SYMPOSIA OF THE
SOCIETY FOR EXPERIMENTAL BIOLOGY

NUMBER XXXXI

PROCEEDINGS OF A MEETING
HELD IN UNIVERSITY OF DURHAM, ENGLAND
10–12 SEPTEMBER 1986

SYMPOSIA OF THE
SOCIETY FOR EXPERIMENTAL BIOLOGY

SYMPOSIA OF THE
SOCIETY FOR EXPERIMENTAL BIOLOGY

NUMBER XXXXI

TEMPERATURE AND ANIMAL CELLS

EDITED BY

K. BOWLER AND B. J. FULLER

Published for the Society for Experimental Biology
by The Company of Biologists Limited,
Department of Zoology, University of Cambridge,
Downing Street, Cambridge CB2 3EJ

Typeset, Printed and Published by The Company of Biologists Limited,
Department of Zoology, University of Cambridge,
Downing Street, Cambridge CB2 3EJ

CONTENTS

Contents

PREFACE

The Symposium that gave rise to this volume was held at the University of Durham from 10–12 September 1986. It was a meeting organized jointly by the Society for Experimental Biology and the Society for Low Temperature Biology. It brought together, in a unique way, an international group of biochemists, biophysicists, cell biologists and cryobiologists to discuss the impact of temperature on animal cells.

The aim was to concentrate on those aspects of cell biology where temperature is likely to have a primary effect. The subject was introduced by contributions on the effect of temperature on the properties and behaviour of the principle molecules that make up animal cells. The effect of temperature in the 'normal' range on a variety of aspects of cell functions was considered with emphasis on acclimatization and adaptation responses. Temperatures outside the normal range damage cells and so special attention was paid to the injurious effects of hyperthermia, and also hypothermia. The practical importance of hyperthermia in tumour therapy, and hypothermia in tissue cold-storage was also dealt with.

It is a pleasure to acknowledge the contributions of many people and organizations involved in ensuring the success of the Symposium meeting and this publication, The Royal Society, The Wellcome Trust and Pergamon Press generously assisted with the expenses of some of our contributors. We particularly appreciate the considerable help from officers of the Society for Experimental Biology at all stages of the planning and execution of the Symposium, and from the committee of the Society for Low Temperature Biology, and of Richard Skaer of The Company of Biologists Limited for guidance through publication. Our greatest thanks must go to Alice Milburn who was called on at the last minute to act as local secretary and did her job superbly.

K. Bowler
B. Fuller
July 1987.

Printed in Great Britain © *Society for Experimental Biology 1987*   1

WATER, TEMPERATURE AND LIFE

FELIX FRANKS

Pafra Ltd, Biopreservation Division, 150 Cambridge Science Park, Cambridge CB4 4GG and Department of Botany, University of Cambridge, Cambridge CB2 3EA

Water in the living cell

Water and life are inseparable, which led Nobel laureate Szent-György to describe water as the 'matrix of life'. Nevertheless, among life scientists the involvement of water in life processes is either ignored or oversimplified, sometimes to the extent that fundamental physical laws are infringed. A limited literature deals with the so-called 'state' of water in living cells, tissues, organs and organisms (Drost-Hansen & Clegg, 1979; Beall, 1983), although rigorous and informative definitions of the term 'state' are hard to find. Under optimum physiological conditions water is the major component of all cell types. Furthermore, each individual cell exists in osmotic equilibrium with its extracellular environment and therefore also with other cells; the *thermodynamic* activities of intracellular and extracellular water are thus identical. Nevertheless, claims have been made that the water in living cells exists in a 'state' which differs from that in the plasma (Ling & Negendank, 1980).

The properties which have been ascribed to water in living systems cover a wide range, from those of bulk water to those of water sorbed at various types of surfaces, such as membrane components, various dissolved macromolecules and cytoskeletal structures. This latter type of water is frequently described as 'bound' (Clegg, 1982), although here, again, binding is rarely defined in a satisfactory manner, either in terms of structure (distances and angles), energetics (strength of binding) or dynamics (lifetimes, exchange rates). Sometimes further distinctions are drawn between tightly and loosely bound water (Finer & Darke, 1974), mainly on the basis of measured diffusion rates and their assignments to the various hypothetical water fractions (Derbyshire, 1982; Beall, 1983). What is usually overlooked is that in living cells, especially under conditions of water stress (e.g. partial freezing), the aqueous phase is likely to be a supersaturated solution the properties of which are dictated by kinetics; equilibrium thermodynamics are inappropriate in the description of interactions in such 'metastable' water (Franks, 1986). In any case, the interactions are always weak, no stronger than hydrogen bonds.

Cell fluids and water stress

A well-defined dividing line can be drawn between chemical and physical rate processes which we associate with life and apparent chemical inertia. Such inertia exists in states of suspended animation, as observed, for instance, in bacterial spores, seeds or cells subjected to cryopreservation at liquid nitrogen temperatures. The dividing line between suspended animation and death is not so well defined, unless death is identified with a range of irreversible chemical reactions with the common feature of decomposition (hydrolysis, oxidation, passive diffusion) of the molecules which are necessary for life.

For optimum functioning, all living cells depend on external and internal aqueous environments the properties of which may only vary within very narrow limits of temperature, pressure, pH, ionic strength and chemical composition. A variation of one or more of these parameters beyond such limits will be perceived by the cell as a stress. In order to survive, the cell must change its chemistry to counter or tolerate the stress. Such alterations require the diversion of energy into the acclimation reactions and will therefore be accompanied by a reduction in the rates of normal growth and reproduction processes. If the cell cannot adequately respond to the applied stress, then symptoms of injury appear which may eventually prove lethal.

There are many different forms of physiological stress conditions induced, for instance, by starvation, radiation, oxygen depletion or changes in light intensity. This review is limited to water stress, that is stress conditions caused by changes in the properties of the intracellular aqueous phase, with an emphasis on effects produced by changes in temperature. In one sense it is artificial to isolate water (osmotic) stress, as produced by drought, freezing or salinity, from other, accompanying, stress conditions. In other words, it is naive to represent a living cell as responding only to osmotic pressure differences. This is, however, a common practice, particularly in microbiology, where the viability of microorganisms is said to be determined by the water activity (a_w), without much regard to the nature of the chemical species involved in producing a given a_w (Griffin, 1981). The whole concept of osmoregulation as a means of countering stress is based on this assumption, although the description of some solutes as 'compatible' (others are presumably toxic) is a tacit admission that a_w alone is not an adequate index of tolerance to water stress (Gould, 1985).

Water as a reactant in the chemistry of life

The involvement of cell water in biochemistry and physiology can be viewed at different levels of resolution, from the role played by discrete and/or specifically placed water molecules to the macroscopic behaviour of water as a transport medium or the lifelong environment of many plants and animals.

The four chemical reactions which are basic to all life processes – hydrolysis, condensation, oxidation and reduction – all involve water, either as a reactant or a product. The detailed mechanisms of some of these reactions are still obscure. An example is the splitting of H_2O during photosynthesis to yield molecular oxygen, a reaction which is not easily visualized in terms of pure chemistry. The subtlety of the role of water, both as a medium and a reactant, is well illustrated by a comparison of the hydrolysis of phosphates, e.g. pyrophosphate (PP_i), ATP or acyl phosphate residues in different environments. In dilute aqueous solution or under conditions as they probably exist in the cytosol, the hydrolysis of PP_i is accompanied by a large change in free energy, typically of the order of $-10\,kJ\,mol^{-1}$ (Flodgaard & Fleron, 1974). On the other hand, the same reaction, catalysed by yeast pyrophosphatase, occurs with $\Delta G = -2\,kJ\,mol^{-1}$ (Springs *et al.* 1981). Because a similar reduction in ΔG is observed from the hydrolysis in aqueous mixtures of dimethyl sulphoxide, ethane diol and polyethylene glycol, some workers have concluded that the water on the enzyme surface and/or in aqueous/organic mixtures has similar structural properties which differ from those of the pure aqueous medium (De Meis *et al.* 1985). Here, again, no details are available about the nature of any such structures or how they affect the hydrolysis reaction.

Still on the subject of water as a reactant in cellular chemical reactions, it must be stressed that ΔG values and other thermodynamic quantities of metabolic reactions quoted in the standard texts refer to *in vitro* conditions and dilute aqueous solutions. Thus, for the hydrolysis of ATP at pH 7 and 37°C, $\Delta G = -31\,kJ\,mol^{-1}$, but under conditions as they exist in the cytosol ΔG is most likely closer to $-45\,kJ\,mol^{-1}$ (Mahler & Cordes, 1967). For an ATPase located in a predominantly lipid membrane environment, the conditions of excess water do not apply and the term pH, as defined, becomes meaningless. Water is likely to be available only at very low concentrations, and dilute solution thermodynamics become inapplicable. The same caveat applies to reaction kinetics. Hydrolysis reactions in dilute solution usually obey pseudo-first-order kinetics, but as the concentration (and availability) of water decreases, so the observed kinetics change to second-order.

Even when water is not a net reactant or product in a reaction sequence, most organic reactions in solution involve one or several steps in which proton transfer to or from the solvent plays a part, e.g. the mutarotation of sugars (Capon & Walker, 1974). The kinetics of such intermediate steps are also very sensitive to the intermolecular details of the aqueous substrate.

Water structure and its dependence on temperature

Much has been written about the structure of water and hypothetical structure making/breaking processes which govern biochemical and physiological phenomena. It must, however, be borne in mind that any specific spatial arrangements which might exist in the liquid state are not permanent but subject to rapid changes and rearrangements, typically on a picosecond time scale.

The current consensus is that in liquid water the H_2O molecules are distributed such that on a time-averaged basis they form an irregular, infinite, hydrogen-bonded network, with each water molecule surrounded on an average by *four*

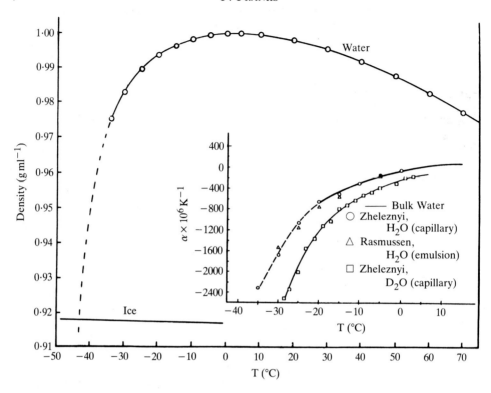

Fig. 1. The density and coefficient of expansion (inset) of liquid water as a function of temperature. The broken line is the extrapolation of a polynomial fit to the density data over the temperature range −35 to +100°C. Reproduced, with permission, from Angell (1982).

other molecules placed approximately at the vertices of an irregular tetrahedron (Dore, 1985). The corresponding *regular* structure, based on the regular tetrahedron, is hexagonal ice (ice-Ih). Such a four-coordinated structure, held by weak bonds, is extremely labile and easily perturbed by temperature, pressure and solutes, especially those containing groups that are themselves able to participate in hydrogen bonding (−OH, >NH, −O−, etc.).

The effects of temperature on the physical properties of water are well documented (Franks, 1971). Until recently, emphasis had been placed on the high-temperature behaviour, but over the past decade the properties of undercooled water (i.e. liquid water below its equilibrium freezing point) have become a subject of intensive study (Angell, 1982). Fig. 1 shows the density (d) of water as a function of temperature (Angell, 1982) which exhibits the well-known maximum at 4°C. However, d(T) is by no means symmetrical about the temperature of maximum density, falling off quite rapidly at subzero temperatures and approaching the density of ice in the neighbourhood of −45°C. Actually, this temperature cannot be reached in practice, because the homogeneous nucleation temperature of ice intervenes at −40°C. If the volume expansion which accompanies the freezing of water is one cause of freeze injury in living tissues, the d(T), as shown in Fig. 1, suggests that if freezing can be achieved under conditions of maximum undercooling, this type of injury would be reduced.

Another property which shows a pronounced temperature sensitivity is the ionization of water. Thus pK_w changes by 5–6 orders of magnitude over the 80 degree temperature range which is associated with life on earth (Franks, 1985). Any process which is particularly sensitive to hydrogen or hydroxyl ion activity must therefore be seriously affected by changes in temperature. A practical example is provided by buffering capacities of acid/salt mixtures. Since the ionization of an acid (or a base) is related to the ionizing power of water, pK_a values are likely to be sensitive to temperature. This is illustrated in Fig. 2 for some common buffering acids. Whereas some physical properties of water show a

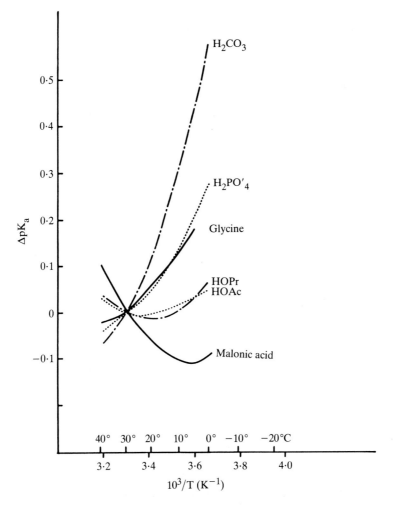

Fig. 2. The effect of temperature on pK_a of acids commonly used in pH buffer mixtures. Reproduced, with permission, from Franks (1985).

pronounced temperature dependence, e.g. the diffusion constant and the compressibility, others such as the dielectric constant and the specific heat are sensibly independent of temperature, except at subzero temperatures, where all properties of liquid water seem to become extremely sensitive to temperature (Angell, 1982).

In terms of changes in the water structure induced by temperature, the above results indicate the direct effect of kinetic energy on the labile, hydrogen-bonded network. Some workers have used such results to estimate the degree of hydrogen boding in water, taking as reference points the freezing and boiling points (Nemethy & Scheraga, 1962). This is misleading, however, because the real temperature range of the liquid extends from the homogeneous nucleation temperature ($-40°C$) to the critical temperature ($374°C$).

The hydration of ions and molecules

The interactions between ions in solution are governed primarily by the laws of electrostatics, as formalized in the Debye–Hueckel theory of electrolytes, according to which activities and interactions are determined by the ionic strength *only*, i.e. by the concentration and valence of the ions present. No explicit allowance is made for hydration, specific ion effects or for the properties of the solvent, except through the assignment of a bulk dielectric constant.

The advent of refined neutron diffraction techniques has made possible the detailed study of ionic hydration shells (Enderby & Nielson, 1979). Fig. 3(A) and (B) shows the typical dispositions of water molecules relative to monatomic cations and anions, respectively. The hydration shells consist of six water molecules arranged octahedrally about the ion. The lifetime of a water molecule in a given position can vary from 10 ns (K^+) to $10\,\mu s$ (Ni^{2+}), but the time-averaged geometry is similar for many ions (except Li^+).

Diffraction methods have not yet been able to yield unequivocal information about the spatial distribution of water molecules beyond the primary hydration shell. However, the octahedral geometry of the primary hydration shell is incompatible with the tetrahedral arrangement characteristic of pure water, and this incompatibility must give rise to a region of structural and dynamic mismatch, as first suggested by Frank & Wen (1957). Any possible significance of such mismatch to biological processes continues to be a subject of some speculation.

Although at low ionic strengths ($I < 0.15$) the behaviour of electrolytes in solution and also their effect on protein solubility and stability is well accounted for by the Debye-Hueckel theory, at higher ionic strengths ion-specific effects begin to dominate the behaviour of aqueous solutions. It is interesting in this context that the isotonic conditions of many living cells correspond to $I = 0.15$.

The ion-specific effects can be classified in terms of salting-in and salting-out and this classification forms the basis of the lyotropic (Hofmeister) series of ions (Hofmeister, 1888), according to which ions can be graded depending on their influence on the solubility of proteins. Thus, SO_4^{2-} and PO_4^{3-} are strongly salting-out, whereas ClO_4^- and CNS^- are strongly salting-in, with other ions taking

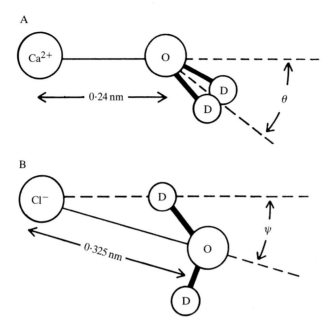

Fig. 3. Geometrical details of the primary hydration shells of (A) monatomic cations and (B) halide ions. The ion-oxygen distance varies with the nature of the ion and the angles of inclination are functions of the salt concentration. After Enderby & Nielson (1979).

intermediate positions. Since it was first reported in 1888, the lyotropic series has been 'rediscovered' on several occasions. It applies equally in its influence on the solubility of rare gases and hydrocarbons, the flocculation of colloidal sols, the micellization of surfactants and the stabilization of native states of proteins and nucleotides, in the sense that salting-out is equivalent to the stabilization of the native state against denaturation (von Hippel & Wong, 1965; Franks & Eagland, 1975). The details of ion hydration probably determine the position of a given ion in the series, but there is as yet no theoretical explanation for the lyotropic series. Furthermore, non-electrolytes can also be placed in the series, although it seems inappropriate to refer to such effects as 'salting-in/out'. This is shown in Fig. 4, where the thermal stability of ribonuclease is expressed as a function of a_w produced by a variety of solutes. There is a wide divergence in behaviour between the strongly destabilizing urea, alkanols and CNS^- and the stabilizing effects of sugar alcohols; a_w is hardly a universal or reliable index of enzyme stability in different media, although for a given solute, the stability is approximately linear in a_w.

The hydration of polar molecules is probably determined mainly by the spatial disposition of hydrogen bonding groups on the solute molecule. This is well illustrated by a comparison of the properties of stereoisomers, such as sugars (Franks, 1983). The peripheral disposition of $-OH$ groups, as perceived by the surrounding water environment, is also sensitive to slight torsional motions of the

8

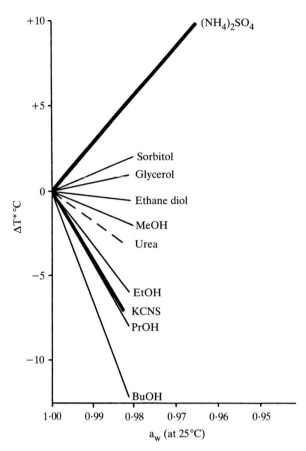

Fig. 4. The effects of additives on the thermal stability of ribonuclease. ΔT^* is the observed shift in the thermal denaturation temperature T^*.

ring, as seen, for instance, in the furanose sugars shown in Fig. 5. Hydrophilic hydration is believed to be of a short range nature, because the interaction energies involved resemble those that exist between water molecules, but also because the water network can tolerate a considerable degree of hydrogen bond distortion. The lifetimes of individual water molecules in the hydration shell are not expected to be lengthened to any major extent (Suggett, 1976), but even apparently minor stereochemical differences can affect the manner in which polyhydroxy compounds in solution interact with water and with each other (Barone et al. 1981; Franks, 1983). This is particularly well demonstrated by the rheological properties of respiratory, gastric and cervical mucus gels which are extremely sensitive to small changes in the water content (Pain, 1982). Since sugar residues (on glycoproteins) fulfil the biological role of recognition, minor changes in conformation, as induced by hydration processes, may well be amplified in solute–solute interactions. Altogether, sugar–sugar interactions and their dependence on the detailed molecular shapes and solvation is a much neglected field of study.

The response of water to the introduction of apolar molecules or residues (e.g. alkyl groups) is generally referred to as hydrophobic hydration. Since water cannot directly interact with such residues by hydrogen bonding, the water molecules rearrange and reorient in order to maintain the maximum degree of water–water hydrogen bonding, thereby creating cavities to accommodate the 'guest' molecule, as shown schematically in Fig. 6. The net effect is to reduce the number of configurational degrees of freedom available to the individual water molecules. This is reflected in a *negative* entropy of mixing. The significance in terms of biochemical processes is that the mixing of water and alkyl derivatives (including apolar amino acid residues in proteins) is thermodynamically unfavour-able ($\Delta G > 0$) *not* because the components repel each other but because many water molecules are forced to adopt specific and constrained hydrogen-bonding configurations so as to provide room for the non-reactive residue (Franks, 1975; Franks & Desnoyers, 1985).

The reverse of hydrophobic hydration is observed on a macroscopic scale as a spontaneous aggregation of apolar residues (usually referred to as hydrophobic interaction), as in micelles, phospholipid liquid crystalline phases in aqueous

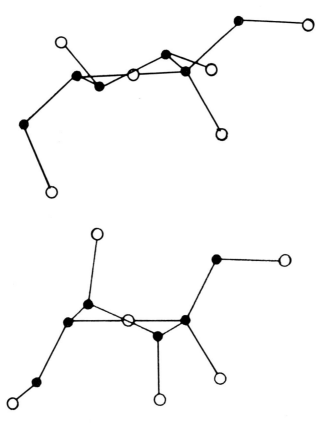

Fig. 5. The effect of furanose sugar ring flexibility on the spatial disposition of the −OH oxygen atoms.

F. FRANKS

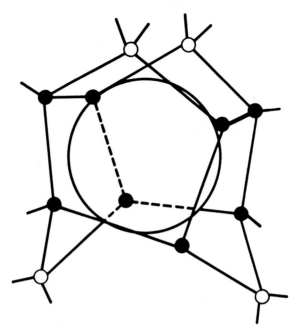

Fig. 6. Example of a concave hydration cage around a hydrophobic residue. The particular cage shown is composed of 12 water molecules of which 8 (filled-in circles) are nearest neighbours and 4 (open circles) are situated further away from the solute.

media and in folded globular proteins. The driving force for such aggregation processes is *not* an attractive interaction energy between apolar groups but a relaxation of the constraints on the water network. The practical result of such entropy-dominated processes is that the temperature dependence of the free energy (driving force) has the opposite sign from that usually associated with equilibrium processes, i.e. hydrophobic association becomes more pronounced at high temperatures and weaker at subambient temperatures.

Little is as yet known about the range of the hydrophobic interaction, although the indications are that the effects of the rearrangement of water molecules around an apolar residue extend well beyond the nearest neighbour water shell (Okazaki *et al.* 1983). Marginal reductions (less than an order of magnitude) in diffusion rates of water molecules in the proximity of apolar residues have been reported (Goldammer & Hertz, 1970), but it is not certain whether any physiological significance can be ascribed to such minor diffusional perturbations.

Water as a macromolecular cement

Most, if not all, biochemical processes in aqueous media occur between hydrated molecules and/or ions. Possibly, trans-membrane processes may involve dehydrated molecular or ionic species, but dehydration energies are large and dehydration is unlikely, unless the water is replaced by another ligand of similar solvation energy, such as occurs during enzyme/substrate binding. Similarly, complex macromolecular structures derive their stability from a combination of intramolecular forces and hydration interactions. The contribution provided by water to the stability of globular proteins in the crystalline state has been a subject of intensive study over the past decade (Finney, 1979). Whereas, in the past, X-ray diffraction could only detect the oxygen atom in the water molecule, neutron diffraction, now being applied on an increasing scale, also makes possible the assignment of spatial coordinates to the hydrogen (actually deuterium) atoms (Savage, 1986).

Water molecules are found to occupy typical sites as follows: (1) as metal atom ligand, e.g. in carbonic anhydrase, (2) as link between a $>C=O$ group on one residue and a $>N-H$ group on another, distant residue, (3) as link between backbone peptide groups and polar side chains, e.g. serine, tyrosine, and (4) in proximity to two or more ionizable residues buried in the interior of the protein structure, e.g. glu or lys. Fig. 7 shows the water electron density in one region of the porcine insulin crystal and also a possible water network, based on the refined crystal structure, calculated water–water potential energy functions and assumed cut-off distances for hydrogen bond lengths (Savage, 1986). It is apparent that the packing of water molecules hardly resembles that in unperturbed liquid water and that the water in the crystal will tolerate appreciable distortions of hydrogen bond angles and lengths. Altogether 340 water sites have been assigned in the insulin crystal, but it is not known how many or which of these water molecules crucially affect the stability of the native protein. It is clear, however, that many water molecules in the environment of the protein must be regarded as forming an integral part of the folded protein structure and that their removal must lead to destabilization of the native state.

Hydration and the stability of complex biological structures

Under physiological conditions the stability margin of a fully functional small (*ca* 100 residues) globular protein, relative to the inactivated state, is small, typically of the order of $<70\,kJ\,mol^{-1}$. This is equivalent to only 3–4 hydrogen bonds, whereas the folded structure actually contains >100 such bonds. The stability, as measured, must therefore be the net resultant of two or more contributions, of which some are stabilizing and others destabilizing. It is difficult to draw up a stability balance sheet, listing the various contributions, because some of them are not amenable to calculation and others might in fact overlap. The balance sheet drawn up by Finney (1982), and shown in Table 1, based on the available experimental evidence and existing theory, highlights the nature of the problem. What is clear, however, is that most of the contributing interactions contain a hydration component.

The native (N) state is stable only within very narrow limits of temperature, pressure, pH, ionic strength and the chemical composition of the medium. Within the context of this review it is pertinent to enquire into the effect of temperature on the various contributions listed in Table 1 and the probable influence on the net

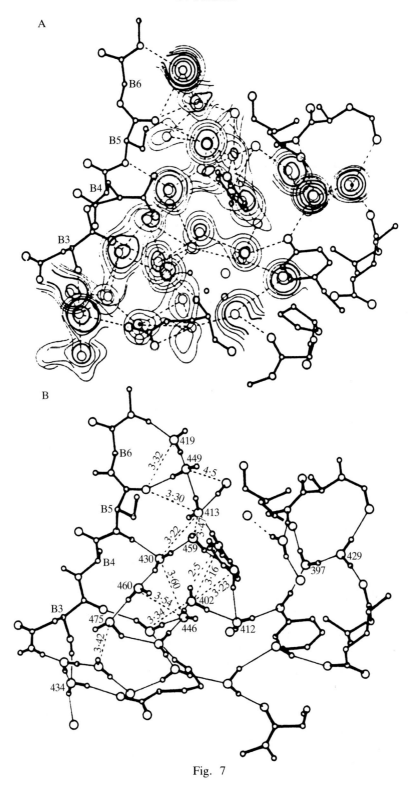

Fig. 7

free energy of stabilization. Intuitively an increase in temperature (kinetic energy) would be expected to destroy the delicately balanced N-state, and thermal denaturation is indeed a well-known and well-studied phenomenon (Franks & Eagland, 1975). Whether it has any physiological or ecological importance is open to doubt, bearing in mind that, with very few exceptions, the upper temperature limit to life is approx. 40°C, whereas denaturation events are usually observed above this temperature.

On the assumption that the denaturation of a small globular protein can be expressed in terms of a simple two-state equilibrium N→D (Brandts, 1964), characterized by an equilibrium constant, it can be shown that if the specific heat of the D-state exceeds that of the N-state, then $\Delta G(T)$ is a curvilinear function, the curvature depending on the magnitude of the difference in the specific heats of the two states. To a first-order approximation $\Delta G(T)$ takes the form of a parabola, suggesting the existence of a cold-induced transition. From a physical point of view, and without consideration of possible effects due to potentially large contributions from changes in hydration, it seems sensible that the above specific heat condition should always apply, since the N-state, being ordered, would be expected to have a lower specific heat than the more flexible (or completely disordered) D-state.

The extensive calorimetric studies of Pfeil & Privalov (1979), some results of which are shown in Fig. 8, confirm the above prediction that there should in general exist a *cold* denaturation limit to protein stability, although for the proteins investigated by these workers the predicted cold denaturation temperatures lie below the equilibrium freezing point of water and are therefore not easily accessible to experiment. Recent investigations with the aid of undercooled solutions of chymotrypsinogen (Franks & Hatley,1985) and lactate dehydrogenase (LDH) (Hatley & Franks, 1986) have established the existence of fully reversible conformational transitions at subzero temperature. Fig. 9 illustrates the effects of temperature cycling on the absorbance at 286 nm of LDH, showing the cold transition, centred on −15°C. It is interesting to note that the low temperature form of LDH still exhibits enzymatic activity (measurable down to −23°C) but the mechanism cannot be the same as that characteristic of native LDH, because of a marked increase in the activation energy at the onset of the cold transition. Such changes in enzyme mechanisms at low temperature have also been reported for phosphofructokinase (PFK) in insects (Storey *et al.* 1981). The cold inactivation of PFK in winter hardy insects (*Eurosta solidaginis*) results in the accumulation of high concentrations (150 μg g^{-1} wet weight) of sorbitol in the haemolymph which acts as natural cryoprotectant. Equivalent (molar) concentrations (240 μg g^{-1}) of glycerol are also produced, by the blockage of pyruvate kinase.

Fig. 7. Water structure over one region of solvent density adjacent to residues B3–B7 in porcine 2 Zn insulin crystals. (A) Solvent electron density at 0·15 nm resolution from X-ray refinement studies of the crystal structure. The refined water positions are included and the broken lines represent 0---0,N distances of <0·34 nm. (B) One of several possible water networks. The broken lines represent non-bonded 0---0 contacts. Reproduced, with permission, from Savage (1986).

Table 1. *Tentative balance sheet of contributions to the free energy change accompanying the unfolding of lysozyme and ribonuclease. n_c and n_w are the respective number of mols of residues and water involved in the unit process (after Finney, 1982)*

Term	Definition	Comments	Estimated magnitude (kJ mol^{-1})
$\Delta G_{N \to D}$	Free energy change on unfolding	Small $\approx$ 3–5 hydrogen bonds	50–100
ΔG_{titr}	pK$_a$ changes on unfolding		−20?
ΔG_{es}	Changes in surface charge–charge interactions	Estimated at neutral pH	0?
ΔG_ϕ	Hydrophobic free energy change	Surface exposure model	500
		Allow for possible long range*	100–2500!
ΔH_ψ	Several hydrogen bonding effects		
	(a) Difference in strength of water–polar group and polar–polar group	Average of several quantum mechanical values	300
	(b) Hydrogen-bond distortion	Using water–water energy surface	−400 to −800
	(c) 'Unsaturated' polar groups		−300
n_c TΔS config	Configurational entropy change per residue	Value of between 8–20 J K^{-1} (residue mol)$^{-1}$ used	−300 to −800
n_wTΔS release	Water forms polar interactions with peptide groups	Assumes 25 % of value for water freezing (underestimate?)	500
ΔH_{VDW}	Changes in van der Waals interactions	Protein denser than water implying some stabilization? Possibly included in G$_\phi$	?
$\Delta H_{\psi - \phi}$	Polar–apolar interactions within protein		?
TΔS$_{vibr}$	Vibrational entropy changes	Potentially large and solvent-related	?

* Evidence is growing that the simple proportionality between buried accessible area and free energy change is probably not valid and that the hydrophobic interaction is of a long-range nature.

The protective effects of sugars and sugar alcohols can be considered as twofold. These compounds markedly raise the viscosity of water, and at the high concentrations at which they occur in the haemolymph the rate of ice crystallization (or evaporation during prolonged periods of high temperature) is markedly reduced (Franks, 1986). As the glass transition is approached, desiccation may be

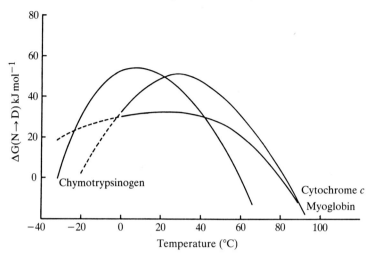

Fig. 8. Free energy of stabilization of some native proteins as function of temperature.

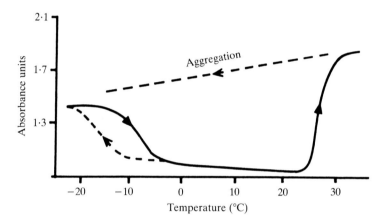

Fig. 9. Cold and heat destabilization of lactate dehydrogenase. The broken lines represent absorbance changes during cooling, and the solid line the corresponding change during heating. After Hatley & Franks (1986).

inhibited completely. Polyhydroxy compounds also 'salt-out' proteins (see Fig. 4) and thus protect proteins and other labile structures against the effects of desiccation (Gekko & Morikawa, 1981).

In vivo *water dynamics*

Within the overall hydrologic cycle involving the constant evaporation and precipitation of water in the ecosphere can be placed two subcycles, relating to the water turnover by plants and animals. The former consists of photosynthesis on the one hand and transpiration and evaporation on the other. Some 10^{11} tonnes of water are turned over annually.

Taking the 'normal human adult' as a standard, his daily water balance involves a net turnover of 2·5 kg, corresponding to approx. 4 % of the total body weight, or 6 % of the total water content. Of the daily water intake, some 300 g is endogenously produced during glycolysis. This water synthesis is accompanied by the generation of 7600 kJ of free energy, and it requires 185 l of oxygen which the lung extracts from air with a 14 % efficiency. (Actually the total daily oxygen requirement is closer to 500 l.) Since air contains 21 % oxygen, the lung processes some 6300 l of air daily in order to generate the necessary supply of oxygen. The oxygen is carried to the tissue cells by the carrier protein haemoglobin which is also linked to the removal of carbon dioxide.

The blood pressure in the arteries is sufficiently high to set up an ultrafiltration system whereby ions, low molecular weight species and water can be transported across the cell membranes. There is a pressure drop along the vascular system, and eventually the osmotic pressure of the plasma exceeds the blood pressure so that tissue fluid is drawn into the capillaries by osmosis. The combined effects of ultrafiltration and the subsequent osmotic reabsorption promote the supply of compounds into the cells and the removal of waste products out of the cells. To supply the cells with enough oxygen for the daily combustion of glucose, the heart has to pump 7000 l of blood around the vascular system.

Several other organs are implicated in maintaining the physiological water balance and quality. Thus, every day the kidneys produce 180 l of diluted urine which is then concentrated to 1·5 l and excreted. The remainder of the water is suitably purified and returned to the body. In this manner the body loses 50–75 g of nitrogenous waste products, mainly in the form of urea. From the point of view of water conservation, birds and insects are rather more efficient, in that they excrete their nitrogen waste in the form of almost insoluble compounds.

The transport of water, whether it is considered as bulk flow, or as diffusion of discrete molecules, occurs as a result of potential gradients across boundaries such as membranes. The effect of temperature is expressed in terms of an Arrhenius activation energy. At the macroscopic level, water flux is measured as the quantity transported across a membrane of unit area. It is determined by the pressure gradient and the hydraulic permeability (L_p) of the membrane.

In plant cells dynamic equilibrium is maintained by the cell wall. Turgor pressure builds up until it equals the difference in osmolality between the intracellular fluid and its environment. Animal cells regulate the osmotic potential of the Gibbs-Donnan ion distribution and thus achieve osmotic equilibrium. Selected ions (or molecules) are transferred against their electrochemical gradients into or out of the cell. In this way osmotically active ions, which, if free to diffuse, would produce osmotic death, are restrained by a continuous expenditure of metabolic energy and are rendered functionally impermeable.

An alternative explanation is based on the proposition that substantial fractions of intracellular water *and* solute are in physical states (what is a 'state'?) which differ from those of the environment (Negendank & Shaller, 1979). Intracellular water is believed to be more organized and able to exclude solutes. The term non-solvent water has been applied to this hypothetical fluid (Horowitz & Paine, 1979). The evidence for non-solvent water or specially constrained water is tenuous. Although in the neighbourhood of biopolymers water does adopt intermolecular geometries which differ markedly from those of water in the bulk (see Fig. 7), there is no evidence that water molecules which participate in these peculiar configurations differ energetically to any marked extent from normal molecules. On the other hand, it must be admitted that the commonly held view of the living cell as a bag filled with water is equally untenable. The known architectures of the cytoskeleton and the extracellular matrix, composed of laminin, fibronectin, collagen IV, etc., strongly suggest that many, if not most, water molecules must be under the influence of anisotropic fields produced by charged macromolecules. Pashley & Israelachvili (1981) have shown that the universal hydration forces which operate in colloidal systems decay over much longer distances ($0\cdot3$–$0\cdot8$ nm) than the typical r^{-6} dependence of van der Waals forces predicts. What is still very much open to speculation is the nature of the signal that could trigger changes in the force fields experienced by intracellular water and solute molecules.

The technique most commonly employed in studies of water diffusion at the microscopic level is nuclear magnetic resonance (nmr). Although much progress has been made in the improvement of sensitivity, resolution and data processing equipment, the translation of experimental data, which refer to the decay of nuclear magnetization, into details of the diffusive motions of molecules in a liquid, let alone in a complex and anisotropic biological matrix, requires refined models and sophisticated experimentation. The details of the spectral output are affected by many different factors, including proton exchange rates, exchange of water molecules between distinguishable sites, the viscosity of the medium, motion of the macromolecular substrate and anisotropy in the diffusional motions performed by the water molecules. Bryant & Halle (1982) have provided a thoughtful summary of the alternative interpretations of nmr spectral data in terms of the motions molecules in biological systems.

Whatever might be the diffusional perturbations suffered by water molecules in a complex biological matrix, they appear to be subtle, second-order effects, because when tissues are cooled to subzero temperatures most of the water freezes with crystallization rates and ice morphologies which resemble those in pure water. The proportion of water which does not freeze is not 'bound' by the matrix but acts as plasticizer to the amorphous (non-crystallizable) matrix components (e.g. connective tissues, cytoskeleton). As the glass transition is approached, so the viscosity increases and the ice growth rate decreases. At the glass temperature, which may lie in the range -40 to -10°C, ice crystallization becomes very slow, *ca* $0\cdot3\,\mu$m year^{-1}. The phenomenon of aqueous glass formation, resulting from the freeze dehydration of cells, may well be implicated in natural freeze tolerance mechanisms in plants and insects.

Membranes possess different permeabilities towards different molecular and ionic species, but all biological membranes appear to have a very high water permeability. Its measurement is beset by problems, because the osmotic pressure gradient (either by freezing or hypertonic conditions) which is required for the measurement of L_p also affects the membrane. For the yeast cell membrane, Schwartz & Diller (1983) have quoted $L_p = 0.0116 \, \mu m \, min^{-1}$ with an activation energy $E^* = 19.4 \, kJ \, mol^{-1}$ at a cooling rate of $9°min^{-1}$. At a cooling rate of $35°min^{-1}$ the corresponding values are $2.11 \, \mu m \, min^{-1}$ and $101 \, kJ \, mol^{-1}$.

Conclusions

The involvement of water in life processes can be viewed at many different levels of resolution, and in any description of a 'state' of water, clear distinctions must be drawn between structural, energetic and dynamic specifications.

Extrapolations to *in vivo* systems from studies of crystals or dilute solutions must be carefully validated. Simplistic concepts, such as water activity or bound water, are unhelpful, if not incorrect, in descriptions of biological dynamics.

Bearing in mind the marginal free energy differences which characterize native structures and reactions, it is likely that the major component of all living cells, water, fulfils an energy-buffering role in which destabilization is almost as important as stabilization.

References

Angell, C. A. (1982). Supercooled water. In *Water – A Comprehensive Treatise*, vol. 7 (ed. F. Franks), pp. 1–82. New York: Plenum Press.

Barone, G., Bove, B., Castronuovo, G. & Elia, V. (1981). Excess enthalpies of aqueous solutions of polyols. *J. Solution Chem.* **10**, 803–810.

Beall, P. T. (1983). States of water in biological systems. *Cryobiology* **20**, 324–334.

Brandts, J. F. (1964). The thermodynamics of protein denaturation. I. The denaturation of chymotrypsinogen. *J. Am. chem. Soc.* **86**, 4291–4301.

Bryant, R. G. & Halle, B. (1982). NMR relaxation of water in heterogeneous systems – consensus views? In *Biophysics of Water* (ed. F. Franks & S. F. Mathias), pp. 389–393. Chichester: John Wiley & Sons.

Capon, B. & Walker, R. B. (1974). Kinetics and mechanism of mutarotation of aldoses. *J. chem. Soc. Perkin Trans.* **2**, 1600–1610.

Clegg, J. S. (1982). Alternative views on the role of water in cell function. In *Biophysics of Water* (ed. F. Franks & S. F. Mathias), pp. 365–383. Chichester: John Wiley & Sons.

De Meis, L., Behrens, M. I., Petretski, J. H. & Pollitt, M. J. (1985). Contribution of water to the free energy of hydrolysis of pyrophosphate. *Biochem.* **24**, 7783–7789.

Derbyshire, W. (1982). The dynamics of water in heterogeneous systems. In *Water – A Comprehensive Treatise*, vol. 7 (ed. F. Franks), pp. 339–450. New York: Plenum Press.

Dore, J. C. (1985). Structural studies of water by neutron diffractions. *Water Sci. Revs.* **1**, 3–92.

Drost-Hansen, W. & Clegg, J. S. (eds) (1979). *Cell-Associated Water*. New York: Academic Press.

Enderby, J. E. & Nielson, G. W. (1979). X-ray and neutron scattering by aqueous solutions of electrolytes. In *Water – A Comprehensive Treatise*, vol. 6 (ed. F. Franks), pp. 1–46. New York: Plenum Press.

Finer, E. G. & Darke, A. (1974). Phospholipid hydration studied by deuteromagnetic resonance spectroscopy. *Chem. Phys. Lipids* **12**, 1–16.

Finney, J. L. (1979). The organization and function of water in protein crystals. In *Water – A Comprehesive Treatise*, vol. 6 (ed. F. Franks), pp. 47–121. New York: Plenum Press.

FINNEY, J. L. (1982). Solvent effects in biomolecular processes. In *Biophysics of Water* (ed. F. Franks & S. F. Mathias), pp. 55–58. Chichester: John Wiley & Sons.

FLODGAARD, H. & FLERON, P. (1974). Thermodynamic parameters for hydrolysis of inorganic pyrophosphate at pH 7·4 as a function of Mg^{2+}, K^+ and ionic strength determined from equilibrium studies of reaction. *J. biol. Chem.* **249**, 3465.

FRANK, H. S. & WEN, W. Y. (1957). Structural aspects of ion-solvent interactions in aqueous solutions: a suggested picture of water structure. *Disc. Faraday Soc.* **24**, 133–140.

FRANKS, F. (ed.) (1971). *Water – A Comprehensive Treatise*, vol. 1. New York: Plenum Press.

FRANKS, F. (1975). The hydrophobic interaction. In *Water – A Comprehensive Treatise*, vol. 4 (ed. F. Franks), pp. 1–94. New York: Plenum Press.

FRANKS, F. (1985). *Biophysics and Biochemistry at Low Temperatures*. Cambridge: Cambridge University Press.

FRANKS, F. (1986). Metastable water at subzero temperatures. *J. Microscopy* **141**, 243–249.

FRANKS, F. DESNOYERS, J. E. (1985). Alcohol-water mixtures revisited. *Water Sci. Revs.* **1**, 171–232.

FRANKS, F. & EAGLAND, D. (1975). Role of solvent interactions in protein conformation. *Crit. Rev. Biochem.* **3**, 165–219.

FRANKS, F. & HATLEY, R. H. M. (1985). Low temperature unfolding of chymotrypsinogen. *Cryo-Lett.* **6**, 171–180.

GEKKO, K. & MORIKAWA, T. (1981). Preferential hydration of bovine serum albumin in polyhydric alcohol-water mixtures. *J. Biochem. (Tokyo)* **90**, 39–50.

GOLDAMMER, E. v. & HERTZ, H. G. (1970). Molecular motion and structure of aqueous mixtures with non-electrolytes. *J. phys. Chem.* **74**, 3734–3746.

GOULD, G. W. (1985). Present state of knowledge of a_w effects on microorganisms. In *Properties of Water in Foods* (ed. D. Simatos & J. L. Multon), pp. 229–245. Dordrecht: Martinus Nijhoff.

GRIFFIN, D. M. (1982). Water and microbial stress. *Adv. microb. Ecol.* **5**, 91–136.

HATLEY, R. H. M. & FRANKS, F. (1986). Denaturation of lactate dehydrogenase at subzero temperatures. *Cryo-Lett.* **7**, 226–235.

HOFMEISTER, F. (1888). Zur Lehre von der Wirkung der Salze. *Arch. exptl Pathol. Pharmakol.* **24**, 247–260.

HOROWITZ, S. B. & PAINE, P,. L. (1979). Reference phase analysis of free and bound intracellular solutes. 2. Cytoplasmic sodium, potassium and water. *Biophys. J.* **25**, 45–62.

LING, G. N. & NEGENDANK, W. (1980). Do isolated membranes and purified vesicles pump sodium – critical review and reinterpretation. *Persp. biol. Med.* **24**, 215–239.

MAHLER, H. R. & CORDES, E. H. (1967). *Biological Chemistry*. New York: Harper & Row.

NEGENDANK, W. & SHALLER, C. (1979). Potassium-sodium distribution in human lymphocytes – description of the association-induction hypothesis. *J. Cell Physiol.* **98**, 95–105.

NEMETHY, G. & SCHERAGA, H. A. (1962). Structure of water and hydrophobic bonding in proteins. *J. chem. Phys.* **36**, 3382–3400.

OKAZAKI, S., NAKANISHI, K. & TOUHARA, H. (1983). Computer experiments on aqueous solutions. *J. chem. Phys.* **78**, 454–469.

PAIN, R. H. (1982). Molecular hydration and biological function. In *Biophysics of Water* (ed. F. Franks & S. F. Mathias), pp. 3–14. Chichester: John Wiley & Sons.

PASHLEY, R. M. & ISRAELACHVILI, J. (1981). Comparison of surface forces and interfacial properties of mica in purified surfactant solutions. *Colloids & Surfaces* **2**, 169–187.

PFEIL, W. & PRIVALOV, P. L. (1979). Conformational changes in proteins. In *Biochemical Thermodynamics* (ed. M. N. Jones), pp. 75–115. Amsterdam: Elsevier.

SAVAGE, H. (1986). Water structure in crystalline solids: ices to proteins. *Water Sci. Revs.* **2**, 67–148.

SCHWARTZ, G. J. & DILLER, K. R. (1983). Osmotic response of individual cells during freezing. II. Membrane permeability analysis. *Cryobiology* **20** (61), 542–552.

SPRINGS, B., WELSH, K. M. & COOPERMAN, B. S. (1981). The kinetics and mechanism in yeast inorganic phosphatase catalysis of inorganic pyrophosphate – inorganic phosphate equilibration. *Biochem.* **20**, 6384–6391.

STOREY, K. B., BAUST, J. G. & STOREY, J. M. (1981). Intermediate metabolism during low temperature acclimation in the overwintering gall fly larva *Eurosta solidaginis. J. comp. Physiol.* **144**, 183–190.

SUGGETT, A. (1976). Molecular motion and interactions in aqueous carbohydrate solutions. *J. Solution Chem.* **5**, 33–46.

VON HIPPEL, P. H. & WONG, K. Y. (1965). On the conformational stability of globular proteins. *J. biol. Chem.* **240**, 3909–3923.

Printed in Great Britain © Society for Experimental Biology 1987   21

TEMPERATURE AND MACROMOLECULAR STRUCTURE AND FUNCTION

ROGER H. PAIN

Department of Biochemistry, University of Newcastle, NE1 7RU, UK

Summary

Stability is frequently a knife-edge phenomenon and it is this aspect which is both essential for the effective involvement of macromolecules in the living cell and also provides the basis for the sensitivity of some macromolecular systems to temperature.

The response of proteins and nucleic acids is relatively simple at the phenomenological level, with rather sharp 'melting' transitions occurring. Since however the thermodynamic stability of both depends on the cooperation of a variety of non-covalent interactions which are qualitatively well understood but quantitatively difficult to assess, the full understanding and prediction of structural stability still evades us.

We shall consider the effects of temperature on the different non-covalent interactions and how far these can account for protein and nucleic acid denaturation at elevated temperatures and also the cold inactivation of proteins. The latter has recently been shown to involve unsuspected complications in terms of protein conformational change. The increased stability of proteins and nucleic acids from thermophiles will be discussed.

Stability is important not only in native folded proteins but also with respect to intermediate structures which occur during the folding of the newly synthesized polypeptide chain into the native, active protein. Through studies of protein folding the molecular basis of the phenomenon of temperature sensitive synthesis has been revealed.

Given a stable molecular structure, its function will frequently be subject to variation with temperature. The deceptively simple temperature dependence of enzyme activity will involve the non-covalent interactions considered above for interaction between enzyme and substrate and for stability of the transition state complex. This complexity again makes the temperature dependence difficult to interpret. Further, the fact that proteins are dynamic structures is becoming recognized as an important feature factor in determining function. A balance has to be struck between on the one hand dynamic mobility which is essential for catalytic activity and on the other thermodynamic stability which holds the molecule in a potentially functional conformation under the given conditions of temperature and pressure. Readjustment of thermostability stability, as in thermophiles (or *vice versa?*), must also involve readjustment of dynamic mobility.

Implicit in the foregoing is the fact that temperature can affect not only stability of the 'finished product', the functional macromolecular structure, but also the

enzyme catalysed synthesis of polypeptides, the structures and interactions involved in gene control and the physicochemical process by which the polypeptide
folds into the functional structure. The life of a macromolecule ends in breakdown, however, and this too relies in part on the carefully selected thermostability
of each structure. The reasons why animal cells are in general finely adapted for
function and survival within a narrow range of temperature must include the
delicate physicochemical balance of stability and dynamics involved in macromolecular structures.

Introduction

The effects of temperature on biological materials are common culinary knowledge arising from everyday experience. Living processes can be slowed down or
suspended by reducing the temperature, under certain conditions with a high
degree of reversibility albeit frequently with hysteresis. Raising the temperature
can again result in stopping living processes, usually with more obvious lack of
reversibility. Changes in physical state seen in boiling an egg may, within severely
proscribed limits, be reversed (Perutz, 1980) but only as a result of understanding
the molecular basis of the events taking place. Molecular studies enable us also to
define those temperature induced changes in the cell which cannot in principle be
reversed. Such studies help to set limits to the principle of self-assembly in cells
and their component structures which in turn focusses attention on those aspects
of the living cell which can only be fully appreciated in terms of their evolutionary
origin.

Even the superficially simple observations of temperature effects raise fundamental questions as soon as explanations at the molecular level are attempted.
Simple kinetic theory would suggest that processes slow down by a factor of two in
rate for every 10°C drop in temperature. Experience shows however that some
'living processes' actually cease with a rather modest drop in temperature. Again,
much use is made in molecular biology of mutants in which the control of gene
expression is sensitive to temperature changes over a very narrow range. Such
sensitivity to temperature is outside the immediate ken of reaction kinetics and
chemical structure and pinpoints the need for detailed investigations of macromolecular structures for clues to such behaviour. The following brief review of some of
the effects of temperature of proteins and nucleic acids indicates therefore one of
the areas of basic knowledge which is essential in beginning to appreciate the
rationale for the fine temperature control of some of the more complex and highly
evolved multicellular organisms such as humans. For reviews see Brandts (1967);
Tanford (1968, 1970); Pace (1975); Privalov (1975); Lapanje (1978); Pain (1979).

Stability of macromolecules

The structure, reactions and hence biological functions of proteins and nucleic
acids depend on their chemically covalent structures being able to take up specific

three-dimensional conformations which persist in time. These conformations are dictated and largely stabilized by intrinsically weak (from the chemist's point of view) non-covalent interactions which are formed within and between macromolecules, usually in a predominantly aqueous environment. The stability of species depends on the stability of the double helix form of DNA and its interactions with proteins in the chromosome and at one and the same time on limitations in stability which allow replication and transcription to occur. Antibodies have three-dimensional conformations sufficiently stable for them to combine specifically with foreign antigens but sufficiently unstable for them to be broken down with a half life sufficiently short for the circulation never to become solid with protein.

Stability has both thermodynamic and kinetic elements. A macromolecule

$$M \underset{k_{-1}}{\overset{k_1}{\rightleftharpoons}} D \overset{k_2}{\rightarrow} X$$

M will in general be in equilibrium with a thermally denatured state D and the associated equilibrium constant and hence free energy change for the reaction will describe the thermodynamic stability. It is likely that state D is more susceptible than state M to aggregation, adsorption to surfaces, proteolytic degradation, oxidation or reduction to states X. If k_2 is large compared with k_{-1} the rate of disappearance of M can be fast even though the equilibrium constant k_1/k_{-1} favours M. Thus, the experiment which measures thermal stability by incubation for a given period of time followed by measure of activity or function cannot *per se* distinguish between kinetic and thermodynamic stability. If there is a significant element of kinetic stability involved the measure will be particularly sensitive to environmental conditions and not easily translated to other conditions, say *in vivo*.

The thermodynamic stability on the other hand, which can usually be measured by some physical parameter which distinguishes M from D under conditions where k_2 is not significant, will not necessarily reflect the persistence of functional M under conditions *in vivo*. Its use in quantitative terms is further limited, particularly for proteins, by the fact that the thermally denatured state can rarely be characterized so that there is no common baseline, e.g. that of a random coil state, against which to measure the stability of state M.

The precise measurement of thermodynamic parameters for the thermal unfolding of proteins by differential scanning calorimetry demonstrates one important principle, that of the cooperativity of protein structures (Privalov & Khechinashvili, 1974). The 'one out, all out' principle results in the thermal denaturation of small, single domain proteins as being a 'two state' process. Thermal perturbation of state M will result in an equilibrium between states M and D with no significant population of intermediate states. This is itself a definition of a single domain protein. Intermediate states which can be detected by n.m.r. for example in general reflect minor perturbations from state M and do not differ from it significantly in heat content. Such experiments on larger proteins such as pepsinogen (Privolov *et al.* 1981) have shown the presence of more than one cooperative structure in the single protein molecule, which correlates well with the known X-ray structure.

Provided still that k_2 is negligible, the thermal denaturation of a protein M is in many cases fully reversible. Exceptions can occur when the protein has been modified after translation and folding for example by proteolytic activation. The modified polypeptide, once denatured, may now not have the information required to direct its folding into the lower free energy native state (Anfinsen, 1967).

How temperature can affect macromolecules

What is loosely described as 'temperature' in this context is actually the level of thermal energy. An increase in thermal energy causes quantized increases in vibration of interatomic bonds, affecting bond lengths, bond angles and the order of the macromolecular system. Non-covalent interactions in the form of hydrogen bonds, van der Waal's interactions and salt bridges will also be subject to increased vibration as the temperature rises until the participating atoms and groups separate with rupture of the interaction. The stacking interactions and hydrogen bonds which play a major role in stabilizing the double helix of DNA will tend increasingly to be dissociated until with a critical time-average proportion of interactions broken the remainder break spontaneously with the cooperative melting of the structure.

With globular proteins a further interaction is of major importance in stabilizing the folded structure. The hydrophobic interaction is responsible for the collapse of the polypeptide chain into a globular form, minimizing the contact of non-polar side chains with water (Tanford, 1968, 1970). Because water molecules cannot hydrogen bond to a non-polar surface, they become orientated in such a way as to maximize the number of hydrogen bonds formed with the surrounding water and in doing so form a less disordered lattice immediately around the non-polar group. Driven by the requirement to satisfy the hydrogen bonding groups there is a decrease in entropy ($\Delta S-$) accompanied by an increase in the heat capacity, (ΔC_p+). When two such non-polar surfaces are juxtaposed so as to exclude water, the above situation in water is reversed and the system has a larger entropy and, due to the relatively small enthalpy change, a lower free energy so that the groups are more stable when together than when apart. Like two Englishmen abroad the non-polar groups stay together, not because they have any particular affinity for each other but because they are constitutionally indisposed to interact with the environment. The hydrophobic interaction is best characterized and defined by the accompanying large negative ΔC_p.

With the $T\Delta S$ term making the dominant contribution to the free energy of 'interaction' ($\Delta G = \Delta H - T\Delta S$), ΔG will exhibit an unexpected temperature dependence, the clustering of hydrophobic groups being more stable as the temperature is raised. For example, in the association of tobacco mosaic virus subunits the free energy of interaction is dominated by the $T\Delta S$ term and as a

Table 1. *Variation of free energy of stabilization of β-lactamase with temperature*

T°C	8	15	30
[a]ΔG^l kJ^{-1} mol^{-1}	6·8	20·5	7·8

[a] Apparent free energies of stabilization were obtained (Thomas, 1981) by extrapolation of urea denaturation data to zero denaturant concentration by the method of Pace (1975). They refer to the inter-domain interaction within the structure.

result marked dissociation can be observed as the temperature is lowered below 30°C (Lauffer, 1966). At higher temperatures still, the thermal vibration in the water lattice will reduce the difference in order between bulk solvent and that immediately surrounding a non-polar group, thus reducing the favourable entropy difference between isolated and clustered groups, so that the contribution of the hydrophobic interaction would be expected to fall off.

Folded protein conformations involve the full range of non-covalent interactions with the hydrogen bond, van der Waal's interactions and electrostatic interactions entailing significant enthalpic contributions and negative temperature coefficients of stability. Although well characterized, the quantitative contribution of these and the hydrophobic interactions is uncertain so that the temperature dependence of stability for a globular protein of known sequence and structure is still unpredictable. A further temperature sensitive factor is the ionization constant of charged groups. Histidine and lysine have appreciably enthalpies of ionization and can therefore shift the isoelectric point of the protein under neutral pH conditions. Despite the uncertainty in the quantitative contribution to stability of the various types of interaction however, it is clear that the overall stabilizing forces are large compared with the actual stability of the protein. The latter takes values characteristically in the range $\Delta G' = 5-12$ kcal mol^{-1} whereas the entropic tendency to unfold the protein has been estimated in terms of a few hundreds of kcal mol^{-1}. The net stability is small, being equivalent in strength to that of two to four hydrogen bonds.

In some of the earlier work on thermostability Brandts (1964) showed that a temperature of maximum stability would be expected for globular proteins, generally between -10°C and 30°C with those below 0°C being experimentally inaccessible. Values of the thermodynamic stability of β-lactamase shown in Table 1 bear this out, showing a marked T_{max} in the region of 15°C. In principle, with a higher T_{max} and a steeper dependence of $\Delta G'$ on temperature below T_{max}, the situation could be envisaged where $\Delta G'$ becomes positive at lower temperatures and the protein unfolds. This possibility has been realized by Franks (1987) using supercooled systems. However many proteins and all allosteric enzymes involve interactions between subunits mediated by the same non-covalent interactions as those that stabilize the subunit conformation. The interfaces between subunits are in many cases made up of a high proportion of non-polar residues so that the interaction between the subunits shows the thermodynamic characteristics of the hydrophobic interaction. In other proteins such as phosphofructokinase, the

subunit interactions are sensitive to pH and, it is suggested, to temperature
sensitive ionization of histidine residues (Bock & Frieden, 1978). Depending on
the relative stabilities of the inter-subunit interactions and of the monomeric
subunit at a given temperature the following equilibria may obtain for a tetrameric
enzyme following temperature perturbation:

$$P_4 \underset{}{\overset{K_1}{\rightleftharpoons}} 4P \underset{}{\overset{K_2}{\rightleftharpoons}} 4D$$

If the dissociation constant K_1 is greater than K_2 for denaturation of the monomer,
intact monomers will be seen; if $K_2 > K_1$ the only populated states will be P_4 and D.

Low temperatures can bring about loss of activity

The phenomenon of cryoinactivation accounts for much of the frustration of
assiduous biochemists who store their enzyme preparation overnight in the cold
room only to find that it has lost activity by the morning. Dissociation has long
been suspected as the cause of cryoinactivation (Irias, Olmsted & Utter, 1969;
Bock & Frieden, 1978) but the degree of subunit dissociation seen on lowering the
temperature is not always sufficient to account for the high losses of activity
experienced. Recent work on lactate dehydrogenase (Weber, 1986; King &
Weber, 1986a,b) has shown that lowering the temperature results in slow inactivation of the enzyme in a way that involves both equilibrium and kinetic factors. A
simplified scheme of the explanation

$$
\begin{array}{ccc}
 & K_1 & \\
T & \rightleftharpoons & 4M \\
\text{slow} \uparrow & & \downarrow \text{slow} \\
 & K_2 & \\
T^* & \rightleftharpoons & 4M^*
\end{array}
$$

given by these authors is shown here. Slight but rapid dissociation of the tetramer
T occurs at reduced temperature (4°C) into monomer M. This is followed by
relatively slow 'conformational drift' into conformationally different monomers
M* which can associate rapidly to tetramers T* which are both inactive and possess
dissociation constants K_2 that are larger than K_1. In this reassociated form a
conformation rearrangement that is kinetically blocked at low temperatures can
occur at higher temperature to give the active tetramer once again.

This explanation, for which King and Weber have produced strong supporting
evidence, shows that a slow but essentially irreversible conformational change or
drift can pull across a dissociation equilibrium under conditions where there is only
a small amount of free monomer.

Evolutionary adaptation to different temperatures

We have shown that the conformation of proteins and nucleic acids can be
perturbed and disrupted by relatively small changes in temperature and that these

transitions are cooperative, occurring over ranges of temperature of a few degrees only. This occurs as a consequence of the delicate balance between relatively large stabilizing and destabilizing forces to give functional conformations of only marginal stability under given conditions. Thus enzymes of one species may be inactive at the temperature at which another species lives – bovine lactate dehydrogenase is inactivated at temperatures (and pressures!) at which coelocanth LDH and the LDH of extreme thermophilic archaebacteria are both presumably active. Animals which have adapted to live over a range of temperatures just below zero to about 50°C must have evolved by pathways which include adaptation of enzymes, nucleic acids and at least some structural proteins to structures which have optimum stability at the appropriate temperature and pressure.

Examination of these different cells and their respective proteins and nucleic acids should throw light on the way in which nature deals with the demands of thermal energy and also lead to a better understanding of the basis of conformational stability and its relation to function. To date there has been no spectacular success in this field but there are some suggestive pointers. In the field of microbial organisms various theories to explain their survival at elevated temperatures have been assessed and reviewed (Amelunxen & Murdock, 1978) and in general discarded to leave explanations to be found in terms of the structure and conformation of proteins and nucleic acids and of course membranes. At first sight the sequences of proteins from thermophilic organisms do not differ markedly from those of the mesophilic counterparts and the number of changes is not greater than the differences between a protein in different species which live at the same temperature. It will be surprising if the same sort of result is not obtained when sufficient three dimensional structures can be compared. From model building studies involving the fitting of sequences of a variety of ferredoxins to the known conformation of ferredoxin from *Micrococcus aerogenes*, Perutz & Raidt (1975) showed that an increased number of salt bridges could be found on the surface of more thermostable ferredoxins. Quantitatively this could readily account for the increased thermostability, so marginal are the requirements. There have been criticisms of this proposal both on the grounds of the basis of the thermostability measurements which did not meet the criteria discussed earlier and also on the expected contribution to stability of surface groups (Amelunxen & Murdock, 1978). The contribution of charged groups to the stabilization of a protein by electrostatic interactions integrated over the total surface is a much more complex calculation (Matthew, 1985) and has not been applied in this situation but may be a more appropriate way of approaching thermostability in ferredoxins rather than considering the amino acid substitutions as introducing 'struts' into the structure. It does seem clear however that only a few amino acid substitutions are necessary to affect substantially the stability.

The basis of thermostability appears somewhat more clear cut in the case of nucleic acids. It has long been recognized that the melting temperatures of DNA from different sources is related to the $(G+C)/(A+T)$ ratio, this being interpreted in terms of there being three hydrogen bonds between G and C, and only two between A and T. Recently a series of transfer RNAs and ribosomal RNAs have been sequenced from mesophilic and thermophilic organisms (Leinfelder, Jarsch & Bock, 1985; Wich & Bock, 1986). In the latter there is a marked increase in the number of G to C base pairs in the regions of the molecules proposed to be involved in structure. Those regions concerned with specificity do not differ between the RNAs from different organisms.

'From being to becoming'

This title of a book on the irreversible thermodynamics of biological systems (Prigogine, 1980) makes the important point that even the macromolecular structures in cells are not permanent but have a time dimension involving their biosynthesis and conformational folding, a degree of conformational mobility enabling them to perform their function, and a well-defined half life for turnover and breakdown. Temperature therefore can affect more than the conformational stability of a protein or nucleic acid.

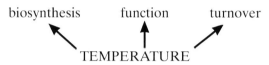

biosynthesis   function   turnover

TEMPERATURE

Biosynthesis

The rate of covalent synthesis of a polypeptide or polynucleotide will be temperature dependent as a natural consequence of being an enzyme catalysed chemical reaction. This will in general be expected to hold over a wide range of temperatures. The folding of the chain and its assembly into more complex structures, both driven and guided by non-covalent interactions, will also be temperature dependent, but can show more dramatic qualitative breaks in that dependence.

Jonathan King and his colleagues (Yu & King, 1984; Smith, Goldenberg & King, 1984) have studied the folding and assembly of the *Salmonella typhimurium* phage tail spike protein using genetic methods. The process is outlined in Fig. 1. Certain mutants have been isolated which either destabilize critical non-covalent interactions early on in the folding pathway or which positively stabilize the folding monomer in a conformation that aggregates spontaneously. Either explanation results in a block in folding so that the trimer state is never attained, the polypeptides being diverted into massive aggregates. The crucial feature is that these blocks are temperature dependent. At lower temperatures they are not operative and the newly synthesized polypeptide can fold and associate into a trimer which finally rearranges into the functional tail spike protein. The product is in each case as stable as or even more stable than the wild type product in which no block operates. At temperatures a few degrees higher the block prevents assembly of any functional product.

Temperature sensitive defects

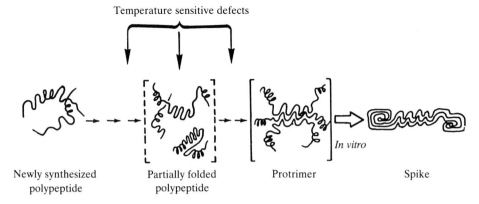

| Newly synthesized polypeptide | Partially folded polypeptide | Protrimer | Spike |

Fig. 1. The folding pathway for phage P22 tail spike endorhamnosidase. The suggested points in the pathway of the mutant proteins at which correct folding is inhibited are marked by arrows. (After Smith *et al.* 1984.)

An example of the inverse of the tail spike protein behaviour is shown by the assembly of ribosomes in yeast (Bayliss & Ingraham, 1974). A mutation resulted in cold sensitivity due to a block in the assembly of ribosomal particles at 15 °C. At 30 °C, the ribosomes are assembled as in the wild type and are then functional at 15 °C. In both these examples the amino acid replacements have no effect on the conformation or function of the assembled product but affect the stability of a critical interaction on the assembly pathway. The phenomenon may be explained in terms of there being several alternative pathways of assembly or folding (Harrison & Durbin, 1985) which would be open or blocked in the wild type protein in a temperature dependent fashion. Only some of the pathways would be blocked by the mutation so that at the permissive temperature assembly occurs by unperturbed pathways.

Temperature dependent production of proteins is important in several systems. A temperature-sensitive repressor can be used for high level expression of proteins in bacteria both by amplification of the plasmid copy number and by turning on a promoter (Remaut, Tsao & Fiers, 1983). The phenomenon of heat shock proteins (Burdon, 1987) involves temperature sensitive control of the production of proteins, again over quite narrow ranges of temperature. All such phenomena will involve interactions of proteins with nucleic acids or proteins and three different sorts of explanation can be advanced. (i) The interactions may be temperature-sensitive. For an interaction involving pure hydrophobic interaction, the effect of temperature on the $T\Delta S$ term is small and likewise on the other extreme of pure hydrophilic interactions. Cooperativity between sites, present in the attachment of regulatory proteins to nucleic acid (Ptashne *et al.* 1980), will amplify small changes in the free energy of binding of the single site, but unless there is a very large amount of cooperativity the final effect will not be large. (ii) A temperature dependent conformational change may be proposed as an explanation – and not as is so often the case, an expression of ignorance. The degree of cooperativity in simple protein systems does not approach that in DNA for example and an adequate shift in equilibrium with temperature is subject to the same questions as in (i). (iii) A temperature dependence in protein folding can exhibit a shift in the production of a functional protein over a narrow temperature range. Unlike the other explanations which depend on the effect of temperature on equilibria, it involves the more sensitive *balance* of kinetic processes such as folding *versus* aggregation. It remains to be seen whether this hypothesis is productive.

Function

In addition to the effect on structures, the temperature dependence of living processes must depend on the effect of temperature on the constituent chemical reactions. The magnitude of the rate constant is determined by the free energy difference between reactants and the transition state while the temperature dependence of the rate constant is dependent on the enthalpy of activation. The role of the enzyme is to stabilize the transition state in ways which are becoming clearer (Fersht *et al.* 1986). Detailed interpretation of the temperature dependence of enzyme catalysed reactions is complex. Interactions between substrate and enzyme take place through the variety of non-covalent interactions already discussed and thus exhibit a varied dependence upon temperature.

An aspect of proteins which is becoming increasingly recognized as being of importance in the functioning and control of enzymes is the dynamic nature of the conformation (Cooper, 1980). A protein structure is mobile as a result of thermal vibrations acting on a system of the folded polypeptide chain in which the constituent atoms are packed rather densely but not so much so as to preclude small but significant 'repacking' from taking place. The vibrational modes are in general greater for residues near the surface while structures such as β-pleated sheets persist and flex more slowly with a rippling motion. An example of modified mobility as an essential complement to function is that of cytochrome *c* (Salemme, 1977). This protein can exist in either the reduced Fe^{2+} or oxidized Fe^{3+} form in which states it will bind preferentially to cytochrome oxidase or to cytochrome reductase respectively. X-ray crystallography shows minor differences in the *average* position of the constituent atoms in the two states but several techniques provide evidence for the dynamic mobility of the two states being substantially different. The temperature dependent mobility of a protein structure occurs however within the intact globular state. Below the temperature of thermal denaturation, the dynamics will change with temperature without increase in exposure of buried groups to solvent (Delepierre *et al.* 1983).

As the distribution of nodes and antinodes on a glass plate set in vibration by a violin bow can be altered qualitatively by touching the edge of the plate, so the pattern of mobility within a globular protein can be modified on binding a substrate or an allosteric activator or inhibitor, thus affecting enzyme rate. Fructose-1,6-diphosphatase from rat liver is controlled by a product AMP acting as a feedback inhibitor. At 46°C the concentration of AMP that results in half-maximal inhibition is $8 \times 10^{-4}\,mol\,l^{-1}$ while at 2°C it is $8 \times 10^{-6}\,mol\,l^{-1}$ (Taketa &

Pogell, 1965). This admittedly non physiological effect may be a result of the temperature dependence of mobility of the enzyme affecting its allosteric properties with the corollary that the degree of mobility has been selected for the *in vivo* temperature to give the level of inhibition required at existing concentrations of AMP. Thus, in discussions of temperature dependence, the mobility must be taken into account, with the changes in enthalpy and entropy which will occur on changing temperature.

If mobility is essential to catalytic function then any form of stabilization, either artificial by protein engineering or natural by evolutionary selection could be detrimental to function. One might expect that stabilization for higher temperatures would have to be balanced by the provision of an adequate amplitude of vibration under the new conditions. It is possible to conduct cell-free protein synthesis using the components isolated from an extreme thermophile at temperatures between 80 and 90°C. Lowering the temperature below 60°C completely stops synthesis (Böck, 1986). This result is in keeping with the presence of a component whose mobility at sub-optimal temperatures is too small, owing to generally increased thermostability, for its biochemical function.

Turnover of macromolecules

The final stage in the transitory existence of macromolecules in the living cell is that of metabolic turnover and breakdown. The half-life of a functional protein must be carefully regulated so that it is removed after exerting its biological effect or in order to ensure localization of its function. One factor that is relevant to this process is the thermodynamic stability of the protein under the prevailing conditions of temperature and pH. A protein that has high stability would be expected to turn over more slowly than another with low stability and greater susceptibility to proteolytic cleavage. Thus human lysozyme should persist longer than phosphoglycerate kinase (Pain, 1979) and this would be in accord with the environment and function of these two proteins.

Conclusions

Much of the precise relationship between stability, dynamics and function has yet to be established in detail and the molecular basis of the relationships requires much more investigation, with closer collaboration between biologists and physical biochemists. One of the most interesting problems in relation to animal cells is their adaptation to function within a narrow temperature range. It seems a potentially productive hypothesis however to propose that the answer may be seen in the context of a balance between *stability* – just great enough to allow persistence for a physiologically appropriate time without being too great to inhibit the requisite turnover rate – and *dynamics* to allow efficient catalysis and control of metabolic activity.

32 R. H. PAIN

References

AMELUNXEN, R. E. & MURDOCK, A. L. (1985). Microbial life at high temperatures: Mechanisms and molecular aspects. In *Microbial Life in Extreme Environments* (ed. D. J. Kushner), pp. 217–278.

ANFINSEN, C. B. (1967). The formation of tertiary structure of proteins. *Harvey Lect.* **61**, 95–116.

BAYLISS, F. T. & INGRAHAM, J. L. (1974). Mutation is *Saccharomyces cerevisiae* conferring streptomycin and cold sensitivity by affecting ribosome formation and function. *J. Bact.* **118**, 319–337.

BOCK, P. E. & FRIEDEN, C. (1978). Another look at the cold lability of enzymes. *Trends biochem. Sci.* **3**, 100–103.

BÖCK, A. (1986). (Personal communication).

BRANDTS, J. F. (1964). The thermodynamics of protein denaturation. I. The denaturation of chymotrypsinogen. *J. Am. chem. Soc.* **86**, 4291–4301.

BRANDTS, J. F. (1967). Heat effects on proteins and enzymes. In *Thermobiology* (ed. A. H. Rose), pp. 25–72. New York & London: Academic Press.

BURDON, R. H. (1987). Thermotolerance and heat shock proteins (this volume).

COOPER, A. (1980). Conformational fluctuation and change in biological macromolecules. *Sci. Progr. Oxf.* **66**, 473–497.

DELEPIERRE, M., DOBSON, C. M., SELVARAJAH, S., WEDIN, R. E. & POULSEN, F. M. (1983). *J. mol. Biol.* **168**, 687–692.

FERSHT, A. R., LEATHERBARROW, R. J. & WELLS, T. N. C. (1986). Binding energy and catalysis: a lesson from protein engineering of the tyrosyl-tRNA synthetase. *Trends biochem. Sci.* **11**, 321–325.

FRANKS, F. (1987). Water, temperature and life (this volume).

HARRISON, S. C. & DURBIN, R. (1985). Is there a single pathway for the folding of a polypeptide chain? *Proc. natn. Acad. Sci. U.S.A.* **82**, 4028–4030.

IRIAS, J. J., OLMSTEAD, M. R. & UTTER, M. F. (1969). Pyruvate carboxylase. Reversible inactivation by cold. *Biochemistry* **8**, 5136–5147.

KING, L. & WEBER, G. (1986a). Conformational drift of dissociated lactate dehydrogenases. *Biochemistry* **25**, 3632–3636.

KING, L. & WEBER, G. (1986b). Conformational drift and cryoinactivation of lactate dehydrogenase. *Biochemistry* **25**, 3637–3640.

LAPANJE, S. (1978). *Physicochemical aspects of protein denaturation.* New York: Wiley.

LAUFFER, M. (1966). Polymerisation – depolymerisation of tobacco mosaic virus protein. VII. A model. *Biochemistry* **5**, 2440–2446.

LEINFELDER, W., JARSCH, M. & BÖCK, A. (1985). The phylogenetic position of the sulfur-dependent archaebacterium *Thermoproteus tenax*: sequence of the 16S rRNA gene. *Syst. appl. Microbiol.* **6**, 164–170.

MATTHEW, J. B. (1985). Electrostatic effects in proteins. *A. Rev. Biophys. Biophys. Chem.* **14**, 387–417.

PACE, C. N. (1975). The stability of globular proteins. *C.R.C. Crit. Rev. Biochem.* **3**, 1–43.

PAIN, R. H. (1979). Conformation and stability of folded proteins. In *Characterisation of Protein Conformation and Function*, pp. 19–36. London: Symposium Press.

PERUTZ, M. F. (1980). In *Protein Folding* (ed. R. Jaenicke), p. 13–14. Amsterdam: Elsevier.

PERUTZ, M. F. & RAIDT, H. (1975). Stereochemical basis of host stability in bacterial ferredoxins and in haemoglobin A2. *Nature, Lond.* **255**, 256–259.

PRIGOGINE, I. (1986). *From being to becoming: time and complexity in the physical sciences.* San Francisco: W. H. Freeman.

PRIVALOV, P. L. (1979). Stability of proteins. *Adv. Protein Chem.* **33**, 167–241.

PRIVALOV, P. L. & KHECHINASHVILI, N. N. (1974). A thermodynamic approach to the problem of stabilisation of globular protein structure: a calorimetric study. *J. mol. Biol.* **86**, 665–684.

PRIVALOV, P. L., MATEO, P. L., KHECHINASHVILI, N. N., STEPANOV, V. N. & REVINA, L. P. (1981). Comparative thermodynamic study of pepsinogen and pepsin structure. *J. mol. Biol.* **152**, 445–464.

PTASHNE, M., JEFFREY, A., JOHNSON, A. D., MAURER, R., MEYER, B. J., PABO, C. O., ROBERTS, T. M. & SAUER, R. T. (1980). How the λ repressor and Cro work. *Cell* **19**, 1–11.

REMAUT, E., TSAO, H. & FIERS, W. (1983). Improved plasmid vectors with a thermoinducible expression and temperature-regulated runaway replication. *Gene* **22**, 103–113.

SALEMME, F. R. (1977). Structure and function of cytochromes *c*. *A. Rev. Biochem.* **46**, 299–329.

SMITH, D. H., GOLDENBERG, D. P. & KING, J. (1984). Use of temperature sensitive mutants to dissect pathways of protein folding and subunit interaction. In *The Protein Folding Problem* (ed. D. B. Wetlaufer), pp. 115–143. Colorado: Westview Press.

TAKETA, K. & POGELL, B. M. (1965). Allosteric inhibition of rat liver fructose 1,6-diphosphatase by adenosine 5'-monophosphate. *J. biol. Chem.* **240**, 651–662.

TANFORD, C. (1968). Protein denaturation. *Adv. Protein Chem.* **23**, 121–282.

TANFORD, C. (1970). Protein denaturation. *Adv. Protein Chem.* **24**, 1–95.

THOMAS, R. (1981). *Folding intermediates in Staphylococcal penicillinase.* Ph.D. Thesis, University of Newcastle upon Tyne.

WEBER, G. (1986). Phenomenological description of the association of protein subunits subjected to coformational drift. Effects of dilution and of hydrostatic pressure. *Biochemistry* **25**, 3626–3631.

WICH, G. & BÖCK, A. (1986). (Personal communication).

YU, M.-H. & KING, J. (1984). Single amino acid substitutions influencing the folding pathway of the phage P22 tail spike endorhamnosidase. *Proc. natn. Acad. Sci. U.S.A.* **81**, 6584–6588.

Printed in Great Britain © *Society for Experimental Biology 1987*

THE EFFECTS OF TEMPERATURE ON BIOLOGICAL MEMBRANES AND THEIR MODELS

DAVID C. LEE and DENNIS CHAPMAN

Department of Biochemistry and Chemistry, Royal Free Hospital School of Medicine (University of London), Rowland Hill Street, London NW3 2PF, UK

Summary

The physical effects of temperature on biological membranes are reviewed. Our current understanding of membrane structure is based on a model in which proteins are inserted in or are attached to a lipid bilayer structure. The lipid composition varies between cell types. Phospholipid bilayers undergo phase transitions at temperatures which are dependent on chain length and the structure of the headgroup. These factors also affect the miscibility of phospholipid types. Cholesterol orders the phospholipid chains above their phase transition temperature but disorders them below this temperature. Cerebrosides, which are important constituents of the myelin membrane, form metastable states under certain conditions. The activities of membrane-bound enzymes may be affected by the physical state of the lipid bilayer which, in turn, is affected by temperature. The interactions between the lipid and protein components of membranes have been investigated by a variety of techniques.

Introduction

Structural and functional studies of biological membranes have long been recognized as important in advancing our understanding of the barriers which delimit the cell and its organelles. The cell membrane is important in regulating ion transport, cell recognition, receptor-mediated processes, energy transduction and as a barrier to diffusion. Models for the structure of membranes have evolved from a rigid lipid bilayer to the more complex and highly variable fluid mosaic model proposed by Singer & Nicolson (1972). The functional asymmetry of the cell membrane is reflected in structural asymmetries. Transport of molecules across the cell membrane must be directional and any receptor, enzyme active site or immunological determinant is found on only one side of the membrane.

Studies of the effects of temperature on cell membranes are important not only for their physiological effects on membrane function but also as a tool for structural perturbation. Thus the phase properties of membrane lipids and the details of the interaction between lipids and proteins have been revealed by following the effects of temperature on these components in model systems. In order to study the function of membrane proteins in detail it has been necessary to purify them from their original environment and reconstitute them into defined lipid vesicles. This review aims to summarize the important advances in our understanding of the effects of temperature on biological membranes and their models which have been made with the assistance of biophysical techniques.

Membrane lipids

Almost all of the lipids found in biological membranes are amphipathic, possessing a hydrophilic polar headgroup and a hydrophobic hydrocarbon region. The structural diversity among membrane lipids arises from variations in each of these regions. The most abundant lipids are the phospholipids, phosphatidyl-choline (PC) being predominant in animal membranes, phosphatidylethanolamine (PE) in bacterial membranes and glycolipids in plant membranes. Glycosphingo-lipids and gangliosides differ from sphingomyelin in having a short carbohydrate chain in place of the normal headgroup. Gangliosides act as receptors for some hormones and toxins on the cell surface. Cholesterol is the major sterol found in mammalian plasma membranes, lysosomal membranes and Golgi membranes, with sitosterol and stigmasterol predominant in plant plasma membranes.

The acyl chains of natural membrane lipids are, in most cases, even numbered. The proportion of odd-numbered fatty acyl chain substituents is rarely above 2 mole percent. Eighty percent of the acyl chains in most membranes are C_{16}, C_{18} and C_{20} substituents. Unsaturated chains are present in some membranes. The $C_{18:1}$, $C_{18:2}$, $C_{18:3}$ and $C_{20:4}$ lipids are the major unsaturated species (Rouser et al. 1968). It was proposed (Chapman et al. 1966) that 'the particular distribution of fatty acyl residues present in a membrane provides the appropriate membrane fluidity at a particular environmental temperature to match the diffusion rate or rate of metabolic processes required for the tissue'.

The effects of temperature on phospholipids

Although biological membranes contain several types of lipid, physico-chemical studies of bilayers of pure lipids have provided much useful information on the structure and dynamics of lipids within membranes.

Pure, anhydrous 1,2-diacylphosphatidylcholines exhibit a capillary melting point at 230 °C (Chapman, 1975) which is largely independent of acyl chain length. Below this temperature phospholipids may exist in several mesomorphic forms similar to those of simple soaps (Chapman et al. 1966, 1967). The physical state of the lipid is dependent on both temperature (thermotropic mesomorphism) and water content (lyotropic mesomorphism). Fully hydrated phospholipids spon-taneously form bilayers which may exist in the gel or crystalline (L_β) state and the fluid, liquid-crystalline (L_α) state. In the former, the fatty acyl chains are fully extended in an all-*trans* conformation and packed in a quasi-hexagonal array. Phosphatidylcholines in the gel state have their acyl chains tilted with respect to the bilayer normal ($L_{\beta'}$ state). The transition to the liquid-crystalline state is readily achieved by raising the temperature. Rotational isomerization about the C-C bonds of the fatty acyl chains results in the formation of *gauche* isomers producing kinks in the chains. These kinks are highly mobile along the length of the chain and occur with increasing probability toward the methyl terminus. The increased disorder of the liquid-crystalline state has been demonstrated by X-ray diffraction (Chapman *et al.* 1967).

The phase transition to the liquid-crystalline state is endothermal (Fig. 1) and the temperature at which it occurs is dependent on both bilayer hydration and the phospholipid species present. The gradual addition of water to the phospholipid reduces the phase transition temperature to a limiting value (T_c) on complete hydration. The phase transition temperature of a phospholipid is also influenced by the combined nature of the polar headgroup and the fatty acyl chains. Longer fatty acyl chain species have higher transition temperatures (Keough & Davis, 1979) but the presence of unsaturated bonds reduces this value. *Cis* double bonds are more effective in disrupting chain packing than *trans* double bonds. Most naturally occurring phospholipids have one saturated and one unsaturated chain with a transition temperature midway between those of the respective phospholipids with identical chains (Keough & Davis, 1979).

The structure and the electrostatic charge of the headgroup influence the phase transition temperature. The structure of the phosphatidylcholine headgroup induces a tilt in the fatty acyl chains reducing the interchain interaction at the methyl terminus. Thus phosphatidylcholines have a reduced transition temperature when compared with phosphatidylethanolamines of identical chain composition. The presence of a negative charge at the headgroup produces charge repulsion between adjacent headgroups and lateral expansion favouring the liquid-crystalline state. The phase transition temperature of negatively charged phospholipids is extremely sensitive to pH, cation concentration and ionic strength (Verkleij *et al.* 1974; Trauble & Eibl, 1974).

The transition occurs over a narrow temperature range, usually 1–2°C. During this process the two phases coexist. A structural model for the transition has been proposed (Lee, 1977). Here, small pools of fluid lipid form below the phase transition temperature and grow larger as the temperature rises, eventually coalescing to form a complete liquid-crystalline phase. On cooling, clusters of gel phase lipid form in a 'sea' of fluid lipid. Clusters in which the hydrocarbon chains are regularly packed constitute the centres from which the complete gel phase is formed. Thus the transition is a highly cooperative event and hysteresis occurs because of the differing paths of the melting and freezing phenomena. This model also implies the presence of a domain of interfacial lipid between the L_α and L_β domains. The free energy of the lipid in this region is high, owing to the mismatch of molecular packing between the adjacent domains. A rapid cooperative transition is promoted to reduce the number of molecules in this state. The presence of highly disordered interfacial lipid may account for the increase in permeability to ions and small molecules which occurs during the phase transition.

Phosphatidylcholines exhibit a second endothermic event, termed the pretransition, a few degrees below the main transition (see Fig. 1). This transition involves reorientation in both the acyl chain and headgroup regions. Early X-ray data (Tardieu *et al.* 1973) were used to deduce a perpendicular-to-tilted reorientation of the chains. In addition, a periodic ripple was observed in freeze-fracture electron micrographs of liposomes quenched from temperatures below, but not above, the main transition (Verkleij *et al.* 1972). Nuclear magnetic resonance (NMR) studies have also indicated a conformational change of the headgroup during the pretransition (Oldfield *et al.* 1971; Salsbury *et al.* 1972). A combined X-ray diffraction and calorimetric study (Janiak *et al.* 1976) has shown that below the pretransition the lipid fatty acyl chains are tilted in the $L_{\beta'}$ phase. The angle of tilt was shown to decrease with increasing temperature. At the pretransition a transformation from a one- to a two-dimensional lattice occurs, the lipid lamellae being distorted by a periodic ripple in the $P_{\beta'}$ phase.

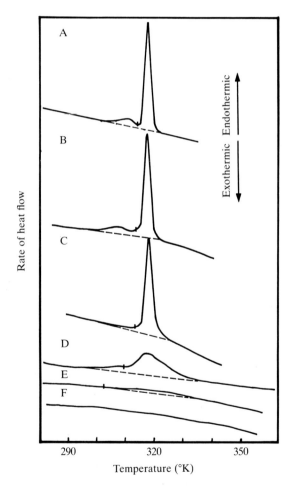

Fig. 1. Differential scanning calorimetry (DSC) curves of 50 wt % dispersions in water of 1,2-dipalmitoylphosphatidylcholine/cholesterol mixtures containing (A) 0·0 mol%, (B) 5·0 mol%, (C) 12·5 mol%, (D) 20·0 mol%, (E) 32·0 mol% and (F) 50·0 mol% cholesterol. (After Ladbrooke *et al.* 1968.)

Phase separations in mixtures of pure phospholipids

Biological membranes, being composed of a mixture of lipid species, do not undergo a single sharp phase transition. However, on heating and cooling, phase separations can occur which result in the coexistence, over a particular temperature range, of distinct lipid phases of differing chemical composition.

A variety of physical techniques, particularly calorimetry and electron spin resonance (ESR), have been used to demonstrate phase separations in binary mixtures of phospholipids. If two lipids have identical headgroups and acyl chains of similar length (for example dimyristoyl- and dipalmitylphosphatidylcholine) then mixing is ideal in both the fluid and solid phases. However, when the chain length or unsaturation of the two components is different, monotectic behaviour is observed (Phillips *et al.* 1970) (see Fig. 2). Immiscibility in the fluid state is rare but has been observed in mixtures of dielaidoylphosphatidylcholine and dipalmitoyl-phosphatidylethanolamine (DPPE) (Wu & McConnell, 1975) along with the expected solid-phase immiscibility.

Phase separations in mixtures of negatively-charged and neutral phospholipids can be induced by changes in pH or cation concentration (Ito & Ohnishi, 1974). A domain of Ca^{2+}-induced, solid-phase acidic phospholipid may exist within a bulk pool of fluid, neutral phospholipid.

The effect of cholesterol

Cholesterol is an important constituent of many biological membranes. The cholesterol:phospholipid ratio has been measured in a variety of animal membranes and is found to vary from 0·11 in liver mitochondria to 1·32 in brain myelin

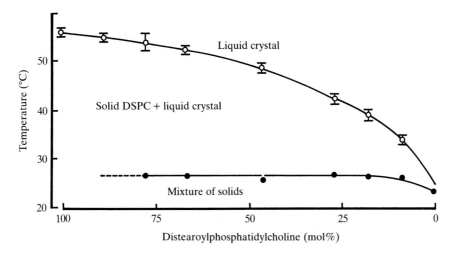

Fig. 2. Temperature-composition diagram for distearoylphosphatidylcholine-dimyristoylphosphatidylcholine mixtures obtained by plotting the (○) onset temperature of the DSC cooling curve and the (●) onset temperature of the DSC heating curve of each mixture in excess water. (After Phillips *et al.* 1970.)

(Ashworth & Green, 1966). Cholesterol influences a wide variety of membrane-mediated cellular processes. Effects on permeability (Chapman, 1975), membrane enzyme activity (Poznansky et al. 1973; Klein et al. 1978) and physiological processes such as cell–cell adhesion, fusion and endocytosis have been demonstrated.

X-ray diffraction experiments have indicated that the lamellar arrangement of the phospholipids is preserved on incorporation of cholesterol (Ladbrooke et al. 1968). Cholesterol is oriented perpendicular to the plane of the bilayer as shown by an increase in the X-ray long-spacing up to a concentration of 7·5 %. Above this concentration a reduction in the long spacing is indicative of a reduction in interchain forces and chain fluidization. The β-hydroxyl group of cholesterol forms hydrogen bonds with the carbonyl oxygen which links the phospholipid acyl chain with the glycerol backbone (Huang, 1977). The sterol ring interacts with the adjacent fatty acyl chain to inhibit the formation of gauche conformers above the phase transition temperature.

Ladbrooke et al. (1968) studied the dipalmitoylphosphatidylcholine (DPPC)-cholesterol-water system by differential scanning calorimetry and X-ray diffraction. They showed that the incorporation of cholesterol lowered the enthalpy of the phase transition and the temperature at which it occurred. Equimolar mixtures of DPPC and cholesterol did not exhibit a transition. This ratio corresponds to the maximum amount of cholesterol which can be introduced into the bilayer before cholesterol precipitation occurs. A later study pointed out that below 20 mole percent cholesterol the calorimetric curves could be resolved into narrow and broad components suggesting the coexistence of two immiscible solid phases (Mabrey et al. 1978). The condensation effect is not restricted to DPPC – it is also seen with unsaturated phosphatidylcholines and the lipid extract of human erythrocyte ghosts (Ladbrooke et al. 1968).

Laser Raman spectroscopy (Lippert & Peticolas, 1971) has shown that cholesterol causes an increase in the proportion of gauche conformers below the phase transition temperature and a decrease in the proportion of gauche conformers above this temperature. This observation has also been observed using difference infrared (IR) spectroscopy (Cortijo & Chapman, 1981) (Fig. 3).

The ordering effect of cholesterol on the acyl chains above the phase transition temperature T_m, and the disordering effect below T_m, has been clearly demonstrated using deuterium NMR (Jacobs & Oldfield, 1979; Rice et al. 1979). Above T_m, the addition of cholesterol at an equimolar ratio caused an increase in the quadrupole splitting from 3·6 kHz to 7·8 kHz for dimyristoylphosphatidylcholine (DMPC) deuterium labelled at C14 (Rice et al. 1979). This is equivalent to an increase in molecular order parameter (S_{mol}) from 0·18 to 0·41. Below T_m, cholesterol caused a decrease in quadrupole splitting. Similar effects were observed using DMPC which was deuterium-labelled at C6, however, the relative increase in order parameter above T_m became smaller toward the top of the hydrocarbon chain.

Electron spin resonance (ESR) has shown that cholesterol inhibits chain motion in liquid-crystalline egg yolk lecithin and increases chain motion in DPPC below its transition temperature (Oldfield & Chapman, 1971).

Cerebrosides

Cerebrosides are the simplest mammalian glycosphingolipids and are found in large quantities in the myelin membrane and in the brush border membrane of the intestinal wall. Their phase transition temperatures are extremely high with respect to body temperature and it is likely that part of their function lies in ordering the membranes of which they form a major part. In aqueous dispersion they form lamellar arrays which undergo complex phase transitions upon heating or cooling. Cerebrosides which contain saturated or predominantly saturated (as in glucocerebrosides from Gaucher's patient spleen) acyl chains and which are void of α-hydroxy residues give metastable states upon fast cooling from the liquid-crystalline state (Freire *et al.* 1980). Upon reheating the metastable state transforms into the stable one producing an exothermic peak at approximately 65 °C. This behaviour is primarily due to head group interactions and is accompanied by changes in hydration which cause further changes in chain packing. X-ray diffraction and calorimetry have shown that this results in two

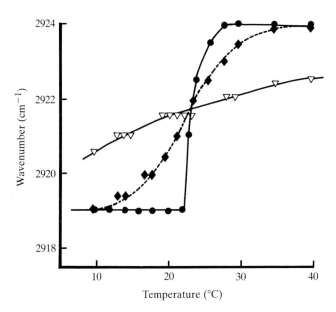

Fig. 3. Temperature dependence of the frequency of the CH_2 asymmetric stretching vibration in DMPC (●) and DMPC/cholesterol mixtures (◆ 22:1 molar ratio; ▽ 1·5:1 molar ratio). The frequency of this C–H stretching vibration reflects the static order (*trans/gauche* isomerization) of the hydrocarbon chains of the lipid. In the pure DMPC system a sharp increase in frequency indicates a loss of order at the phase transition. This effect is progressively removed by addition of increasing amounts of cholesterol. (After Costijo & Chapman, 1981.)

forms in the gel state: a stable tightly packed form and a looser metastable form. Our recent infrared studies have shown that inter-molecular hydrogen bonding between neighbouring amide groups is important in the stable state (Lee *et al.* 1986). Reduced hydrogen bonding to the amide groups was found in the metastable state.

Lipid–protein interactions

Significance in membranes

The fluid mosaic model of membrane structure, specifically the proposed insertion of membrane proteins within the hydrophobic core of the lipid bilayer, has instigated much recent research in the field of lipid–protein interactions. Research has been directed toward the influence of integral proteins on the structure and dynamics of the bilayer and, conversely, the effect of the physical state of the bilayer on the structure and function of integral proteins. These problems have been approached using calorimetric, spectroscopic and biochemical techniques. Much of the current controversy in this field has arisen from the difficulty in summarizing conceptually the results obtained from these diverse approaches.

It is known that the activities of many membrane enzymes are related to the physical state of the bilayer (Hesketh *et al.* 1976). Thus, lipid–protein interactions in natural membranes are important in both sealing integral proteins into the bilayer while maintaining the permeability barrier, and in controlling the conformation of integral proteins in the fluid matrix.

Calorimetric studies

Early work by Chapman *et al.* (1974) demonstrated that the interaction of cytochrome *c* with phosphatidylserine (PS) bilayers caused a reduction in the main transition temperature. Interaction with this protein is mainly electrostatic and there is little penetration into the bilayer.

The effects of a variety of proteins on the thermotropic properties of phospholipid bilayers were examined by Papahadjopoulos *et al.* (1975). They defined three types of interaction according to the effects on the main transition. Cytochrome *c* and A1-basic myelin protein decreased T_c and the transition enthalpy of phosphatidylglycerol (PG) bilayers but not of PC bilayers. This, again, suggested an electrostatic interaction with limited bilayer penetration. Myelin proteolipid N-2 apoprotein and gramicidin A had almost no effect on T_c but caused a linear decrease in enthalpy with increasing protein concentration. They concluded that these proteins are embedded within the hydrocarbon core of the bilayer perturbing a limited number of adjacent lipid molecules. Finally, the water-soluble proteins ribonuclease and polylysine caused an increase in enthalpy with or without an increase in T_c. In these cases interaction takes place between positively charged protein residues and negatively charged lipids, stabilizing the bilayer without penetration.

The length of the phospholipid fatty acyl chains is important in determining the perturbation of the bilayer. Myelin basic protein raises the T_c of di-C_{12}-PG (Verkleij *et al.* 1974) but lowers the T_c of PG with longer chains (Papahadjopoulos *et al.* 1975).

The number of lipid molecules perturbed by the presence of an integral protein or polypeptide can be estimated by plotting the transition enthalpy against the mole ratio of protein to lipid and extrapolating to zero enthalpy (Fig. 4). In the case of bacteriorhodopsin, 19 molecules of DMPC and 22 molecules of DPPC per molecule of protein were removed from the cooperative transition (Alonso *et al.* 1982). For Ca^{2+}/Mg^{2+}-ATPase this figure was found to be 45 molecules of DMPC and 42 molecules of DPPC (Gomez-Fernandez *et al.* 1980). A decrease in transition enthalpy together with no change or an increase in T_c has been detected for a wide range of integral proteins (Bach, 1983). The precise number of lipid

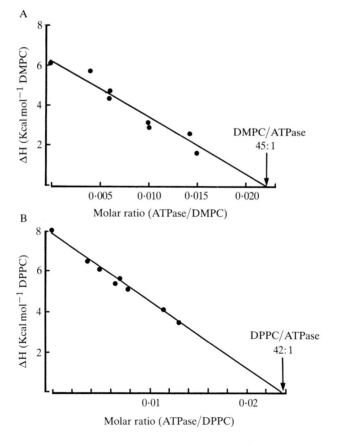

Fig. 4. Plots of the enthalpy change (ΔH) of the main calorimetric endotherm as a function of the molar ratio in (A) DMPC/Ca^{2+}-ATPase and (B) DPPC/Ca^{2+}-ATPase recombinants. (After Gomez-Fernandez *et al.* 1980.)

molecules perturbed by insertion of the protein depends on both the protein and the type of lipid studied.

Electron spin resonance studies

ESR spectra of lamellar phospholipids containing fatty acid spin labels have provided information on the mobility of the lipid acyl chains. When incorporated into bilayers, spin labels with nitroxide groups attached at various positions along the fatty acid hydrocarbon chain are sensitive to the rotational mobility of the adjacent C-C bonds of the phospholipids.

Jost et al. (1973a,b) using 16-doxylstearate and 5-doxylstearate spin labels in vesicular preparations of cytochrome oxidase, recorded spectra which could be resolved into two components. The relative intensities of these components altered as the phospholipid to protein ratio was varied. The mobile component was shown to be similar to the ESR spectra of isolated lipids in the lamellar form, reflecting the fluid, bulk lipid. The motionally restricted component, which was stronger at low lipid to protein ratios, was thought to reflect a layer of boundary lipid immobilized at the protein surface. Aqueous bilayer systems containing the relatively simple polypeptide gramicidin A also display an immobilized component in ESR spectra (Chapman et al. 1977a) suggesting that this effect is not restricted to proteins which require lipid for enzyme activity.

ESR has also been used to demonstrate that cytochrome oxidase preferentially interacts with lipids containing specific types of head group (Knowles et al. 1981). Various headgroup derivatives of 14-doxyl phospholipid spin labels were incorporated in DMPC-cytochrome oxidase complexes with identical lipid to protein ratio. A stronger immobilized component was observed with a cardiolipin spin label than with PC, PE, PG and phosphatidylserine spin labels suggesting a specificity of protein-lipid interactions.

Nuclear magnetic resonance studies (NMR)

In contrast to ESR, which requires the addition of a spin label to the system under study, NMR spectroscopy provides a non-perturbing method for analysing the structure of membranes and model systems.

Deuterium NMR has provided much useful information. The position of the reporter ^2H nuclei in the lipid fatty acyl chains may be selected and the low natural abundance of ^2H nuclei ensures that the observed resonances derive entirely from the chosen reporter. The residual deuterium quadrupole splitting $\Delta\nu_Q$, provides an estimate of the orientational order in the environment of the C-^2H bond. High quadrupole splitting values are characteristic of a high degree of order whereas rapid tumbling in isotropic solution reduces the spectrum to a single line. In contrast to an earlier study (Dahlquist et al. 1977), Oldfield et al. (1978) found no evidence for the presence of ordered boundary lipid, above T_c, with a variety of membrane proteins and polypeptides in reconstituted systems. This work was extended to a comparison of the effects of integral proteins and cholesterol on lipid bilayers which were deuterated at a variety of positions along the acyl chains (Rice *et al.* 1979). Again these workers found no evidence for chain ordering above the main lipid transition temperature on incorporation of integral proteins, in contrast to the effect of cholesterol. The quadrupole splitting was slightly reduced in the presence of integral proteins compared with the pure fluid bilayer, suggesting that some disordering occurs at the lipid–protein interface. Exchange between bulk lipid and protein-associated lipid was fast on the deuterium NMR time scale ($>10^3 \, \text{s}^{-1}$).

Deuterium NMR studies of intact membranes have been reported. In *Escherichia coli*, grown on media supplemented with deuterated fatty acids, the orientational order of the outer membrane was greater than that of the cytoplasmic membrane at each temperature studied (Davis *et al.* 1979). Later, Gally *et al.* (1980) showed that the fatty acyl chain order of both membranes was less than that of a bilayer of total lipid extract. Distortion of the acyl chains probably occurs so as to accommodate the irregular surfaces of integral proteins.

Fluorine-19 NMR has indicated that below T_c, both rigid crystalline and protein-perturbed lipid regions occur in *E. coli* cell membranes (Gally *et al.* 1981). Above T_c, only one narrow resonance was detected indicating that rapid exchange between bulk and boundary lipid occurs. The conformation of the phospholipid glycerol backbone is also not altered by the presence of high amounts of integral proteins in *E. coli* (Gent & Ho, 1978).

^{31}P-NMR has provided both structural and motional information on phospholipid headgroups in model systems (Skarjune & Oldfield, 1979; Seelig *et al.* 1981). The phosphorus chemical shielding anisotropy is related to phospholipid headgroup structure and motional information is obtained by measurement of spin-lattice relaxation times (T_1). Oldfield and co-workers (1981) have found an increase in ^{31}P-NMR linewidth and a decrease in both T_1 and T_2 (spin-spin relaxation time) for liquid-crystalline bilayers containing a variety of proteins. These results suggest that the phospholipid headgroup is immobilized by integral proteins.

Infrared spectroscopic studies

Infrared (IR) spectroscopy is a non-perturbing technique sensitive to molecular vibrations on a time-scale which is complimentary to the ESR and NMR methods. Perturbations to the hydrophobic acyl chain region may be studied *via* the C-H asymmetric and symmetric stretching frequencies. These parameters are sensitive to the static order (*trans/gauche* isomerization) of the acyl chains. Cortijo *et al.* (1982) have reported the effects of the intrinsic proteins Ca^{2+}-ATPase and bacteriorhodopsin and the intrinsic polypeptide gramicidin A on the acyl chain order of phosphatidylcholine bilayers. They found that below T_m these molecules behaved in a similar manner to cholesterol; that is they caused an increase in the proportion of *gauche* conformers. Above T_m, the perturbation of the acyl chains differs from that observed with cholesterol. At high lipid:protein molar ratios (e.g. DMPC/Ca^{2+}-ATPase 150:1) a reduction in the proportion of *gauche* conformers with respect to the pure lipid bilayer was observed. However, when the concentration of the intrinsic protein or polypeptide was increased this effect was removed and the static order of the acyl chains was essentially the same as in the pure lipid system. This IR study also provided further evidence for a reduction in the cooperativity of the phase transition in the presence of these intrinsic molecules.

More recently, we have investigated lipid–protein interactions using diperdeuteriomyristoylphosphatidylcholine, an analogue of DMPC in which the acyl chain hydrogen atoms are completely replaced by deuterium atoms (Lee *et al.* 1984). This enabled lipid C-^{2}H stretching frequencies to be measured without interference from the absorption of amino acid side-chain absorptions. It was shown that above the lipid phase transition temperature, low concentrations of gramicidin A and alamethicin caused a small ordering of the lipid chains while bacteriorhodopsin had no effect.

IR spectroscopy has also been used to monitor lipid–protein interactions in endogenous biomembranes. Intact and deproteinated plasma membranes of *Acholeplasma laidlawii* were studied after the biosynthetic incorporation of diperdeuteriopalmitoylphosphatidylcholine (Casal *et al.* 1980). Plots of the temperature dependence of the C^2H_2 symmetric stretching frequency indicated a broad phase transition, centred at the growth temperature for the organism, which was interpreted as a regulatory mechanism for cell growth. The effect of the endogenous protein is to stabilize the acyl chains in their preferred conformation at a particular temperature.

In subsequent studies (Cameron *et al.* 1983, 1985) these workers investigated the thermotropic behaviour of *A. laidlawii* grown on a fatty acid-depleted medium supplemented with perdeuterated fatty acids and an inhibitor of endogenous fatty acid synthesis. Analysis of the C^2H stretching frequencies revealed that the isolated membranes are predominantly in the gel state at the growth temperature. However, 95–100 % of the lipids in the live cell membranes are in the liquid-crystalline phase at 37 °C.

Protein distribution

An important consequence of the insertion of protein molecules into a phospholipid bilayer is the generation of lateral packing faults. Model studies have shown that packing faults radiate from an inserted protein when it does not occupy the space of an integral number of phospholipid molecules (Chapman *et al.* 1977*b*). As the amount of inserted protein is increased these packing faults become comparable in size to the inter-protein spacing. The observed reduction in the cooperativity of the phospholipid phase transition as the protein to lipid ratio is increased (Bach, 1983) supports this model.

When the protein concentration is not too high the bulk phospholipid can form a gel phase below its transition temperature. Under these circumstances protein molecules are forced to segregate into protein-rich regions which contain small domains of trapped lipid. This effect has been observed in freeze-fracture electron micrographs (see Fig. 5). When Ca^{2+}-ATPase is reconstituted into vesicles of dipalmitoylphosphatidylcholine the micrographs show that there is a random distribution of protein molecules (revealed as intramembranous particles) within the phospholipid matrix. However, when samples are flash-frozen from temperatures below the phase transition discrete areas of crystalline lipid and clusters of protein are revealed.

Protein diffusion

Physical techniques have been developed for measuring both lateral and rotational movement of proteins within membranes. Early experiments on the lateral diffusion of rhodopsin in the rod outer segment membrane were made by Poo & Cone (1974) using the technique of fluorescence recovery after photo-bleaching (FRAP). The rotational motion of proteins has been measured by saturation transfer electron spin resonance (Thomas *et al.* 1976) and laser-induced dichroism (Razi-Naqvi *et al.* 1973).

The freedom of integral proteins to diffuse laterally within the plane of the bilayer varies from one cell type to another. Rhodopsin has a lateral diffusion coefficient of $4\times10^{-9}\,cm^2\,s^{-1}$ in rod outer segments (Poo & Cone, 1974) whereas there is little, if any, lateral movement of proteins in the erythrocyte membrane (Peters *et al.* 1974). Erythrocyte membranes do not undergo a phase transition upon cooling to $0\,°C$ (Ladbrooke *et al.* 1968) and do not show low-temperature phase separations (Wunderlich *et al.* 1974). The cytoskeletal framework plays an important role in restricting protein diffusion and aggregation. The high cholesterol content of some mammalian membranes is important in preventing phase separation of lipids and proteins at low temperatures (Ferber *et al.* 1972).

Biochemical studies

Biochemical approaches to the study of lipid–protein interactions have mainly concentrated on the measurement of membrane enzyme activities in endogenous, delipidated and reconstituted environments.

In reconstituted systems, Ca^{2+}/Mg^{2+}-ATPase required a minimum of around 30 lipid molecules to maintain maximal activity (Warren *et al.* 1973). Activity was completely inhibited below $30\,°C$ in DPPC-ATPase complexes and below $25\,°C$ in DMPC-ATPase complexes, the latter corresponding to the phospholipid phase transition temperature. This work was extended to show that ATPase activity was maintained down to $30\,°C$ over a wide range of DPPC:ATPase ratios, up to 1500:1 (Hesketh *et al.* 1976). Again, a minimum of about 35 lipid molecules were required for activity. It was shown that at least 30 lipid molecules interact directly with the

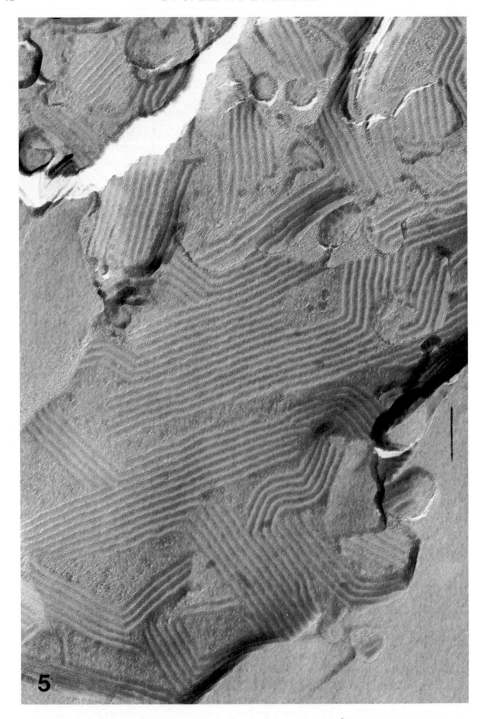

Fig. 5. Freeze-fracture electron micrograph of a DMPC/Ca^{2+}-ATPase recombinant (237:1 molar ratio) quenched from 18 °C. This shows a replica of the surface of a vesicle at a temperature between the pre-transition and the main transition for this lipid. The rippling of the lipid lamellae in the $P_{\beta'}$ phase is shown as is the clustering of the intra-membranous particles (which correspond to proteins). Scale bar represents 0·2 μm. (After Gomez-Fernandez et al. 1980.)

ATPase and that these molecules cannot undergo a phase transition at 41 °C. However, a transition at this temperature was detected in the presence of excess lipid corresponding to an endothermic event in the bulk lipid phase.

The increase in ATPase activity at 30 °C in DPPC bilayers is coincident with a sudden increase in rotational motion (Hoffman *et al.* 1980). Both phenomena occur approximately 10 °C below the main transition temperature. Hoffman *et al.* (1980) invoked protein aggregation to interpret these effects. As noted above, discrete areas of crystalline lipid together with protein patches are observed in freeze-fracture electron micrographs obtained below the phase transition temperature. The increase in protein rotation and enzymatic activity is due to the protein patch melting whilst the bulk lipid melts at higher temperatures.

Conclusions

Variations in temperature produce a variety of effects on biological membranes. Temperature-dependent phase transitions and phase separations affect the fluidity of the membrane and may lead to protein aggregation. The activity of membrane-bound enzymes is, in some cases, strongly dependent on membrane fluidity.

The authors thank the Wellcome Trust for financial support.

References

ALONSO, A., RESTALL, C. J., TURNER, M., GOMEZ-FERNANDEZ, J. C., GONI, F. M. & CHAPMAN, D. (1982). Protein–lipid interactions and differential scanning calorimetric studies of bacteriorhodopsin reconstituted lipid-water systems. *Biochim. biophys. Acta* **689**, 283–289.

ASHWORTH, L. A. E. & GREEN, C. (1966). Plasma membranes: phospholipid and sterol content. *Science* **151**, 210–211.

BACH, D. (1983). Calorimetric studies of model and natural membranes. In *Biomembrane Structure and Function* (ed. D. Chapman), pp. 1–41. Topics in Molecular and Structural Biology, vol. 4. London: MacMillan.

CAMERON, D. G., MARTIN, A. & MANTSCH, H. H. (1983). Membrane isolation alters the gel to liquid-crystal transition of *Acholeplasma laidlawii* B. *Science* **219**, 180–182.

CAMERON, D. G., MARTIN, A., MOFFAT, D. J. & MANTSCH, H. H. (1985). Infrared spectroscopic study of the gel to liquid-crystal phase transition in live *Acholeplasma laidlawii* cells. *Biochemistry* **24**, 4355–4359.

CASAL, H. L., CAMERON, D. G., SMITH, I. C. P. & MANTSCH, H. H. (1980). *Acholeplasma laidlawii* membranes: a Fourier transform infrared study of the influence of protein on lipid organisation and dynamics. *Biochemistry* **19**, 444–451.

CHAPMAN, D. (1975). Phase transitions and fluidity characteristics of lipids and cell membranes. *Q. Rev. Biophysics* **8**, 185–235.

CHAPMAN, D., BYRNE, P. & SHIPLEY, G. G. (1966). Solid state and mesomorphic properties of some 2,3-diacyl-D1-phosphatidylethanolamines. *Proc. Roy. Soc.* **A290**, 115–142.

CHAPMAN, D., CORNELL, B. A., ELIASZ, A. W. & PERRY, A. (1977*a*). Interactions of helical polypeptide segments which span the hydrocarbon region of lipid bilayers. Studies of the gramicidin A-lipid-water system. *J. molec. Biol.* **113**, 517–538.

CHAPMAN, D., CORNELL, B. A. & QUINN, P. J. (1977*b*). Phase transitions, protein aggregation and a new method for modulating membrane fluidity. *FEBS Symposium* **42** (ed. G. Semenaz & E. Carafoli), pp. 72–85. Berlin: Springer-Verlag.

CHAPMAN, D., URBINA, J. & KEOUGH, K. H. (1974). Biomembrane phase transitions. Studies of lipid-water systems using differential scanning calorimetry. *J. biol. Chem.* **249**, 2512–2521.

Chapman, D., Williams, R. M. & Ladbroke, B. D. (1967). Physical studies of phospholipids. VI. Thermotropic and lyotropic mesomorphism of some 1,2-diacyl-phosphatidylcholines (lecithins). *Chem. Phys. Lipids* **1**, 445–475.

Cortijo, M., Alonso, A., Gomez-Fernandez, J. C. & Chapman, D. (1982). Intrinsic protein–lipid interactions. Infrared spectroscopic studies of gramicidin A, bacteriorhodopsin and Ca-ATPase in biomembranes and reconstituted systems. *J. molec. Biol.* **157**, 597–618.

Cortijo, M. & Chapman, D. (1981). A comparison of the interactions of cholesterol and gramicidin A with lipid bilayers using an infrared data station. *FEBS Lett.* **131**, 245–248.

Dahlquist, F. W., Muchmore, D. C., Davis, J. H. & Bloom, M. (1977). Deuterium magnetic resonance studies of the interaction of lipids with membrane proteins. *Proc. natn. Acad. Sci. U.S.A.* **74**, 5435–5439.

Davis, J. H., Nichol, C. P., Weeks, G. & Bloom, M. (1979). Study of the cytoplasmic and outer membranes of *Escherichia coli* by deuterium magnetic resonance. *Biochemistry* **18**, 2103–2112.

Ferber, E., Resch, K., Wallach, D. F. H. & Imm, W. (1972). Isolation and characterisation of lymphocyte plasma membranes. *Biochim. biophys. Acta* **266**, 494–504.

Freire, E., Bach, D., Correa-Freire, M., Miller, I. R. & Barenholz, Y. (1980). Calorimetric investigation of the complex phase behaviour of glucocerebroside dispersons. *Biochemistry* **19**, 3662–3664.

Gally, H. U., Pluschke, G., Overath, P. & Seelig, J. (1980). Structure of *Escherichia coli* membranes. Fatty acyl chain order parameters of inner and outer membranes and derived liposomes. *Biochemistry* **19**, 1638–1643.

Gally, H. U., Pluschke, G., Overath, D. & Seelig, J. (1981). Structure of *Escherichia coli* membranes. Glycerol auxotrophs as a tool for the analysis of phospholipid head group region by deuterium magnetic resonance. *Biochemistry* **20**, 1826–1831.

Gent, M. P. N. & Ho, C. (1978). Fluorine-19 nuclear magnetic resonance studies of lipid phase transitions in model and biological membranes. *Biochemistry* **17**, 3023–3038.

Gomez-Fernandez, J. C., Goni, F. M., Bach, D., Restall, C. J. & Chapman, D. (1980). Protein–lipid interaction. Biophysical studies of (Ca+Mg)-ATPase reconstituted systems. *Biochim. biophys. Acta* **598**, 502–516.

Hesketh, T. R., Smith, G. A., Houslay, M. D., McGill, K. A., Birdsall, N. J. M., Metcalfe, J. C. & Warren, G. B. (1976). Annular lipids determine the ATPase activity of a calcium transport protein complexed with dipalmitoyllecithin. *Biochemistry* **15**, 4145–4151.

Hoffmann, W., Sarzala, G. M., Gomez-Fernandez, J. C., Goni, F. M., Restall, C. J., Chapman, D., Heppeler, G. & Kreutz, W. (1980). Protein rotational diffusion and lipid structure of reconstituted systems of Ca-activated adenosine triphosphatase. *J. molec. Biol.* **141**, 119–132.

Huang, C. H. (1977). A structural model for the cholesterol-phosphatidylcholine complexes in bilayer membranes. *Lipids* **12**, 348–356.

Ito, T. & Ohnishi, S.-I. (1974). Ca-induced lateral phase separations in phosphatidic acid-phosphatidylcholine membranes. *Biochim. biophys. Acta* **352**, 29–37.

Jacobs, R. & Oldfield, E. (1979). Deuterium nuclear magnetic resonance investigation of dimyristoyllecithin-dipalmitoyllecithin and dimyristoyllecithin-cholesterol mixtures. *Biochemistry* **18**, 3280–3285.

Janiak, M. J., Small, D. M. & Shipley, G. G. (1976). Nature of the thermal pretransition of synthetic phospholipids: dimyristoyl- and dipalmitoyl lecithin. *Biochemistry* **15**, 4575–4580.

Jost, P. C., Griffith, O. H., Capaldi, R. A. & Vanderkooi, G. (1973a). Identification and extent of fluid bilayer regions in membranous cytochrome oxidase. *Biochim. biophys. Acta* **311**, 141–152.

Jost, P. C., Griffith, O. H., Capaldi, R. A. & Vanderkooi, G. (1973b). Evidence for boundary lipid in membranes. *Proc. natn. Acad. Sci. U.S.A.* **70**, 480–484.

Keough, K. M. W. & Davis, P. J. (1979). Gel to liquid-crystalline phase transitions in water dispersions of saturated mixed-acid phosphatidylcholines. *Biochemistry* **18**, 1453–1459.

Klein, I., Moore, L. & Pastan, I. (1978). Effect of liposomes containing cholesterol on adenylate cyclase activity of cultured mammalian fibroblasts. *Biochim. biophys. Acta* **506**, 42–53.

Knowles, P. F., Watts, A. & Marsh, D. (1981). Spin-label studies of head-group specificity in the interaction of phospholipids with yeast cytochrome oxidase. *Biochemistry,* **20**, 5888–5894.

LADBROOKE, B. D., WILLIAMS, R. M. & CHAPMAN, D. (1968). Studies on lecithin-cholesterol-water interactions by differential scanning calorimetry and X-ray diffraction. *Biochim. biophys. Acta* **150**, 333–340.

LEE, A. G. (1977). Lipid phase transitions and phase diagrams. I. Lipid phase transitions. *Biochim. biophys. Acta* **472**, 237–281.

LEE, D. C., DURRANI, A. A. & CHAPMAN, D. (1984). A difference infrared spectroscopic study of gramicidin A, alamethicin and bacteriorhodopsin in perdeuterated dimyristoylphosphatidylcholine. *Biochim. biophys. Acta* **769**, 49–56.

LEE, D. C., MILLER, I. R. & CHAPMAN, D. (1986). An infrared spectroscopic study of metastable and stable forms of hydrated cerebroside bilayers. *Biochim. biophys. Acta* **859**, 266–270.

LIPPERT, J. L. & PETICOLAS, W. L. (1971). Laser Raman investigation of the effect of cholesterol on conformational changes in dipalmitoyllecithin multilayers. *Proc. natn. Acad. Sci. U.S.A.* **68**, 1572–1576.

MABREY, S., MATEO, P. L. & STURTEVANT, J. M. (1978). High-sensitivity scanning calorimetric study of mixtures of cholesterol with dimyristoyl- and dipalmitoylphosphatidylcholines. *Biochemistry* **17**, 2464–2468.

OLDFIELD, E. & CHAPMAN, D. (1971). Effects of cholesterol and cholesterol derivatives on hydrocarbon chain mobility in lipids. *Biochem. biophys. Res. Commun.* **43**, 610–616.

OLDFIELD, E., GILMORE, R., GLAZER, M., GUTOWSKY, H. S., HSHUNG, J. C., KANG, S. Y., KING, T. E., MEADOWS, M. & RICE, D. (1978). Deuterium nuclear magnetic resonance investigation of the effects of proteins and polypeptides on hydrocarbon chain order in model membrane systems. *Proc. natn. Acad. Sci. U.S.A.* **75**, 4657–4660.

OLDFIELD, E., JANES, N., KINSEY, R. A., KINTANAR, A., LEE, R. W. K., ROTHGEB, T. M., SCHRAMM, S., SKARJUNE, R., SMITH, R. L. & TSAI, M.-D. (1981). Protein crystals, membrane proteins and membrane lipids. Recent advances in the study of their static and dynamic structures using NMR techniques. *Biochem. Soc. Symp.* **46**, 155–181.

OLDFIELD, E., MARSDEN, J. & CHAPMAN, D. (1971). Proton NMR relaxation study of mobility in lipid water systems. *Chem. Phys. Lipids* **7**, 1–8.

PAPAHADJOPOULOS, D., MOSCARELLO, M., EYLAR, E. H. & ISAC, T. (1975). Effects of proteins on thermotropic phase transitions. *Biochim. biophys. Acta* **401**, 317–335.

PETERS, R., PETERS, J., TEWS, K. H. & BAHR, W. (1974). A microfluorometric study of translational diffusion in erythrocyte membranes. *Biochim. biophys. Acta* **367**, 282–294.

PHILLIPS, M. C., LADBROOKE, B. D. & CHAPMAN, D. (1970). Molecular interactions in mixed lecithin systems. *Biochim. biophys. Acta* **196**, 35–44.

POO, M. M. & CONE, R. A. (1974). Lateral diffusion of rhodopsin in the photoreceptor membrane. *Nature, Lond.* **247**, 438–441.

POZNANSKY, M., KIRKWOOD, D. & SOLOMON, A. K. (1973). Modulation of red cell K^+ transport by membrane lipids. *Biochim. biophys. Acta* **330**, 351–355.

RAZI-NAQVI, K., GONZALEZ-RODRIGUEZ, J., CHERRY, R. J. & CHAPMAN, D. (1973). Spectroscopic technique for studying protein rotation in membranes. *Nature New Biol.* **245**, 249–251.

RICE, D. M., MEADOWS, M. D., SCHEINMAN, A. O., GONI, F. M., GOMEZ-FERNANDEZ, J. C., MOSCARELLO, M. A., CHAPMAN, D. & OLDFIELD, E. (1979). Protein–Lipid Interactions. A nuclear magnetic resonance study of sarcoplasmic reticulum Ca, Mg-ATPase, lipophilin and proteolipid apoprotein-lecithin systems and a comparison with the effects of cholesterol. *Biochemistry* **18**, 5893–5903.

ROUSER, G., NELSON, G. J., FLEISCHER, S. & SIMON, G. (1968). Lipid composition of animal cell membranes, organelles and organs. In *Biological Membranes* (ed. D. Chapman), pp. 5–69. London: Academic Press.

SALSBURY, N. J., DARKE, A. & CHAPMAN, D. (1972). Deuteron magnetic resonance studies of water associated with phospholipids. *Chem. Phys. Lipids* **8**, 142–151.

SEELIG, J., TAMM, L., HYMEL, L. & FLEISCHER, S. (1981). Deuterium and phosphorus nuclear magnetic resonance and fluorescence depolarisation studies of functional reconstituted sarcoplasmic reticulum membrane vesicles. *Biochemistry* **20**, 3922–3932.

SINGER, S. J. & NICOLSON, G. L. (1972). The fluid mosaic model of the structure of cell membranes. *Science* **175**, 720–731.

Skarjune, R. & Oldfield, E. (1979). Physical studies of cell surface and cell membrane structure. Determination of phospholipid head group organisation by deuterium and phosphorus nuclear magnetic resonance spectroscopy. *Biochemistry* **18**, 5903–5909.

Tardieu, A., Luzatti, V. & Remen, F. C. (1973). Structure and polymorphism of the hydrocarbon chains of lipids: A study of lecithin-water phases. *J. molec. Biol.* **75**, 711–731.

Thomas, D. D., Dalton, L. R. & Hyde, J. S. (1976). Rotational diffusion studied by passage saturation transfer electron paramagnetic resonance. *J. Chem. Phys.* **65**, 3006–3024.

Trauble, H. & Eibl, H. (1974). Electrostatic effects on lipid phase transitions: Membrane structure and ionic environment. *Proc. natn. Acad. Sci. U.S.A.* **71**, 214–219.

Verkleij, A. J., de Kruyff, B., Ververgaert, P. H. J., Tocanne, J. F. & Van Deenen, L. L. M. (1974). The influence of pH, Ca and protein on the thermotropic behaviour of the negatively charged phospholipid, phosphatidyl glycerol. *Biochim. biophys. Acta* **339**, 432–437.

Verkleij, A. J., Ververgaert, P. H. J., Van Deenen, L. L. M. & Elbers, P. F. (1972). Phase transitions of phospholipid bilayers and membranes of *Acholeplasma laidlawii* B visualised by freeze-fracture electron microscopy. *Biochim. biophys. Acta* **288**, 326–332.

Warren, G. B., Toon, P. A., Birdsall, N. J. M., Lee, A. G. & Metcalfe, J. C. (1974). Reversible lipid titrations of the activity of pure adenosine triphosphatase-lipid complexes. *Biochemistry* **13**, 5501–5507.

Wu, S. H.-W. & McConnell, H. M. (1975). Phase separations in phospholipid membranes. *Biochemistry* **14**, 847–854.

Wunderlich, F., Batz, W., Speth, V. & Wallach, D. F. H. (1974). Reversible thermotropic alteration of nuclear membrane structure and nucleocytoplasmic RNA transport in *Tetrahymena. J. Cell Biol.* **61**, 633–640.

Printed in Great Britain © *Society for Experimental Biology 1987*

TEMPERATURE EFFECTS ON RED CELL MEMBRANE TRANSPORT PROCESSES

J. C. ELLORY and A. C. HALL

University Laboratory of Physiology, Parks Road, Oxford OX1 3PT

It is generally, but not universally true that chemical and biological processes slow down with decreasing temperature. One might therefore expect that temperature effects on membrane transport would be predictable, and perhaps unexciting. In practice, some surprises and some fundamental information have emerged from investigating the influence of temperature on specific pathways.

Most studies involving temperature effects in biology fall into two categories. In one case temperature is used as a fundamental thermodynamic variable in the hope of gaining an insight from its effects on transport into the mechanisms of transport function. Alternatively, temperature is of interest in the context of cells, tissues or animals surviving in different thermal conditions, where membrane adaptation can be an essential part of the response. In the present paper we describe the effects of temperature on a wide variety of membrane transport systems concentrating on the erythrocyte as a model system and showing that the response of a given pathway to decreasing temperature is often surprisingly complex. In particular, although there have been recent improvements in methodology which have allowed the study of discrete transport systems, there can be several pitfalls which await the unwary experimenter.

Transport studies as a function of temperature are usually best performed on whole cells. Although there is an extensive literature on temperature effects on soluble enzymes (e.g. Hochachka & Somero, 1973), red cell transport studies have indicated that there are discrepancies between transport in whole cells compared with measurement of transport-related biochemical parameters e.g. Na/K-activated ATPase in isolated membranes (Ellory & Willis, 1976). For intact cells, there are at least three levels at which temperature can exert its effects. The most obvious is by a direct action on the properties of the transporter molecule itself. An interesting recent example of this is the temperature dependence of Na channel activation in skeletal muscle membranes where there is evidence that the break in the Arrhenius curve of channel activity seen on cooling below 6°C probably reflects an effect on the channel protein itself, rather than on the lipid environment (Kirsch & Sykes, 1987). Similar conclusions have also been drawn from experiments on the sarcoplasmic reticulum Ca-ATPase (see below). However, for most transport systems their function is critically dependent on the properties of the membrane with which the transporter is associated. Much attention has been paid to the role of membrane lipids as a control mechanism and has produced the concept of 'homeoviscous adaptation' involving changes in

membrane lipids, affecting the fluidity of the membrane. This approach has been
most successful for prokaryotes (e.g. *Acholeplasma laidlawii*; *Tetrahymena*) and
poikilotherms (e.g. goldfish) (see Cossins, 1981, this volume). Finally, temperature may have indirect effects on membrane transport pathways *via* changes in the
physico-chemical characteristics of biological systems (e.g. pN of water, pK of
protein charged groups), and also cellular metabolism (e.g. alterations in ATP
levels, or intracellular pH or Ca).

Perhaps it is relevant at this stage to justify the importance of studying
membrane transport in the context of temperature. Selective permeation of cell
membranes is a fundamental property of living systems. There is an obvious
requirement for movement of metabolically relevant molecules (glucose, amino
acids, lactate). Additionally, active transport systems for extruding Na, Ca and
protons or for accumulating K are critical for maintenance of cell volume and
intracellular composition. Changing temperature will affect membrane permeation systems differentially; active transport processes have a higher temperature sensitivity than 'simple' passive permeation. This means that the steady-state
may only be maintained over a limited temperature range so that extremes of
temperature will ultimately lead to gross changes in cell composition which are
lethal.

Types of membrane transport processes

Membrane transport pathways may be classified into four distinct categories,
(Table 1). The simplest of these is the basal or ground state permeability which is
defined theoretically as the 'simple' electrodiffusive leak, and operationally as the
residual flux which remains after all known transport pathways have been
inhibited. This flux should follow Fick's law of diffusion and be linearly dependent
on the concentration (i.e. not be saturable). The next order of complexity is
facilitated transport *via* a specific system operating either as a channel or carrier.
Although such systems are capable of mediating enormous fluxes, (e.g. 3 moles of
Cl/1. cells/min *via* Band 3), they are gradient-driven i.e. non-concentrative.
Secondary active transport utilizes the inward sodium gradient as the energy
source for the translocation of biologically important molecules. Examples include
the symport of amino acids and glucose, and the antiport of Ca and H. Simple cotransport has a stoichiometry of 1 Na to transported species, whilst more complex
carriers can either transport several species simultaneously (e.g. Na–K–2 Cl cotransport) or have different coupling ratios (e.g. 3 Na × Ca exchange). Finally,
primary active transport is driven by ATP hydrolysis and is responsible for
pumping ions across membranes and maintaining the appropriate intracellular
concentrations of Na, K and Ca. Thus, to describe the temperature characteristics
of membrane transport at the cellular level requires the consideration of the
properties of each of these types of transport systems in detail. Before discussing
this, a final concern is the methods for identifying discrete systems. By far the most
useful criterion is the use of specific pharmacological inhibitors (e.g. ouabain for

Table 1. *Classification of membrane transport processes*

1. Electrodiffusive leak	
True passive permeability	
2. Facilitated transport	
Carriers	Channels
glucose transporter	Ca-activated K
nucleoside transporter	channel (Gardos)
amino acid transporters (L,y+)	
anion transporter (Band 3, capnophorin)	
3. Secondary active transport	
Cotransport	Countertransport
Simple	Simple
sodium–amino acid (system ASC)	sodium–hydrogen
	exchange
Complex	Complex
sodium, potassium cotransport	sodium–calcium
sodium–glycine transport	exchange
4. Primary active transport	
Na/K pump	
Ca pump	

All pathways shown are found in mammalian red cells. For further details and definitions see Ellory & Tucker (1983); Stein (1986).

the Na/K pump, bumetanide for the (Na+K) cotransport system). In some cases we do not yet have ideal inhibitors and therefore other less specific criteria have to be used. Alternatives include the application of enzyme kinetics (Michaelis-Menten substrate dependence; competitive inhibition studies), metabolic depletion to inhibit primary active transport, and the use of ionophores and uncouplers to control transmembrane gradients.

(i) Basal membrane permeability

Diffusion of a wide range of electrolytes and non-electrolytes in solution even including high molecular weight species such as albumin (60 000 D) has a simple temperature dependence characterized by low energies of activation (Table 2), and obeying the Stokes-Einstein equation (Stein, 1986). When a membrane barrier is introduced, however, non-Stokesian behaviour can become an important and complicating factor. For the simplest model system (e.g. phospholipid membranes prepared as liposomes) with selected substrates (notably non-electrolytes) over certain temperature ranges the permeability changes monotonically with temperature with low activation energies (Table 2). For ions (e.g. Na, K) even for this simple paradigm, complex behaviour is often observed particularly at the so-called 'critical' (or phase transition) temperature of the lipids. This is presumably due to phase separation and the mismatch between

Table 2. *Values for the apparent activation energies (E_a) for the permeation of various electrolytes and non-electrolytes through water, across artificial lipid membranes and erythrocyte membranes*

Self diffusion	E_a (kJ mol^{-1})	References
H_2O	20	1,2,3,4
K, Na	16	
Urea	19	
Glucose	21	
Alanine	20	
Bovine serum albumin	21	
Liposomes		
H_2O	60	5
Na	complex; increases at T_c then decreases	6,7
Na	55 (DPPG with equimolar cholesterol)	7
K	23 (+gramicidin A)	8
Red cells		
'Pores'		
H_2O (total)	26	9
H_2O (+PCMBS)	60	
K	14–52 (<18·5 °C) (depends on species)	
	22–79 (>18·5 °C) (depends on species)	10
*Ca-sensitive K channel	>200	11
'Carriers'		
Glucose	84 (<20 °C)	12
	30–42 (37 °C)	
L-leucine	71 (>20 °C)	13
	197 (<20 °C)	
Urea	34 (total at 1 M urea)	14
	59 (basal)	
*Band 3 (Capnophorin)	84 (>15 °C)	15
	126 (<15 °C)	
*(Na+K) co-transport	74 (>5 °C)	16
Na/K pump	104 (>15 °C)	17
Ca pump	60–142 (range of values from various red cell preparations)	18

References
1. Wang *et al.* 1953; 2. Wang, 1965; 3. Bruins, 1929; 4. Stein, 1986; 5. Cass & Finkelstein, 1967; 6. Papahadjopoulos *et al.* 1973; 7. Singer, 1982; 8. Hladky & Haydon, 1972; 9. Brahm, 1982; 10. Hall & Willis, 1986; 11. Simons, 1976; 12. Sen & Widdas, 1962; 13. Hoare, 1972; 14. Brahm, 1983; 15. Brahm, 1977; 16. Hall *et al.* 1982; 17. Ellory & Willis, 1982; 18. Rega & Garrahan, 1986.
* Activation energies for some transport systems have been calculated without taking into account kinetic effects, i.e. under non-saturating substrate concentrations. These values should, therefore be treated as underestimates. (Abbreviation: DPPG, dipalmitoylphosphatidylglycerol.)

adjacent phases (Antonov *et al.* 1980; Bohein *et al.* 1980). Increasing liposome complexity using lipid mixtures including cholesterol added in equimolar amounts, has been shown to reduce Na diffusion rates at all temperatures, and yield linear Arrhenius plots in marked contrast to results obtained when single lipids are studied. Thus, cholesterol present in the same ratio as that in erythrocyte membranes inhibits the phase-transition-induced changes in Na permeability (Papahadjopoulos *et al.* 1973).

Liposome permeability may be dramatically increased by the addition of specific ionophores, and these mediated fluxes can show even greater temperature sensitivity (in the case of nystatin a change of $\times 10^4$ can occur between 26° and 36°C (Cass *et al.* 1970)), although in other cases, a relatively low temperature dependence has been reported (gramicidin A; Hladky & Haydon, 1972). It is obvious that increasing the complexity of model membranes by using mixed lipids, asymmetric bilayers and incorporating proteins to approach the real biological situation will complicate further the analysis of the effects of temperature on liposome permeability. We therefore intend to treat this situation empirically and consider the case for red cell membranes directly.

Fig. 1 shows the basal K uptake in red cells from various species as a function of temperature. In these experiments the Na/K pump is inhibited by ouabain and the (Na+K) co-transport system by bumetanide (see Hall & Willis, 1986). Under these conditions at 37°C, K uptake is a linear function of the external K concentration which is indirect support for the notion that no carrier-mediated mechanisms are involved. It can be seen that lowering temperature can have a variety of effects. In the simplest case (hedgehog, Fig. 1(A)), there is a monotonic fall in K permeability. In contrast, in guinea pig cells (Fig. 1(B)) such simple behaviour only occurs in the absence of Ca added to the medium, or in the presence of quinine. The paradoxical increase in K flux seen on cooling in guinea pig cells in the presence of Ca can be simply explained as a failure of the Ca pump

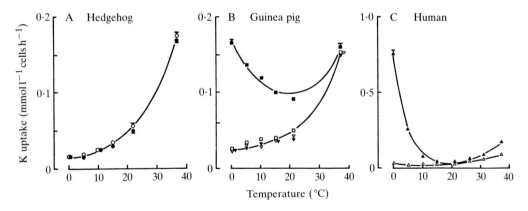

Fig. 1. The temperature dependence of passive K uptake in red cells of different species under various conditions. Passive K influx (using ^{86}Rb as a congener) was studied using standard techniques in the presence of ouabain and bumetanide to inhibit Na/K pump and (Na+K) co-transport systems respectively (see Hall & Willis, 1984*b*, 1986, for further experimental details). The symbols represent the following conditions; hedgehog control (○), +Ca (1·5 mM) (●); guinea pig control (□), +Ca (1·5 mM) (■), control + quinine (0·5 mM) (▽), Ca + quinine (▼); human control (△), +NMDG (▲). Results are means ± S.E.M., $n = 3$, with errors not shown where they were smaller than the symbols.

as temperature is reduced, leading to an increased level of intracellular Ca and subsequent stimulation of the Ca-sensitive K channel (Gardos effect (Gardos, 1958; Hall & Willis, 1984a,b)). The apparent absence of the Gardos effect in hedgehog red cells can be explained by the persistence of active Ca pump activity in the cold which maintains intracellular Ca at low levels in this hibernator. Finally, human erythrocytes represent a real exception to this situation (Fig. 1(C)). Even when the Ca concentration is at micromolar levels, there is a minimum for K uptake at about 12°C, below which there is a paradoxical increase in K uptake (Stewart *et al.* 1980). If extracellular Na is replaced by a variety of cations (e.g. NMDG (*n*-methyl *d*-glucamine) as illustrated), a marked enhancement of this effect is observed. This can be inhibited by small monovalent cations (e.g. Na, Li) at relatively low concentrations (1 mM) (Blackstock & Stewart, 1986). Similar paradoxical effects on Na and K permeability are observed when chloride is replaced by various anions (e.g. SCN; Weith, 1970). On the other hand ghosting of red cells abolishes this paradoxical increase at reduced temperature (Simons, 1976) resulting in a linear temperature profile for both Na and K permeability, with low activation energies (25 kJ mol^{-1}).

Of the 20 species which have been studied to date, only human and primate red cells exhibit this 'paradoxical' increase in residual cation permeability as temperature is reduced. The K permeability of the other species shows a continuous decrease in flux with cooling (Fig. 1; and Hall & Willis, 1986). These data have also been expressed as the Arrhenius relationship, where for some species apparently bimodal behaviour was observed. Over the lower temperature range (i.e. below about 18·5°C) values for the E_a are very low and for the red cells of some species are close to the values for free ionic diffusion through bulk water (Table 2). This has led to the suggestion that under these conditions K transport across the erythrocyte membrane of these species is mediated by water-filled 'pores' (Hall & Willis, 1986). Thus, after the inhibition of all known K transport systems including the elimination of the Ca effect (Fig. 1(B)), most species' red cells respond to cooling by a modest continuous decrease in K permeation. The complex effects of temperature on the 'residual' permeability in human (and primate) but not in other species' erythrocytes, mean that it is difficult to avoid the conclusion that some cryptic (and as yet poorly understood) pathways remain, which under certain conditions are unmasked. Similar complex activating effects on cation permeability have been observed by reducing the ionic strength of incubation media (Bernhardt *et al.* 1984).

(ii) Diffusional water permeability

Brahm (1982) has studied the influence of temperature on the diffusional water permeability of human red cells and their ghosts. The apparent activation energies for water exchange in human red cells, ghosts and ghosts treated with 1 mM PCMBS (*p*-chloro-mercuribenzene sulphonate) were 21, 30 and 60 kJ mol^{-1} respectively with no evidence of a change in slope over the temperature range investigated (2·5–38 °C). The increased activation energy observed after the addition of PCMBS was explained by the inhibition of a water transport pathway associated with the intermediate phase between integral membrane proteins and their surrounding lipids. Thus, in PCMBS-treated cells the permeability of the membrane for water approaches that expected in artificial lipid membranes (Cass & Finkelstein, 1967; Table 2). Additionally, this ground state permeability exhibits a temperature dependence which is very similar to the theoretically and experimentally verified activation energy of about 55 kJ mol^{-1} for water diffusion through the hydrocarbon phase of a lipid membrane (Redwood & Haydon, 1969; Price & Thompson, 1969). Significantly, the temperature dependence of water permeability in the absence of PCMBS is similar in magnitude to that for the self-diffusion of water (Table 2).

(iii) Mediated passive transport

Facilitated 'passive' transport which relies on specific membrane carriers or channels represents the next level of complexity. Conventionally, these two transport systems have been differentiated on kinetic grounds (see Stein, 1986 for review). In fact for the red cell only limited data are available on channels, but Simons (1976) in human erythrocyte ghosts, and Hall & Willis (1984*b*) using intact rodent red cells have shown that the Ca-activated K channel is relatively temperature-sensitive. For comparison, both early (Hodgkin *et al.* 1952; Franken-hauser & Moore, 1972) and recent (Kirsch & Sykes, 1987) experiments on gating of Na channels in excitable membranes (nerve, muscle) have indicated a high temperature sensitivity.

The effects of temperature on carrier-mediated transport in red cells have been studied in some detail, in particular for amino acid (L-leucine) (Hoare, 1972), glucose (Sen & Widdas, 1962; Lowe & Walmsley, 1986), aldose (Fisher & Nimmo, 1973), urea (Brahm, 1983) and anion (*via* Band 3) transport (Brahm, 1977) (Table 2). In all cases, the rate of transport is highly temperature-sensitive, however it is not sufficient to compare transport rates at different temperatures without also considering the influence of cooling on the kinetic properties of the transporter. The recent approach of Lowe & Walmsley (1986) has indicated that the temperature dependence of the entry, exit and exchange fluxes of these different transport systems in human erythrocytes can be successfully analysed in terms of a simple carrier model. The best example in this context is the glucose transporter.

(iv) Glucose transport

The conventional carrier model for the red cell glucose transporter is shown in Fig. 2(B). The carrier can exist in either the loaded (CG) or unloaded forms (C) at either face of the membrane, and loaded or unloaded carriers can relocate to the opposite side of the membrane giving either exchange or net transport. From the

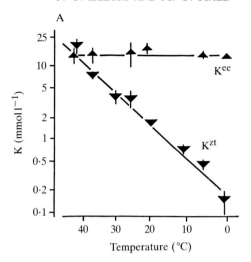

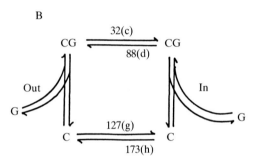

Fig. 2. The effects of temperature on glucose transport. Fig. 2(A) shows the temperature dependences of D-[^{14}C]-glucose transport under equilibrium exchange (▲) (K^{ee}), and zero-trans (▼) (K^{zt}) conditions. Fig. 2(B) shows schematically the carrier model for glucose transport (Lowe & Walmsley, 1986; Stein, 1986). 'Out' and 'in' represent the outside and inside of the cell, and the letters in parentheses, the conventional rate constants. The activation energies (E_a) in kJ mol^{-1} for the appropriate rate constants are given; the corresponding values for the rate constants (c), (d), (g) and (h) are 1113, 90, 12 and 0·7 (s^{-1}) respectively, with all determinations being made at 0 °C. These figures have been redrawn and modified after Lowe & Walmsley (1986).

rate constants for these reactions it is possible to write kinetic equations describing the dependence of fluxes on the concentrations of substrates on either side of the membrane (Stein, 1986). The equations have defined two particularly useful conditions, equilibrium exchange (i.e. identical substrate concentrations at either side of the membrane) or zero-trans (i.e. no substrate present on the opposite side of the membrane), which can be used experimentally to attempt to validate the carrier model, and then in the present context to define the temperature-sensitive steps in the transport process.

A striking difference between the effects of temperature on these two fluxes emerges when the results are analysed kinetically. The constant, K^{ee}, (which can be considered closely related to the Michaelis constant at the outer binding site for equilibrium exchange) is almost constant over the temperature range studied. Thus, the apparent affinity for glucose outside only changes from 12 to 17 mM from 0° to 43°C (Fig. 2(A)). In marked contrast, the analagous constant for net transport under zero-trans conditions, K^{zt}, is reduced by 100-fold (from 19·6 mM at 43°C to 0·15 mM at 0°C) yielding an E_a of 74 kJ mol^{-1} (Fig. 2(A)). Analysis of glucose flux experiments in terms of the carrier model can yield the temperature dependence of the individual rate constants for the overall reaction scheme (Fig. 2(B)). The legend to Fig. 2 also gives the values for the rate constants for glucose transport mediated by the individual reactions. From these data it can be seen that the slowest and most temperature-sensitive step is the return of the unloaded carrier to the outside surface of the membrane, and this therefore accounts for the marked effect of cooling on the zero-trans flux (K_{oi}^{zt}) relative to that of equilibrium exchange (Fig. 2(A)). Therefore, to assess fully the overall effect of temperature on the glucose transporter, account must be taken not only of the temperature-sensitivity of the individual steps of the reaction sequence, but also of their transport capacity.

The V_{max} of the glucose carrier decreases markedly with falling temperature so that the maximal transport rate at 0°C is less than 0·1 % of the rate at 37°C. Under some circumstances this can be partially offset by the increased affinity of the carrier at low temperatures, a phenomenon referred to by others as 'positive thermal modulation' (e.g. Klein, 1982). If the substrate concentration is at about the K_m of the carrier, (which often occurs physiologically) transport will tend to be stabilized on cooling since the decrease in K_m results in an increased loading of the carrier. Although this may be biologically desirable because this tends to minimize the effects of temperature on transport, experimentally this situation should be avoided, i.e. transport should be studied with the carrier maximally saturated with substrate, or be taken account of by kinetic analysis.

(v) Active transport; Na/K pump, Ca pump

It is well established that under certain conditions the Na/K pump functions differently from its normal mode of operation (uptake of 2 K, efflux of 3 Na with hydrolysis of 1 ATP molecule), to perform exchange (Na×Na; K×K), electrogenic (Na×O) or reversal (uptake of 3 Na, efflux of 2 K) reactions with the equivalent modified biochemical reactions (ATP–ADP exchange, ATP synthesis) (see e.g. Glynn & Karlish, 1975). These so-called 'partial reactions' are thought to represent various combinations of the intermediate steps in the overall transport process. The reaction sequence for the Na/K pump may be considered as Na-preferring (E_1) or K-preferring (E_2) forms of the enzyme. It is possible to classify broadly certain partial reactions; thus, Na–Na exchange and Na–ATPase phosphorylation from ATP+Na are E_1-reactions, and K–K exchange, K-occlusion, K-pNPPase and ouabain binding with Mg and P_i are E_2-reactions (see Glynn & Karlish, 1975; Stekhoven & Bonting, 1981). By analogy with the simple carrier systems described in the preceding section, the more complex reaction scheme of the Na/K pump could be analysed for the temperature-sensitive steps, in particular the possibility that the E_1/E_2 transition is involved. Several workers have found that reducing temperature tends to favour an E_2 form of the Na/K ATPase (Barnett & Palazzotto, 1974; Swann & Albers, 1979, 1981; Kaplan & Kenny, 1985), suggesting that this might be a reason for pump failure at low temperature. However, Ellory & Willis (1982) have observed that in erythrocytes from both cold-sensitive and cold-tolerant rodents there is a *symmetrical* increase in cation affinity of the pump with reduced temperature, and furthermore, the number of ouabain binding sites did not change significantly with cooling. These results argue against a change in the E_2/E_1 ratio, and suggest that the pump is not blocked preferentially in one form but that the rate of the overall reaction is reduced. Thus, although the partial reactions which have been studied appear to have differential temperature-sensitivity, at present we are unable to identify clearly the temperature-sensitive steps in the Na/K pump reaction sequence as precisely as has been achieved for the glucose carrier.

The other primary active transport system whose importance in the cell was emphasized in earlier sections is the Ca pump. Failure of this system at low temperature can lead to the intracellular Ca concentration rising, and this leads to the activation of the Ca-sensitive K channel. Mechanistically, the Ca pump is quite similar to the Na/K pump, in terms of its overall reaction sequence and phosphorylated intermediates (e.g. Garrahan, 1986). The recent results of J. D. Cavieres (personal communication) on the alterations in Mg-sensitivity of ATP–ADP exchange in this system at low temperature indicate that indeed there may be particular temperature-sensitive steps to be identified in this system and offer the possibility of eludicating the reactions at which the pump fails in the cold. In temperature-resistant species, (e.g. 'true' mammalian hibernators) both the Na/K (Willis *et al.* 1980) and Ca (Hall *et al.* 1986) pumps may be modified so that these critical reaction steps are less temperature-dependent. Such a modification tends to favour a specific change in the transport molecule rather than alterations to lipid environment. This concept receives indirect support from recent work on intact sarcoplasmic reticulum vesicles. When most of the vesicle lipids are replaced experimentally by detergents, the temperature dependence of Ca transport is very similar to that of the natural membrane. This has also been taken as support for the notion that in the sarcoplasmic reticulum at least, lipid–protein interactions are not primarily responsible for the breaks seen in Arrhenius plots, but that there are temperature-dependent changes which are intrinsic to the enzyme reaction (De Meis, 1981).

Physiological considerations

Cell viability is critically dependent on membrane transport systems. Obvious roles for such systems include the selective transport of metabolically important substrates into and out of the cells. Further, the balance between the passive leaks and the active pumps which maintains the intracellular ionic composition and hence cell volume constant. In this paper so far, we have considered the effects of changing temperature on the properties of discrete membrane transport systems, but clearly the overall cellular response to cooling will be the net result of changes in all pumps and leaks. The failure of pumps at low temperature relative to leaks is well known because with cooling, the cells of many species gain Na, Ca and lose K at reduced temperature. For the three species given in Fig. 1, the maximum Na/K pump activity which remains at 5 °C compared with 37 °C is 3·2, 0·3 and 0·25 % for hedgehog, guinea pig and human red cells respectively (Willis *et al.* 1980). In the case of the two temperature-sensitive species, this activity is not sufficient to balance the leak flux at low temperature. In addition, as shown in Fig. 1(B), there can be significant amplification of the leak fluxes at low temperatures caused by the activation of the Gardos channel when the Ca pump fails. Furthermore, even a modest rise in $[Ca]_i$ can have other dramatic effects on cell survival for example, interaction with cytoskeletal proteins, activation of kinases etc.

In the discussion of the changes in the properties of the transporter systems on cooling, we have exclusively considered cases where fluxes are measured under experimental conditions with the composition of the extracellular medium controlled within narrow limits. *In vivo* there may be a complex series of effects, relating to the internal and external environment at the tissue or whole animal level which may profoundly influence transport function. An obvious example would be the availability of intracellular ATP, which is crucial for the functioning of pumps. If metabolism is affected by temperature so that ATP production becomes rate-limiting, there will be immediate and serious consequences for the Ca and Na pumps. Accumulation of ADP or P_i, or changes in *trans*-concentrations of substrates can also dramatically affect pump activity, particularly if transport is not being measured under saturating substrate conditions. Other, perhaps less obvious effects should also be considered. For example cooling alters the physico-chemical properties of water and ionizable groups on proteins. Most experimenters choose to accommodate the temperature shifts of the pK of buffers by maintaining pH constant while temperature varies. However, as pointed out by Reeves (1972, 1977), the pK of alpha-imidazole groups, (pK_{im}), which is quantitatively the most important buffer group of most proteins, increases as temperature is lowered. Cold-adapting ectotherms therefore tend to alter their blood acid–base regulation with temperature to keep a constant $pH-pK_{im}$ difference, rather than a constant pH. Clearly, in terms of physiological function, such an effect in compensating temperature-dependent changes in transport should be noted.

These considerations emphasize the experimental limitations for the apparently simple manoeuvre of changing temperature. Even under well-defined experimental conditions, there can be pitfalls in analysing data in physico-chemical terms. For example, many workers have expressed their results in terms of the Arrhenius relationship to derive information about thermodynamic properties, principally the activation energies and 'transition temperatures' of transport systems. However, there are a number of serious kinetic and thermodynamic problems with this approach, and caution should be exercised (Silvus & McElhaney, 1981; Klein, 1982).

Conclusions

We have attempted to describe the behaviour of all classes of membrane transporter when temperature is reduced, using red cells as a model system. Our results emphasize the different effects of temperature on active and passive transport, and earlier we discussed some of the further variables introduced when considering the *in vivo* situation. Can we now draw any conclusions with regard to temperature adaptation and transport processes at the molecular level? Temperature adaptation of membrane transport systems could be accommodated *via* changes in the lipid environment of the transport system (see Cossins, this volume), or could involve changes in the transport protein itself. Although transport proteins are highly conserved in terms of structure (e.g. consider the homology in the bacterial and mammalian sugar transporters (Maiden *et al.* 1987), it is nevertheless likely that some instances of temperature adaptation for example hibernation will involve structural changes in the transport molecule itself. Now that an increasing number of transporters have been cloned and sequenced, it should be possible to use the genetic approaches applied successfully in the context of temperature and bacterial growth to gain new insights into the temperature-sensitivity of mammalian transport systems.

References

Antonov, V. F., Petrov, V. V., Molnar, A. A., Predvoditelev, D. A. & Ivanov, A. A. (1980). The appearance of single-ion channels in unmodified lipid bilayer membranes at the phase transition. *Nature, Lond.* **283**, 585–586.

Barnett, R. E. & Palazzotto, J. (1974). Mechanism of the effects of lipid phase transitions on the Na,K ATPase and the role of protein conformational change. *Ann. N.Y. Acad. Sci.* **242**, 69–75.

Bernhardt, I., Donath, E. & Glaser, R. (1984). Influence of surface charge and transmembrane potential on rubidium-86 efflux of human red blood cells. *J. Membr. Biol.* **78**, 249–255.

Blackstock, E. J., Ellory, J. C. & Stewart, G. W. (1986). K influxes in red cells from two primate species. *J. Physiol., Lond.* **374**, 34P.

Blackstock, E. J. & Stewart, G. W. (1986). The dependence on external cation of sodium and potassium fluxes across the human red cell membrane at low temperatures. *J. Physiol., Lond.* **375**, 403–420.

Boheim, G., Hanke, W. & Eibl, H. (1980). Lipid phase transition in planar bilayer membranes and its effect on carrier- and pore-mediated ion transport. *Proc. natn. Acad. Sci. U.S.A.* **77**, 3403–3407.

Brahm, J. (1977). Temperature-dependent changes of chloride transport kinetics in human red cells. *J. gen. Physiol.* **70**, 283–306.

Brahm, J. (1982). Diffusional water permeability of red cells and their ghosts. *J. gen. Physiol.* **79**, 791–819.

Brahm, J. (1983). Urea permeability of red cells. *J. gen. Physiol.* **82**, 1–23.

BRUINS, H. R. (1929). Coefficients of diffusion in liquids. In *International Critical Tables*, vol. 5 (ed. E. W. Washburn), pp. 63–76. New York: McGraw-Hill.

CASS, A. & FINKELSTEIN, A. (1967). Water permeability of thin lipid membranes. *J. gen. Physiol.* **50**, 1765–1784.

CASS, A., FINKELSTEIN, A. & KRESPI, V. (1970). The ion permeability induced in thin lipid membranes by the polyene antibiotics nystatin and amphotericin B. *J. gen. Physiol.* **56**, 100–124.

COSSINS, A. R. (1981). The adaptation of membrane dynamic structure to temperature. In *Effects of Low Temperatures on Biological Membranes* (ed. G. J. Morris & A. Clarke), pp. 83–106. London: Academic Press.

DE MEIS, L. (1981). The sarcoplasmic reticulum. Transport and energy transduction. In *Transport in the Life Sciences*, vol. 2 (ed. E. E. Bittar). Chapter 8. New York: John Wiley & Sons.

ELLORY, J. C. & TUCKER, E. M. (1983). Cation transport in red blood cells. In *Red Cells of Domesticated Animals* (ed. N. S. Agar & P. G. Board), pp. 291–312. Amsterdam: Elsevier Biomedical Press.

ELLORY, J. C. & WILLIS, J. S. (1976). Temperature dependence of membrane transport function. Disparity between active potassium transport and (Na+K)ATPase activity. *Biochim. biophys. Acta* **443**, 301–305.

ELLORY, J. C. & WILLIS, J. S. (1982). Kinetics of the sodium pump in red cells of different temperature sensitivity. *J. gen. Physiol.* **79**, 1115–1130.

FISHER, R. B. & NIMMO, I. A. (1973). The stereo-specificity and temperature-dependence of the permeability of human erythrocytes to aldoses. *Q. Jl exp. Physiol.* **58**, 153–161.

FRANKENHAUSER, B. & MOORE, L. E. (1963). The effect of temperature on the sodium and potassium permeability changes in myelinated nerve fibres of *Xenopus laevis*. *J. Physiol., Lond.* **169**, 431–437.

GARDOS, G. (1958). The function of calcium in the potassium permeability of human erythrocytes. *Biochim. biophys. Acta* **30**, 653–654.

GARRAHAN, P. J. (1986). Partial reactions of the Ca-ATPase. In *The Ca Pump of Plasma Membranes* (ed. A. F. Rega & P. J. Garrahan), pp. 105–125. Boca Raton, Florida: CRC Press.

GLYNN, I. M. & KARLISH, S. J. D. (1975). The sodium pump. *A. Rev. Physiol.* **37**, 13–55.

HALL, A. C., ELLORY, J. C. & KLEIN, R. A. (1982). Pressure and temperature effects on human red cell cation transport. *J. Membr. Biol.* **68**, 47–56.

HALL, A. C. & WILLIS, J. S. (1984a). Activation of the Ca-sensitive K channel in erythrocytes at low temperatures. *Biochem. Soc. Trans.* **12**, 312–314.

HALL, A. C. & WILLIS, J. S. (1984b). Differential effects of temperature on three components of passive permeability to potassium in rodent red cells. *J. Physiol., Lond.* **348**, 629–643.

HALL, A. C. & WILLIS, J. S. (1986). The temperature dependence of passive potassium permeability in mammalian erythrocytes. *Cryobiology* **23**, 395–405.

HALL, A. C., WOLOWYK, M. W., WANG, L. C. H. & ELLORY, J. C. (1986). The effects of temperature on Ca transport in red cells from a hibernator (*Spermophilus richardsonii*). *J. Therm. Biol.* (in press).

HLADKY, S. B. & HAYDON, D. A. (1972). Ion transfer across lipid membranes in the presence of Gramicidin A. *Biochim. biophys. Acta* **274**, 294–312.

HOARE, D. G. (1972). The temperature dependence of the transport of L-leucine in human erythrocytes. *J. Physiol., Lond.* **221**, 331–348.

HOCHACHKA, P. W. & SOMERO, G. N. (1973). *Strategies of Biochemical Adaptation.* Philadelphia: Saunders.

HODGKIN, A. L., HUXLEY, A. F. & KATZ, B. (1952). Measurement of current–voltage relations in the membrane of the giant axon of *Loligo*. *J. Physiol., Lond.* **116**, 424–448.

KAPLAN, J. H. & KENNY, L. J. (1985). Temperature effects on sodium pump phosphoenzyme distribution in human red blood cells. *J. gen. Physiol.* **85**, 123–137.

KIRSCH, G. E. & SYKES, J. S. (1987). Temperature dependence of Na currents in rabbit and frog muscle membranes. *J. gen. Physiol.* **89**, 239–251.

KLEIN, R. A. (1982). Thermodynamics and membrane processes. *Q. Rev. Biophysics* **15**, 667–757.

Kregenow, F. M. (1981). Osmoregulatory salt transporting mechanisms: Control of cell volume in anisotonic media. *A. Rev. Physiol.* **43**, 493–505.

Lowe, A. G. & Walmsley, A. R. (1986). The kinetics of glucose transport in human red blood cells. *Biochim. biophys. Acta* **857**, 146–154.

Maiden, M. C. J., Davios, E., Baldwin, S. A., Moore, D. C. M. & Henderson, P. J. F. (1987). Mammalian and bacterial sugar transport proteins are homologous. *Nature, Lond.* **325**, 641–643.

Papahadjopoulos, D., Jacobson, K., Nir, S. & Isac, T. (1973). Phase transitions in phospholipid vesicles. Fluorescence polarization and permeability measurement concerning the effect of temperature and cholesterol. *Biochim. biophys. Acta* **311**, 330–348.

Price, H. D. & Thompson, T. E. (1969). Properties of lipid bilayer membranes separating two aqueous phases: temperature dependence of water permeability. *J. molec. Biol.* **41**, 443–457.

Redwood, W. R. & Haydon, D. A. (1969). Influence of temperature and membrane composition on the water permeability of lipid bilayers. *J. theor. Biol.* **22**, 1–8.

Reeves, R. B. (1972). An imidazole alphastat hypothesis for vertebrate acid–base balance: tissue carbon dioxide content and body temperature in bullfrogs. *Respir. Physiol.* **14**, 219–236.

Reeves, R. B. (1977). The interaction of body temperature and acid–base balance in ectothermic vertebrates. *A. Rev. Physiol.* **39**, 559–586.

Rega, A. F. & Garrahan, P. J. (1986). The Ca pump of plasma membranes. Boca Raton, Florida: CRC Press.

Sen, A. K. & Widdas, W. F. (1962). Determination of the temperature and pH dependence of glucose transfer across the human erythrocyte membrane measured by glucose exit. *J. Physiol., Lond.* **160**, 392–403.

Silvus, J. R. & McElhaney, R. N. (1981). Non-linear Arrhenius plots and the analysis of reaction and motional rates in biological membranes. *J. theor. Biol.* **88**, 135–152.

Simons, T. J. B. (1976). The preparation of human red cell ghosts containing Ca buffers. *J. Physiol., Lond.* **256**, 203–225.

Singer, M. (1982). Permeability of bilayers composed of mixtures of saturated phospholipids. *Chemistry and Physics of Lipids* **31**, 145–159.

Stein, W. D. (1986). *Transport and Diffusion Across Cell Membranes.* London: Academic Press.

Stekhoven, F. S. & Bonting, S. L. (1981). Transport adenosine triphosphatases: Properties and functions. *Physiol. Rev.* **61**, 1–76.

Stewart, G. W., Ellory, J. C. & Klein, R. A. (1980). Increased human red cell cation permeability below 12 °C. *Nature, Lond.* **286**, 403–404.

Swann, A. C. & Albers, R. W. (1979). (Na+K) adenosine triphosphatase of mammalian brain. *J. biol. Chem.* **254**, 4540–4544.

Swann, A. C. & Albers, R. W. (1981). Temperature effects on cation affinities of the (Na,K)ATPase of mammalian brain. *Biochim. biophys. Acta* **644**, 36–40.

Wang, J. H. (1965). Self-diffusion coefficients of water. *J. Phys. Chem.* **69**, 4412.

Wang, J. H., Robinson, C. V. & Endleman, I. S. (1953). Self-diffusion and structure of liquid water. III. Measurements of the self-diffusion of liquid water with ^{2}H, ^{3}H and ^{18}O as tracers. *J. Am. Chem. Soc.* **75**, 466–470.

Weith, J. O. (1970). Paradoxical temperature dependence of sodium and potassium fluxes in human red cells. *J. Physiol., Lond.* **207**, 563–580.

Willis, J. S., Ellory, J. C. & Wolowyk, M. W. (1980). Temperature sensitivity of the sodium pump in red cells from various hibernator and non-hibernator species. *J. comp. Physiol.* **138**, 43–47.

Printed in Great Britain © *Society for Experimental Biology 1987*

TEMPERATURE ACCLIMATION AND METABOLISM IN ECTOTHERMS WITH PARTICULAR REFERENCE TO TELEOST FISH

IAN A. JOHNSTON and JEFF DUNN

Gatty Marine Laboratory, Department of Physiology and Pharmacology, University of St Andrews, St Andrews, Fife, Scotland, KY16 8LB

Summary

As body temperature decreases, changes in the physical chemistry of the cell produce a reduction in metabolic activity. In temperate fish, cold water temperatures either lead to dormancy or else trigger a range of homeostatic responses which serve to offset the passive effects of reduced temperature. Compensatory adjustments to temperature occur with time courses ranging from less than a second to more than a month. Although swimming performance may increase with cold-acclimation, active metabolic rate remains significantly below that for warm-acclimated fish. Compensatory and dormancy responses are not mutually exclusive and sometimes occur in the same species depending on the temperature.

Cold-acclimation results in significant increases in the density of mitochondria and capillaries in skeletal muscle. This serves to reduce diffusion distances and increase the capacity for aerobic ATP production relative to fish acutely exposed to low temperature. There is evidence that cold acclimation has differential effects on the synthesis and degradation rates of mitochondrial proteins leading to a net increase in their concentration. In contrast, the activities of enzymes associated with glycolysis and phosphocreatine hydrolysis show no consistent changes with thermal acclimation suggesting that flux through these pathways is modulated by factors other than enzyme concentration.

Higher mitochondrial densities have also been reported for the liver, brain and gill tissue of cold compared with warm acclimated fish. In spite of their increased concentration, the activities of aerobic enzymes remain much lower at cold than warm temperatures. Acclimation temperature affects hepatosomatic index, the concentration of energy reserves, and the relative importance of glucose and fatty acid catabolism in liver. The fraction of glucose oxidized by the hexose monophosphate shunt (HMPS) pathway also increases with cold acclimation in some species. It is likely that many of the changes in liver metabolism with temperature acclimation reflect associated changes in feeding behaviour and/or diet, and other energetic demands (e.g. gametogenesis).

Possible mechanisms underlying alterations in pathway utilization with temperature acclimation are discussed. They include changes in factors influencing enzyme structure and activity (e.g. pH, substrate/modulator concentrations, phosphorylation state, membrane composition), and effects of temperature on gene expression.

Experiments with fish

Whole animal studies

An estimate of total energy production can be obtained from measurements of oxygen consumption provided there is no net contribution from anaerobic pathways. This is generally true for levels of activity which can be sustained for long periods. A number of measures of metabolic rate are in common usage (Fry, 1957; Doudoroff & Shumway, 1970). Standard or basal metabolism represents the oxygen uptake of animals in the post-absorbtive state and in the complete absence of activity. This provides an estimate of maintenance energy requirements but is difficult to measure and interpret. The term 'routine metabolic rate' is often applied to measurements in which there is a contribution from spontaneous activity. However, measurements of active metabolic rate and the aerobic scope for activity are more useful. These terms refer to steady state oxygen consumption at the maximum sustainable swimming speed, and the difference between the active and standard metabolic rates, respectively. Sustainable speeds are usually taken as those that can be maintained for periods of 30–60 min (Brett, 1972). Higher levels of performance require a contribution from anaerobic ATP production. The time to fatigue shortens dramatically as the level of effort increases, often falling to less than a second at maximum speed (Bainbridge, 1962).

Species clearly differ in their ability to swim at different temperatures. For example, the minimum temperature for locomotion is below 0°C for antarctic fish compared with around 15°C for most tropical species. Such differences are genetically fixed and must be taken into account in any discussion on the effects of temperature on energy metabolism. How does adaptation temperature influence basal and maximal energy flux? This question is difficult to answer largely because of interspecific variation associated with differences in body-shape and activity patterns. For example, even among temperate fish active metabolic rate varies some 5-fold between sluggish and athletic species (Fig. 1). This review will therefore only be concerned with metabolic responses to acute and seasonal temperature change.

Activity produces a greater effect on total energy production than changes in acclimation temperature (Fig. 1). For example, in the sockeye salmon (*Oncorhynchus nerka*), an increase in temperature from 5°C to 15°C produces a 4-fold

Fig. 1. Effects of acclimation temperature on maximum sustained swimming speed, standard and active metabolic rates for three species of teleost fish. The dotted line between the standard and active rates indicates the scope for aerobic activity. The data were obtained from the following sources: lemon sole (*Microstomus kitt*) (~220 g bodyweight) (Duthie & Houlihan, 1982); largemouth bass (*Micropterus salmoides*) (150 g bodyweight) (Beamish, 1970) and sockeye salmon (*Oncorhynchus nerka*) (~50 g bodyweight) (Brett, 1964).

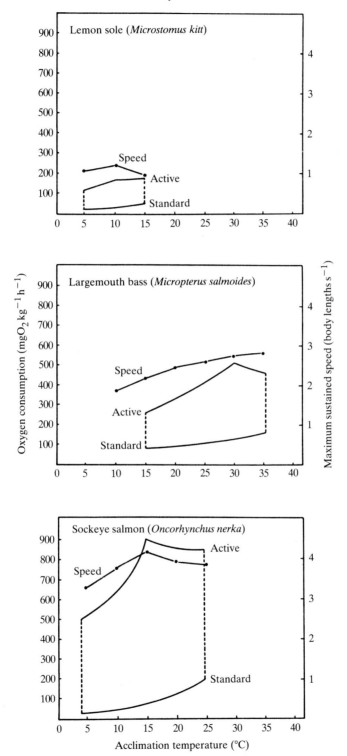

Fig. 1

increase in basal metabolic rate and a 2-fold increase in active metabolism (Brett, 1964). This compares with a 15-fold increase in energy production over basal rates at the maximum sustained swimming speed (Brett, 1964). At temperatures above 15°C the active rate declines resulting in a reduced aerobic scope for activity (Fig. 1). Largemouth bass (*Micropterus salmoides*) prefer warmer waters to the salmon and in this species aerobic scope increases between 15 and 30°C but is lower at 35 than 30°C (Beamish, 1970). It has been suggested that the temperature at which the aerobic scope starts to decline represents the point at which the energy demands of ventilation and associated circulation become excessive, restricting increased oxygen supply to the tissues (Jones, 1971).

The behaviour and metabolism of many fish varies with the length of time they are exposed to a particular temperature. The effects of temperature acclimation are most marked for species living in environments which exhibit a marked seasonal variation in temperature, e.g. inland waterways and shallow seas. Two main types of behaviour are observed in response to the low water temperatures experienced during the winter months. Some species become relatively inactive and cease feeding as temperature is reduced and are thought to enter a torpid or dormant state (Crawshaw, 1984). Winter dormancy is characterized by extremely low metabolic rates which serve to spare food reserves until prey numbers and water temperatures increase in the spring. In other species, cold-acclimation results in improvements in swimming performance relative to fish acutely exposed to low temperature. This phenomenon which may take several weeks to complete is often referred to as temperature compensation. These two kinds of response are not mutually exclusive and sometimes occur in the same species depending on the temperature. For example, acclimated largemouth bass, *Micropterus salmoides*, become torpid below 7°C, yet maintain a similar level of spontaneous activity over the range 7–30°C (Lemons & Crawshaw, 1985). Feeding in this species is inhibited at higher temperatures (10°C) than are required to suppress locomotion (Lemons & Crawshaw, 1985). This suggests that not all physiological systems respond in the same manner to a given change in body temperature (Fig. 2).

Goldfish (*Carassius auratus*) is the only species showing a significant temperature compensation response for which both standard and active metabolic rates have been determined (Kanungo & Prosser, 1959; Beamish & Mookherjii, 1964; Fry & Hochachka, 1970). Since fish acutely exposed to low temperature are stressed and perform badly it is difficult to quantify accurately the overall effects of acclimation on total energy production. However, it is clear that even in fully acclimated individuals standard and active metabolic rate decrease with a Q_{10} of 1·6–2·0 as water temperature is reduced (Fry & Hochachka, 1970) (see also Fig. 1).

Metabolism of tissues

Measurements of enzyme activities have often been employed to investigate metabolic responses to temperature acclimation at the tissue level (Hazel & Prosser, 1974). The assumption behind such studies is that maximal enzyme

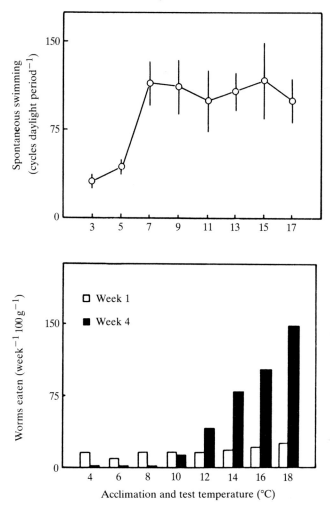

Fig. 2. Spontaneous swimming activity and food intake of largemouth bass (*Micropterus salmoides*) at various acclimation temperatures. Values given represent mean ± SEM for 6–9 fish (adapted from Lemons & Crawshaw, 1985).

activities measured *in vitro* can provide information on the relative capacities of different metabolic pathways *in vivo*. Enzymes catalysing non-equilibrium reactions generally have low activities and are often subject to allosteric modulation (Newsholme & Start, 1973). Newsholme and co-workers have suggested that the activities of non-equilibrium enzymes can be used to provide a quantitative index of the flux through a pathway (Zammit & Newsholme, 1979; Newsholme & Paul, 1983). It has been argued, that the activities of equilibrium enzymes also exert a degree of control over total flux (Kacser & Burns, 1979). Precht (1958) has classified changes in rate functions (capacity adaptations) according to five types: (1) overcompensation, (2) perfect compensation, the acclimated rate being the same at the two temperatures, (3) partial compensation, (4) no compensation and (5) inverse or paradoxical change, the acclimated rate in the cold being lower than the rate on direct transfer from warm to cold. This terminology is purely descriptive and requires that enzyme activities are assayed at all the acclimation temperatures investigated. Another common experimental design involves acclimating animals to either high or low temperatures and assaying enzyme activities at some intermediate temperature. This approach is often used to make inferences about changes in the relative concentrations of enzymes (Sidell, 1980; Jones & Sidell, 1982).

Measurements of enzyme activities cannot provide an insight into metabolic flux under different physiological states. However, this kind of information can be obtained by measuring the rates of utilization of radio-labelled substrates in experiments using either whole animals, tissue homogenates/slices, or isolated cells. The interpretation of isolated tissue experiments is often complicated by either cell damage and/or by differences between *in vivo* and *in vitro* assay conditions. For example, few studies on isolated hepatocytes have considered the effects of hormones on metabolism.

Since the metabolic response of individual tissues to temperature change varies with their function and requirements for ATP production, experiments on liver, muscle, brain etc. are initially considered separately.

Liver

It is likely that the metabolic responses of liver to changes in acclimation temperature vary with nutritional factors, e.g. lipogenesis is very sensitive to the composition of the diet (Henderson & Sargent, 1981). Differences can also be expected between species that cease feeding at low temperatures and exhibit winter dormancy (e.g. eel, *Anguilla rostrata*; Walsh, Foster & Moon, 1983), and those that remain active and continue feeding (e.g. striped bass, *Morone saxatilis*; Nichols & Miller, 1967; Stone & Sidell, 1981). For a particular species the exact temperature of laboratory acclimation may also be critical since a slightly lower temperature may inhibit feeding and trigger the winter dormancy response (Walsh *et al.* 1983; Lemons & Crawshaw, 1985). Many of the discrepancies and contradictory findings in the literature may result from a failure to take these factors into account.

The liver is the major reservoir for glycogen in fish and an important site for gluconeogenesis from lactate, amino acids and glycerol (Walton & Cowey, 1982; Foster & Moon, 1986). In some species the liver is also an important site for the synthesis and storage of triacylglycerol (Henderson & Sargent, 1981; Walton & Cowey, 1982). Stone & Sidell (1981) acclimated striped bass to either 5, 15 or 25 °C for at least 4 weeks. All fish were intestinally postabsorptive at the time of sacrifice. They found that the relative weight of the liver was almost two times higher in the 5 °C-acclimated compared to the 25 °C-acclimated group. Similar though smaller increases in hepatosomatic index have been reported for 10 °C-acclimated relative to 20 °C European eels (*Anguilla anguilla*) (Jankowsky, Hotopp & Seibert, 1984). Average cell size is higher at low acclimation temperatures in goldfish (Das, 1967) but not eels (Jankowsky *et al*. 1984). In the latter species the increase in hepatosomatic index appears to be due to an increase in cell numbers (Jankowsky *et al*. 1984). Lipid content of livers from 5 °C-acclimated striped bass is 40 % of tissue dry weight compared to 25 % of tissue dry weight for 25 °C-acclimated fish (Stone & Sidell, 1981). Thus striped bass acclimated to the colder temperature have four times more hepatic lipid stores than the same sized fish acclimated to warmer temperatures. The glycogen and protein content of striped bass liver was unaffected by acclimation temperature (Stone & Sidell, 1981). A different pattern of energy storage is observed in the livers of American eels which cease feeding below 8 °C, burrow in the mud and become dormant (Walsh *et al*. 1983). Six months starvation at 10 or 15 °C results in substantial decreases in hepatosomatic index, lipid stores, and enzyme activities. In contrast, metabolic reserves and enzyme activities of eels fasted at 5 °C are similar to those of well fed eels acclimated to 15 °C (Walsh *et al*. 1983).

Oxygen uptake by liver slices from striped bass acclimated to 15 °C were found to be 2·75 times higher than slices from 5 °C-acclimated fish (Stone & Sidell, 1981). However, when assayed at a constant 15 °C oxygen consumption was higher for slices from 5 °C- than 25 °C-acclimated individuals (Fig. 3). Similar differences in respiration rate between cold- and warm-acclimated fish have been obtained for homogenates from goldfish (Kanungo & Prosser, 1959) and brook trout (Hochachka & Hayes, 1962) and for hepatocytes from rainbow trout (Hazel & Sellner, 1979). These changes are accompanied by relative increases in the activities of tricarboxylic acid cycle enzymes (Campbell & Davies, 1978; Kleckner

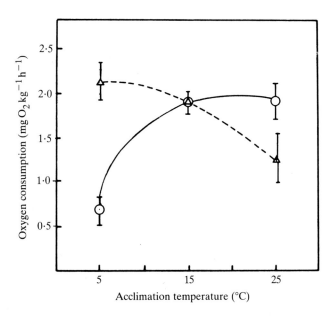

Fig. 3. Rate of oxygen utilization by liver slices from temperature acclimated striped bass (*Morone saxatilis*). Values represent mean ± SEM for 6–8 fish. Adapted from Stone & Sidell (1981).

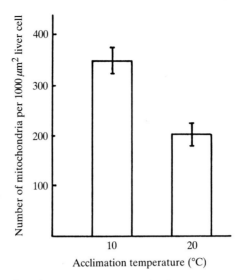

Fig. 4. Effects of acclimation temperature on mitochondrial number in liver cells from the blenny, *Blennius pholis*. Values represent mean ± SEM. Data taken from Campbell & Davies (1978).

& Sidell, 1985) and are thought to reflect an increase in enzyme concentration following cold-acclimation (Sidell, 1983). Campbell & Davies (1978) have reported a 70 % increase in the number of mitochondria per liver cell in blennies (*Blennius pholis*) acclimated to 10°C compared to 20°C (Fig. 4). This type of metabolic re-structuring of the cell presumably increases the maximum potential for ATP production relative to non-acclimated individuals. Note, however, that absolute respiration rates or aerobic enzyme activities are always found to be lower at the cold temperature. Relative increases in the aerobic capacity of the liver are not generally observed for species such as eels that undergo winter dormancy at low temperatures (Walsh *et al.* 1983; Jankowsky *et al.* 1984).

Studies with inhibitors provide evidence that the relative importance of carbohydrate, amino acid and lipid catabolism to the total aerobic ATP flux is critically dependent on acclimation temperature (Hazel & Prosser, 1974). For example, in striped bass, oxygen consumption by liver slices is inhibited by iodoacetate (a blocker of glyceraldehyde-3-P dehydrogenase) to a much greater extent in 5°C than either 15°C or 25°C fish (Fig. 5). Inhibitor experiments with liver homogenates from goldfish suggest a relatively increased importance of carbohydrate catabolism with cold acclimation (Kanungo & Prosser, 1959) and brook trout (Hochachka & Hayes, 1962). Stone & Sidell (1981) found that $^{14}CO_2$ production from uniformly labelled glucose by striped bass liver slices was not significantly different in fish acclimated and assayed at 5, 15 and 25°C. In contrast, $^{14}CO_2$ production form ^{14}C-L-palmitate is markedly temperature dependent and 4 times higher at 25 than 5°C (Stone & Sidell, 1981). This is consistent with carbohydrates accounting for a higher proportion of hepatic energy metabolism as acclimation temperature decreases (Fig. 6). Hochachka & Hayes (1962) found

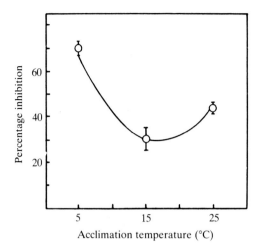

Fig. 5. Percentage inhibition by iodoacetate of oxygen utilization by liver slices from temperature acclimated striped bass. Values represent mean ± SEM of 4–8 fish. Assays were performed at the acclimation temperatures. Adapted from Stone & Sidell (1981).

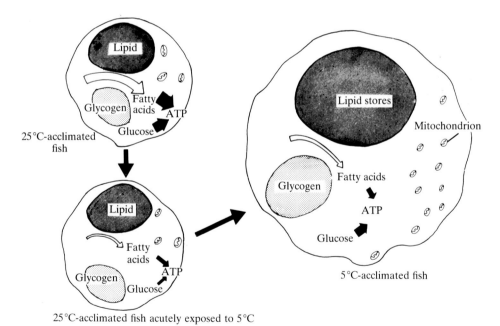

Fig. 6. Diagrammatic representation of the effects of acclimation temperature on energy stores and ATP production in liver cells from striped bass. The width of arrows provides an indication of the relative importance of fatty acid and glucose catabolism at each temperature. Based on the experiments of Stone & Sidell (1981). It is assumed that the higher oxygen consumption of liver slices of 5°C compared with 25°C fish (assayed at 15°C) is due, in part, to an increase in mitochondrial volume density and that this results in an increased rate of ATP production relative to fish acutely exposed to the same temperature.

that acetate-L-^{14}C oxidation was lower, but incorporation into fat was higher in liver homogenates from 4°C-acclimated compared with 15°C-acclimated brook trout. Thus in both striped bass and brook trout cold acclimation appears to shift the balance of dietary lipids assimilated from catabolism to storage (Hochachka & Hayes, 1962; Stone & Sidell, 1981) (Fig. 6). Stone & Sidell (1981) have suggested that this may reflect a need to increase lipid stores for gonad development and for the energetic demands of migration when the water temperature rises in the spring.

A different pattern of fuel utilization and storage is observed in the liver of temperature acclimated mummichogs *Fundulus heteroclitus* (Moerland & Sidell, 1981). The livers of fish acclimated to 5°C have much higher concentrations of glycogen and significantly lower lipid reserves than fish acclimated to either 15 or 25°C (Fig. 7). The relative rates of utilization of ^{14}C-labelled palmitate and glucose also differ significantly between acclimation temperatures (Fig. 7). It would appear that the higher energetic demands of hepatocytes isolated from 15 *versus* 5°C-acclimated fish are met to a greater degree by increased lipid than by increased carbohydrate catabolism. In 25°C-acclimated individuals lipid catabolism declines relative to 15°C-acclimated fish whereas there is a dramatic increase

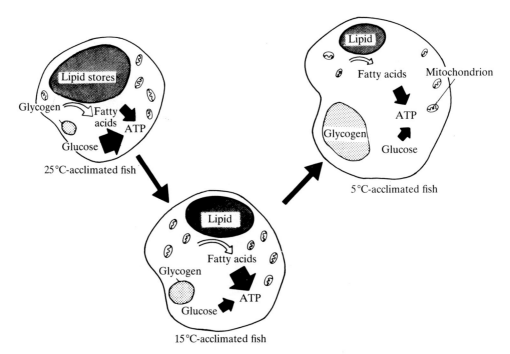

Fig. 7. Diagrammatic representation of the effects of acclimation temperature on energy stores and ATP production in liver cells from the mummichog *Fundulus heteroclitus*. The width of arrows provides an indication of the relative importance of fatty acid and glucose catabolism at each temperature. Based on the experiments of Moerland & Sidell (1981). It is assumed that the higher activities of cytochrome oxidase in liver tissue of 15°C relative to 5°C and 25°C acclimated fish (assayed at 15°C) is due, in part, to an increase in mitochondrial volume density.

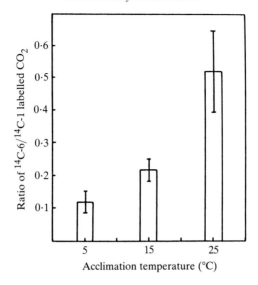

Fig. 8. Ratio of $^{14}CO_2$ released from ^{14}C-6/^{14}C-L-labelled glucoses by liver tissue from temperature acclimated striped bass. Values represent mean SEM for four fish. Tissues were assayed at the acclimation temperature. Adapted from Stone & Sidell (1981).

in glucose utilization. Moerland & Sidell (1981) suggested that the decreased reliance on fatty acids as fuel in 25 °C-acclimated animals may reflect a partitioning of available lipid into gametogenesis triggered by the increase in water temperature. During the winter months a marked serum hyperglycemia enables *Fundulus* to survive periods of sub-zero temperatures in a super-cooled state (Parker, 1972; Umminger, 1972). Since this largely involves hepatic glycogenolysis the marked increase in glycogen stores in 5 °C-acclimated fish may represent an adaptive strategy for freezing avoidance.

There is evidence for an increased importance of the hexose monophosphate shunt pathway (HMPS) with cold acclimation in the liver of a number of species, including brook trout (Hochachka & Hayes, 1962; Yamauchi, Stegeman & Goldberg, 1975), goldfish (Kanungo & Prosser, 1959), blenny (Campbell & Davies, 1978) and striped bass (Stone & Sidell, 1981). For example, the ratio of ^{14}C-6-O_2/^{14}C-1-O_2 released from labelled glucose by striped bass liver slices is lower at cold- than warm-acclimation temperatures (Fig. 8). This is consistent with an increased participation of the pentose shunt as acclimation temperature decreases. Similarly, brook trout acclimated to 4 °C had greater activities of liver glucose-6-phosphate dehydrogenase, hexose-6-phosphate dehydrogenase and 6-phosphogluconate dehydrogenase (assayed at 25 °C) than those acclimated to either 10 or 15 °C (Yamauchi *et al.* 1975). An important function of the HMP in liver is thought to be the provision of NADPH for fatty acid biosynthesis. It is well established that there are major changes in the turnover rates and composition of fatty acid moieties in liver membranes following thermal acclimation (Cossins, Kent & Prosser, 1980; Hazel & Neas, 1982). There is also evidence for changes in the relative activities of individual enzymes associated with the turnover and processing of membrane phospholipids. For example, phospholipase A_2 activity is more than two times higher at 5°C in liver microsomes from 5 than 20°C-acclimated rainbow trout (Neas & Hazel, 1985). Indeed, the activities of LPA_2 and enzymes associated with the desaturation and elongation reactions of $n-3$ fatty acids are similar for warm- and cold-acclimated fish when assayed at their respective acclimation temperatures (Hazel & Neas, 1982; Neas & Hazel, 1985). The re-structuring of membrane lipids following thermal acclimation is thought to require several weeks (Hazel, 1979). Yamauchi et al. 1975 found that rainbow trout maintained at 5°C without feeding for more than 2 months showed a decrease in G6PD and PGD activities (the normal starvation response). They concluded that the increase in HMPS activity was a transitory phenomenon related to the extra requirement for fatty acid biosynthesis during the early stages of thermal acclimation. However, this does not appear to be the case for striped bass since HMPS activity remained enhanced in fish kept for up to 10 months at 5°C (Stone & Sidell, 1981). By contrast, in Fundulus heteroclitus no increase in HMPS activity was observed in hepatocytes from 5°C- compared with 25°C-acclimated fish (Moerland & Sidell, 1981). It has been suggested that in this species carbohydrate is partitioned away from catabolic outlets (i.e. HMPS and glycolysis) and into glycogen reserves at low temperatures (Moerland & Sidell, 1981).

In hepatocytes from rainbow trout, triacylglycerol synthesis is much more temperature dependent than either fatty acid synthesis or cholesterogenesis (Voss & Jankowsky, 1986). These authors found that glycogen storage levels in hepatocytes affected both the distribution and rate of incorporation of C-2 units from [^{14}C] lactate into the main lipid classes. However, similar rates of lipogenesis were obtained for trout acclimated to either 10°C or 20°C at assay temperatures from 5 to 20°C (Voss & Jankowsky, 1986). This contrasts with the study of Hazel & Sellner (1979) who found higher rates of fatty acid synthesis in hepatocytes from cold- relative to warm-acclimated trout. In another study with trout hepatocytes, Hazel & Prosser (1979) investigated the incorporation of [^{14}C] acetate and found evidence for an elevated rate of sterol lipogenesis in cold-acclimated fish but no effect of temperature acclimation on fatty acid synthesis. The sensitivity of lipogenesis in isolated hepatocytes to carbohydrate storage levels, dietary and/or hormonal factors may explain some of the apparently contradictory results reported by different investigators. An altered balance between the synthesis and catabolism of lipids could result in higher lipid storage levels at low temperature in the absence of acclimatory changes in lipogenesis.

Protein and RNA synthesis rates are higher in hepatocytes isolated from common carp acclimatized in outdoor pools to summer (16–18°C) relative to winter (9–10°C) conditions at assay temperatures of 10, 20 and 30°C (Saez et al. 1982). Although protein synthesis is reduced at low temperature there is evidence for certain compensatory changes following thermal acclimation. Rates of incorporation of labelled amino acids into protein at low temperatures are generally higher in the liver of cold- than warm-acclimated goldfish (Das & Prosser, 1967), rainbow trout (Dean & Berlin, 1969) and toadfish (Haschemeyer,

1969). In toadfish liver the average time required for the assembly of a polypeptide chain of average length (50 000 daltons) decreased from 6·6 to 3·9 min following cold acclimation (8–10°C), equivalent to synthetic outputs of $16 \, \mu g$ protein g^{-1} liver min^{-1} and $27 \, \mu g \, g^{-1}$ liver min^{-1}, respectively (Haschemeyer, 1969). Relative increases in protein synthesis are associated with changes in the concentration of low molecular weight elongation factors EF_1 (Nielsen, Plant & Haschemeyer, 1977). Below 7°C, liver protein synthesis in the toadfish liver appears to be inhibited (Mathews & Haschemeyer, 1978). However, it should be stressed that absolute rates of protein synthesis are lower at cold than warm temperatures (Kent & Prosser, 1980). Liver from winter adapted carp generally show lower levels of nuclear RNA, rRNA and tRNA, and a polysomal population which corresponds to a state of depressed protein synthesis compared to summer adapted carp (Zuvic, Brito, Villanueva & Krauskopf, 1980).

Muscle

Sustained swimming activity is largely supported by the recruitment of a relatively small volume of aerobic slow muscle fibres (Bone, 1978; Johnston, 1981). Electromyographical studies have shown that in common carp there is a sequential activation of slow > fast oxidative glycolytic > fast glycolytic motor units as speed increases (Johnston, Davison & Goldspink, 1977). On reducing the water temperature from 20 to 10°C, the threshold speed for recruitment of fast glycolytic fibres decreases from 2·6 to 1·4 bodylength/s (Rome, Loughna & Goldspink, 1984). This suggests that at low temperatures slow oxidative fibres can no longer provide all the power necessary to swim even at moderate sustained speeds, necessitating the recruitment of faster fibre types (Fig. 9). However, following several weeks acclimation to 8°C the recruitment threshold for fast glycolytic fibres has been shown to have increased (Rome, Loughna & Goldspink, 1985). After cold acclimation carp are able to swim faster with their aerobic muscle and have higher sustained swimming speeds than acutely cooled fish (Rome *et al.* 1985).

Changes in swimming performance and central patterns of muscle fibre recruitment with cold acclimation are associated with a major remodelling of the skeletal muscle. In goldfish (Johnston & Lucking, 1978) and striped bass (Jones & Sidell, 1982) the volume of aerobic fibre types is significantly higher in cold- than warm-acclimated fish. For example, in goldfish the number of fast oxidative glycolytic fibres/myotome was found to be 1700 for fish acclimated to 31°C compared with over 3000 for fish acclimated to 3°C (Johnston & Lucking, 1978). In addition, studies in a variety of species, including goldfish (Smit *et al.* 1974; Sidell, 1980), green sunfish (Shaklee, Christiansen, Sidell & Prosser, 1977), common carp (Johnston, Sidell & Driedzic, 1985), striped bass (Jones & Sidell, 1982), chain pickerel (Kleckner & Sidell, 1985) and flounder (Johnston & Wokoma, 1986) have shown that when assayed at an intermediate temperature the activity of aerobic enzymes is significantly higher in cold- than warm-acclimated fish. This results in part from a higher density of mitochondria in the muscles of cold-acclimated populations (Fig. 10). For example, mitochondria occupied 25 % of slow fibre volume in crucian carp acclimated to 2°C, compared with only 14 % for fish maintained at 28°C (Johnston & Maitland, 1980). Tyler & Sidell (1984) measured the harmonic mean of mitochondrial spacing in slow muscles from goldfish acclimated for 5 weeks to either 5 or 25°C. They found a 23 % shorter diffusion path length in cold- as opposed to warm-acclimated fibres. Several studies have also reported that aerobic enzyme activity is higher in muscle mitochondria isolated from cold- than warm-acclimated fish (Hazel, 1972a; Wodtke, 1981). However, mitochondria isolated from the slow muscle of 10°C-acclimated common carp contain a similar amount of cytochrome c oxidase per mg of protein as 30°C-acclimated fish (Wodtke, 1981). This suggests that cold acclimation does not change the density of enzymes on the inner mitochondrial membrane. This conclusion is supported by the stereological analysis of Tyler & Sidell (1984), which found no difference in the surface densities of inner mitochondria membrane between warm- and cold-acclimated goldfish muscle fibres. Another

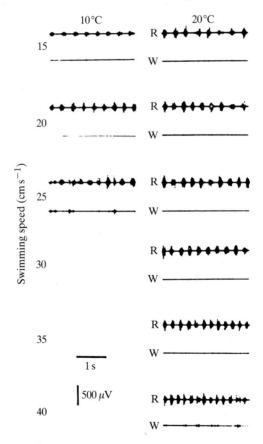

Fig. 9. Effects of acute temperature change on the recruitment of slow (R) and fast (W) muscles of the carp (*Cyprinus carpio*). Adapted from Rome, Loughna & Goldspink (1984).

possibility is that the changes in mitochondrial enzyme activity occur as a result of the changes in membrane phospholipid composition that accompany thermal acclimation (Hazel, 1972a,b; Van den Thillart & Modderkolk, 1978; van den Thillart & Bruin, 1981).

Cold acclimation in crucian carp has also been shown to result in higher surface and volume densities of muscle capillaries relative to warm acclimated fish (Johnston, 1982). The ultrastructural evidence (Fig. 10) suggests that prolonged exposure to low temperature increases the potential for aerobic ATP production in muscle relative to acutely exposed fish. Assayed at intermediate temperatures the rate of oxygen utilization of muscle fibres is higher for cold- than warm-acclimated fish. This is likely to result in a higher active metabolic rate and an increase in the range of swimming speeds that can be sustained at low temperatures (Fry & Hochachka, 1970; Rome *et al.* 1985). Although, the absolute capacity for aerobic ATP production usually remains significantly lower in cold than warm-acclimated fish, some exceptions have been reported (Fig. 11).

The power to swim at maximum speed largely comes from the recruitment of the fast muscle system (Johnston, 1981). The initial fuel for sprint activity is derived from the hydrolysis of phosphocreatine stores. This is rapidly followed by the activation of glycogenolysis resulting in the accumulation of lactic acid (see Hochachka, 1985). In contrast to the results for aerobic enzymes, the activities, and hence concentration, of glycolytic enzymes in muscle is generally unchanged by temperature acclimation (green sunfish, Shaklee *et al.* 1977; goldfish, Sidell

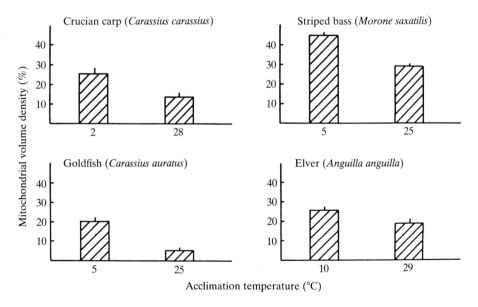

Fig. 10. Effects of acclimation temperature on volume density of mitochondria in teleost slow muscle fibres. Data are from the following sources: crucian carp (*Carassius carassius*) (Johnston & Maitland, 1980); goldfish (*Carassius auratus*) (Tyler & Sidell, 1984); striped bass (*Morone saxatilis*) (Egginton & Sidell, 1986) and elvers (*Anguilla anguilla*) (Egginton & Johnston, 1984).

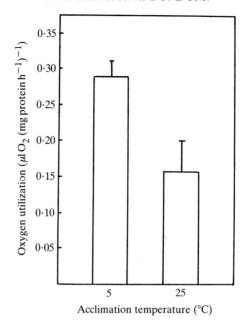

Fig. 11. Oxygen consumption by slow muscle from temperature acclimated striped bass. Values represent mean $\dot{V}O_2 \pm$ SEM measured at the acclimation temperature. $N = 5$. Adapted from Jones & Sidell (1982).

(1980); common carp, Johnston *et al.* 1985; chain pickerel, Kleckner & Sidell (1985)). Kleckner & Sidell (1985) compared maximal enzyme activities in tissues from 5 and 25°C-acclimated chain pickerel with those from winter- and summer-acclimatized fish to determine whether laboratory studies of thermal acclimation can be extrapolated to the natural environment. They found the general pattern of metabolic response to be similar in acclimated and acclimatized groups although enzyme activities were generally higher for laboratory held than wild fish. An exception was creatine phosphokinase (CPK) activity which increased in the slow muscle of winter-acclimatized compared with summer acclimatized fish but did not change between acclimated groups. The opposite result was found for CPK activities in fast muscle. Kleckner & Sidell (1985) suggested that these differences were related to differences in swimming behaviour between fish in the wild and fish in the laboratory.

Johnston, Davison & Goldspink (1975) found that in goldfish the myofibrillar ATPase activity of fast muscle fibres increased at low temperatures following a period of cold-acclimation. Furthermore the ATPase isolated from cold-acclimated goldfish was more susceptible to thermal denaturation than the enzyme from warm-acclimated fish (Johnston *et al.* 1975). This was the first demonstration that the myofibrillar proteins of some species underwent a re-modelling with cold-acclimation. Similar changes have now been reported for the slow oxidative fibres in goldfish (Sidell, 1980) and for the fast muscles of common carp, tench and roach (Heap *et al.* 1985). Johnston *et al.* (1985) acclimated common carp to either 7°C or 23°C for 2 months and studied the mechanical properties of single isolated fibres from fast and slow muscles. Fast muscle fibres from cold-acclimated fish developed 2-times more tension and had twice the unloaded contraction velocity at 7°C of fibres from warm-acclimated fish. These changes in contractile proteins with thermal acclimation take 4–6 weeks to complete and are not observed in starved fish in which protein synthesis is reduced to a very low level (Heap, Watt & Goldspink, 1986). It seems likely that myofibrillar re-modelling increases the demand for ATP in cold-acclimated fish and contributes to observed increases in swimming performance. However, in many eurythermal fish the thermal characteristics of the myofibrillar proteins appear to be genetically fixed and are not altered by acclimation. This is the case for muscle tissue from brook trout (Walesby & Johnston, 1981), rainbow trout (Penney & Goldspink, 1981), mummichogs (Sidell *et al.* 1983), flounder (Johnston & Wokoma, 1986) and striped bass (Moerland & Sidell, 1986).

Temperature acclimation also affects the concentrations of energy stores in muscle. Glycogen concentrations are more than 2-times higher in muscle fibres from 2°C than 28°C-acclimated crucian carp (Johnston & Maitland, 1980). In striped bass the volume density of lipid droplets in slow muscle is 15-fold higher in 5°C-acclimated (7·9 %) than 25°C-acclimated fish (0·6 %) (Egginton & Sidell, 1986).

There have been very few studies of muscle metabolism in species that undergo winter dormancy. Egginton & Johnston (1984) found that the density of mitochondria was slightly higher in the slow muscle of elvers of the European eel acclimated to 10°C (26 %) than those acclimated to 29°C (18·8 %). The lower acclimation temperature studied was slightly above that reported as triggering dormancy (8°C) and indicates a limited amount of metabolic compensation. However, American eels acclimated to 5°C were found to have similar or lower activities of most enzymes measured (assayed at 15°C) than fish acclimated to 15°C (Walsh *et al.* 1983).

Brain

Several studies have reported a relative increase in aerobic metabolism of brain tissue following a period of cold acclimation (goldfish: Evans, Purdie & Hickman, 1962; Caldwell, 1969; chain pickerel: Kleckner & Sidell, 1985). Nervous tissue is largely dependent on the aerobic catabolism of glucose for its energy supply. Breer & Jeserich (1980) found that on reducing the water temperature of goldfish from 20°C to 7°C there was a 5-fold increase in brain glucose concentration over a 2 h period. Following a further 2 weeks acclimation to this low temperature, brain glucose and glycogen concentrations were 2-times higher than for fish maintained at 20°C. Breer & Jeserich (1980) provided evidence that increased glycogen storage levels with cold acclimation resulted from a decreased activity of glycogen phosphorylase rather than an increased incorporation of glucose into glycogen.

Of particular interest are studies concerned with aspects of metabolism specific to the function of nervous tissue. Hazel (1969) found that, assayed at 5°C, acetylcholinesterase activity in homogenates of goldfish brain was lower in 5°C- than in 25°C-acclimated fish. These changes, which suggest a reduction of nervous activity at low temperatures, were complete in approximately 3 weeks (Hazel, 1969). In contrast, no acclimation effects were observed for brain acetylcholin-esterase activity in *Fundulus heteroclitus* maintained at either 5, 15 or 25°C (Hazel, 1969). Baldwin & Hochachka (1970) obtained evidence for 'warm' and 'cold' isozymes of acetylcholinesterase in the brain of rainbow trout (*Salmo gairdneri*). Electrophoresis of brain extracts from 12°C-acclimated trout revealed two bands. However, the anodic component was only found in brain extracts from 17°C-acclimated fish and the cathodic band was only found in extracts from fish acclimated to 2°C (Baldwin & Hochachka, 1970). The relationship between the apparent Km for acetylcholine and temperature was found to differ for the 'warm' and 'cold' isozymes. In each case the Km was at a minimum at temperatures corresponding to the acclimation temperature. Hochachka & Somero (1973) have suggested that such selective adjustments in Km are of adaptive significance in providing some temperature independence of substrate and ligand binding properties.

Other tissues

Caldwell (1969) acclimated goldfish for 2–4 weeks to either 10°C or 30°C and found that activities of electron transport enzymes (cytochrome oxidase, succinate cytochrome-c reductase, NADH cytochrome-c reductase) in gill tissue were higher at all temperatures between 10–40°C in the cold-acclimated group. Interestingly, the concentrations of cytochromes aa_3, b and $c+c_1$ did not differ between the two groups. This suggests that in contrast to liver and muscle relative increases in aerobic metabolism with cold acclimation do not appear to involve a significant increase in the mitochondrial content of gill cells (Caldwell, 1969).

Mechanisms underlying temperature compensation phenomena in ectotherms

The processes or mechanisms by which cellular activities are altered when the temperature changes may be grouped into two categories. Passive effects parallel the time course of thermal change and are the result of changes in the free energy of ions and molecules. Other metabolic changes may be regarded as an attempt to maintain cellular homeostasis in the face of the direct or passive effects of temperature change. The time course of these 'active' changes ranges from seconds (e.g. altered patterns of muscle fibre recruitment) to several weeks (e.g. changes in mitochondrial density and pathway utilization). Some of the possible mechanisms underlying these longer term changes in metabolism are shown in Table 1. The degree of activation of individual enzymes is determined by the concentrations of substrates and modulators and by the rate of product removal. In goldfish, the concentration of certain adenylates and sugar phosphates changes following transfer (over 30 min) form 15°C to either 5°C or 25°C, reaching new

Table 1. *Mechanisms underlying changes in enzyme activity and pathway utilization with temperature acclimation*

1.	Factors influencing enzyme microenvironment (e.g. pH, substrate/modulator concentrations, membrane composition)
2.	Conversion inactive to active form enzyme (e.g. phosphorylation)
3.	Changes in enzyme concentration
4.	Conformational changes ('instantaneous isozymes')
5.	Altered gene expression – isozymes
6.	Altered postranslational processing

steady state levels after around 7 h (Freed, 1971). Walesby & Johnston (1980) acclimated brook trout for 10 weeks to either 4°C or 25°C. Both groups were exercised at 2·5 bodylengths s^{-1} for 2 weeks prior to sacrifice in order to control for differences in spontaneous activity. They found that parameters of metabolic control including; adenylate energy charge ([ATP]+0·5 [ADP])/([ATP]+[ADP]+[AMP]), phosphorylation state ratio [ATP]/([ADP].[Pi]), and the ratios [ATP]/[ADP] and [ATP]/[AMP] were all significantly lower in the muscles of trout acclimated to 4°C. The observed changes in metabolic control ratios are consistent with a greater degree of activation of mitochondrial respiration and glycolysis at cold acclimation temperatures. It has also been postulated that temperature may have differential effects on the regulatory properties of enzymes that lie at the branchpoints of metabolism. For example, glucose-6-phosphate is a branch-point metabolite which can either be catabolized *via* the hexose monophosphate shunt or glycolysis or alternatively stored as glycogen. Hochachka (1968) incubated liver slices from the tropical fish *Symbranchus marmoratus* and *Lepidosiren paradoxa* at different temperatures in the presence of radio-labelled substrates. He found that as temperature was increased the rate of glycogen synthesis and HMPS activity increased relative to the rate of glycolysis.

In addition, certain regulatory steps are under hormonal control, e.g. glycogen phosphorylase. However, almost nothing is known about changes in endocrine status with temperature acclimation. Thyroid hormones are thought to be involved in regulating turnover rates of mitochondria and tissue aerobic capacity in amphibia (Jankowsky, 1960). In contrast, similar patterns of oxygen uptake with temperature acclimation were found for goldfish following various treatments designed to alter thyroid status (radiothroidectomy, TSH, ACTH thiourea treatment) (Klicka, 1965). Similar results have been obtained for hypophysectomized mud-minnows *Umbra limi* (Hanson & Stanley, 1970). However, in this study hypophysectomy was found to cause a reduction in oxygen consumption, but had no effect on metabolic temperature compensation between 5°C and 22°C.

In goldfish, the content of cytochrome *c* oxidase was shown to be 66% greater in skeletal muscle of fish acclimated to 5°C than in those acclimated to 25°C (Wilson, 1973). Similarly, the concentration of cytochrome *c* in the skeletal muscle of the green sunfish *Lepomis cyanellus* increases with cold acclimation. There is evidence

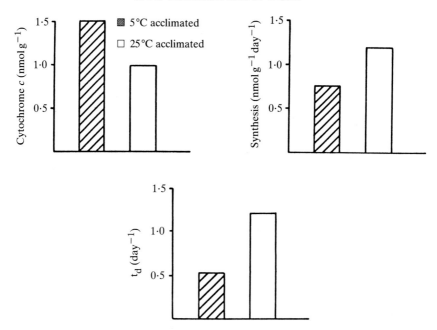

Fig. 12. Concentration, and rate constants of synthesis and degradation for muscle cytochrome c in temperature acclimated green sunfish *Lepomis cyanellus*. Data from Sidell (1977).

that such changes in enzyme concentration result from different thermal sensitivities of the rates of synthesis and degradation (Sidell, 1977). Sidell (1977) used [δ-^{14}C] aminolaevulinic acid, a non-reutilizable precursor of haem, to determine the loss of specific radioactivity of cytochrome c during acclimation of green sunfish from 5 to 25°C. He found that on reducing the temperature the rate of degradation of cytochrome c decreased less than its rate of synthesis resulting in a net increase in protein concentration (Fig. 12). This suggests that changes in tissue mitochondrial density with temperature acclimation could result, at least in part, from direct effects of temperature on mitochondrial turnover. The net effect of higher mitochondrial densities in cold-acclimated animals is to increase the rate of aerobic ATP production relative to that for acutely cooled tissues. Higher mitochondrial densities may also offset reduced diffusion rates at low temperatures. The diffusion coefficient for oxygen in frog skeletal muscle has been shown to decrease more than 40 % between the temperatures of 23°C and 0°C (Mahler, 1978). Similarly, the diffusion constant of lactate in fish muscle is 1·7 times lower at 5°C than 25°C (Hazel & Sidell, unpublished results cited in Tyler & Sidell, 1984). Tyler & Sidell (1984) have calculated that the increase in surface and volume density of mitochondria in 5°C compared with 25°C goldfish could compensate for a 3·1 to 3·4 fold decrease in diffusivity constants of metabolites.

Changes in enzyme microenvironment provide another mechanism for altering the rate of substrate utilization (Table 1). For example, the cytoplasm becomes more alkaline as temperature is reduced and this can have important consequences for enzyme function (Somero, 1981; Busa & Nuccitelli, 1984). The effects of temperature on acid–base balance varies between species and tissue compartments and with the time course of thermal change (Reeves, 1972; Heisler, 1978; Heisler, 1984; Walsh & Moon, 1982; Walsh *et al.* 1983). When 3°C acclimated largemouth bass were placed in a thermal gradient they moved to a temperature of 28°C over 18 h (Crawshaw *et al.* 1982). As temperature was increased neither the pH nor the total CO_2 content of the blood exhibited major changes suggesting that neither was an important determinant of the behavioural response (Crawshaw *et al.* 1982; Crawshaw, 1984). Other examples of changes in enzyme microenvironment with thermal acclimation include the modification of membrane phospholipid composition. These effects are reviewed in a companion chapter in this volume (Cossins).

Altered isoenzyme expression provides another potential mechanism for changing enzyme activities with temperature acclimation. It would appear that the 'on-off' switching of particular genes, which results in products which are only observed over a particular temperature range, is rare (e.g. trout brain acetylcholinesterase, Baldwin & Hochachka, 1970). More common are changes in the subunit composition of enzymes. For example, cold acclimation has been shown to result in a relative increase in the activities of M to H sub-units of lactate dehydrogenase in tissues from goldfish (Hochachka, 1965) and various amphibia including, *Xenopus laevis* (Tsugawa, 1980) and painted frog, *Discoglossus pictus* (DeCosta, Alonso-Bedate & Fraile, 1981). Tsugawa (1980) found that acclimation of *Xenopus laevis*, to either 14°C or 25°C, did not change the temperature dependence of the apparent Km of liver LDH for pyruvate. However, the H_4 and H_3M isozymes, isolated by affinity chromatography, had a higher Q_{10} for lactate oxidation than that of the M_4 isozyme. This, together with the relative decline in the proportions of H_4 and H_3M isozymes with cold acclimation, would be expected to favour pyruvate reduction at low temperatures. Acclimation temperature also affects the relative proportions of A and B subunits of malate dehydrogenase in cardiac and skeletal muscles of the teleost *Leiostomus xanthurus* (Fig. 13). The relative increase in AA compared with BB isozymes in 30°C-acclimated fish is thought to have some adaptive significance because the A subunit is more heat stable (Schwantes & Schwantes, 1982).

Temperature-mediated changes in protein conformation may provide another mechanism for producing enzyme variants at different acclimation temperatures. Somero (1969) provided kinetic evidence for two forms of pyruvate kinase in the king crab (*Paralithodes camtschatica*). At 15°C the (phospho-enol pyruvate) saturation curve is sigmoidal with a minimum $S_{0.5}$ (maximum PEP affinity) at 12°C. However, lower assay temperatures promote a shift from sigmoidal to hyperbolic kinetics. The hyperbolic form of the enzyme displays maximal PEP affinity at 5°C (Somero, 1969). There are very few other examples of 'instantaneous' isozymes (see Hochachka & Somero, 1984). Other potential mechanisms for modifying enzyme activity with temperature acclimation include changes in

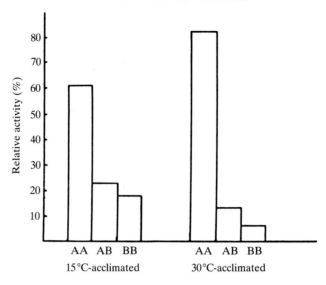

Fig. 13. Relative activities of malate dehydrogenase isozymes from temperature acclimated spot (*Leiostomus xanthurus*). Adapted from Schwantes & Schwantes (1982).

phosphorylation and post-translational modification of proteins (Table 1). The importance of these factors has yet to be explored.

References

BAINBRIDGE, R. (1962). Training, speed and stamina in trout. *J. exp. Biol.* **35**, 109–133.

BALDWIN, J. & HOCHACHKA, P. W. (1970). Functional significance of isoenzymes in thermal acclimatization: acetylcholinesterase from trout brain. *Biochem. J.* **116**, 883–887.

BEAMISH, F. W. H. (1970). Oxygen consumption of largemouth bass, *Micropterus salmoides*, in relation to swimming speed and temperature. *Can. J. Zool.* **48**, 1221–1228.

BEAMISH, F. W. H. & MOOKHERJII, P. S. (1964). Respiration of fishes with special emphasis on standard oxygen consumption. I. Influence of weight and temperature on respiration of goldfish, *Carassius auratus* L. *Can. J. Zool.* **42**, 161–175.

BONE, Q. (1978). Locomotor Muscle. In *Fish Physiology* (ed. W. S. Hoar & D. J. Randall), pp. 361–424. New York, London: Academic Press.

BREER, H. & JESERICH, G. (1980). Temperature effect on carbohydrate metabolism in fish brain. In *Proc. Europ. Soc. Comp. Physiol. Biochem*, vol. 1. pp. 107–108. Cambridge: Cambridge University Press.

BRETT, J. R. (1964). The respiratory metabolism and swimming performance of young sockeye salmon. *J. Fish. Res. Bd. Canada* **21**, 1183–1226.

BRETT, J. R. (1972). The metabolic demand for oxygen in fish, particularly salmonids, and a comparison with other vertebrates. *Resp. Physiol.* **14**, 151–170.

BUSA, W. B. & NUCCITELLI, R. (1984). Metabolic regulation via intracellular pH. *Am. J. Physiol.* **246**, R409–R438.

CALDWELL, R. S. (1969). Thermal compensation of respiratory enzymes in tissues of the goldfish (*Carassius auratus* L.). *Comp. Biochem. Physiol.* **31**, 79–93.

CAMPBELL, C. M. & DAVIES, P. S. (1978). Temperature acclimation in the teleost *Blennius pholis*: changes in enzyme activity and cell structure. *Comp. Biochem. Physiol.* **61B**, 165–167.

COSSINS, A. R., KENT, J. & PROSSER, C. L. (1980). A steady state and differential polarised phase flourimetric study of the liver microsomal and mitochondrial membranes of the thermally-acclimated green sunfish (*Lepomis cyanellus*). *Biochim. biophys. Acta* **599**, 341–358.

CRAWSHAW, L. I. (1984). Low-temperature dormancy in fish. *Am. J. Physiol.* **246**, R479–R486.

CRAWSHAW, L. I., ACKERMAN, R. A., WHITE, F. N. & HEATH, M. E. (1982). Metabolic and acid–base changes during selection of warmer water by cold-acclimated fish. *Am. J. Physiol.* **242**, R157–R161.

DAS, A. B. (1967). Biochemical changes in tissues of goldfish acclimated to high and low temperatures. II. Synthesis of protein and RNA of subcellular fractions and tissue composition. *Comp. Biochem. Physiol.* **21**, 469–485.

DAS, A. B. & PROSSER, C. L. (1967). Biochemical changes in tissues of goldfish acclimated to high and low temperature. I. Protein synthesis. *Comp. Biochem. Physiol.* **21**, 449–467.

DEAN, J. M. & BERLIN, J. D. (1969). Alterations in hepatocyte function of thermally acclimated rainbow trout (*Salmo gairdneri*). *Comp. Biochem. Physiol.* **29**, 307–312.

DE COSTA, J., ALONSO-BEDATE, M. & FRAILE, A. (1981). Temperature acclimation in amphibians: changes in lactate dehydrogenase activities and isoenzyme patterns in several tissues from adult *Discoglossus pictus pictus*. *Comp. Biochem. Physiol.* **70B**, 331–340.

DOUDOROFF, P. & SHUMWAY, D. L. (1970). *Dissolved oxygen requirements of freshwater fishes.* F.A.O. (United Nations) Fisheries Technical Paper No. 86.

DUTHIE, G. & HOULIHAN, D. F. (1982). The effect of single step and fluctuating temperature changes on oxygen consumption of flounders *Platichthys flesus* (L.): lack of temperature adaptation. *J. Fish. Biol.* **21**, 215–226.

EGGINTON, S. & JOHNSTON, I. A. (1984). Effects of acclimation temperature on routine metabolism, muscle mitochondrial volume density, and capillary supply in the elver (*Anguilla anguilla*). *J. therm. Biol.* **9**, 165–170.

EGGINTON, S. & SIDELL, B. D. (1986). Changes in mitochondrial spacing in fish skeletal muscle induced by environmental temperature. *J. Physiol., Lond.* **373**, 78.

EVANS, R. M., PURDIE, F. C. & HICKMAN, C. P., JR (1962). The effect of temperature and photoperiod on the respiratory metabolism of rainbow trout (*Salmo gairdneri*). *Can. J. Zool.* **40**, 107–118.

FOSTER, G. D. & MOON, T. W. (1986). Cortisol and liver metabolism of immature American eels, *Anguilla rostrata* (LeSueur). *Fish Physiol. Biochem.* **1**, 113–124.

FREED, J. M. (1971). Properties of muscle phosphofructokinase of cold and warm acclimated *Carassius auratus*. *Comp. Biochem. Physiol.* **39B**, 747–764.

FRY, F. E. J. (1957). The aquatic respiration of fish. In *The Physiology of Fishes*, vol. 1 (ed. A. H. Rose), pp. 1–63. London: Academic Press.

FRY, F. E. J. & HOCHACHKA, P. W. (1970). Fish. In *Comparative Physiology of Thermoregulation*, vol. 1 (ed. G. C. Wittow), pp. 97–134. New York: Academic Press.

HANSON, R. C. & STANLEY, J. G. (1970). The effects of hypophysectomy and temperature acclimation upon the metabolism of the central mudminnow, *Umbra limi* (Kirtland). *Comp. Biochem. Physiol.* **33**, 871–879.

HASCHEMEYER, A. E. V. (1969). Rates of polypeptide chain assembly in liver *in vivo*: relation to the mechanism of temperature acclimation in *Opsanus tau*. *Proc. natn. Acad. Sci. U.S.A.* **62**, 128–135.

HAZEL, J. R. (1969). The effect of thermal acclimation upon brain cholinesterase activity of *Carassius auratus* and *Fundulus heteroclitus*. *Life Sci.* **8**, 775–784.

HAZEL, J. R. (1972*a*). The effect of temperature acclimation upon succinic dehydrogenase activity from the epaxial muscle of the common goldfish (*Carassius auratus* L.). I. Properties of the enzyme and the effect of lipid extraction. *Comp. Biochem. Physiol.* **43B**, 837–861.

HAZEL, J. R. (1972*b*). The effect of temperature acclimation upon succinic dehydrogenase activity from the epaxial muscle of the common goldfish (*Carassius auratus* L.). II. Lipid reactivation of the soluble enzyme. *Comp. Biochem. Physiol.* **43B**, 863–882.

HAZEL, J. R. (1979). Influence of thermal acclimation on membrane lipid composition of rainbow trout liver. *Am. J. Physiol.* **236**, R91–R101.

HAZEL, J. R. & NEAS, N. P. (1982). Turnover of the fatty acyl and glycerol moieties of membrane lipids from liver, gill and muscle tissue of thermally acclimated rainbow trout, *Salmo gairdneri*. *J. comp. Physiol.* **149**, 11–18.

HAZEL, J. R. & PROSSER, C. L. (1974). Molecular mechanisms of temperature compensation in poikilotherms. *Physiol. Rev.* **54**, 620–677.

HAZEL, J. R. & PROSSER, C. L. (1979). Incorporation of I-^{14}C-acetate into fatty acids and sterols by isolated hepatocytes of thermally acclimated rainbow trout (*Salmo gairdneri*). *J. comp. Physiol.* **134**, 321–329.

HAZEL, J. R. & SELLNER, P. A. (1979). Fatty acid and sterol synthesis by hepatocytes of thermally acclimated rainbow trout (*Salmo gairdneri*). *J. exp. Zool.* **209**, 105–113.

HEAP, S. P., WATT, P. W. & GOLDSPINK, G. (1985). Consequences of thermal change on the myofibrillar ATPase of 5 freshwater teleosts. *J. Fish. Biol.* **26**, 733–738.

HEAP, S. P., WATT, P. W. & GOLDSPINK, G. (1986). Myofibrillar ATPase activity in the carp *Cyprinus carpio*: interactions between starvation and environmental temperature. *J. exp. Biol.* **123**, 373–382.

HEISLER, N. (1978). Bicarbonate exchange between body compartments after changes of temperature in the larger spotted dogfish (*Scyliorhinus stellaris*). *Resp. Physiol.* **33**, 145–160.

HEISLER, N. (1984). Role of ion transfer processes in acid–base regulation with temperature changes in fish. *Am. J. Physiol.* **246**, R441–R451.

HENDERSON, R. J. & SARGENT, J. R. (1981). Lipid biosynthesis in rainbow trout fed diets of different fat content. *Comp. Biochem. Physiol.* **69C**, 31–37.

HOCHACHKA, P. W. (1965). Isozymes in metabolic adaptation of a poikilotherm: subunit relationships in lactic dehydrogenases of isolated hepatocytes from the European eel *Anguilla anguilla* L. *J. therm. Biol.* **6**, 201–208.

HOCHACHKA, P. W. (1968). Action of temperature on branch points in glucose and acetate metabolism. *Comp. Biochem. Physiol.* **25**, 107–118.

HOCHACHKA, P. W. (1985). Fuels and pathways as designed systems for support of muscle work. *J. exp. Biol.* **115**, 149–164.

HOCHACHKA, P. W. & HAYES, F. R. (1962). The effect of temperature acclimation on pathways of glucose metabolism in the trout. *Can. J. Zool.* **40**, 261–270.

HOCHACHKA, P. W. & SOMERO, G. (1973). *Strategies of biochemical adaptation*. Philadelphia, London: W. B. Saunders Co.

HOCHACHKA, P. W. & SOMERO, G. (1984). *Biochemical Adaptation*. Princeton: Princeton University Press.

JANKOWSKY, H. D. (1960). Uber die hormonale beeinflussing der temperatureadaptation beim grasfrosch (*Rana temporaria* L.). *Z. Vergl. Physiol.* **43**, 392–410.

JANKOWSKY, H. D., HOTOPP, H. & SEIBERT, H. (1984). Influence of thermal acclimation on glucose production and ketogenesis in isolated eel hepatocytes. *Am. J. Physiol.* **246**, R471–R476.

JOHNSTON, I. A. (1981). Structure and function of fish muscles. *Symp. zool. Soc. London.* **48**, 71–113.

JOHNSTON, I. A. (1982). Capillarisation, oxygen diffusion distances and mitochondrial content of carp muscles following acclimation to summer and winter temperatures. *Cell. Tiss. Res.* **222**, 325–337.

JOHNSTON, I. A., DAVISON, W. & GOLDSPINK, G. (1975). Adaptations in Mg^{2+}-activated myofibrillar ATPase activity induced by temperature acclimation. *FEBS. Lett.* **50**, 293–295.

JOHNSTON, I. A., DAVISON, W. & GOLDSPINK, G. (1977). Energy metabolism of carp swimming muscles. *J. comp. Physiol.* **114**, 203–216.

JOHNSTON, I. A. & LUCKING, M. (1978). Temperature induced variation in the distribution of different types of muscle fibre in the goldfish (*Carassius auratus*). *J. comp. Physiol.* **124**, 111–116.

JOHNSTON, I. A. & MAITLAND, B. (1980). Temperature acclimation in crucian carp (*Carassius carassius* L.); morphometric analysis of muscle fibre ultrastructure. *J. Fish. Biol.* **17**, 113–125.

JOHNSTON, I. A., SIDELL, B. D. & DRIEDZIC, W. R. (1985). Force-velocity characteristics and metabolism of carp muscle fibres following temperature acclimation. *J. exp. Biol.* **119**, 239–250.

JOHNSTON, I. A. & WOKOMA, A. (1986). Effects of temperature and thermal acclimation on contractile properties and metabolism of skeletal muscle in the flounder (*Platichthys flesus* L.). *J. exp. Biol.* **120**, 119–130.

JONES, D. R. (1971). Theoretical analysis of factors which may limit the maximum oxygen uptake of fish: the oxygen cost of the cardiac and branchial pumps. *J. theor. Biol.* **32**, 341–349.

Jones, P. L. & Sidell, B. D. (1982). Metabolic responses of striped Bass (*Morone saxatilis*) to temperature acclimation. II. Alterations in metabolic carbon sources and distributions of fibre types in locomotory muscle. *J. exp. Zool.* **219**, 163–171.

Kacser, H. & Burns, J. A. (1980). *The control of flux.* Symp. Soc. Exp. Biol. XXVII, pp. 65–105. Cambridge: Cambridge University Press.

Kanungo, M. S. & Prosser, C. L. (1959). Physiological and biochemical adaptation of goldfish to cold and warm temperatures. II. Oxygen consumption of liver homogenate; oxygen consumption and oxidative phosphorylation of liver mitochondria. *J. cell. comp. Physiol.* **54**, 265–274.

Kent, J. & Prosser, C. L. (1980). Effects of incubation and acclimation temperatures on incorporation of carbon-14 uniformly labelled glycine into mitochondrial protein of liver cells and slices from green sunfish *Lepomas cyanellus*. *Physiol. Zool.* **53**, 293–304.

Kleckner, N. W. & Sidell, B. D. (1985). Comparison of maximal activities of enzymes from tissues of thermally acclimated and naturally acclimated chain pickerel (*Esox niger*). *Physiol. Zool.* **58**, 18–28.

Klicka, J. (1965). Temperature acclimation in goldfish: lack of evidence for hormonal involvement. *Physiol. Zool.* **38**, 177–189.

Lemons, D. E. & Crawshaw, L. I. (1985). Behavioral and metabolic adjustments to low temperatures in the largemouth bass (*Micropterus salmoides*). *Physiol. Zool.* **58**, 175–180.

Mahler, M. (1978). Diffusion and consumption of oxygen in the resting frog sartorius muscle. *J. gen. Physiol.* **71**, 533–557.

Mathews, R. W. & Haschemeyer, A. E. V. (1978). Temperature dependency of protein synthesis in toadfish liver *in vivo*. *Comp. Biochem. Physiol.* **61B**, 479–484.

Moerland, T. S. & Sidell, B. D. (1981). Characterization of metabolic carbon flow in hepatocytes isolated from thermally acclimated killifish *Fundulus heteroclitus*. *Physiol. Zool.* **54**, 379–389.

Moerland, T. S. & Sidell, B. D. (1986). Biochemical responses to temperature in the contractile protein complex of striped bass *Morone saxatilis*. *J. exp. Zool.* **238**, 287–295.

Neas, N. P. & Hazel, J. R. (1985). Phospholipase A_2 from liver microsomal membranes of thermally acclimated rainbow trout (*Salmo gairdneri*). *J. exp. Zool.* **233**, 51–60.

Newsholme, E. A. & Paul, J. M. (1983). The use of *in vitro* enzyme activities to indicate the changes in metabolic pathways during acclimatisation. In *Cellular Acclimatisation to Environmental Change* (ed. A. Cossins & P. Sheterline), pp. 81–101. Cambridge: Cambridge University Press.

Newsholme, E. A. & Start, C. (1973). *Regulation in Metabolism.* London: John Wiley & Sons.

Nichols, P. R. & Miller, R. V. (1967). Seasonal movements of striped bass, *Roccus saxatilis* (Walbaum), tagged and released in the Potomac River, Maryland, 1959–1961. *Chesapeake Sci.* **8**, 102–124.

Nielsen, J. B. K., Plant, P. W. & Haschemeyer, A. E. V. (1977). Control of protein synthesis in temperature acclimation. II. Correlation of elongation factor 1 activity with elongation rate *in vivo*. *Physiol. Zool.* **50**, 22–30.

Parker, J. (1972). Spatial arrangement of some cryoprotective compounds in ice lattices. *Cryobiology* **9**, 247–250.

Penny, D. & Goldspink, G. (1981). Short term temperature acclimation in myofibrillar ATPase of a stenotherm *Salmo gairdneri* and a eurytherm *Carassius auratus*. *J. Fish. Biol.* **18**, 715–722.

Precht, H. (1958). Concepts of the temperature adaptation of unchanging reaction systems of cold-blooded animals. In *Physiological adaptation* (ed. C. L. Prosser), pp. 50–78. Washington D.C.: American Physiological Society.

Reeves, R. B. (1972). An imidazole alphastat hypothesis from vertebrate acid–base regulation: Tissue carbon dioxide content and body temperature in bullfrogs. *Resp. Physiol.* **14**, 219–236.

Rome, L. C., Loughna, P. T. & Goldspink, G. (1984). Muscle fiber activity in carp as a function of swimming speed and temperature. *Am. J. Physiol.* **247**, R272–R279.

Rome, L. C., Loughna, P. T. & Goldspink, G. (1985). Temperature acclimation: improved sustained swimming performance in carp at low temperatures. *Science* **228**, 194–196.

Saez, L., Goicoichea, O., Amthaver, R. & Krauskopf, M. (1982). Behavior of RNA and protein synthesis during the acclimatization of carp: studies with isolated hepatocytes. *Comp. Biochem. Physiol.* **72B**, 31–38.

SCHWANTES, M. L. B. & SCHWANTES, A. R. (1982). Adaptive features of ectothermic enzymes: 2. The effects of acclimation temperature on the malate dehydrogenase of the spot, *Leistomus xanthurus*. *Comp. Biochem. Physiol.* **72B**, 59–64.

SHAKLEE, J. B., CHRISTIANSEN, J. A., SIDELL, B. D., PROSSER, C. L. & WHITT, G. S. (1977). Molecular aspects of temperature acclimation in fish: contributions of changes in enzymic activities and isoenzyme patterns to metabolic reorganisation in the green sunfish. *J. exp. Zool.* **201**, 1–20.

SIDELL, B. D. (1977). Turnover of cytochrome c in skeletal muscle of green sunfish (*Lepomas cyanellus* R.) during thermal acclimation. *J. exp. Zool.* **199**, 233–250.

SIDELL, B. D. (1980). Response of goldfish (*Carassius auratus* L.) muscle to acclimation temperature: alterations in biochemistry and proportions of different fibre types. *Physiol. Zool.* **53**, 98–107.

SIDELL, B. D. (1983). Cellular acclimatisation to environmental change by quantitative alterations in enzymes and organelles. In *Cellular Acclimatisation to Environmental Change* (ed. A. Cossins & P. Sheterline), pp. 103–120. Cambridge: Cambridge University Press.

SIDELL, B. D., JOHNSTON, I. A., MOERLAND, T. S. & GOLDSPINK, G. (1983). The eurythermal myofibrillar protein complex of the mummichog (*Fundulus heteroclitus*): adaptation to a fluctuating thermal environment. *J. comp. Physiol.* **153**, 167–173.

SMIT, H., VAN DEN BERG, R. J. & KIJN-DEN KARTOG, I. (1974). Some experiments on thermal acclimation in the goldfish (*Carassius auratus* L.). *Neth. J. Zool.* **24**, 32–49.

SOMERO, G. N. (1969). Pyruvate kinase varients of the Alaska king crab: evidence for a temperature-dependent interconversion between two forms having distinct and adaptive kinetic properties. *Bioch. J.* **114**, 237–241.

SOMERO, G. N. (1981). pH-temperature interactions on proteins: principles of optimal pH and buffer system design. *Mar. Biol. Lett.* **2**, 163–178.

STONE, B. B. & SIDELL, B. D. (1981). Metabolic response of striped bass (*Morone saxatilis*) to temperature acclimation. I. Alterations in carbon sourses for hepatic energy metabolism. *J. exp. Zool.* **218**, 371–379.

TSUGAWA, K. (1980). Thermal dependence in kinetic properties of lactate dehydrogenase from the African clawed toad, *Xenopus laevis*. *Comp. Biochem. Physiol.* **66B**, 459–466.

TYLER, S. & SIDELL, B. D. (1984). Changes in mitochondrial distribution and diffusion distances in muscle of goldfish (*Carassius auratus*) upon acclimation to warm and cold temperatures. *J. exp. Zool.* **232**, 1–10.

UMMINGER, B. L. (1978). Physiological studies on supercooled killifish (*Fundulus heteroclitus*). III. Carbohydrate metabolism and survival at subzero temperatures. *J. exp. Zool.* **173**, 159–174.

VAN DEN THILLART, G. & DE BRUIN, G. (1981). Influence of environmental temperature on mitochondrial membranes. *Biochim. biophys. Acta* **640**, 439–447.

VAN DEN THILLART, G. & MODDERKOLK, J. (1978). The effect of acclimation temperature on the activation energies of state III respiration and on the unsaturation of membrane lipids of goldfish mitochondria. *Biochim. biophys. Acta* **510**, 38–51.

VOSS, B. & JANKOWSKY, H. D. (1986). Temperature dependence of lipogenesis in isolated hepatocytes from rainbow trout (*Salmo gairdneri*). *Comp. Biochem. Physiol.* **83B**, 13–22.

WALESBY, N. J. & JOHNSTON, I. A. (1980). Temperature acclimation in brook trout muscle: adenine nucleotide concentrations, phosphorylation state and adenylate energy charge. *J. comp. Physiol.* **139**, 127–133.

WALESBY, N. J. & JOHNSTON, I. A. (1981). Temperature acclimation of Mg^{2+} Ca^{2+} myofibrillar ATPase from a cold-selective teleost, *Salvelinus fontinalis*: a compromise solution. *Experientia* **37**, 716–718.

WALSH, P. J., FOSTER, G. D. & MOON, T. (1983). The effects of temperature on metabolism of the american eel *Anguilla anguilla* Rostrata (Le Sueur): Compensation in the summer and torpor in the winter. *Physiol. Zool.* **56**, 532–540.

WALSH, P. J. & MOON, T. (1982). The influence of temperature on extracellular and intracellular pH in the American eel, *Anguilla rostrata* (Le Sueur). *Resp. Physiol.* **50**, 129–140.

WALTON, M. J. & COWEY, C. B. (1982). Aspects of intermediary metabolism in salmonid fish. *Comp. Biochem. Physiol.* **73B**, 59–79.

WILSON, F. R. (1973). *Quantitating changes of enymes of the goldfish (*Carassius auratus *L.) in response to temperature acclimation. An immunological approach.* PhD thesis, University of Illinois.

WODTKE, E. (1981). Temperature adaptation of biological membranes. Compensation of the molar activity of cytochrome C oxidase in the mitochondrial energy-transducing membrane during thermal acclimation of the carp (*Cyprinus carpio* L.). *Biochim. biophys. Acta* **640**, 710–720.

YAMAUCHI, T., STEGEMAN, J. J. & GOLDBERG, E. (1975). The effects of starvation and temperature acclimation on pentose phosphate pathway dehydrogenases in brook trout liver. *Archs Biochem. Biophys.* **167**, 13–20.

ZAMMIT, V. A. & NEWSHOLME, E. A. (1979). Activities of enzymes of fat and ketone body metabolism and effects of starvation on blood concentrations of glucose and fat fuels in teleost and elasmobranch fish. *Biochem. J.* **184**, 313–322.

ZUVIC, T., BRITO, M., VILLANUEVA, J. & KRAUSKOPF, M. (1980). *In vivo* levels of aminoacyl-t.RNA species during acclimatization of the carp *Cypinus carpio. Comp. Biochem. Physiol.* **67B**, 167–170.

Printed in Great Britain © Society for Experimental Biology 1987

ADAPTIVE RESPONSES OF ANIMAL CELL MEMBRANES TO TEMPERATURE

A. R. COSSINS *and* R. S. RAYNARD

Department of Zoology, University of Liverpool, PO Box 147, Liverpool L69 3BX, UK

Introduction

Great advances have been made over the past 15 years in our understanding of the structure–function relationships of biological membranes. Yet little is known about the way in which cells produce a diverse and highly differentiated collection of membranes or regulate their respective chemical compositions. One obvious way of revealing such control mechanisms is to disturb the *status quo* and look for corrective responses. This strategy has recently been used to great advantage in biological membranes where temperature has been used as the disturbing influence.

Historically it has been known for many years that exposure of microorganisms, plants and animals to low temperatures leads to the incorporation of increased proportions of unsaturated fatty acids in the storage lipids (i.e. triglycerides) and in membrane phospholipids (Hazel, 1984). On the basis of physical studies this trend was interpreted as an adaptive response which offsets the direct effects of cooling upon the physical state of the lipid structures. Thus phospholipids containing unsaturated fatty acids displayed greater cross-sectional areas (i.e. a greater expansion of a monolayer), a greater molecular flexing motion and lower phase transition temperatures (i.e. melting temperatures) than lipids containing saturated fatty acids.

A direct test of this hypothesis was only possible with the development of techniques for the direct investigation of membrane physical structure. This was first demonstrated in bacteria using electron spin resonance spectroscopy by Sinensky (1974), who showed that the 'fluidity' and phase state of membranes from *Escherichia coli* was maintained constant at each of several growth temperatures. He coined the term 'homeoviscous adaptation' to emphasize both the homeostatic nature of the response and its adaptive importance. The hypothesis of homeoviscous adaptation states that the chemical composition of membranes is regulated to maintain the physical structure or 'fluidity' of the bilayer within tolerable limits. Fluidity is a term which describes the relative conformational freedom and mobility of membrane constituents.

The significance of homeoviscous regulation relates to the direct effects of bilayer physical properties upon the functional properties of the membrane since bilayer lipids provide the effective solvent environment of the functional components, namely integral membrane-bound proteins. Thus by conserving the physical structure of the bilayer within certain limits the functional properties of the membrane are thought to be protected from the direct effects of environmental change. Although homeoviscous adaptation was originally restricted to temperature adaptations there is accumulating evidence for similar regulatory responses to other influences which also disturb membrane fluidity, such as anaesthetics (Littleton, 1983), dietary manipulations (Cossins & Sinensky, 1984) and hydrostatic pressure (Macdonald & Cossins, 1985). Therefore it probably represents a response of general and fundamental significance rather than applying only to a restricted group of animals suffering from a specific thermal stress.

This short review discusses some of the evidence for adaptations of both fluidity and phase structure, changes in lipid composition and the enzymatic bases for these changes and, finally, how the biosynthetic apparatus is adaptively regulated. Most studies of homeoviscous adaptations have used poikilothermic organisms such as bacteria, the protozoan *Tetrahymena* and teleost fish. This is not only because the cells of these organisms are routinely exposed to temperature variations and hence disturbed fluidity, but also because they display a suite of physiological adjustments to temperature variations which collectively offset the disturbance that this might otherwise cause. Some species have evolved in specific thermal environments and comparison of these species reveals genotypic adaptations to variations in temperature.

Membrane fluidity, membrane order and phase structure

Textbooks of biochemistry invariably illustrate the structure of biological membranes with diagrams of phospholipids aligned in a parallel and highly ordered manner. Although these membrane constituents are indeed organized as a bilayer they exhibit, at physiological temperatures, a considerable degree of disorder and molecular mobility. These motions range from a rapid flexing of the hydrocarbon chain by rotations about carbon-carbon bonds, to the wobbling motion of the entire molecule and to the lateral displacement of the molecule along the plane of the membrane. Because of this motion biological membranes have a liquid-like character and are thus described as being fluid. However, the properties and physical structure of the hydrocarbon interior of phospholipid bilayers is quite distinct from that of bulk hydrocarbon fluids, such as paraffin, and concepts of bulk viscosity and fluidity (i.e. reciprocal of viscosity) cannot have the same precise meaning. For one thing the bilayer is highly anisotropic and semi-ordered and for another the ΔH values associated with the gel-to-liquid crystalline phase transition are significantly lower than for the melting of hydrocarbon fluids (Phillips *et al.* 1969). Finally, biological membranes are heterogeneous systems in that a diverse chemical composition may lead to microdomains in the plane of the membrane as well as different physical properties in each monolayer.

In practice, membrane fluidity is operationally defined and quantified using one of several biophysical techniques which measure the rotational characteristics of spectroscopic nuclei or molecules ('probes') which by chemical synthesis or by partitioning can be intercalated within the bilayer interior. The basic assumptions in this approach are first that the measured motional characteristics of the probe are sensitive to the dynamical motion or order of the surrounding hydrocarbon chain and second that the normal physical structure of the bilayer is not greatly disturbed by the probe. Each technique provides information specifically on the type of motion which affects the spectroscopic property in question so that different techniques report on different aspects of the 'fluid' condition. Thus ^{2}H- and ^{13}C-nuclear magnetic resonance spectroscopy provides rather specific information on the dynamics of hydrocarbon flexing motion and by the chemical synthesis of fatty acids with ^{2}H or ^{13}C nuclei positioned at specific sites along the hydrocarbon chain a detailed understanding of bilayer structure can be obtained (Seelig & Seelig, 1980). By contrast, fluorescence polarization spectroscopy of a rigid, rod-shaped fluorescent molecule such as 1,6-diphenyl-1,3,5-hexatriene (DPH) indicates the extent of wobbling motion of the probe over the nanosecond timescale. This is influenced by bilayer structure over much of the length of the fatty acid at least equivalent to the length of the DPH molecule (*ca* 13 Å) rather than at a specific segment. Given these differences it is necessary to use several different techniques to characterize fully membrane fluidity or to use probes with different rotational properties. Each type of motion occurs over a rather different time domain and changes in one may not correlate with changes in another (Kleinfeld *et al.* 1981). Any of these methods are useful in a comparative experiment though attempts to convert derived motional parameters to absolute units of viscosity by comparisons of probe behaviour in hydrocarbon solvents of different bulk viscosities have been severely criticized largely because of the qualitative differences in physical structure of the bilayer and bulk solvents.

Technical and theoretical advances have led to an improved understanding of these spectroscopic techniques and their interpretation. Nowhere is this more true than with fluorescence polarization spectroscopy where early interpretations assumed that the rotational motion of rod-like probes was unhindered so that polarization of fluorescence was viewed as an index of rotational rate. More recent time-resolved studies have demonstrated more complex rotational properties in that free rotations of the probe were constrained by the anisotropic structure of the bilayer (van Blitterswijk *et al.* 1981). Thus DPH polarization is now thought to reflect primarily the degree of hindrance to free rotation rather than a rotational rate. The dynamic concept of fluidity has been superseded by the static concept of membrane order.

Finally, it is necessary to recognize that probe techniques generally provide average information on fluidity/order depending upon the distribution of probe positions within the membrane. The membrane is thus treated as though it were a single hydrophobic phase with discrete and definable properties, even though it is probably made up of a number of distinct microdomains each with different physical properties. The more common probe techniques, such as fluorescence polarization using DPH or electron spin resonance spectroscopy (ESR) using nitroxy-labelled fatty acids, are not able to differentiate between these different microdomains and yield a weighted average of the probe positions. This problem may be overcome to some extent by using probes with well-defined positions in the so-called 'fluidity gradient' though at present there are no techniques to probe separately microdomains with subtly different properties. From a practical point of view this means that it is frequently not possible to link causally a change in average membrane fluidity either to changes in all molecules in a population of probes, to changes in the distribution of probe between different microdomains or to changes in the properties in some of the microdomains but not others.

Phenotypic membrane adaptations

Fig. 1 compares the order of membranes isolated from cold- and warm-acclimated fish, as measured using the fluorescence polarization technique with DPH as probe. DPH polarization was much lower in cold-acclimated fish compared to warm-acclimated fish, indicating a greater probe wobbling motion in cold-acclimated fish and by inference a reduced lipid order. In that cooling causes a general ordering of the acyl chains this difference offsets or compensates for at

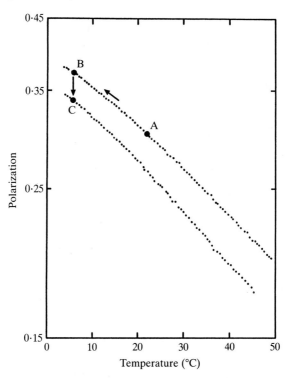

Fig. 1. The effects of thermal acclimation of goldfish upon the 'fluidity' of a brain synaptic membrane preparation. Membrane fluidity was estimated using the fluorescence polarization technique with DPH as probe. A low polarization indicates a greater degree of probe rotation and by inference a greater fluidity of its immediate environment. Cooling of 25°C-acclimated fish to 5°C results in an increase in polarization (point A to B), but acclimation at 5°C for 10–14 days results in a reduction of polarization (B to C) which offsets about half of the temperature-induced increase (M. Behan, G. Jones, K. Bowler & A. R. Cossins, unpublished observations).

least part of this direct cooling effect. However, because DPH polarization in cold-acclimated fish was higher than in warm-acclimated fish *when measured at their respective acclimation temperatures*, the difference was not sufficient to overcome the effects of cooling completely.

The extent of the compensatory response can be simply calculated as the ratio of the observed shift of the polarization/temperature curve along the temperature axis to the shift required for complete or perfect compensation. This measure has been termed 'homeoviscous efficacy' and lies between 0·2 and 0·5 or 20 and 50 % of a perfect compensation for a broad range of membranes in various species of fish and the protozoan *Tetrahymena* (Cossins & Sinensky, 1984). Only in the case of fish muscle sarcoplasmic reticulum was no compensatory difference found between cold- and warm-acclimated fish (Cossins, Christiansen & Prosser, 1978).

Homeoviscous responses by an individual are not limited to fish or even to poikilothermal animals since under the appropriate conditions mammalian tissues display equivalent responses. Lagerspetz & Laine (1986) have found differences in DPH polarization of a synaptic membrane preparation of neonatal and adult mice which corresponded with a difference in body temperature of neonate and adult of approximately 11 °C. This difference might, of course, be due to tissue differentiation during growth and maturation of the tissue (Hitzemann & Harris, 1984). However, differences in DPH polarization were not observed between neonatal and adult guinea pigs, a species in which the neonate is euthermic. Homeoviscous adaptation has also been observed in cultured mammalian cells (Anderson *et al.* 1981).

Studies on the time-course of homeoviscous adaptation in fish have given conflicting results. In brain synaptosomes of goldfish, the change in DPH polarization during cold-acclimation took 30–40 days whilst the corresponding period during warm-acclimation was 10–14 days (Cossins, Friedlander & Prosser, 1977). By contrast, in liver microsomes of the carp, warm-acclimation was achieved within 3–4 days with a half-time of 2 days (Wodtke & Cossins, to be published). In that the changes in DPH polarization depend upon changes in lipid composition these different rates in brain and liver probably reflect differences in their respective rates of lipid turnover. In unicellular organisms the time-course of homeovisous response is comparatively rapid. In *Tetrahymena*, for example, DPH polarization and ESR order parameters change within 4h after a reduction in culture temperature from 39·5 °C to 15 °C (Martin & Thomspon, 1978; Ohki *et al.* 1984) and recent studies have shown even more rapid responses (Dickens & Thompson, 1981). The rapid responses of bacteria and protozoans in culture may be due to a very rapid synthesis of membrane components of cells in the logarithmic growth phase.

Genotypic homeoviscous adaptations

The ability to adjust physiology to suit seasonally-varying temperatures is most apparent in species such as carp and goldfish which have evolved in mid-latitude, continental climatic zones. Other species live in relatively constant environmental temperatures, such as antarctic fish, or have adopted thermoregulatory strategies to maintain constant body temperature, such as mammals and birds. These species have a more limited ability to acclimatize physiologically and it is interesting to enquire whether the adaptation of species over evolutionary time to more or less constant temperatures also results in adaptive shifts in membrane order and whether the thermal specialization of species enables a more complete homeoviscous adaptation than observed during seasonal acclimatization. Implicit in the comparison of different species is the broad similarity of the compared membrane fractions in all respects other than body temperature; namely in their functional properties, tissue morphology and membrane-type compositions of the resulting membrane preparations. These conditions are arguably best satisfied in the brain where neuronal structure is relatively constant, certainly within the teleost fish and perhaps also within the vertebrates as a whole.

Interspecific and apparently adaptive differences in DPH polarization have been reported for a crude 'synaptosomal' preparation from the brain of different fish species (Cossins & Prosser, 1978). Fig. 2 shows a similar comparison of DPH

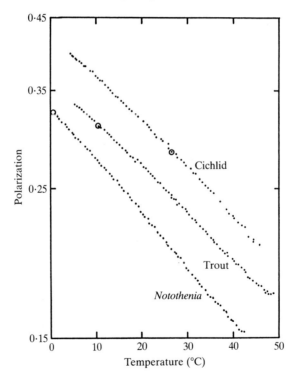

Fig. 2. A comparison of membrane fluidity of brain synaptic membrane preparations from an Antarctic fish (*Notothenia*, 0°C), trout (10°C) and an East African cichlid (28°C). Again fluidity was estimated using the fluorescence polarization technique with DPH as probe. The open circles represent DPH polarization at the respective body temperatures for each species (M. Behan, G. Jones, K. Bowler & A. R. Cossins, unpublished observations).

polarization of several fish species but for a more refined 'synaptic' preparation. Polarization increases at a single measurement temperature in the order *Notothenia*, trout, and cichlid, which correlates with their normal environmental temperatures of 0°C, 14°C and 28°C, respectively. These interspecific differences were sufficiently large to provide similar values of polarization *at their respective body temperatures*. DPH polarization measured for the rat at 37°C is also rather similar to that measured in the different species of fish at their respective body temperatures. These remarkable observations suggest first that thermal specialization of fish species results in a near-complete homeoviscous adaptation and second that the concept of evolutionary homeoviscous adaptation can be extended to include other vertebrate classes, such as mammals and birds. The large interspecific and interclass differences in membrane order become explicable when the differences in body temperatures are taken into account.

Phase structure and 'homeophasic' adaptations

As temperature is gradually reduced, phospholipid bilayers undergo a reversible thermotropic transition from a fluid-like, liquid-crystalline state at high temperatures to a more rigidified gel state at low temperatures (Melchior & Steim, 1976). Similar transitions have been demonstrated calorimetrically in biological membranes, though because of their chemical heterogeneity the transitions occur over a broad range of temperatures (10–30°C) compared to the restricted, well-defined transition in pure phospholipid bilayers (2–5°C, Melchior & Steim, 1976). This means that within the phase transition of biological membranes the liquid-crystalline and gel phases coexist, producing a lateral differentiation of the bilayer into microdomains (Shimshick & McConnell, 1973). There are also suggestions in the literature of more subtle rearrangements of lipids. For example, Lee *et al.* (1974) have suggested the formation of 'clusters' of quasi-crystalline lipid in the bilayer at the onset of a bulk transition. Wu & McConnell (1975) have shown how under certain conditions fluid domains may become separated.

In that thermotropic transitions involve rather dramatic changes in physical structure they pose a potentially more serious problem for maintained membrane function in the cold than altered fluidity. Indeed, it may be that homeoviscous adaptation is, in reality a manifestation of the regulation of membrane phase state rather than a 'fine-tuning' of a liquid-crystalline membrane *per se* (Silvius & McElhaney, 1980; Cossins & Sinensky, 1984). One might expect that thermal acclimation would also result in adaptive shifts in transition temperatures especially in situations where the transition occurs over the physiological temperature range. Such 'homeophasic' adaptations have been claimed in studies using probe techniques both in *E. coli* (Sinensky, 1974) and *Bacillus stearothermophilus* (Esser & Souza, 1974). A more satisfactory and less equivocal technique of establishing transition temperatures is by scanning calorimetry. McElhaney & Souza (1976) have shown a shift in phase transition temperatures in response to

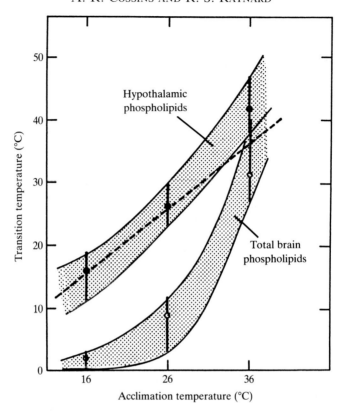

Fig. 3. The effect of thermal acclimation of the garden lizard, *Calotes versicolor*, upon the temperatures for the thermotropic phase transitions of extracted brain and hypothalamic phospholipids. The symbols represent the midpoint temperatures and the bars represent the extent of the upper and lower boundary for the calorimetric transition. The shaded area shows how the temperatures at which gel and liquid-crystalline state coexist varies with acclimation temperature. For comparison the dashed line represents the body temperature. (Data from Durairaj & Vijayakamur, 1984.)

variations in growth temperature of *B. stearothermophilus*, such that the upper boundary of this transition occurred just below the growth temperature.

In a remarkable recent study, Durairaj & Vijayakuma (1984) have observed some sharp calorimetric transitions in extracted phospholipids of lizard hypothalamus. Fig. 3 shows how these transitions were shifted to lower temperatures as acclimation temperature was reduced, maintaining the phase state constant at each acclimation temperature. A similar and no less impressive demonstration has been provided in *Tetrahymena* using semi-quantitative freeze-fracture electron microscopy (Martin *et al.* 1976). Unfortunately, there is at present no information regarding the phase state or microdomain structure of fish membranes though judging from their highly unsaturated fatty acid composition it is likely that the main phase transition occurs well below the physiological range of temperatures.

Adaptive changes in lipid composition

Underlying the compensatory changes in membrane physical structure are changes in lipid composition. There are several types of compositional adjustment which are reviewed in detail by Hazel (1984). These include changes in acyl chain composition, their positional distribution on phospholipids, the phospholipid headgroup composition and finally changes in cholesterol content. On the basis of experiments with artificial membranes most of these adjustments have been shown to have clear-cut effects on membrane physical structure which may act to offset the direct influence of temperature variations. For example, the incorporation of greater proportions of unsaturated fatty acids gives phospholipids with lower melting point temperatures and greater cross-sectional areas, all of which leads to a greater conformational freedom in membranes of cold-acclimated animals. Similarly, the observed increase in phosphatidylethanolamine in the cold may decrease bilayer order because its wedged shape disturbs the otherwise close packing of the hydrocarbon chains.

Of these various compositional adjustments most attention has been paid to the mechanisms accounting for the change in acyl chain composition. Whilst these adjustments in composition have been explored in some considerable detail there is only a poor understanding of their physical significance. This is largely because changes in lipid composition are complex with simultaneous changes in several structural features and second because the contribution of the various types of acyl chain to the fluidity of the bilayer has not been established in any quantitative way. General indices of fatty acid composition such as mole % saturated fatty acids or the unsaturated index (double bond index) do not adequately take account of the important structural features of fatty acids (Cossins & Lee, 1985). Lands & Davis (1984) have devised a semi-empirical formula in which the overall fluidity of a mixture of fatty acids was estimated by a linear sum of individual contributions of the individual components to the total. These workers have demonstrated good agreement between the calculated fluidity and measured properties of bacterial and yeast membranes, and with growth of these organisms under conditions where fatty acid supplementation was necessary.

Adaptive changes in lipid biosynthesis

Because the changes in lipid saturation that occur on cold-acclimation vary in a detailed way between different tissues, between different membrane fractions and even between different phospholipid classes of the same fraction, the underlying mechanisms of change are probably complex and highly specific. Nevertheless, there are a number of regulatory possibilities in common.

Fig. 4 displays in simplified form the metabolic pathways involved in the regulation of the acyl chain composition of membrane phospholipids. This shows three specific points where regulation is likely to occur:

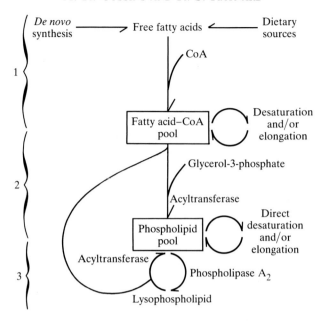

Fig. 4. A simplified scheme for the pathways involved in the biosynthesis of membrane phosphoglycerides. The numbers refer to the potential sites of adaptive regulation as described in the text. (Modified after Hazel, 1984.)

(i) The provision of the fatty acyl-CoA pool. This is established by the dietary intake of fat and by *de novo* synthesis, but is then subject to modification by desaturases and chain elongation enzymes.

(ii) The selective incorporation of fatty acids from the pool during *de novo* phospholipid synthesis and their selective placement on the sn-1 and sn-2 positions of the glycerol moiety.

(iii) The direct modification of the phospholipid pool either by the selective turnover of phospholipids through the deacylation-reacylation cycle or by the direct desaturation of acyl chains on the phospholipid.

(i) *Fatty acid desaturases*

Distinct enzymes are thought to account for desaturation at different positions on the acyl chain and adaptive changes in lipid composition may result from complex changes in the activity of some or all of these enzymes. The situation is at its simplest in microorganisms where the acyl group composition of membrane lipids is, by comparison with multicellular organisms, very simple. In *Bacillus megaterium* the activity of the Δ^5-desaturase is regulated by temperature in two ways. Firstly, total activity is dramatically increased through an increased rate of enzyme biosynthesis due to the absence of a postulated temperature-sensitive modulator protein that normally represses protein synthesis (Fujii & Fulco, 1977). Secondly, the enzyme is labile at high temperatures and so becomes progressively inactivated at higher temperatures.

In the protozoan *Tetrahymena* there are two possible means of altering acyl-CoA desaturase. The first is by the induced synthesis of more enzyme at the lower temperatures; thus chilling leads to a rapid and substantial increase in *in vitro* palmitoyl-CoA and stearoyl-CoA desaturase activity. This increase is inhibited in the presence of cycloheximide. The second mechanism is a direct modulation of specific activity of the enzyme by the fluidity of the lipids in its immediate microenvironment. Both mechanisms are thought to operate within the same cell though their relative contributions are unclear (Thompson & Nozawa, 1984).

Wodtke and his colleagues have demonstrated an increased desaturase activity on cold exposure in the endoplasmic reticulum of carp liver on cold exposure (Schünke & Wodtke, 1983). Fig. 5 shows the time-course of desaturase activity following transfer of 30°C-acclimated fish to 10°C. There was a sudden increase in activity within 1–3 days after cold transfer from very low levels followed by a reduction after day 3, and a second increase between 7–10 days. The first increase in desaturase activity coincided well with changes in lipid saturation of endoplasmic reticulum phospholipids and with a decrease in DPH polarization (Wodtke & Cossins, unpublished data). The half-times for the changes in desaturase activity, lipid saturation and DPH polarization were 1·31, 1·98 and 2·16 days, respectively (Wodtke, 1986). The second peak of desaturase activity was without any effect upon DPH polarization and may, therefore, reflect a secondary 'retailoring' of acyl group composition with no clear-cut structural effects as judged by the polarization technique. Because the desaturase reaction displayed a normal temperature dependence (Schünke & Wodtke, 1983) the enzyme is clearly not cold-activated. It is likely, therefore, that the transient changes in desaturase rate constant reflect changes in the number of active enzymes.

The study by Schünke & Wodtke was also notable for examining the effects of altered dietary saturation upon desaturase activity. Not surprisingly, feeding of 30°C-acclimated fish with a more saturated diet led to higher desaturase activities. However, transfer of these animals to 10°C did not elicit the violent oscillations in desaturase activity that were noted in fish fed a more unsaturated diet, although the time-course of changes in lipid saturation were broadly similar with both dietary regimes. Thus transient changes in desaturase activity far exceed those observed between long-term acclimated animals, a point that has also been made by Hazel (1984) in reference to other studies. Desaturase activities in fish at constant temperatures are more dependent upon the dietary source of lipids than temperature *per se* and transient alterations in desaturase activity reflect the changes required to elicit the adaptive changes in lipid composition. Farkas *et al.* (1980) have also found that lipid composition and lipid biosynthesis are adjusted according to diet as well as temperature.

The modification of activity of different desaturases during thermal acclimation may be highly specific. Sellner & Hazel (1982) found complete compensation for Δ^6-desaturation, and overcompensation for Δ^4- and Δ^5-desaturation acting on linolenic acids (18:3) in trout hepatocytes. The compensatory patterns for desaturation of 18:1 and 18:2 were rather different suggesting highly complex

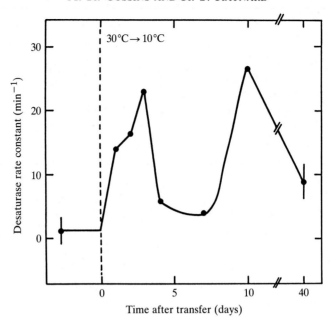

Fig. 5. The time-course of changes in desaturase activity of liver microsomal membranes following transfer of 30°C-acclimated carp to 10°C. (Modified after Schunke & Wodtke, 1983.)

changes in fatty acid metabolism when temperature is altered. The patterns observed may well be highly species- and tissue-specific.

(ii) *Phospholipid synthesis*

Fatty acids may be selectively incorporated into membrane phospholipids. There are two principal steps where this may occur, namely the acyltransferase (the sequential addition of fatty acids to glycerol-3-phosphate and to lysophosphatidate) and the phospho- and CDP-diglyceride transferases where phosphatidic acid is broken down to form either diglyceride or CDP-diglyceride and when specific molecular species may be preferentially used in phospholipid synthesis.

Sellner & Hazel (1982) have demonstrated, again in trout hepatocytes, a selective incorporation of fatty acids into phospholipids, and the modification of this pattern by both assay and acclimation temperature. This applied particularly to the long-chain polyunsaturates such as 20:4, 20:5 and 20:6, where incorporation was selectively increased in cold-acclimated fish. These modifications may be due to acclimation-dependent changes in the substrate preference of enzymes involved in phospholipid biosynthesis. Thus, incorporation of radiolabelled CDP-choline into phosphatidylcholine was diverted in the cold into molecular species of phosphatidylcholine containing polyunsaturated fatty acids and away from molecular species containing mono- and diunsaturated fatty acids (see Hazel, 1984). By contrast, the acyltransferases of trout liver microsomes do not display selectivity when pre-tested with different 18-carbon fatty acids (Hazel *et al.* 1983).

(iii) *Phospholipid retailoring*

The effective regulation of phospholipid saturation during *de novo* biosynthesis of phospholipids and fatty acids is feasible only in membranes with a significant rate of turnover of phospholipids. A more economical and potentially more rapid tactic is to regulate selectively the incorporation of fatty acids during acylation of lysophospholipids formed during membrane turnover.

The evidence in favour of this process is circumstantial rather than direct. For example, the activity of phospholipase A_2 in liver microsomes of cold-acclimated trout was twice that of warm-acclimated trout such that at their respective acclimation temperatures the activity was similar (Neas & Hazel, 1985). Similarly, incorporation of radiolabelled fatty acids into membrane phospholipids was greater in cold- than in warm-acclimated trout (Hazel, 1983). Finally, the rates of incorporation of radiolabelled glycerol into liver phosphatidylcholine was slower than the incorporation of radiolabelled fatty acid in the gills of cold-acclimated trout. In warm-acclimated trout the rates were similar (Neas & Hazel, 1982). This suggests that cold activates a deacylation-reacylation pathway.

In *Tetrahymena*, sudden chilling from 39 to 15 °C induced a substantial reorganization of the positional distribution of fatty acids on the glycerol moiety. This was achieved within one hour of the shift in temperature which was well before any detectable change in overall lipid saturation (Watanabe *et al.* 1981; Dickens & Thompson, 1981). The inference is that temperature causes changes in acyltransferase and/or in phospholipase specificity such that saturated fatty acids in the sn−1 position were replaced with linoleic acid. Conventional measures of lipid saturation would not detect this rearrangement.

Control mechanisms

The nature and extent of homeoviscous adaptation to temperature is now becoming clear as are the compositional adjustments that accompany and presumably underlie this response. Perhaps the most important unresolved issue is how the biosynthetic apparatus is regulated to produce the appropriate response. Underlying this is the central question of to what extent the response is induced by signals originating from an inappropriate membrane fluidity. Control theory formalizes this situation as a negative feedback loop in which temperature acts as a disturbance to the controlled variable, which in this case is membrane fluidity. Deviations of fluidity from a preferred value or 'set-point' initiate the appropriate corrective response by the biosynthetic apparatus (see Cossins, 1983). The alternative control mechanism is one where the biosynthetic adjustments are made without reference to the level of fluidity to give an open control system. In this case temperature acts as a stimulus to a mechanism that is programmed with temperature-specific properties to elicit the adaptive response.

Physiologists are used to postulating negative feedback mechanisms because of their inherent self-regulatory or homeostatic properties. An additional attraction in this case is the greater flexibility offered by a mechanism that detects a deviation of fluidity from a preferred value rather than temperature *per se*. Thus disturbance by factors other than temperature (pressure, fluidizing drugs) should also elicit the adaptive response. Similarly, changes in the saturation of dietary fats can easily be accommodated into the model by assuming that these also act by perturbing the *status quo*. An open control system must be separately 'programmed' for each perturbing modality.

Translating these principles into actual mechanisms is not straightforward because it is not clear whether the controlling elements operate at the membrane level or alternatively involve induced protein synthesis. In microorganisms there is evidence for both membrane level and nuclear regulation. For example, in *E. coli* it seems that the biosynthetic enzymes exhibit a temperature-specific control. Enzyme preparations from cells grown at different temperatures exhibited the same temperature-specific pattern of incorporation of saturated and unsaturated fatty acids suggesting that the degree of membrane unsaturation has little or no influence on the products of the acyl-transferase reaction (Cossins & Sinensky, 1984). As already discussed, *Bacillus stearothermophilus* exhibits regulation by a cold-induced derepression of the appropriate gene(s) (Fulco, 1972; Fulco & Fujii, 1980).

In *E. coli* it seems that an open control loop operates whilst in *B. stearothermophilus* the role of fluidity in causing the derepression of desaturase synthesis in the cold or the repression in the warm is not known. Induction of desaturase activity during cold-acclimation has been observed in several other studies in microorganisms and animal cells (Schünke & Wodtke, 1983; Nozawa & Kasci, 1978) and it seems likely that nuclear regulation is also involved.

At present there are two other hypotheses which satisfy the negative feedback scheme. First, in *Tetrahymena* the specific activity of the palmitoyl-CoA desaturase was increased four-fold directly after culture temperature was reduced from 39·5°C to 15°C even when protein synthesis was inhibited (Fukushima *et al.* 1979; Kasei *et al.* 1985). This has led to the suggestion that the specific activity or perhaps the substrate specificity of the desaturase is enhanced by a direct effect of lipid fluidity (see Kates *et al.* 1984). The proposed direct effect of fluidity on enzymatic activity, therefore, acts as the feedback element in a closed control loop and there is some experimental evidence for cold-activation of stearoyl-CoA desaturase of rat liver reconstituted in phosphatidylcholine liposomes. However, because there is no clearly defined mechanism by which activity is enhanced in the cold the idea remains controversial. One potential mechanism is by the vertical displacement of the enzyme out of the bilayer as described by Shinitzky (1984) and a second is by the restriction of protein conformational freedom.

A second and quite different mechanism has been proposed by Melchior & Steim (1977) to account for the temperature-specific pattern of fatty acid incorporation seen in the mycoplasma *Acholeplasma*. They noted that the ratio of 16:0 to 18:1 incorporated could be altered by changing the phase state of the membrane. Thus the incoporation ratio was low and constant in the gel state but progressively increased through the phase transition. A similar pattern was observed with the binding of 16:0 and 18:1 to artificial bilayers, suggesting that the temperature-specificity is simply a function of the selective binding or solubilization of fatty acids by the bilayer. The temperature-programmed pattern of incorporation is thus seen as a cooperative property of the bilayer itself rather than of the biosynthetic enzymes. Thus enzymes have little or no ability to distinguish between various fatty acids or alter their substrate specificity with changing temperature. Selective binding of fatty acids may provide a rather simple homeostatic mechanism which operates at the membrane level.

References

ANDERSON, R. L., MINTON, K. W., LI, G. C. & HAHN, G. M. (1981). Temperature-induced homeoviscous adaptation of Chinese hamster ovary cells. *Biochim. Biophys. Acta* **641**, 334–348.

COSSINS, A. R. (1983). The adaptation of membrane structure and function to changes in temperature. In *Cellular Acclimatisation to Environmental Change* (ed. A. R. Cossins & P. S. Sheterline), pp. 3–32. Cambridge: Cambridge University Press.

COSSINS, A. R., CHRISTIANSEN, J. R. & PROSSER, C. L. (1978). Adaptation of biological membranes to temperature. The lack of homeoviscous adaptation in the sarcoplasmic reticulum. *Biochim. Biophys. Acta* **511**, 442–454.

COSSINS, A. R., FRIEDLANDER, M. J. & PROSSER, C. L. (1977). Correlations between behavioural temperature adaptations of goldfish and the viscosity and fatty acid composition of their synaptic membranes. *J. comp. Physiol.* **120**, 109–121.

COSSINS, A. R. & LEE, J. A. C. (1985). The adaptation of membrane structure and lipid composition to cold. In *Circulation, Respiration and Metabolism* (ed. R. Gilles), pp. 543–552. Berlin: Springer-Verlag.

COSSINS, A. R. & PROSSER, C. L. (1978). The evolutionary adaptation of membranes to temperature. *Proc. natn. Acad. Sci. U.S.A.* **75**, 2040–2043.

COSSINS, A. R. & SINENSKY, M. (1984). Adaptation of membranes to temperature, pressure and exogenous lipids. In *Physiology of Membrane Fluidity*, vol. II (ed. M. Shinitzky), pp. 1–20. Boca Raton, Florida: C.R.C. Press.

DICKENS, B. F. & THOMPSON, G. A., JR (1981). Rapid membrane response during low-temperature acclimation. Correlation of early changes in the physical properties and lipid composition of *Tetrahymena* microsomal membranes. *Biochim. Biophys. Acta* **664**, 211–218.

DURAIRAJ, G. & VIJAYAKUMAR, I. (1984). Temperature acclimation and phospholipid phase transition in hypothalamic membrane phospholipids of garden lizard, *Calotes versicolor*. *Biochim. Biophys. Acta* **770**, 7–14.

ENOCH, H. G., CATELA, A. & STRITTMATTER, P. (1976). Mechanism of rat liver microsomal stearoyl-CoA desaturase. *J. biol. Chem.* **251**, 5095–5103.

ESSER, A. F. & SOUZA, K. A. (1974). Correlation between thermal death and membrane fluidity in *Bacillus stearothermophilus*. *Proc. natn. Acad. Sci. U.S.A.* **71**, 7111–7115.

FARKAS, T., CSENGERI, I., MAJOROS, F. & OLAH, J. (1983). Metabolism of fatty acid in fish. III. Combined effects of environmental temperature and diet on formation and deposition of fatty acids in the carp, *Cyprinus carpio*. *Aquaculture* **20**, 29–40.

FULCO, A. J. (1972). The biosynthesis of unsaturated fatty acids by bacilli. IV. Temperature-mediated control mechanisms. *J. biol. Chem.* **247**, 3511–3519.

FULCO, A. J. & FUJII, D. K. (1980). Adaptive regulation of membrane lipid biosynthesis in bacilli by environmental temperature. In *Membrane Fluidity: Biophysical Techniques and Cellular Regulation* (ed. M. K. Kates & A. Kuksis), pp. 77–98. Clifton, New Jersey: Humana Press.

FUJII, D. K. & FULCO, A. J. (1977). Biosynthesis of unsaturated fatty acids by bacilli. Hyperinduction and modulation of desaturase synthesis. *J. biol. Chem.* **252**, 3660–3670.

Fukushima, H., Nagao, S. & Nozawa, Y. (1979). Further evidence of changes in the level of palmitoyl-CoA desaturase during thermal acclimation in *Tetrahymena pyriformis*. *Biochim. Biophys. Acta* **572**, 178–182.

Hazel, J. R. (1983). The incorporation of unsaturated fatty acids of the $n-9$, $n-6$ and $n-3$ families into individual phospholipids by isolated hepatocytes of thermally acclimated rainbow trout, *Salmo gairdneri*. *J. exp. Zool.* **227**, 167–179.

Hazel, J. R. (1984). Effects of temperature on the structure and metabolism of cell membranes of fish. *Am. J. Physiol.* **246**, R460–R470.

Hazel, J. R., Sellner, P. A., Hagar, A. F. & Neas, N. P. (1983). Thermal adaptation in biological membranes: The acylation of glycerol-3-phosphate by liver microsomes of thermally-acclimated rainbow trout, *Salmo gairdneri*. *Mol. Physiol.* **4**, 125–140.

Hitzemann, R. J. & Harris, R. A. (1984). Developmental changes in synaptic membrane fluidity: A comparison of DPH and TMA-DPH. *Devl Brain Res.* **14**, 113–120.

Kasei, R., Kitajima, Y., Martin, C. E., Nozawa, Y., Skriver, L. & Thompson, G. A., Jr (1976). Molecular control of membrane properties during temperature acclimation. Membrane fluidity regulation of fatty acid desaturase action. *Biochemistry* **15**, 5228–5233.

Kasei, R., Yamada, T., Hasegawa, I., Muto, Y., Yoshioka, S., Nakamara, T. & Nozawa, Y. (1985). Regulatory mechanism of desaturation activity in cold acclimation of *Tetrahymena pyriformis*, with special reference to quick cryoadaptation. *Biochim. Biophys. Acta* **836**, 397–401.

Kates, M., Pugh, E. L. & Ferrante, G. (1984). Regulation of membrane fluidity by lipid desaturases. In *Biomembranes* **14** (ed. M. Kates & L. A. Manson), pp. 379–395. New York: Plenum Press.

Kleinfeld, A. M., Dragsten, P., Klausner, R. D., Pjura, W. J. & Matayoshi, E. D. (1981). The lack of relationship between fluorescence polarisation and lateral diffusion in biological membranes. *Biochim. Biophys. Acta* **649**, 471–480.

Lagerspetz, K. Y. H. & Laine, A. (1986). The functional significance of homeoviscous adaptation of cell membranes in multicellular animals. In *Biona Report* No. 4 (ed. H. Laudien), pp. 101–108. Stuttgart: Gustav Fischer.

Lands, W. E. M. & Davis, F. S. (1984). Fluidity of membrane lipids. In *Biomembranes* **14** (ed. M. Kates & L. A. Manson), pp. 475–516. New York: Plenum Press.

Lee, A. G., Birdsall, N. J. M., Metcalfe, J. C., Toon, P. A. & Warren, G. B. (1974). Clusters in lipid bilayers and the interpretation of thermal effects in biological membranes. *Biochemistry* **13**, 3699–3705.

Littleton, J. M. (1983). Membrane reorganisation and adaptation during chronic drug exposure. In *Cellular Acclimatisation to Environmental Change* (ed. A. R. Cossins & P. S. Sheterline), pp. 145–160. Cambridge: Cambridge University Press.

Macdonald, A. G. & Cossins, A. R. (1985). The theory of homeoviscous adaptation of membranes applied to deep-sea animals. In *Experimental Biology at Sea*. Symp. Soc. Exp. Biol., vol. 34 (ed. M. S. Laverack), pp. 301–322. Cambridge: Cambridge University Press.

Martin, C. E., Miramatsu, K., Kitajima, Y., Nozawa, Y., Skriver, L. & Thompson, G. A., Jr (1976). Molecular control of membrane properties during temperature acclimation. Fatty acid desaturase regulation of acclimating *Tetrahymena* cells. *Biochemistry* **15**, 5218–5227.

Martin, C. E. & Thompson, G. A., Jr (1978). Use of fluorescence polarisation to monitor intracellular membrane changes during temperature acclimation. Correlation with lipid compositional and ultrastructural changes. *Biochemistry* **17**, 3581–3586.

McElhaney, R. N. & Souza, K. A. (1976). The relationship between environmental temperature, cell growth and the fluidity and physical state of the membrane lipids in *Bacillus stearothermophilus*. *Biochim. Biophys. Acta* **443**, 348–359.

Melchior, D. L. & Steim, J. M. (1976). Thermotropic transitions in biomembranes. *A. Rev. Biophys. Bioeng.* **5**, 205–238.

Melchior, D. L. & Steim, J. M. (1977). Control of fatty acid composition of *Acholeplasma laidlawii* membranes. *Biochim. Biophys. Acta* **466**, 148–159.

Neas, N. P. & Hazel, J. R. (1985). Phospholipase A_2 from liver microsomal membranes of thermally acclimated rainbow trout. *J. exp. Zool.* **233**, 51–60.

NOZAWA, Y. & KASEI, R. (1978). Mechanisms of thermal adaptation of membranes lipids in *Tetrahymena pyriformis* NT-1. Possible evidence for temperature-mediated induction of palmitoyl-CoA desaturase. *Biochim. Biophys. Acta* **529**, 54–66.

OHKI, K., GOTO, M. & NOZAWA, Y. (1984). Thermal adaptation of *Tetrahymena* membranes with special reference to mitochondria. II. Preferential interaction of cardiolipin with specific molecular species of phospholipid. *Biochim. Biophys. Acta* **769**, 563–570.

PHILLIPS, M. C., WILLIAMS, R. M. & CHAPMAN, D. (1969). On the nature of hydrocarbon chain motion in lipid liquid crystal. *Chem. Phys. Lipids* **3**, 234–244.

SCHÜNKE, M. & WODTKE, E. (1983). Cold induced increase of Δ^9 and Δ^6-desaturase activities in endoplasmic reticulum membranes of carp liver. *Biochim. Biophys. Acta* **734**, 70–75.

SEELIG, J. & SEELIG, A. (1980). Lipid conformation in model membranes and biological membranes. *Quart. Rev. Biophys.* **13**, 19–61.

SELLNER, P. A. & HAZEL, J. R. (1982). Desaturation and elongation of unsaturated fatty acids in hepatocytes from thermally-acclimated rainbow trout. *Archs Biochem. Biophys.* **213**, 58–66.

SHIMSHICK, E. J. & McCONNELL, H. M. (1973). Lateral phase separations in phospholipid membranes. *Biochemistry* **12**, 2351–2360.

SILVIUS, J. R. & McELHANEY, R. N. (1980). Membrane lipid physical state and modulation of the Na^+, Mg^+-ATPase activity in *Acholeplasma laidlawii* B. *Proc. natn. Acad. Sci. U.S.A.* **77**, 1255–1259.

SINENSKY, M. (1974). Homeoviscous adaptation – a homeostatic process that regulates viscosity of membrane lipids in *Escherichia coli*. *Proc. natn. Acad. Sci. U.S.A.* **71**, 522–525.

THOMPSON, G. A., JR & NOZAWA, Y. (1984). The regulation of membrane fluidity in *Tetrahymena*. In *Membrane Fluidity* (ed. M. Kates & L. A. Manson), pp. 397–450. New York: Plenum Press.

VAN BLITTERSWIJK, W. J., VAN HOEVEN, R. P. & VAN DER MEER, B. W. (1981). Lipid structural order parameters (reciprocal of fluidity) in biomembranes derived from steady-state fluorescence polarisation measurements. *Biochim. Biophys. Acta* **644**, 323–332.

WATANABE, T., FUKUSHIMA, H., KASEI, R. & NOZAWA, Y. (1981). Studies on thermal adaptation in *Tetrahymena* lipids. Changes in positional distribution of fatty acids in diacylphospholipids and alkyl-acyl-phospholipids during temperature acclimation. *Biochim. Biophys. Acta* **665**, 66–73.

WODTKE, E. (1986). Adaptation of biological membranes to temperature: Modifications and their mechanisms in the eurythermic carp. In *Temperature Relations in Animals and Man* (ed. H. Laudien), pp. 129–138. Stuttgart: Gustav Fischer.

WU, S. H. & McCONNELL, H. M. (1975). Phase separations in phospholipid membranes. *Biochemistry* **147**, 847–854.

Printed in Great Britain © *Society for Experimental Biology 1987*   113

TEMPERATURE AND ANIMAL CELL PROTEIN SYNTHESIS

ROY H. BURDON

Department of Bioscience & Biotechnology, Todd Centre, University of Strathclyde,
Glasgow G4 0NR, Scotland, UK

Summary

A predominant feature of brief heat stress to animal cells is the vigorous but transient activation of a small number of specific genes, previously either silent, or active at low levels. New mRNAs are actively transcribed from these genes and are translated into proteins, known collectively as the heat shock proteins, or hsps.

The number of different types of hsp varies considerably in different organisms and cell types but in all cases proteins of approximately 84 and 70 kDa are amongst the most prominent. A dramatic feature that emerges from a study of their genes is that these proteins have been highly conserved during evolution. Gene sequence data reveal specific nucleotide sequences upstream of the transcription start sites that are essential for induction. These are known as 'heat shock elements' and are believed to be the region to which activated 'heat shock transcription factors' bind to facilitate hsp gene transcription. A recent model suggests that transcriptional regulation is based on competition between *abnormal* intracellular proteins and such a labile regulatory factor. Other experiments suggest such hsp inducers as heat, ethanol, arsenite, or oxygenation after anoxia, may cause protein damage through oxygen-derived free radical action.

Although heat shock can cause very considerable changes in transcriptional patterns, effects specifically on translational control are no less dramatic. Different organisms achieve a rapid change in different ways. In *Drosophila*, heat shock promotes the translation of hsp mRNAs. In yeast there is no mechanism for sequestering pre-existing mRNAs from translation. Instead these mRNAs simply disappear rapidly from the cell. Mammalian cells are different in yet another way. There is neither sequestration of pre-existing mRNAs nor their removal from the cell.

Superimposed upon the transient heat induced activation of hsp genes there are now clear indications of developmental regulation. The mechanism for the specific developmental control of hsp expression is not yet known. However, as oncogene products are known to induce hsp synthesis it may be that hsps are involved in eukaryotic growth control.

Hypothermia causes loss of protein synthetic activity in cultured animal cells. No specific proteins are however induced and recovery at 37°C is rapid.

Introduction

Animal life is mainly limited to a narrow range of temperatures, from a few degrees below the freezing point of water to approximately 50°C. Animals nevertheless differ in the range of temperatures that they can tolerate. Temperature tolerance may however change with time and a certain degree of adaptation is possible. The limits of temperature tolerance for a given animal are not fixed. Indeed it has been known for some time that exposure to a near lethal temperature often leads to a degree of adaptation so that a previously lethal temperature is tolerated. This particular response has attracted considerable attention over the last decade and has resulted in a rapid accumulation of data providing considerable insights into stress physiology in general. The response to heat shock is now known to occur in bacteria, in plants as well as in animals, and is a rapid, but transient reprogramming of cellular activities (a) to ensure survival during the stress period, (b) to protect essential cell components against heat damage and (c) to permit a rapid resumption of normal cellular activities during the recovery period.

A main feature of the heat shock response from bacteria to man (Schlesinger *et al.* 1982) is the vigorous but *transient* activation of a small number of specific genes previously either silent, or active at low levels. New mRNAs are actively transcribed from these genes and translated into proteins which are collectively referred to as the heat shock proteins, or hsps. These are listed in Table 1 along with their apparent intracellular locations.

The optimum temperature range of hsp induction varies considerably with the organism, but relates to the physiological range of supraoptimal temperatures within which active adaptation is observed. For example this is around 40–50°C for birds and mammals, 35–37°C for *Drosophila*, 33–35°C for yeast and 35–40°C for plants. However it has been found that the optimum can vary between different cell types of a single organism between individual hsps from even one cell type.

Heat shock proteins

Heat shock proteins (hsps) were first reported in *Drosophila* by Tissières and his colleagues (Tissières *et al.* 1974). The exact number of different types of hsp varies considerably in different organisms, and cell types, but in all cases proteins of approximately 84 and 70 kDa (hsp 84 and hsp 70) are amongst the most prominent. A dramatic feature is that these particular proteins have been highly conserved in evolution. For example the hsp 70 proteins of *Drosophila* and yeast have a 72% amino acid identity (Ingolia *et al.* 1982), and their hsp 84 proteins have a 63% identity (Farrelly & Finkelstein, 1984). The derived protein sequence of a human hsp 70 gene is found to be 73% homologous to *Drosophila* hsp 70 (Hunt & Morimoto, 1985).

Table 1. *Heat shock proteins of animal cells*

Class	Species	Size range (kDa)	Covalent modifications	Intracellular location	Comment
hsp 110	Mammals	110		Nucleolus	Normal nucleolar component
hsp 95	Vertebrates	92–105	Phosphorylation	Golgi	Normal cell component identical to membrane proteins induced by glucose deprivation
hsp 84	Vertebrates	83–90	Methylation phosphorylation ADP-ribosylation	Cytosol	Normal cell component associates with steroid receptors
	Drosophila	84			
hsp 70	all	68–74	Phosphorylation	Cytoplasm, cytoskeleton; migration to nucleus and nucleolus after heat shock.	Mainly constitutive cell proteins (mammalian hsp 72 however may be only heat inducible) *E. coli* equivalent is the Dna K protein.
Small hsps	Vertebrates	23–30	Glycosylation	Nuclear matrix after heat shock	
	Drosophila	22 23 26 28			

Most organisms also produce small hsps of 15 to 30 kDa. Whilst *Drosophila* cells induce four very closely related proteins of 22, 23, 26 and 28 kDa, yeast appears to produce only one at 26 kDa. These smaller hsps are not so well conserved in evolutionary terms, although nucleic acid sequence analysis has demonstrated certain homology amongst the small hsps of insects, vertebrates and nematodes (Bienz, 1985) as well as a part homology to α-crystallin (Ingolia & Craig, 1982a).

A problem when evaluating the hsp spectrum of an organism by two-dimensional electrophoresis has been the complexity due to the presence of 'isoforms' of the major hsps (Slater *et al.* 1981; Kioussis *et al.* 1981, Welch *et al.* 1983). This would appear to be the result of multigene families as well as extensive covalent modifications. An additional problem revealed recently in studies on hsp 70 class proteins is *in vitro* degradation (Mitchell *et al.* 1985). This occurs even during electrophoresis and does not appear to be mediated by a general protease, rather the hsp 70s have a slow proteolytic action upon themselves.

During investigation of hsp induction it became clear, in mammalian cells, that particular hsps such as hsp 28, hsp 72, hsp 23, hsp 84, hsp 90, hsp 100 and hsp 110 were all *transiently* synthesised at an elevated rate after heat stress. However with the possible exception of hsp 72, all were present at a low level in unstressed cells at 37°C (Welch *et al.* 1983). Thus in the mammalian 'hsp 70 class' a distinction can be drawn between the 'constitutive' hsp 73 and the 'inducible' hsp 72. Peptide mapping shows that the hsp 73 and hsp 72 proteins are similar but not identical (Welch & Feramisco, 1985).

A third type of protein, that also appears to be related to the 'hsp 70 class', was detected in mouse cells. Proteins of this type are synthesised constitutively but are *not* elevated upon heat shock (Lowe & Moran, 1984). Such proteins have been referred to as 'heat shock cognate proteins' and are also detectable in other mammalian cells as well as in *Drosophila* (Ingolia & Craig, 1982b). The possibility that the 'constitutive' group also contains these 'cognate' species remains to be determined. Present separation techniques do not yet permit a clear distinction.

Mechanisms underlying hsp gene transcription

We find that the induced synthesis of hsps in mammalian cells can in fact be brought about by a variety of agents other than heat (see Table 2). This makes it difficult to be clear about the mechanisms responsible for triggering the induction process. Nevertheless the mechanism of transcriptional induction would appear to be highly conserved during evolution. For instance when cloned *Drosophila* hsp genes are transfected into mammalian cells they are transcribed only when the recipient cells are heated (Pelham, 1982, 1985). *In vitro* deletion analysis of the sequences upstream from the *Drosophila* hsp 70 gene has identified a short consensus sequence of nucleotides necessary for the heat induced transcription of the gene (Pelham, 1985). This sequence C--GAA--TTC--G is some 20 nucleotides upstream of the 'TATA-box', a conserved sequence element found in all eukaryotic promoters. The consensus sequence is referred to as the heat shock

Table 2. *Hsp synthesis in mammalian cells*

Inducers	Non-inducers	
Heat	cAMP	Hydroxyurea
Ethanol (5 %)	Butyrate	Bleomycin
Arsenite	DMSO	NaCN
Cadmium	Phorbol esters	NaF
Amino acid	Colchicine	DNP
analogues	Cytochalasin B	Antimycin A
Glucose starvation	Cycloheximide	Oligomycin
followed by refeeding	Azacytidine	
Anoxia followed	Ouabain	
by oxygenation	Calcium ionophore	

element (HSE) and recent evidence indicates that it serves, in heated cells, as a binding site for a protein (Wu, 1984). This protein is specifically required for the transcription *in vitro* of cloned hsp genes (Parker & Topol, 1984) and is referred to as the heat shock transcription fctor (HSTF) and is required along with a second protein which normally stimulates transcription and binds to the 'TATA-box' region (see Fig. 1). The HSTF can be extracted from both heat shocked and untreated cells. However, it appears to be less active when extracted from the latter (Parker & Topol, 1984). On this basis it has been speculated that HSTF might be modified in some way and thereby activated in response to heat and other

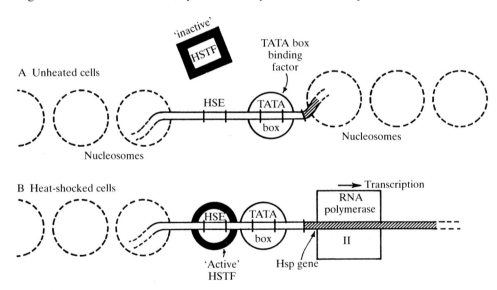

Fig. 1. A possible mechanism involving transcription factors whereby hsp genes might be activated. The regulatory DNA sequences, HSE (heat shock element) and TATA box, lie upstream from the hsp gene in a region free of nucleosomes. In unheated cells (A) nuclease sensitivity studies indicate that the TATA box binding factor is bound to the TATA box sequence whereas the HSE element is free of protein. After heat shock (B) there is some specific structural alteration to the heat shock transcription factor (HSTF) that allows it to bind to the HSE sequence. This allows the transcription of the hsp gene by RNA polymerase II.

stresses mentioned in Table 2. Certainly inducers such as heat, ethanol, cadmium, arsenite and amino acid analogues (when incorporated into protein) might be expected to denature, or alter cellular proteins. On the other hand sudden increases in metabolic activity, such as would follow refeeding of glucose after starvation, or reoxygenation following hypoxia, might conceivably cause protein damage through increased metabolism and subsequent oxygen related free radical activity. A recent view is that the intracellular accumulation of abnormal or damaged proteins *per se* may trigger hsp gene transcription. For example the presence of abnormal, truncated actin molecules in cells of certain flightless *Drosophila* mutants results in the synthesis of hsps (Karlik *et al.* 1984; Hiromi & Hotta, 1985) as does the injection of denatured proteins into frog oocytes (Ananthan *et al.* 1986).

Another important feature relating to heat induced transcriptional changes is that the transcriptional activation of hsp genes is a *transient* phenomenon (Di Domenico *et al.* 1982; Slater *et al.* 1981). In heat shocked *Drosophila* cells it is clear that transcriptional regulation of hsp genes is dependent on accumulated levels of hsps themselves (Di Domenico *et al.* 1982). A specific accumulation of functional hsp appears to repress hsp gene transcription. How this is achieved is not clear, but a number of studies indicate the migration of hsps to nuclei following heat shock (Velasquez & Lindquist, 1984; Welch & Feramisco, 1984). Proteins of the hsp 70 class have also been found in association with heterogeneous nuclear ribonucleo-proteins as well as with messenger ribonucleoproteins (Kloetzel & Bautz, 1983). Such associations may have an important role in the autoregulation phenomenon, at least eukaryotes.

Translational control following hyperthermia

Although heat shock can cause very considerable changes in transcriptional patterns, effects specifically on translational control are no less dramatic. On the other hand it is clear that different organisms achieve rapid shifts in protein synthesis in quite different ways.

Perhaps the most well studied organism in this connection is *Drosophila* where heat shock induces a mechanism of translational control which both promotes the translation of hsp mRNAs and specifically represses the translation of pre-existing mRNAs (McKenzie *et al.* 1975; Lindquist, 1980; Lindquist, 1981). The pre-existing mRNAs however remain stable and translatable in other *in vitro* systems, or *in vivo* following recovery from heat shock (Kruger & Benecke, 1981, Storti *et al.* 1980). From experiments with whole cells, and with *in vitro* translation systems, two points of regulation appear important. Firstly there appears to be a change in the translation mechanism that permits the exclusive recognition of hsp mRNAs, and secondly, a requirement for some specific structural feature present in hsp mRNAs that allow them to be recognised as such. Scott & Pardue (1981), from work on *in vitro* translation systems from heat shocked *Drosophila* cells, found that addition of ribosomes from normal cells restored normal protein synthesis. A key question was whether the dephosphorylation of ribosomal protein S6 observed after heat shock was the reason for the change in translational specificity. However experiments with whole cells suggested that there was no correlation between the two phenomena (Olsen *et al.* 1983). More recent work with mRNA-dependent *in vitro* translation systems from normal and heat shocked *Drosophila* cells that can be fractionated into ribosomes and supernatant fractions, as well as reconstituted from these fractions, suggests that the specificity for the protein synthetic patterns after heat shock is due to supernatant factors, rather than ribosomes *per se* (Sanders *et al.* 1986). With regard to special structural features of hsp mRNAs, it has been known for sometime that 5′-untranslated regions of *Drosophila* hsp mRNAs are quite long. For example in mRNAs for hsp 70 such a region is 244 nucleotides. Moreover this sequence is unusually rich in adenine residues (46 %) (Ingolia & Craig, 1981). In recent experiments (Klemenz *et al.* 1985, McGarry & Lindquist, 1985) various fusion genes were constructed using elements of the hsp 70 gene of *Drosophila* and the gene for alcohol dehydrogenase. From transfection experiments using various gene fusions the specific feature responsible for the preferential translation of hsp mRNAs is not yet clear, but nevertheless resides somewhere in the 5′-untranslated region.

In mammalian cells the situation is quite different. There is no extensive sequestration of pre-existing mRNAs from translation nor do pre-existing mRNAs disappear (Slater *et al.* 1981; Kioussis *et al.* 1981). When Hela cells for example are subject to hyperthermia an immediate effect is a marked inhibition of total protein synthesis and decay of polysomes. Nevertheless when the heat treatment is *continuous* at temperatures *less* than 42°C, polysomes reform and protein synthesis recovers to reach levels higher than in untreated cells (Fig. 2). After acute, but brief (10–15 min) hyperthermia (43–46°C), the polysome profile is also restored and protein synthesis recovers (to levels between 100–200 %) if the temperature is returned to 37°C (Burdon *et al.* 1982). These recovery processes are *partly* inhibited if low levels of actinomycin D ($0.5\,\mu\mathrm{g\,ml}^{-1}$) are added (see Burdon, R. H. this volume).

Heat-induced proteins synthesis inhibition

Possible reasons for the initial inhibition and loss of polysomes include (a) the inactivation of initiation factor eIF-4F function (Panniers *et al.* 1985), (b) the dephosphorylation of ribosomal protein S6 and the phosphorylation of protein L14 which are observed during the initial phase of heat shock (Kennedy *et al.* 1984), (c) the activation of a protein kinase analogous to the haem-controlled repressor of reticulocytes and the subsequent phosphorylation of initiation factor, eIF-2 (de Benedetti & Baglioni, 1986), or other initiation factors (Duncan & Hershey, 1984). Whilst eIF-2 phosphorylation appears to limit mRNA translation in heat shocked cells (de Benedetti & Baglioni, 1986) we find the addition of purified eIF-2 to lysates from heat shocked cells has a marked stimulatory effect on *in vitro* protein synthesis (Burdon *et al.* 1986). However, addition of fresh ribosomal subunits to such lysates also had restorative effects. Thus it may well be that the phosphorylation of eIF-2 as well as alteration to other initiation factors and ribosome sub-units are all contributory to the initial inhibition effects.

The problem of recovery from heat shock also requires consideration. During the recovery period in Hela cells, not only is the synthesis of hsps dramatically enhanced (Slater *et al.* 1981) but the phosphorylation of S6 ribosomal protein is partly restored whilst that of L14 ribosomal protein remains unchanged (Kennedy *et al.* 1984). In addition phosphorylation of eIF-2 is reduced (de Benedetti & Baglioni, 1986). However these observations do not account for the partial requirement for RNA suggested from the experiments using low levels of actinomycin D. Thus whilst a readjustment of covalent modifications in the protein synthetic apparatus may be of considerable importance, there may be other damage and a need to replace some affected ribosomes and/or initiation factors.

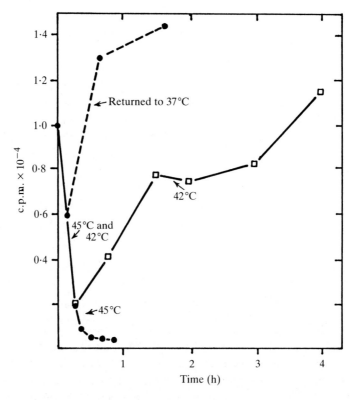

Fig. 2. Effect of various hyperthermic protocols on labelled amino acid incorporation into Hela cell protein. The Hela cell cultures were grown as monolayers (0.5×10^6 cells per 2·5 cm diam. dish) for 24 h in Eagles MEM medium supplemented by 10% calf serum. Triplicate cultures were held at 45°C and labelled with ^{35}S-methionine for 10 min periods at various times (●——●). Some cultures were removed after 10 min at 45°C, returned to 37°C for various times and then labelled for 10 min at 37°C (●----●). Other cultures were held at 42°C for various times and labelled for 10 min periods at 42°C (□).

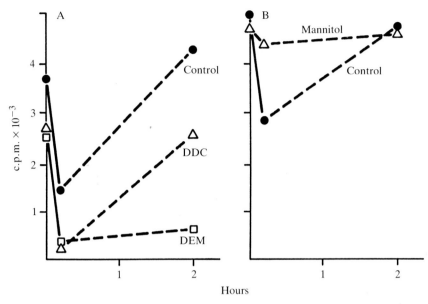

Fig. 3. Effects of diethyldithiocarbamate, diethylmaleate and mannitol on the heat induced inhibition of ^{35}S-methionine incorporation into protein. (A) Hela cells cultures (see Fig. 2) were pretreated with 1 mM diethyldithiocarbamate (DDC) for 2 h at 37° before subjecting them to 45°C for 10 min (△———△) and returning them at 37°C for 2 h (△----△). Untreated control cells at 45° for 10 min (●———●) and then at 37°C for 2 h (●----●). Some cells were exposed to 0·5 mM diethylmaleate (DEM) in a similar fashion and then subject to hyperthermia at 45° for 10 min (□———□) and then 37° for 2 h (□----□). (B) Similar to (A) but cells pretreated with 5 mM mannitol (△) and untreated (●). The incorporation of ^{35}S-methionine into protein after the above treatments is measured over 30 min at 37°C (see Burdon *et al.* 1982).

In order to explore the nature of other possible damage, monolayer cultures of Hela cells were pretreated with diethyl dithiocarbamate (DDC) (to deplete cells of superoxide dismutase) (Evans *et al.* 1983) or with diethymaleate (DEM) (to deplete cells of non-protein thiol) (Freeman *et al.* 1985). After such treatments the initial inhibitory effects of heat were more pronounced (Fig. 3(A)). However, if the pretreatment was with 5 mM-mannitol (a scavenger of hydroxyl radicals) then the inhibition was far less marked (Fig. 3(B)). Another means of reducing the initial inhibitory effect of heat on cellular protein synthesis, was the inclusion in the medium of the calcium-chelator, EGTA (0·5 mM), for 1 h beforehand (Fig. 4(A)). Inclusion of the iron-chelator desferrioxamine (Keberle, 1964) on the other hand does not have any significant effect on the initial inhibition (Fig. 4(B)).

To test whether the initial inhibition of protein synthesis was related to intracellular ATP levels, the effect of cyanide was tested. 5 mM KCN pretreatment for 2 h resulted in no enhancement of the heat-induced inhibition; however recovery was poorer than normal (Fig. 5(A)). Sodium azide treatment on the other hand led to a reduction of the initial inhibitory effect of heat (Fig. 5(B)). The effects of pH were also examined. A reduction to pH 6·8 caused a slight increase in the inhibitory effect of heat (Fig. 6(A)) but in conditions of acute acidosis at pH 6·0 there was a considerable reduction in the inhibitory effect of heat (Fig. 6(B)). Indeed at this pH there appears to be a noteably high level of constitutive hsp 70 synthesis which is not increased upon hyperthermic treatment.

The initial heat-induced inhibition can also be modulated by the serum component of the growth medium. Normally heat treatment is carried out on cells that have been in full medium for at least 24 h. In one experiment heating such cells at 45 °C for 10 min led to an inhibition of 12 %. Replacement of the medium with fresh medium immediately prior to the heat treatment however resulted in the complete elimination of the inhibitory effect of heat. On the other hand replacement of the medium with fresh medium, but without the serum component, resulted in a much greater inhibitory effect of heat (36 % inhibition).

From these experiments it is clear that the causes of the initial inhibition may be complex. Whilst covalent modification of initiation factors and/or ribosomal proteins may be important other possible explanations include mechanisms that involve Ca^{2+}, and possibly oxygen derived free radical activity, but yet are affected by pH and serum factors.

In view of the possible free radical involvement, the levels of lipid peroxidation in Hela cells exposed to hsp inducers was examined (Burdon et al. 1986). In Table 3

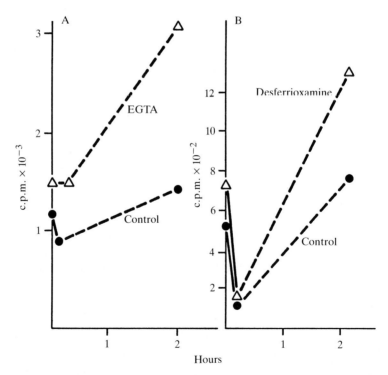

Fig. 4. Effects of EGTA and desferrioxamine on the heat induced inhibition of ^{35}S-methionine incorporation into protein. The procedure used was similar to that employed in Fig. 3. In (A) cells pretreated with 0·5 mM EGTA for 15 min ($\triangle$) and untreated control cells ($\bullet$). In (B) cells pretreated with 0·4 mM desferrioxamine for 15 min ($\triangle$) and untreated control cells ($\bullet$).

the lipid peroxidation is presented as the levels of 2-thiobabituric acid (TBA) reactive material generated (Gutteridge, 1986). Both ethanol and arsenite treatment of the monolayer cultures resulted in notable increases in lipid peroxidation. A significant response to heat however was difficult to detect with this technique, even at the cell densities that were employed. Nevertheless a significant increase in intracellular lipid peroxidation was observable if the serum component of the medium was removed prior to the heat treatment. Under such conditions the response could be diminished if mannitol or EGTA were included in the medium. Lowering of pH to 6·8 also causes increased lipid peroxidation (Table 3).

Because radicals generated in lipid peroxidation reactions can cause damage to cell protein (Wolff *et al.* 1986), they could be a cause of the initial heat-induced protein synthesis-inhibition by virtue of effects on ribosomes and/or initiation factors. A more general question is whether such increases in lipid peroxidation might elicit general cell protein damage. Table 4 shows that heat, as well as other hsp inducers such as arsenite and ethanol, certainly lead to detectable loss of cell protein *in vivo*. Such losses are however not inhibited by the addition of fluoride to the medium, suggesting that they are mediated by degradative systems other than the ATP-dependent ubiquitin proleolytic pathway (Hershko & Ciechanover, 1982). This does however not rule out the occurrence of free radical mediated damage to protein. It has been suggested that the ubiquitin system might simply become overloaded where damage is excessive (Munroe & Pelham, 1985). In this connection it should be noted that both mannitol and EGTA addition stimulate

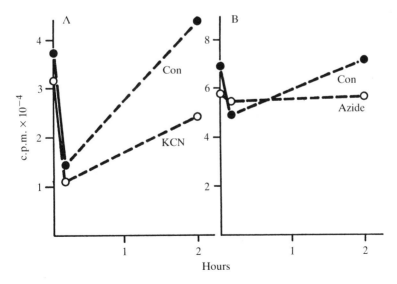

Fig. 5. Effects of potassium cyanide and sodium azide on the heat induced inhibition of ^{35}S-methionine incorporation into protein. The procedure used was similar to that described for Fig. 3. In (A) cells pretreated with 5 mM KCN for 2 h (○) and untreated control cells (●). In (B) cells are pretreated with sodium azide (5 mM) for 30 min (○) and untreated control cells (●).

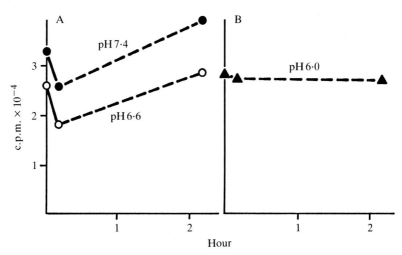

Fig. 6. Effect of pH on heat induced inhibition of ^{35}S-methionine incorporation into protein. The procedure used was similar to that described for Fig. 3. In (A) cells were incubated in media containing HEPES buffer (in place of the normal bicarbonate buffer of Eagles MEM medium) at the pH indicated for 3 h before the heating at 45°C and recovery at 37°C. Protein synthesis was subsequently measured with ^{35}S-methionine in Eagles medium at pH 7·4 as previously described by Burdon *et al.* 1982).

the breakdown of cell protein, possibly by reducing the 'load' on the cellular degradative systems that are either dependent or independent of ubiquitin (Hershko & Ciechanover, 1982). Clearly the situation is complex and multiple degradative systems may operate following hyperthermia.

We have also recently reported extensive losses of stable RNA in cells treated with heat (or arsenite or ethanol) (Burdon *et al.* 1986). As was the case with the protein losses, the RNA losses were not blocked by mannitol, or EGTA, and are thus unlikely to be the direct outcome of lipid peroxidation; rather lysosomal activity maybe the explanation. Indeed increases in lysosomal ribonuclease activity brought about by hyperthermia was reported by Lwoff (1969).

The recovery of protein synthesis after hyperthermia

Whilst the extensive losses of RNA and protein may be due to the activation of certain cellular degradative systems rather than molecular damage *per se*, the possibility that free radical activity contributes to the initial heat induced inhibition of protein synthesis should not be overlooked. This is important if the *recovery* process is to be understood. So far we find that several agents prevent optimal recovery after hyperthermia. These include diethylmaleate (which depletes cells of non-protein thiols), cyanide (which indicates an energy requirement), as well as low levels of actinomycin D (0·05 μg ml^{-1}, which preferentially inhibits ribosomal RNA synthesis). An important factor however that influences recovery at 37°C is the duration of the initial acute hyperthermia. As can be seen from Fig. 7, the efficiency of recovery declines markedly if the hyperthermia at 45°C is prolonged much more than 10 min. This tends to support the view that in addition to covalent

Table 3. *Effect of hsp inducers pH and low temperature on lipid peroxidation in Hela cells*

Expt	Treatment	nmol TBA reactive product per 10^{10} cells
1	Fresh medium 37°C, 3 h	$327·6 \pm 40·9$
	Fresh medium 37°C, 3 h then 45°C, 1 h	$450·5 \pm 204·8$
	Fresh medium minus serum 37°C, 3 h	$368·7 \pm 81·9$
	Fresh medium minus serum 37°C, 2 h, then 45°C, 1 h	$2027·5 \pm 245·7$
2	Fresh medium minus serum at 37°C, 3 h	$266·2 \pm 20·5$
	Fresh medium minus serum 37°C, 2 h then 45°C, 1 h	$409·6 \pm 20·5$
	Fresh medium minus serum 37°C, 2 h then 45°C, 1h with mannitol at 50 mM	$40·9 \pm 61·4$
	Fresh medium minus serum 37°C, 2 h then 45°C, 1 h with EGTA at 0·5 mM	$286·7 \pm 40·9$
3	37°C 1 h	$470·4 \pm 40·9$
	Ethanol (5 %) 37°C, 1 h	$1147·2 \pm 40·9$
	Arsenite (50 μM) 37°C, 1 h	$1290·4 \pm 81·9$
4	37°C 4 h	$339·9 \pm 39·9$
	4°C 4 h	$409·6 \pm 39·9$
5	37°C 16 h	$544·7 \pm 60·8$
	37°C 16 h, pH 6·8	$716·8 \pm 81·6$

Monolayer cultures of Hela cells were established with 10^7 cells in 20 ml medium, in flat plastic culture vessels of 175 cm^2. The cells were allowed to grow to $2·5 \times 10^7$ (3 days) before exposure to the various experimental conditions described.

After treatment, each monolayer culture was washed three times with 0·9 % NaCl (w/v) and then the cells scraped off into 2 ml 0·9 % NaCl for malonaldehyde measurement. Trichloroacetic acid was added to a final concentration of 10 % (w/v) and the suspension centrifuged at 200 g for 5 min. 1 ml of the supernatant was removed and added to 1 ml 0·75 % 2-thiobarbituric acid. Samples were heated at 90°C for 20 min and then spun at 10 000 g for 10 min. The levels of thiobarbituric acid reactive products in each supernatant fraction were determined by measuring the difference between absorbances at 532 and 580 nm, and using malonaldehyde as standard (see Gutteridge, 1986; Walls *et al.* 1976).

modifications molecular damage, and loss, is significant and it may become progressively severe with time so that the Hela cells at 60 min are presented with impossible recovery problems.

Recovery is likely to be a complex process. We find in Hela cells that there is a partial rephosphorylation of ribosomal protein S6 although the heat induced phosphorylation of ribosomal protein L14 remains (Kennedy *et al.* 1984). In addition the phosphorylation of initiation factor eIF-2 is reduced (de Benedetti & Baglioni, 1986). These observations however do not account for the major requirement of energy producing metabolism and RNA synthesis sensitive to low levels of actinomycin D. When such RNA synthesis is examined it appears to be quite heat sensitive compared with the RNA synthesis that is resistant to low levels of the drug (Fig. 8). Nevertheless it recovers after hyperthermia to reach levels similar to that in unheated cells. Recovery of low actinomycin D sensitive RNA synthesis also takes place in cells exposed to continuous hyperthermia at 42°C (see

Table 4. *The effect of hsp inducers on loss of cellular protein*

Additions	Duration of experiment (min)	Temperature (°C)	^{35}S-remaining in total cell protein (cpm)
None	0	37	345
None	60	37	339
None	60	45	308
Mannitol (50 mM)	60	45	194
EGTA (0·5 mM)	60	45	225
Arsenite (50 μM)	60	37	171
Ethanol (5 %)	60	37	266

Hela cell cultures of 0.5×10^6 per dish were grown overnight then labelled for 24 h with 2 μC L-^{35}S-methionine (634·8 Ci mmol^{-1}). After this period, the monolayer cultures were washed two times with fresh non-radioactive medium and then returned to 37°C for a further 24 h. At this point triplicate cultures were taken for determination of ^{35}S-radioactivity in protein as previously described (Burdon *et al.* 1982). These cultures served as zero-time controls. Other cultures were then incubated at 37°C, or 45°C with, or without, various additions, to determine the loss of ^{35}S-radioactivity from intracellular protein.

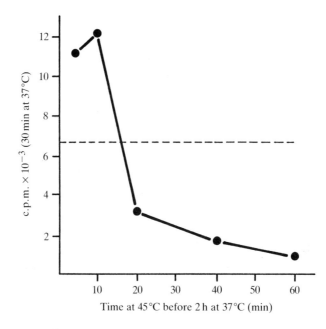

Fig. 7. The effect of length of hyperthermia at 45°C on the ability to recover amino acid incorporation at 37°C. Cells were treated at 45°C for varying lengths of time and then returned to 37°C. After 2 h at 37°C their ability to incorporate ^{35}S-methionine into protein over a 30 min period was determined. The dotted line represents the level of ^{35}S-methionine incorporated into protein by untreated cultures at 37°C.

Burdon, R. H. this volume). Such observations suggest a role for ribosomal RNA synthesis in the recovery process since low levels of actinomycin D are known to preferentially inhibit ribosomal RNA gene transcription (Perry, 1967). Certainly when the level of stable RNA in Hela cell cytoplasm after hyperthermia is followed (Table 5) there is a very noticeable influx from the nucleus during the recovery period. However to decide whether this represents increased levels of

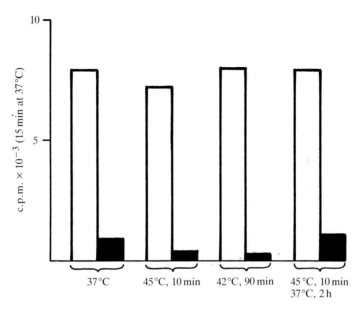

Fig. 8. The effect of hyperthermia on the incorporation of ^{3}H-uridine into RNA by Hela cells. Duplicate monolayer cultures (see Fig. 2) were treated to various hyperthermic protocols in the presence, or absence of 0·05 μg ml^{-1} actinomycin D as indicated. They were then labelled with ^{3}H-uridine (2 μC plate^{-1}, 41 Ci mmol) for 15 min at 37 °C. To determine the level of ^{3}H-radioactivity in RNA, the cultures were washed three times with 5 % trichloroacetic acid (ice-cold) and then the cell monolayer dissolved in 880 μl 0·4 M NaOH, neutralised with 20 μl glacial acetic acid and the radioactivity determined after addition of scintillation fluid. The solid histograms represent incorporation into RNA that is sensitive to the low concentration of actinomycin D, whereas the open histograms are a measure of the insensitive incorporation.

Table 5. *Effects of hyperthermia on cytoplasmic RNA of 'prelabelled' Hela cells*

Expt	Treatment	^{3}H-radioactivity (cpm) in cytoplasmic RNA
1	None	16 770
	45°, 10 min	15 773
2	None	44 393
	45°, 10 min, then 37° for 2 h	115 483

Monolayer cultures of Hela cells in 80 oz rotating bottles were grown for 24 h at 37 °C from 5 × 10^7 cells in Eagles MEM plus 10 % calf serum. They were labelled as described by Burdon *et al.* (1977) with 0·5 mCi amounts of ^{3}H-uridine for a further 24 h at 37 °C. The medium was then replaced with non-radioactive medium and after 1 h the indicated heating protocols were carried out. Cytoplasmic RNA was isolated and its level of ^{3}H-radioactivity determined as described by Burdon *et al.* (1977).

A

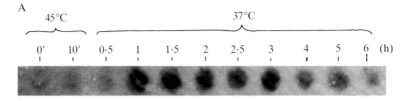

B

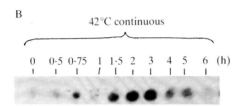

Fig. 9. Analysis by dot-blot hybridisation of 28S ribosomal RNA sequences in Hela cell cytoplasmic RNA. These Hela cell cultures were originally seeded at 5×10^6 cells in flat Roux tissue culture bottles and grown for 2 days in Eagles MEM supplemented with 10 % calf serum. After various treatments cytoplasm was prepared from samples of these cells by the method of White & Bancroft (1982) and after incubation, with 7·4 % formaldehyde, equal aliquots equivalent to approx 5×10^5 cells were dotted onto nitrocellulose as described by these authors. The 'dotted' samples were then hybridised with ^{32}P-nick-translated plasmid pHS 2 (Munro *et al.* 1986) and then autoradiographed as described by White & Bancroft (1982). (A) Cells treated at 45 °C for 10 min followed by recovery at 37 °C (B) cells treated at 42 °C continuously.

ribosomal RNA in the cytoplasm, 'dot blot' analyses of cytoplasmic RNA were carried out using 'nick-translated' plasmid pHS2 which contains an insert corresponding to nucleotides 3627–4105 of the human 28S ribosomal RNA gene (Munro *et al.* 1986). As can be seen from Fig. 9(A), there is a marked rise in the cytoplasmic level of 28S ribosomal RNA sequences during recovery at 37 °C from treatment of Hela cells at 45 °C for 10 min. An increase in these ribosomal RNA sequences also occurs in cells exposed to continuous hyperthermia at 42 °C, the maximum levels being attained after 1·5 h to 3 h (Fig. 9(B)). These data may indicate that in addition to changes in the covalent modification of components of the protein synthetic system there may also be a replacement of damaged components such as ribosomes for the survival of protein synthesis and cell viability. The non-protein thiols (e.g. glutathione) in this process have not yet been clarified, but when cell-free protein synthesis systems are prepared from Hela cells treated at 45 °C for 10 min, it is not possible to restore their activity to normal levels by direct addition of glutathione (Burdon *et al.* 1986). The most likely explanation is that the intracellular levels of non-protein thiols must be maintained as we find there is an *irreversible* loss of protein synthetic activity even when unheated Hela cells are incubated with diethylmaleate (75 % loss over 4 h at 37 °C).

The effect of hypothermia on protein synthesis

When Hela cells are placed at 4°C there is a gradual decline in their ability to synthesise protein when this is determined on return to 37°C (Fig. 10). Such acute cold stress however does not appear to induce the synthesis of any specific proteins. Moreover if the cells are first subject to hyperthermic protocols that will induce hsp synthesis, this appears to have no protective effect against acute cold. Rather it seems to increase the deleterious effects of cold exposure (Fig. 10). Return of the cells to 37°C after a 2h exposure at 4°C nevertheless permits recovery to normal levels of protein synthesis. However the protein synthesis of these cells shows no increased cold resistance when reexposed to to 4°C.

In an attempt to elucidate the possible molecular basis for the decline in protein synthesis on cold exposure we have examined levels of lipid peroxidation using the thiobarbituric acid (TBA) method (Gutteridge, 1986). After 4h there was really not a significant increase in TBA reactive material. Moreover the decline in protein synthetic rate could not be arrested by inclusion of mannitol (50 mM) or desferrioxamine (400 mM), or EGTA (0·5 mM), or cystamine (1 mM) or ascorbic acid (10 mM) in the medium during cold exposure. Such data suggest that lipid peroxidation is unlikely to be a direct cause of loss of protein synthetic function after hypothermic exposure. Another possibility is loss of cytoskeletal integrity. At 4°C cultured cells tend to become more spherical. This change is paralleled by

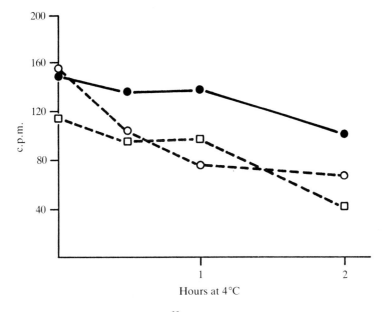

Fig. 10. Effect of hypothermia on ^{35}S-methionine incorporation into Hela cells protein. Triplicate Hela cells monolayer cultures were prepared as in Fig. 2. Some of these (●) were then exposed to 4°C for various times before assay for ability to incorporate ^{35}S-methionine at 37° as described by Burdon *et al.* (1982). In some cases cells were either treated at 42°C for 90 min (□) or 45°C for 10 min and 37°C for 2h (○), before the treatment at 4°C was initiated.

the sequential dismantling of the internal structures of the cell. Initially the microtubules disassemble followed by the microfilaments (Porter & Tucker, 1981). Recently cytochalasin B has been shown to cause the release of mRNA from the cytoskeleton framework and the inhibition of protein synthesis (Ornelles *et al.* 1986).

It must be conceded that such studies are limited and may only be relevant to mammalian cells in culture. Nevertheless a decline in muscle tissue protein synthetic rate is also observed when certain fish species are placed at temperatures lower than normal. However in eurythermal species long term acclimation at these lower temperatures led to a small increase in muscle protein synthetic rate (Loughna & Goldspink, 1985). On the other hand these authors observed no such changes was observable upon acclimation of stenothermal species (Loughna & Goldspink, 1985).

Hyperthermia, hypothermia and hsps

Exposure of cultured human cells to extremes of hot and cold thus elicit quite different effects. Most current knowledge concerns hyperthermia, and the function of the hsps remain obscure. Considerable speculation nevertheless concerns their role in the generation of thermotolerance and other stress resistance. This is discussed elsewhere (See Burdon, R. H., this volume), but it is their impressive evolutionary conservation that suggests even more profound biological roles. Recent data show them to be *constitutively* synthesised at specific stages in the normal development of insect, frog and mouse embryos (see Bienz, 1985). The mechanism for this specific developmental control of hsp expression are not known. There may of course be specific periods of stress during embryogenesis, but there may be more subtle controls of hsp gene expression. For instance it is known that the expression of hsp 70 can be induced by the oncogene products the E1A gene of the DNA tumour virus adenovirus 5 (Nevins, 1982), as well as by the rearranged c-*myc* mouse oncogene (Kingston *et al.* 1984). Such oncogenes are implicated in the immortalisation of primary cells in culture and it may be that hsp 70 has a critical role to play in animal cell growth control. Thus the continued study of temperature and stress related adaptive mechanisms may yield considerable insights relevant to fundamental biological phenomena.

The work is partly supported by grants from the Cancer Research Fund and from the Scottish Home & Health Department.

References

ANANTHAN, J., GOLDBERG, A. L. & VOELLMY, R. (1986). Abnormal proteins serve as eukaryotic stress signals and trigger the activation of heat shock genes. *Science* **232**, 522–524.
BIENZ, M. (1985). Transient and developmental activation of heat-shock genes. *Trends Biochem. Sci.* **10**, 157–161.
BURDON, R. H., GILL, V. & RICE-EVANS, C. A. (1986). Oxidative stress and heat shock protein induction in human cells. *Free Radical Res. Commun.* **3**, 129–139.

BURDON, R. H., SHENKIN, A., DOUGLAS, J. T. & SMILLIE, E. J. (1977). Poly(A)-binding RNAs from nuclei-polysomes of BHK-21 cells. *Biochim. Biophys. Acta* **474**, 254–267.

BURDON, R. H., SLATER, A., MCMAHON, M. & CATO, A. C. B. (1982). Hyperthermia and the heat shock proteins of HeLa cells. *Br. J. Cancer* **45**, 953–963.

DE BENEDETTI, A. & BAGLIONI, C. (1986). Activation of hemin-regulated initiation factor-2 kinase in heat shocked Hela cells. *J. biol. Chem.* **261**, 338–342.

DI DOMENICO, B. J., BUGAISKY, G. E. & LINDQUIST, S. (1982). The heat shock response is self-regulated at both the transcriptional and post-transcriptional levels. *Cell* **31**, 593–603.

DUNCAN, R. & HERSHEY, J. W. B. (1984). Heat shock induced translational alterations in Hela cells. *J. biol Chem.* **259**, 11 882–11 889.

EVANS, R. G., NIELSEN, J., ENGEL, C. & WHEATLEY, C. (1983). Enhancement of heat sensitivity and modification of repair of potentially lethal heat damage in plateau-phase cultures of mammalian cells by diethyldithiocarbamate. *Radiat. Res.* **93**, 319–325.

FARELLY, F. W. & FINKELSTEIN, D. B. (1984). Complete sequence of the heat shock inducible HSP 70 gene of *Saccharomyces cerevisiae. J. biol. Chem.* **259**, 5741–5751.

FREEMAN, M. L., MALCOLM, A. W. & MEREDITH, M. L. (1985). Role of gluthathione in cell survival after hyperthermic treatment of Chinese hamster ovary cells. *Cancer Res.* **45**, 6308–6313.

GUTTERIDGE, J. M. (1986). Aspects to consider when detecting and measuring lipid peroxidation. *Free Radical Res. Commun.* **1**, 173–184.

HERSHKO, A. & CIECHANOVER, A. (1982). Mechanisms of intracellular protein breakdown. *A. Rev. Biochem.* **51**, 335–364.

HIROMI, Y. & HOTTA, Y. (1985). Actin gene mutations in *Drosophila*, heat shock activation in the indirect flight muscles. *EMBO J.* **4**, 1681–1687.

HUNT, C. & MORIMOTO, R. I. (1985). Conserved features of eukaryotic hsp 70 genes revealed by comparison with the nucleotide sequence of human hsp 70. *Proc. natn Acad. Sci. U.S.A.* **82**, 6455.

INGOLIA, T. D. & CRAIG, E. A. (1981). Primary sequence of the 5'-flanking region of the *Drosophila* heat shock genes in chromosome subdivision 67B. *Nucleic Acids Res.* **9**, 1627–1642.

INGOLIA, T. D. & CRAIG, E. A. (1982a). Four small *Drosophila* heat shock proteins are related to each other and the mammalian crystallin. *Proc. natn Acad. Sci U.S.A.* **79**, 2360–2364.

INGOLIA, T. D. & CRAIG, E. A. (1982b). *Drosophila* gene related to the major heat shock-induced gene is transcribed at normal temperatures but not induced by heat shock. *Proc. natn. Acad. Sci. U.S.A.* **79**, 525–529.

INGOLIA, T. D., SLATER, M. R. & CRAIG, E. A. (1982). *Saccharomyces cereviaiae* contains a complex multigene family related to the major heat shock inducible gene of *Drosophila. Mol. Cell Biol.* **2**, 1388–1398.

KARLIK, C. C., COUTU, M. D. & FYRBERG, E. A. (1984). A nonsense mutation within the Act 88F gene disrupts myofibril formation in *Drosophila* indirect flight muscles. *Cell* **38**, 711–719.

KEBERLE,H. (1964). The biochemistry of desferrioxamine and its relation to iron metabolism. *Ann. N.Y. Acad. Sci.* **119**, 758–768.

KENNEDY, I. M., BURDON, R. H. & LEADER, D. P. (1984). Heat shock causes diverse changes in the phosphorylation of ribosomal protein of mammalian cells. *FEBS Letts* **169**, 267–273.

KINGSTON, R. E., BALDWIN, A. S. & SHARP, P. A. (1984). Regulation of heat shock protein 70 gene expression by c-myc. *Nature, Lond.* **312**, 280–282.

KIOUSSIS, J., CATO, A. C. B., SLATER, A. & BURDON, R. H. (1981). Polypeptides encoded by polyadenylated and non polyadenylated messenger RNAs from normal and heat shocked Hela cells. *Nucl. Acids Res.* **9**, 5203–5214.

KLEMENZ, R., HULTMARK, D. & GEHRING, W. J. (1985). Selective translation of heat shock mRNA in *Drosophila melanogaster* depends on sequences information in the leader. *EMBO J.* **4**, 2053–2060.

KLOETZEL, P. & BAUTZ, X. (1983). Heat-shock proteins are associated with hn RNP in *Drosophila melanogaster* tissue culture cells. *EMBO J.* **2**, 705–710.

KRUGER, C. & BENECKE, B.-J. (1981). *In vitro* translation of *Drosophila* heat shock and non-heat shock mRNAs in heterologous and homologous cell-free systems. *Cell* **23**, 595–603.

LINDQUIST, S. (1980). Translational efficiency of heat-induced messages in *Drosophila melanogaster* cells. *J. molec. Biol.* **137**, 151–158.

LINDQUIST, S. (1981). Regulation of protein synthesis during heat shock. *Nature, Lond.* **293**, 311–314.

LOUGHNA, P. T. & GOLDSPINK, G. (1985). Muscle protein synthesis rates during temperature acclimation in a eurythermal (*Cyprinus carpio*) and a stenothermal (*Salmo gairdneri*) species of teleost. *J. exp. Biol.* **118**, 267–276.

LOWE, D. G. & MORAN, L. A. (1984). Proteins related to the mouse L-cell major heat shock protein are synthesised in the absence of heat shock gene expression. *Proc. natn. Acad. Sci. U.S.A.* **81**, 2317–2321.

LWOFF, A. (1969). Death and transfiguration of a problem. *Bact. Rev.* **33**, 390–403.

McGARRY, T. J. & LINDQUIST, S. (1985). The preferential translation of *Drosophila* hsp 70 mRNA requires sequences in the untranslated leader. *Cell* **42**, 903–911.

McKENZIE, S. L., HENIKOFF, S. & MESELSON, M. (1975). Localization of RNA from heat induced polysomes at puff sites ion *Drosophila melanogaster*. *Proc. natn. Acad. Sci. U.S.A.* **72**, 1117–1121.

MITCHELL, H. K., PETERSEN, N. S. & BUZIN, C. H. (1985). Self-degradation of heat shock proteins. *Proc. natn Acad. Sci. U.S.A.* **82**, 4969–4973.

MUNROE, S. & PELHAN, H. R. B. (1985). What turns on heat shock genes. *Nature, Lond.* **317**, 477–478.

MUNRO, J., BURDON, R. H. & LEADER, D. P. (1986). Characterisation of a human orphon 28S ribosomal DNA. *Gene* **48**, 65–70.

NEVINS, J. R. (1982). Induction of the synthesis of a 70.000 dalton mammalian heat shock protein by adenovirus E1A gene product. *Cell* **29**, 913–919.

OLSEN, A. S., TRIEMER, D. F. & SANDERS, M. M. (1983). Dephosphorylation of S6 and expression of the heat shock response in *Drosophila melanogaster*. *Molec. Cell Biol.* **3**, 2017–2027.

ORNELLES, D. A., FEY, E. G. & PENMAN, S. (1986). Cytochalasin releases mRNA from the cytoskeleton framework and inhibits protein synthesis. *Molec. Cell Biol.* **6**, 1650–1562.

PANNIERS, R., STEWART, E. B., MERRICK, W. C. & HENSHAW, E. C. (1985). Mechanism of inhibition of polypeptide chain initiation in heat shocked Ehrlich cells involves reduction of eukaryotic initiation factor 4B activity. *J. biol. Chem.* **260**, 9648–9653.

PARKER, C. S. & TOPOL, J (1984). A *Drosophila* RNA polymerase 11 transcription factor binds to the regulatory site of an hsp 70 gene, *Cell* **37**, 273–283.

PELHAM, H. R. B. (1982). A regulatory upstream promoter element in the *Drosophila* hsp 70 heat shock gene. *Cell* **30**, 517–528.

PELHAM, H. R. B. (1985). Activation of heat shock genes in eukaryotes. *Trends Genetics* **1**, 31–35.

PERRY, R. P. (1967). The nucleolus and the synthesis of ribosomes. *Prog. Nucl. Acids. Res. molec. Biol.* **6**, 219–248.

PORTER, K. E. & TUCKER, J. B. (1981). The ground substance of the living cell. *Scient. Am.* **244**, (3) 41–51.

SANDERS, M. M., TRIEMER, D. F. & OLSON, A. S. (1986). Regulation of protein synthesis in heat shocked *Drosophila* cells. Soluble factors control translation *in vitro*. *J. biol. Chem.* **261**, 2189–2196.

SCHLESINGER, M. J., ASHBURNER, M. & TISSIÈRES, A. (1982). *Heat shock: from Bacteria to Man*. Cold Spring Harbor Lab. New York: Cold Spring Harbor.

SCOTT, M. P. & PARDUE, M. L. (1981). Translational control in lysates of *Drosophila melanogaster* cells. *Proc. natn. Acad. Sci. U.S.A.* **78**, 3353–3357.

SLATER, A., CATO, A. C. B., SILLAR, G. M., KIOUSSIS, J. & BURDON, R. H. (1981). The pattern of protein synthesis induced by heat shock of Hela cells. *Eur. J. Biochem.* **117**, 341–346.

STORTI, R. V., SCOTT, M. P., RICH, A. & PARDUE, M. L. (1980). Translational control of protein synthesis in response to heat shock in *D. melanogaster* cells. *Cell* **22**, 825–834.

TISSIÈRES, A., MITCHELL, H. K. & TRACY, V. M. (1974). Protein synthesis in salivary glands of *D. melanogaster*. Relation to chromosome puffs. *J. molec. Biol.* **84**, 389–398.

WALLS, R., KUMAR, K. S. & HOCHSTEIN, P. (1976). Aging of human erythrocytes. *Archs Biochem. Biophys.* **174**, 463–468.

WELCH, W. J. & FERAMISCO, J. R. (1984). Nuclear and nucleolar localization of the 72,000-dalton heat shock protein in heat shocked mammalian cells. *J. biol. Chem.* **259**, 4501–4513.

WELCH, W. J. & FERAMISCO, J. R. (1985). Rapid purification of mammalian 70,000-dalton stress protein: affinity of the proteins for nucleotides. *Molec. Cell Biol.* **5**, 1229–1237.

WELCH, W. J., GARRELS, J. I., THOMAS, G. P. LIM, J. C. & FERAMISCO, J. R. (1983). Biochemical characterisation of the mammalian stress proteins and identification of two stress proteins as glucose and calcium ionophore regulated protein. *J. biol. Chem.* **258**, 7102–7111.

WHITE, B. A. & BANCROFT, F. C. (1982). Cytoplasmic dot hybridisation. *J. biol. Chem.* **257**, 8569–8572.

WOLFF, S. P., GRANER, A. & DEAN, R. T. (1986). Free radicals, lipids and protein degradation. *Trends in biochem. Sci.* **11**, 27–31.

WU, C. (1984). Activating protein factor binds *in vitro* to upstream control sequences in heat shock gene chromatin. *Nature, Lond.* **311**, 81–84.

VELAZQUEZ, J. M. & LINDQUIST, S. (1984). Hsp 70: nuclear concentration during environmental stress and cytoplasmic storage during recovery. *Cell* **36**, 655–662.

Printed in Great Britain © Society for Experimental Biology 1987   135

A TEMPERATURE-COMPENSATED ULTRADIAN CLOCK EXPLAINS TEMPERATURE-DEPENDENT QUANTAL CELL CYCLE TIMES

D. LLOYD AND F. KIPPERT

Department of Microbiology, University College, Newport Road, Cardiff CF2 1TA, Wales, UK and Institut für Tierphysiologie, Justus-Liebig-Universität, Wartweg 95, D-6300, Giessen, FRG

Summary

The effects of sublethal heat pulses on cell division have provided insights into possible molecular mechanisms. Thus Zeuthen's findings of 'set-backs' up to a transition point provides the basis for the idea that the continuous accumulation of a compound needed for cell division spans a major portion of the cell cycle. The accumulating substance is a 'division protein' which forms part of a structure which is unstable until completely assembled at the transition point. Experiments showing phase resetting of mammalian cells by temperature perturbation indicate limit-cycle oscillator control of the cell cycle with a phase-response curve with a repeat interval equal to the period of the clock. As well as providing a method for establishing synchronized cultures these observations have found application in the selective effects of hyperthermia as an antitumour agent.

Circadian rhythms display several unique features distinguishing them from other periodic processes. Only recently has it been recognized that some of these characteristics may be properties of ultradian rhythms as well. The probably most striking feature of circadian timekeeping, i.e. independence of ambient temperature, was found for ultradian rhythmicity even at the level of the unicellular organization.

Synchronous cultures of some lower eukaryotes were prepared by centrifugal size selection methods. Experiments with asynchronous control cultures substantiated the view that the conditions employed were such as to minimize any perturbative effects: most importantly the organisms were never removed from their culture medium. Whereas the control cultures showed smoothly increasing respiration rates, total RNA, total protein, enzyme activities and enzyme protein (e.g. for cytochrome aa_3, ATPase, catalase), in synchronous cultures all these parameters showed oscillatory behaviour. Different periods were observed in different organisms: thus in *Acanthamoeba castellanii* the period was about 70 min, in *Tetrahymena pyriformis* strain ST it was about 50 min, in *T. pyriformis* AII it was 30 min, and in *Candida utilis* it was about 30 min (all measurements at 30°C). In *A. castellanii* the periods of both the oscillations in rate of respiration and the total cell protein were hardly affected by the temperature of growth over the range 20 to 30°C. The oscillations show no damping during experiments lasting 12 h: these properties suggest that we are observing temperature-compensated endogenous rhythms which presumably serve a timekeeping function in cells undergoing growth and division. The cell cycle of *A. castellanii* has a Q_{10} of 2; at 20°C the division interval is 16 h, whereas at 30°C it is only 8 h. A common phase reference point exists between the cell cycle and the rhythms, so that at any one temperature reproducible cell cycle maps of the timings of maxima (e.g. of respiration or total protein content) can be constructed. Thus at any fixed temperature there are an integral number of subcycles per cell cycle. This suggests that cell cycle times increase by discrete time intervals (equal to the temperature-compensated ultradian period) as temperature decreases.

In *Paramecium tetraurelia* a motility rhythm with a period of 70 min can be followed in single cells. Between 18°C and 33°C total activity increased by 70 % while the Q_{10} of the period over the entire temperature interval was 1·04. The pattern, however, was more complex with the shortest period at 27°C and slightly prolonged periods at temperatures above as well as below 27°C. Determination of generation times of individual cells revealed clustered values separated by time intervals corresponding to the ultradian clock period.

These results provide evidence of temperature compensation associated with ultradian rhythms which is as accurate as that found for their circadian counterparts. Additional comparative studies on circadian and ultradian timekeeping may reveal more about similarities in the clock mechanisms which have been suggested to be the result of a common evolutionary history. A model is proposed for timed cell cycle progression to a threshold that triggers a key cell cycle process.

Introduction

Temperature affects the nutritional requirements for growth, the rates of biochemical reactions, the pathways traversed in metabolism, and hence the composition of cells. Growth rates increase with temperature up to an optimum, and over most of this range the temperature coefficient (Q_{10}) has a value of about 2, that is a two-fold increase in growth rate occurs per 10°C rise in temperature (Pirt, 1975). At 10 to 25°C below the optimum temperature, growth rates approach zero; the limits for growth vary greatly from one organism to another. The temperature range for the growth of some fungi extends over more than 30°C, but for mammalian cells over only about 12°C. That the specific heat of water is at its minimum value of 4·1779 J g^{-1} °C^{-1} at 35°C may explain why the majority of homoiothermic animals maintain their body temperature during non-hibernation within a few degrees of 36°C. It is at this temperature that minimal heat production or dissipation is necessary, for the maintenance of body temperature (Paul, 1986). The approximate upper temperature survival limits reported are as follows: fish and other aquatic vertebrates 38°C, insects 50°C, crustaceans 50°C, protozoa 56°C, algae 60°C and fungi 62°C; some extremely thermophilic prokaryotes live at temperatures higher than 100°C (Brock, 1986).

The upper temperature limit compatible with growth in any species is set by the thermolability of a vital constituent. Advantage has been taken of this characteristic for the experimental arrest of cell growth on shifting the temperature from a favourable ('permissive') to an unfavourable ('non-permissive') condition. The earliest method for the synchronization of cell division within a population of organisms used this principle (Zeuthen & Scherbaum, 1954). The temperature-sensitive cell cycle mutants, selected for a particular thermolabile gene function, arrest at a specific cell cycle stage. Investigations of these genotypes have provided new insights into the cell cycles of yeast (Hartwell, 1974; Nurse *et al.* 1976; Dickinson, 1984), *Tetrahymena pyriformis* (Frankel *et al.* 1980), and mammalian cells (Baserga, 1985; Marcus *et al.* 1985). The observation that tumour cells may be more sensitive to killing by increased temperature (Cavaliere *et al.* 1967) initiated an extension of interest in heating effects into clinical medicine.

All these aspects of the biological effects of heating may be termed perturbative; an extensive literature describes the responses of cell cycle processes. Rather less attention has been brought to bear on steady-state cell cycle progression at different temperatures. A key question is, what controls the duration of the cell cycle at different temperatures? Intuitively one might expect a smooth change in generation time (GT) with changing temperature. However, we show here that this is not always the case. Increases in cell cycle times at lower temperatures occur by discrete increments; generation times are quantized. These discrete units correspond with the temperature-compensated period of an oscillator. This clock provides timekeeping, one component essential for the coordination and control of cell division.

Effects of heat pulses on the cell division cycle

Pioneering studies by Zeuthen & Scherbaum (1954) with the amicronucleate GL strain of *Tetrahymena pyriformis* showed that cell division could be synchronized by exposing organisms to 30 min periods at 34°C, a sublethal temperature, alternating with similar periods at the optimal growth temperature (29°C). After eight cycles of heat treatment, three successive synchronous divisions were observed. Although subsequent refinement and modification eventually enabled development of a less perturbing procedure in which the cell cycle becomes entrained by providing a single heat pulse once per cycle (Fig. 1; Zeuthen, 1971; Kramhøft & Zeuthen, 1971), it was the distortion (Zeuthen, 1978) of 'normal' cell cycle progression observed in the early system which led to numerous experimental and theoretical studies. These aimed at elucidation of the control mechanisms involved. It was shown in *T. pyriformis* that heat shock delays cell division; newly divided cells showed little delay, and the extent of delay increased progressively, until at 25 min before division heat shock became ineffective. Similar observations had previously been described for sea-urchin eggs (Swann, 1957). Zeuthen proposed that synthesis and assembly of a protein-containing structural element, thermolabile only until the transition point, was essential for cell division.

Disassembly back to the starting-point explained progressive set-backs. Pulses of amino acid analogues (Rasmussen & Zeuthen, 1962), or protein synthesis inhibitors such as cycloheximide (Frankel, 1969), gave effects similar to those evoked by heat shocks. In the fission yeast *Schizosaccharomyces pombe*, entrainment of the cell cycle to cyclic temperature changes (Kramhøft & Zeuthen, 1971) gave a method for producing continuously synchronous cultures and suggested that fundamentally similar features of temperature sensitivity exist in different types of organisms. In this yeast, however, several different kinds and periods of sensitivity to heat shocks and inhibitor pulses were distinguished (Polanshek, 1977). In Chinese hamster cells initiation of DNA synthesis is set-back by cycloheximide or puromycin, and this also suggests that accumulation of different kinds of initiators may occur to threshold levels at different stages of the cell cycle (Schneiderman *et al.* 1971; Highfield & Dewey, 1972). The naturally synchronous mitoses in plasmodia of *Physarum polycephalum* can be delayed by heat pulses (Brewer & Rusch, 1968), and evidence for a cytoplasmic initiator is extensive (Rusch *et al.* 1966; Sachsenmaier *et al.* 1972; Sudbery & Grant, 1976). Fusion experiments, in which plasmodia at different stages in their mitotic cycles give fused pairs in which the timing of mitosis is altered, provide evidence for control by an oscillator (Kauffman, 1974; Kauffman & Wille, 1977; Fig. 2). Mathematical modelling of results obtained in phase-synchronization and heat-pulse experiments encourages the concept of oscillator control, and recent controversy (Tyson & Sachsenmaier, 1978) debates whether the oscillator is a continuous one or of the relaxation-type (the nuclear sites-titration model of Sachsenmaier *et al.* 1972). In V79 Chinese hamster cells subjected to 45°C for 10 min, a twice repeated pattern of phase shifts was observed in a cell cycle of 8·5 h (Klevecz *et al.* 1980). This important observation clearly indicates that in some cell-types the period of the oscillator is not that of the cell cycle itself.

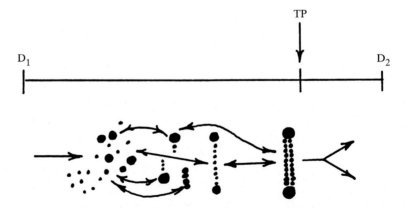

Fig. 1. The division protein model of Zeuthen. At the transition point an assembled protein-containing structure is no longer thermolabile. Before that time heat shock leads to its complete disassembly and 'set-back' to the initiation point for its synthesis.

Applied aspects of the effects of heat treatment

Selective effects of heat on tumour cells have produced an interest in hyperthermia as an approach to the treatment of cancer (Cavaliere *et al.* 1967; Pettigrew *et al.* 1974; Henle & Dethlefsen, 1978; Overgaard, 1984). Synergism between cytotoxic drugs or radiation and hyperthermia has also become an exploitable phenomenon (Robinson *et al.* 1974) but the development of thermotolerance (Kamura *et al.* 1982) may occur. Cell cycle stage dependence of sensitivity to hyperthermia has been measured in normal and malignant cells, with late S and mitotic cells most sensitive (Dewey *et al.* 1971; Bhuyan *et al.* 1977). Blocked progression through the S/G_2 boundary is reversible under favourable growth conditions (Selawry *et al.* 1957). Long division delays are produced in Chinese hamster ovary cells, even with heat treatments which are mild in terms of killing (>20 % survival). Chromosome damage occurs if heat treatment is given during S-phase (Dewey *et al.* 1971), and this effect is probably responsible for the potentiation of the lethal effects of subsequently applied X-rays (Dewey *et al.* 1978; Tomasovic *et al.* 1979; Bichel *et al.* 1979). It is claimed that centrosome structure is damaged by hyperthermia (Barrau *et al.* 1978). But in the fission yeast *Schizosaccharomyces pombe*, DNA damage is not the dominant cause of cell killing by heat, as growth with 2-phenylethanol advances the timing of S-phase from just before cell division without altering the time of greatest heat sensitivity (Bullock & Coakley, 1976, 1978, 1979).

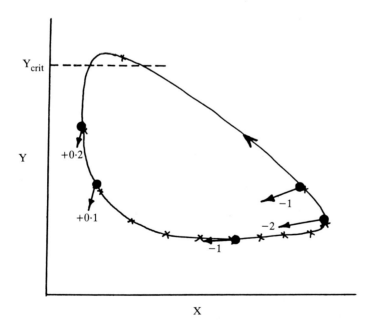

Fig. 2. The limit cycle model. A critical level (threshold) of Y triggers mitosis. X is an inactive precursor. Equal time intervals are shown along the trajectories (x). The effects of 20 % destruction of X and Y from five phases are shown indicating magnitude of delay (−) or advance (+) of the next mitosis.

Since hyperthermia leads to a delay in cell cycle traverse (Zielke-Temme & Hopwood, 1982; Rice *et al.* 1984*b*) it is unclear whether the development and decay of observed thermotolerance are strictly related to cell cycle phase (Majima & Gerweck, 1983). Both G_1- and S-phase cells can develop thermotolerance (Read *et al.* 1983, 1984*a*; Rice *et al.* 1984*b*), but elucidation of the kinetics of its decay is complicated by cell cycle delays. In this context it is worth mentioning that the levels of various modulators of cellular function which have also been ascribed important roles in cell cycle progression show altered levels after heat shock. Relevant findings include changes in pH (Drummond *et al.* 1986), Ca^{2+} and inositol trisphosphate (Stevenson *et al.* 1986), and poly(ADP-ribose) (Juarez-Salinas *et al.* 1984). *In vitro* both heat-induced cell cycle delay and the rates of development and decay of thermotolerance can be slowed by lowering the pH of incubation media (Holahan & Dewey, 1986). Ca^{2+} (Malhotra *et al.* 1986) and calmodulin (Wiegant *et al.* 1985) have been implicated in the heat shock response and the roles they play in heat shock and cell cycle regulation may well interfere with one another.

Studies of the cell cycle under steady-state conditions of temperature

(a) *In* Acanthamoeba castellanii

In order to study time-dependent processes of the cell cycle it is necessary to examine single cells or synchronous populations; observations on ordinary growing populations conceal the time-structure of the individuals (Lloyd *et al.* 1982*a*). It is also essential that steady-state growth be examined as non-invasively as possible (i.e. that the organisms are not removed from their growth media, even for a moment, and thus do not experience transients of nutrient deprivation, anoxia or temperature). Other potential disturbances include osmotic or hydrodynamic stress and rapid pressure fluctuations.

Two methods have been developed which enable size-selection synchrony with minimum perturbation; these are continuous-flow centrifugation size-selection (Lloyd *et al.* 1975) used for large cultures (>21) and centrifugation of the culture (Chagla & Griffiths, 1978) for smaller cultures (25 ml–1 litre). In those cell types where cell cycle dependent changes in density are not pronounced, selection of the most slowly sedimenting subpopulation gives the smallest cells of the culture. Their subsequent growth gives a highly synchronous culture.

Fig. 3 compares the respiration rates of a synchronous population with that of an asynchronous (control) culture which has experienced identical centrifugal forces during the experimental manipulation. In both cultures respiration rates double over one generation time, but only the synchronous culture shows an oscillation of oxygen consumption.

Similar respiratory oscillations have been revealed in a number of protozoa and yeasts (Table 1); the periods all lie in the epigenetic time domain (Goodwin, 1963; Lloyd & Edwards, 1984), and are characteristic of different species (Lloyd &

Edwards, 1986*a*). In *A. castellanii* (Edwards & Lloyd, 1978) and *C. utilis* (Lloyd *et al.* 1981) the biochemical mechanism responsible has been defined as *in vivo* mitochondrial respiratory control. Although the easiest observable, respiration is only an enslaved oscillator, coupled by way of the adenine nucleotide system to the changing demands of energy-utilizing processes (Edwards & Lloyd, 1980). The major energetic sink in the growing cell is protein synthesis, and accumulated levels of cellular protein oscillate too (Edwards & Lloyd, 1980). This observation, on a bulk measurement, indicates that either a few predominant protein species show this phenomenon, or rather that it is a characteristic of the majority of

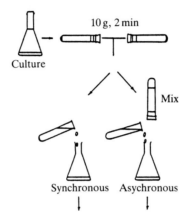

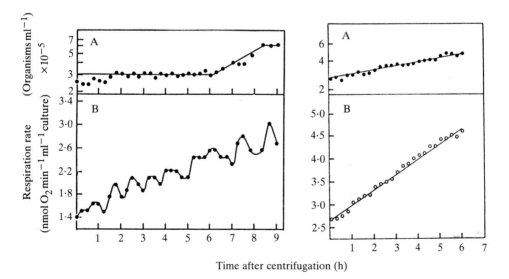

Fig. 3. A method for obtaining synchronous cell division of cultures of *Acanthamoeba castellanii* (Chagla & Griffiths, 1978). A short low-speed centrifugation of the culture is followed by decantation of the most slowly sedimenting organisms to give a synchronous culture. Remixing of the contents of one tube gives an asynchronous control culture. Respiratory oscillations are seen only in the synchronous culture (Edwards & Lloyd, 1978).

proteins; 10 oscillatory enzymes have been demonstrated (Lloyd & Edwards, 1984). The implication for rates of protein turnover is that the dynamic state of cell constituents has been consistently underestimated (Lloyd & Edwards, 1986b).

At least three possible functions can be proposed for these cellular rhythms: (a) they may serve to provide temporal compartments for incompatible reactions; (b) they may facilitate a more speedy response when environmental circumstances alter, or (c) they may have a timekeeping role.

The effect of temperature provides an unambiguous solution to their functional role (Lloyd et al. 1982b). Fig. 4 shows the growth of A. castellanii in synchronous cultures at 30°, 27° and 25°C. The cell counts show an increase in cell cycle time from 8h to 12h over this 5 degree temperature decrease. *But the period of the respiratory oscillation is hardly affected, i.e. it is temperature compensated.* Furthermore, coupling of the rhythm of respiration to that of accumulated cell protein is maintained with close phase-correspondence over a 10°C range (Fig. 5; Marques et al. 1987) so that the periods of both variables appear temperature compensated (Fig. 6). A rhythm of this type must be part of or coupled to a cellular clock (Fig. 7). Selection of the smallest cells of the culture gives not only synchrony of cell division, but also synchrony with respect to the rhythm (i.e. reveals phase-correspondence between cell cycle and subcycles of respiration, so as to give reproducible 'cell cycle maps' of the timing of maxima in replicate experiments at any one temperature (Edwards & Lloyd, 1980). Thus cell cycle time is an integer multiple of the period of the respiratory rhythm, and one of the essential timekeeping functions of the clock is used for the timing of cell division. By analogy with the circadian clock, that in rapidly growing cells may be termed the ultradian timer.

(b) *In* Tetrahymena pyriformis

Tetrahymena pyriformis has been shown to display endogenous rhythms of oxygen uptake with both circadian (Dobra & Ehret, 1975) and ultradian (Lloyd et al. 1978) periods, depending on growth conditions. The circadian type occurs

Table 1. *Respiratory oscillations in synchronous cultures of lower eukaryotes*

Organism	Temperature (°C)	Cell cycle time (h)	Respiratory maxima/ cell cycle	Reference
Crithidia fasciculata	30	5·5	5	Edwards et al. (1975)
Tetrahymena pyriformis ST	30	2·5	3	Lloyd et al. (1978)
Acanthamoeba castellanii	30	7·8	7	Edwards & Lloyd (1978)
Dictyostelium discoideum	22	6·0	6	Woffendin & Griffiths (1985)
Candida utilis				
(glucose)	30	1·5	3 ⎫	
(glycerol)	30	1·5	3 ⎬	Kader & Lloyd (1979)
(acetate)	30	2·0	3 ⎭	

when cells are growing slowly (GT ~24h), or not growing at all, the ultradian type in rapidly cycling cells (GT a few hours).

In order to obtain *Tetrahymena* cultures synchronous with respect to an ultradian rhythmicity, the cells were grown to stationary phase and after two more days diluted into fresh medium. In that medium, after a short lag phase, cells entered cell cycle again with a GT of a few hours. At the same time, this transition to favourable growth conditions induced rhythmic oxygen uptake with a period of about 0·5h which persists throughout many cycles without pronounced damping (Kippert, 1986).

The view that this rhythm is not due to any kind of metabolic perturbation but rather reflects an endogenous property of the cells which are only synchronized by

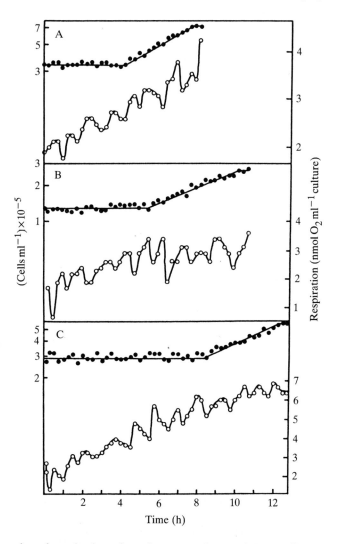

Fig. 4. Growth and respiration of synchronous cultures of *A. castellanii* at 30°C (A), 27°C (B) and 25°C (C). ●, Growth; ○, respiration rate (Lloyd *et al.* 1982*b*).

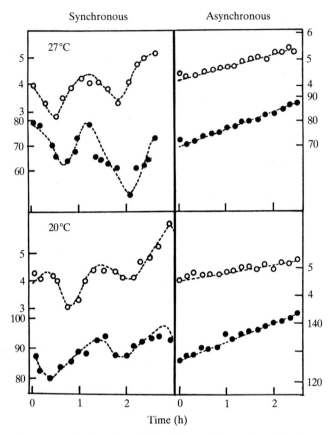

Fig. 5. Protein accumulation ($\mu g\,ml^{-1}$ culture, ●) and respiration rates (nmol $O_2\,min^{-1}\,ml^{-1}$, ○) in synchronous cultures of *A. castellanii* at 27°C and 20°C (Lloyd *et al.* 1982*b*). Curves were fitted after time series analyses.

the reactivation procedure is substantiated by the facts that (i) the rhythmic activity always shows the same pattern and resumes with a definite phase relationship to the moment of dilution of the culture (therefore, traces from different sets of experiments can all be plotted together); (ii) even more important, the period of this rhythm is independent of steady-state ambient temperature. The rhythm was monitored at various temperatures between 19°C and 33°C (four of these temperatures represented in Fig. 8) and period length increased only marginally from 30·3 min at 33°C to 35·2°C min at 19°C. This results in a Q_{10} of 1·12 which is in strong contrast to the Q_{10} of the mean respiratory activity which, showing a value of 2·35, is in the range expected for such physiological functions.

(c) *In* Paramecium tetraurelia

A decisive step towards excluding any perturbation induced by even the mildest synchronization procedure was to look for a relationship between endogenous rhythms and cell cycle on a single cell basis. *Paramecium tetraurelia* proved ideal because an ultradian rhythm of motility and the relationship of this rhythm to the cell cycle can most easily be studied in this large ciliate (Kippert, 1985, 1986). A clear-cut circadian rhythm of motility had been shown for stationary phase cultures of another *Paramecium* species (Hasegawa & Tanakadate, 1984). Screening individual cells of *Paramecium tetraurelia* under optimal growth conditions (GT about 6 h at 27°C) for variations of their motile behaviour revealed an ultradian rhythm of motility which is very similar to its circadian counterpart. Both

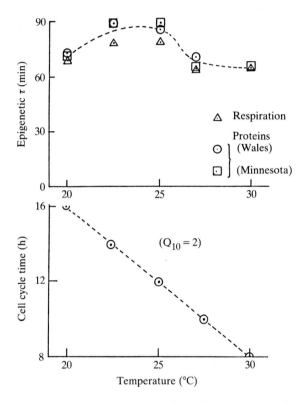

Fig. 6. Temperature independence of the periods of respiratory and protein accumulation rhythms and temperature dependence of cell cycle time (Lloyd & Edwards, 1987).

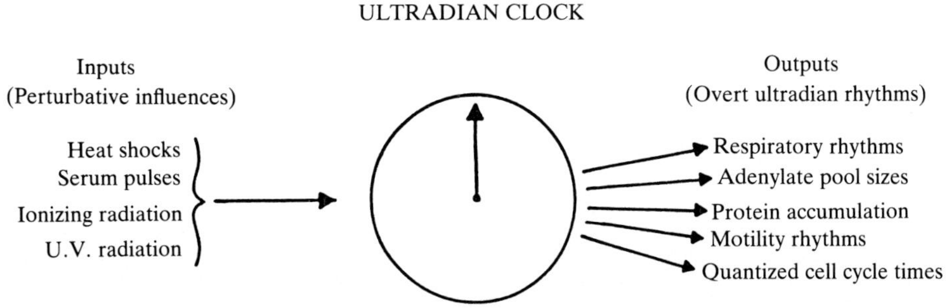

ULTRADIAN CLOCK

Inputs		Outputs
(Perturbative influences)		(Overt ultradian rhythms)

Heat shocks
Serum pulses
Ionizing radiation
U.V. radiation

Respiratory rhythms
Adenylate pool sizes
Protein accumulation
Motility rhythms
Quantized cell cycle times

Fig. 7. The coupling of rhythms to the ultradian clock.

swimming speed and the frequency of changes in direction vary in a rhythmic
fashion. Since these two components of motility are 180° out of phase, a
pronounced overall rhythm of motile behaviour results which can easily be
monitored by just counting the number of field coordinates crossed by an
individual cell within 5 min intervals.

The period of this rhythm was unaffected by the steady-state temperature,
showing a Q_{10} of just 1·04. Registration at different temperatures between 18°C
and 33°C (at 3 temperatures shown in Fig. 9) revealed a biphasic temperature
behaviour with the shortest period at 27°C (68·9 min) and slightly prolonged
periods at temperatures below (up to 74·4 min at 78°C) and above (up to 70·2 min
at 33°C). Again this typical clock-associated Q_{10} of around 1 was in sharp contrast
to the temperature dependence of the overall motility of the cells which increased
by 70 % over the experimental temperature range.

To study the relationship between the putative clock underlying this rhythm on
the one hand and cell cycle timing on the other hand, the GT of individual cells was
monitored. Fig. 10, showing the GT of a total of more than 600 cells at different

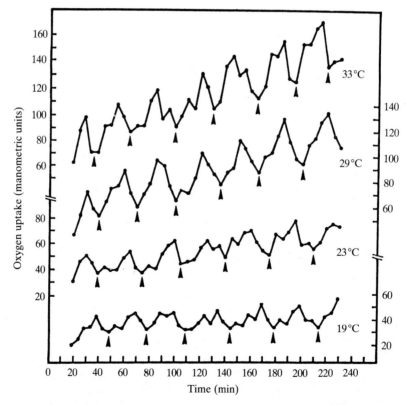

Fig. 8. Ultradian rhythm of oxygen uptake in *Tetrahymena pyriformis* at 4 different
growth temperatures. Stationary phase cultures were transferred into fresh medium at
time zero. Each point represents the mean oxygen uptake of a number of experiments:
56 at 33°C, 147 at 29°C, 36 at 23°C, 32 at 19°C. Minima are indicated by arrows for
better comparison of period length (Kippert, 1986).

growth temperatures, clearly indicates that they are not distributed in a random fashion. Rather, they appear in clusters which are separated by a time span which very closely corresponds to the period of the motility rhythm: when one calculates the length of these quantal elements from the data in Fig. 10, one obtains values of 70·0 min at 33°C, 69·2 min at 27°C, and 71·8 at 21°C, which are almost

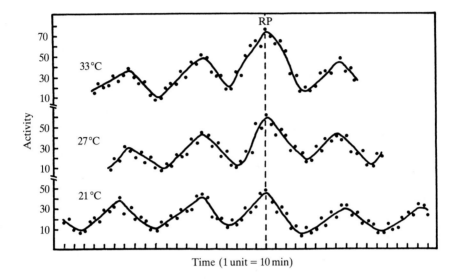

Fig. 9. Ultradian rhythm of motility in *Paramecium tetraurelia*. Activity was registered as the number of coordinates crossed by the individual cell during successive 5 min intervals. Data plotted from one characteristic point (pronounced maximum) into both directions on the time scale. Number of registrations was 17 at 33°C, 16 at 27°C and 11 at 21°C (Kippert, 1986).

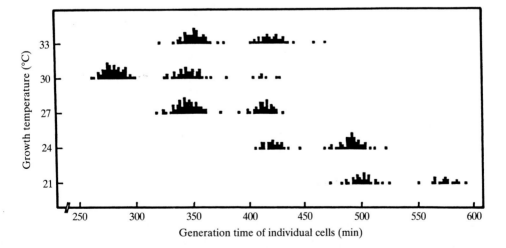

Fig. 10. Distribution of generation times of individual *Paramecium tetraurelia* cells at various steady-state growth temperatures. Each square represents the time span between two successive cell divisions of an isolated cell.

indistinguishable from the periods of the motility rhythm at the respective temperatures.

These results unequivocally show that there is a temperature-compensated clock working in these cells. Coupled to this clock are cellular functions as diverse as motility behaviour and cell cycle events. As was shown in the case of *Acanthamoeba*, cell cycle timing in *Paramecium tetraurelia* is dependent on two distinct features: (i) cellular growth which is strongly dependent on temperature ($Q_{10} = 1·87$) and (ii) a temperature-compensated timekeeping mechanism ($Q_{10} = 1·04$). As a result GT is restricted to a discrete multiple of the clock's period and is prolonged stepwise under less favourable conditions of temperature or nutrient supply. Further investigation of this phenomenon has shown that division occurs only in a certain phase relationship to the rhythm of cell motility (Kippert, 1985). Photoentrainment of ultradian rhythms in *Tetrahymena* is restricted to times defined by modal generation times (Readey, 1985).

Experimental evidence for components oscillating with a period appropriate for cell cycle control is not confined to the systems described here. Klevecz & King (1982) report a temperature-compensated periodicity of 4 h timing the mammalian cell cycle. Confirmation of an ultradian rhythm in *Tetrahymena pyriformis* comes from the experiments of Thiel *et al.* (1985) and Michel & Hardeland (1985). Most importantly they showed that the ultradian rhythm proceeds even when growth is stopped. The only other examples of temperature-compensated ultradian rhythms are those observed by Loh & Golenhofen (1970) and Kyriacou & Hall (1980) in the minute range. Of all the several hundreds of other *metabolic* oscillators described, none have been shown to possess any clock-like properties (Lloyd & Edwards, 1984).

Models of quantized cell cycle

There has been no shortage of hypothetical models for oscillatory control of the cell cycle. Sel'Kov's (1970) limit cycle sees thiols and disulphides synthesized and degraded, oxidized and reduced. Gilbert (1974) developed this type of control system and related it (Gilbert, 1978) to the transition probability concept (Brooks, 1985), and other models (Kauffman & Wille, 1975; Gilbert, 1981) showed how a limit cycle mechanism could explain heat shock delay, phase compromise and several other experimentally observed features. Further discussion of these and other models is to be found in Edmunds (1984).

Almost all these models, however, envisage an oscillator, the period of which is identical to that of the cell cycle itself and they therefore cannot apply to the phenomenon of quantized cell cycles. The question then arises of what is the nature of the underlying mechanism in a pattern of cell cycle timing as shown in Fig. 10.

Primarily, of course, the temperature-compensated clock has to be explained; although several very different models of the clock mechanism have been suggested, experimental evidence is lacking. Besides the long period of these rhythms when compared to metabolic oscillations, probably most difficult to explain is the striking independence of ambient steady-state temperature (as will be discussed in the following section).

More easily explained is how various physiological functions (irrespective of their nature) could be coupled to the central timer (whatever its mechanism): (i) rhythms of protein accumulation (Edwards & Lloyd, 1980) may stem from alterations in the capacity of the translational machinery and/or the activity of protein breakdown pathways. (ii) Rhythms of respiration are probably due to alterations of energy demand defined in i (Edwards & Lloyd, 1978). (iii) Rhythms of motility may be caused by alterations in the number and/or activity of ion channels in the plasma membrane.

But what about cell cycle regulation by the endogenous clock? As is obvious from the results for GT in *Paramecium*, cell division (preceded by mitosis) appears to be 'gated' by the underlying rhythm. Another important step in the cell cycle, the G_1/S transition, will be timed in a similar way. And we know from studies on circadian systems that other clearly distinguishable stages of the cell cycle are only permitted at certain stages of the circadian period (Sweeney, 1982).

We would like to propose here a very generalized scheme of quantized cell cycles which is applicable whatever the molecular mechanism of the underlying clock or the nature of the cell cycle event may be. Threshold phenomena have frequently been proposed to be involved in cell cycle regulation (Kauffman & Wille, 1975). Such a threshold has to be reached before a certain stage in the cell cycle (DNA synthesis, mitosis and cell division) can be initiated. Most workers have suggested that an essential protein is accumulated to a threshold level. We now suggest that such a protein may be either accumulated or modified in a rhythmic fashion.

Fig. 11 shows that, while the length of the subcycle's period is not affected by the temperature, it takes progressively more cycles to reach the threshold as temperature is decreased. The scheme indicates that, in order to initiate a cell cycle event, two requirements have to be fulfilled: (1) a certain threshold has to be reached, and (2) the endogenous rhythm has to be at its maximum. An easily visualized example for such a mechanism would be the accumulation (that could be either continuous or discontinuous) of an essential protein and its rhythmic modification by a phosphorylation/dephosphorylation cycle.

Only an amount of this protein which is above the threshold level and at the same time maximally modified would be effective in regulation. Thus the critical modulation of cellular function is both concentration-dependent and state-dependent. This model, however, is by no means restricted to the afore-mentioned example, but is equally applicable to any other variable implied in cell cycle regulation, be it pH, Ca^{2+}, the concentration of any other ion, or the relationship between the concentrations of different ions.

We suggest that this kind of cell cycle regulation by a temperature-compensated clock may be widespread if not almost universal in eukaryotic cells. With regard to this suggestion, it has to be stressed that such subtle mechanisms of cell cycle regulation can only be studied when: (i) highly synchronous cultures are obtained with respect to both cell cycle and ultradian rhythm, and (ii) parameters are followed continuously or at least sampling is with a frequency high enough to determine periodicities which are only a fraction of the cell cycle.

Future studies

One of the intriguing aspects of this work is the mechanism of temperature compensation. In contrast, most biological processes have a Q_{10} within the range 2–4, e.g. for microbial growth (Pirt, 1975) or for nerve membrane conductance (Hodgkin & Huxley, 1952), and exceptionally may reach 11 (Jaslove & Brink, 1986). Temperature compensation is an unsolved problem in the circadian field too, where many measurements have been made (Sweeney & Hastings, 1960), and although there have been many proposals there are a few experimental studies. Possible mechanisms include: (i) Temperature-compensated product inhibition (i.e. a biochemical analogue of a RC time-constant electronic circuit) (Pittendrigh & Bruce, 1957; Chance *et al.* 1967); (ii) a relatively temperature-independent diffusion-controlled rate-limiting step (Ehret & Trucco, 1967); (iii) adaptation

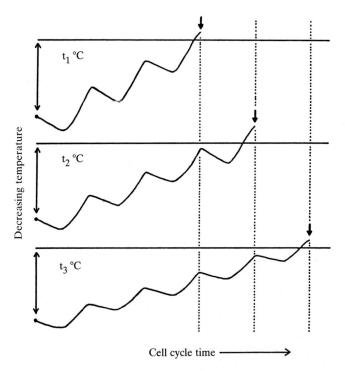

Fig. 11. A model for quantized cell cycles. Attainment of a constant threshold is indicated at three different temperatures by way of a rhythm which has a temperature-compensated period. Cell cycle time increases by 1 quantal increment as the temperature of growth is decreased. Attainment of the threshold has to occur at a maximum of the rhythm for triggering of a cell cycle process (e.g. DNA replication, mitosis or cell division).

until a lowered enzyme content compensates for increased enzyme activities at high temperatures (Pavlidis & Kauzmann, 1969); (iv) the direct effect of temperature on membrane fluidity is opposed by compensatory adapatation which alters unsaturated fatty acid composition (Njus *et al.* 1974); (v) compensation for the temperature coefficient of protein synthesis >1 by a membrane assembly step with a Q_{10} <1 (Schweiger & Schweiger, 1977); (vi) transcription loop length is selected to compensate for temperature (Edmunds & Adams, 1981); (vii) a temperature-dependent threshold (O. E. Rössler, cited by Engelmann & Schrempf, 1980).

One approach to this problem requires studies of transient behaviour after temperature shifts. Here again the temperature effects may give key clues to clock mechanisms, but it might be anticipated that final answers may be as elusive as has proved the case in the circadian time domain (Rosbash & Hall, 1985; Johnson & Hastings, 1986).

References

BARRAU, M. D., BLACKBURN, G. R. & DEWEY, W. C. (1978). Effects of heat on the centrosomes of Chinese hamster ovary cells. *Cancer Res.* **38**, 2290–2294.

BASERGA, R. (1985). *The Biology of Cell Reproduction*, 245pp. Cambridge Massachusetts: Harvard Univ. Press.

BHUYAN, B. K., DAY, K. J., EDGERTON, C. E. & OGUNBASE, O. (1977). Sensitivity of different cell lines and of different phases in the cell cycle to hyperthermia. *Cancer Res.* **37**, 3780–3784.

BICHEL, P., OVERGAARD, J. & NIELSEN, O. S. (1979). Synergistic cell cycle kinetic effect of low doses of hyperthermia and radiation on tumour cells. *Eur. J. Canc.* **15**, 1191–1196.

BREWER, E. N. & RUSCH, H. P. (1968). Effects of elevated temperature shocks on mitosis and on the initiation of DNA replication in *Physarum polycephalum*. *Expl Cell Res.* **49**, 79–86.

BROCK, T. D. (1986). Life at high temperatures. *Science* **230**, 132–138.

BROOKS, R. F. (1985). The transition probability model: successes, limitations and deficiencies. In *Temporal Order* (ed. L. Rensing & N. I. Jaeger), pp. 304–314. Berlin: Springer-Verlag.

BULLOCK, J. G. & COAKLEY, W. T. (1976). Effect of heat on the viability of *Schizosaccharomyces pombe* 972 h⁻ growing in synchronous cultures. *Expl Cell Res.* **103**, 447–449.

BULLOCK, J. G. & COAKLEY, W. T. (1978). Investigation into the mechanisms and repair of heat damage to *Schizosaccharomyces pombe* growing in synchronous cultures. *J. therm. Biol.* **3**, 159–162.

BULLOCK, J. G. & COAKLEY, W. T. (1979). The cell cycle thermal-inactivation stage of *Schizosaccharomyces pombe* is independent of 2-phenylethanol-induced changes in S-phase location. *Expl Cell Res.* **121**, 441–445.

CAVALIERE, R., CIOCATTO, E. C., GIOVANELLA, B. C., HEIDELBERGER, C., JOHNSON, R. O., MARGOTTINI, M., MONDOVI, B., MORICCA, G. & ROSSI-FANELLI, A. R. (1967). Selective heat insensitivity of cancer cells. *Cancer* **20**, 1351–1381.

CHAGLA, A. H. & GRIFFITHS, A. J. (1978). Synchronous cultures of *Acanthamoeba castellanii* and their use in the study of encystation. *J. gen. Microbiol.* **108**, 39–43.

CHANCE, B., PYE, K. & HIGGINS, J. (1967). Wave-form generation by enzymatic oscillators. *IEEE Spectrum* **4**, 79–87.

DEWEY, W. C., SAPARETO, S. A. & BETTEN, D. A. (1978). Hyperthermic radiosensitization of synchronous Chinese hamster cells: Relationship between lethality and chromosome aberrations. *Radiat. Res.* **76**, 48–59.

DEWEY, W. C., WESTRA, A., MILLER, H. H. & NAGASAWA, H. (1971). Heat-induced lethality and chromosome damage in synchronized chinese hamster cells treated with 5-bromodeoxyuridine. *Int. J. Radiat. Biol.* **20**, 505–520.

DICKINSON, J. R. (1984). The biochemical genetics of cell cycle control in eukaryotes. *Trends biochem. Sci.* **9**, 269–271.

DOBRA, K. W. & EHRET, C. F. (1975). Circadian regulation of glycogen, tyrosine aminotransferase, and several respiratory parameters in solid agar cultures of *Tetrahymena pyriformis. Proc. XII Int. Comf. Soc. Chronobiol.,* Il Ponte, Milano, pp. 589–594.

DRUMMOND, I. A., McCLURE, S. A., POENIE, M., TSIEN, R. Y. & STEINHARDT, R. A. (1986). Large changes in intracellular pH and calcium observed during heat shock are not responsible for the induction of heat shock proteins in *Drosophila melanogaster. Mol. Cell Biol.* **6**, 1767–1775.

EDMUNDS, L. N. (ed.) (1984). *Cell Cycle Clocks,* 616pp. New York, Basel: Marcel Dekker.

EDMUNDS, L. N. & ADAMS, K. J. (1981). Clocked cell cycle clocks. *Science* **211**, 1002–1013.

EDWARDS, S. W. & LLOYD, D. (1978). Oscillations of respiration and adenine nucleotides in synchronous cultures of *Acanthamoeba castellanii*: Mitochondrial respiratory control *in vivo. J. gen. Microbiol.* **108**, 197–204.

EDWARDS, S. W. & LLOYD, D. (1980). Oscillations in protein and RNA content during synchronous growth of *Acanthamoeba castellanii*: Evidence for periodic turnover of macromolecules during the cell cycle. *FEBS Lett.* **109**, 21–26.

EDWARDS, C., STATHAM, M. & LLOYD, D. (1975). The preparation of large-scale synchronous cultures of the trypanosomatid, *Crithidia fasciculata*, by cell-size selection: Changes in respiration and adenylate charge through the cell-cycle. *J. gen. Microbiol.* **88**, 141–152.

EHRET, C. F. & TRUCCO, E. (1967). Molecular models for the circadian clock. *J. theor. Biol.* **15**, 240–262.

ENGLEMANN, W. & SCHREMPF, M. (1980). Membrane models for circadian rhythms. *Photochem. Photobiol. Rev.* **5**, 49–86.

FRANKEL, J. (1969). The relationship of protein synthesis to cell division and oral organelle development in synchronized *Tetrahymena pyriformis* GL-C: an analysis employing cycloheximide. *J. Cell Physiol.* **74**, 135–148.

FRANKEL, J., MOHLER, J. & FRANKEL, A. K. (1980). The relationship between the excess-delay phenomenon and temperature-sensitive periods in *Tetrahymena thermophila. J. Cell Sci.* **43**, 75–91.

GILBERT, D. A. (1974). The nature of the cell cycle and the control of cell replication. *Biosystems* **5**, 197–206.

GILBERT, D. A. (1978). The relationship between the transition probability and oscillator concepts of the cell cycle and the nature of commitment to replication. *Biosystems* **10**, 235–240.

GILBERT, D. A. (1981). The Cell Cycle 1981. One or more limit cycle oscillations? *S. Afr. J. Sci.* **77**, 541–546.

GOODWIN, B. C. (1963). *Temporal Organization in Cells.* New York: Academic Press.

HARTWELL, L. H. (1974). *Saccharomyces cerevisiae* cell cycle. *Bact. Rev.* **38**, 164–198.

HASEGAWA, K. & TANAKADATE, A. (1984). Circadian rhythm of locomotor behaviour in a population of *Paramecium multimicronucleatum*: its characteristics as derived from circadian changes in the swimming speed and the frequencies of avoiding response among individual cells. *Photochem. Photobiol.* **40**, 105–112.

HENLE, K. J. & DETHLEFSEN, L. A. (1978). Heat fractionation and thermotolerance A review. *Cancer Res.* **38**, 1843–1851.

HIGHFIELD, D. P. & DEWEY, W. C. (1972). Inhibition of DNA synthesis in synchronized chinese hamster cells treated in G_1 or early S-phase with cycloheximide or puromycin. *Expl Cell Res.* **75**, 314–320.

HODGKIN, A. L. & HUXLEY, A. F. (1952). The components of membrane conductance in the giant axon of *Loligo. J. Physiol., Lond.* **116**, 473–496.

HOLAHAN, P. K. & DEWEY, W. C. (1986). Effect of pH and cell cycle progression on development and decay of thermotolerance. *Radiat. Res.* **106**, 111–121.

JASLOVE, S. W. & BRINK, P. R. (1986). The mechanism of rectification at the electronic motor giant synapse of the crayfish. *Nature, Lond.* **323**, 63–65.

JOHNSON, C. H. & HASTINGS, J. W. (1986). The elusive mechanism of the circadian clock. *Am. Scient.* **74**, 29–36.

JUAREZ-SALINAS, H., DURAN-TORRES, G. & JACOBSON, M. K. (1984). Alteration of poly (ADP-ribose) metabolism by hyperthermia. *Biochem. biophys. Res. Commun.* **122**, 1381–1388.

KADER, J. & LLOYD, D. (1979). Respiratory oscillations and heat evolution in synchronous cultures of *Candida utilis*. *J. gen. Microbiol.* **114**, 455–461.

KAMURA, T., NIELSON, O. S., OVERGAARD, J. & ANDERSEN, A. H. (1982). Development of thermotolerance during fractionated hyperthermia in a solid tumor *in vivo*. *Cancer Res.* **42**, 1744–1748.

KAUFFMAN, S. A. (1974). Measuring a mitotic oscillator: the arc discontinuity. *Bull. math. Biol.* **36**, 171–182.

KAUFFMAN, S. A. & WILLE, J. J. (1975). The mitotic oscillator in *Physarum polycephalum*. *J. theor. Biol.* **55**, 47–93.

KAUFFMAN, S. A. & WILLE, J. J. (1977). Evidence that the mitotic 'clock' in *Physarum polycephalum* is a limit cycle oscillator. In *The Molecular Basis of Circadian Rhythms* (ed. J. W. Hastings & H. G. Schweiger), pp. 421–431. Berlin: Dahlem Konferenzen.

KIPPERT, F. (1985). Evidence for the concept of quantized cell cycles: Generation time of individual *Paramecium* cells is a discrete multiple of ultradian sub-cycles. *Eur. J. Cell Biol.* **38** *Suppl* (9), 16.

KIPPERT, F. (1986). Temperature-compensation of ultradian rhythms in ciliates. *Proc. II Annual Meeting Eur. Soc. Chronobiol.* (in press).

KLEVECZ, R. R. & KING, G. A. (1982). Temperature compensation in the mammalian cell cycle. *Expl Cell Res.* **140**, 307–318.

KLEVECZ, R. R., KROS, J. & KING, G. A. (1980). Phase response to heat shock as evidence for a timekeeping oscillator in synchronous animal cells. *Cytogenet. Cell Genet.* **26**, 236–243.

KRAMHØFT, B. & ZEUTHEN, E. (1971). Synchronization of cell divisions in the fission yeast *Schizosaccharomyces pombe* using heat shocks. *C.r. Trav. Lab. Carlsberg* **38**, 351–368.

KYRIACOU, C. P. & HALL, J. C. (1980). Circadian rhythm mutations in *Drosophila melanogaster* affect short-term fluctuations in the male's courtship song. *Proc. natn. Acad. Sci. U.S.A.* **77**, 6729–6733.

LLOYD, D. & EDWARDS, S. W. (1984). Epigenetic oscillations during the cell cycles of lower eukaryotes are coupled to a clock: Life's slow dance to the music of time. In *Cell Cycle Clocks* (ed. L. N. Edmunds), pp. 27–46. New York, Basel: Marcel Dekker.

LLOYD, D. & EDWARDS, S. W. (1986a). Ultradian rhythms in lower eukaryotes: timers for cell cycles. *Proc. II Annual Meeting Eur. Soc. Chronobiol.* (in press).

LLOYD, D. & EDWARDS, S. W. (1986b). Temperature-compensated ultradian rhythms in lower eukaryotes: periodic turnover coupled to a timer for cell division. *J. interdiscipl. Cycle Res.* **77**, 321–326.

LLOYD, D. & EDWARDS, S. W. (1987). Temperature-compensated ultradian rhythms in lower eukaryotes: timers for cell cycles and circadian events? In *Advances in Chronobiology* (ed. J. E. Pauly & L. E. Scheving). New York: Alan Liss. (in press).

LLOYD, D., EDWARDS, S. W. & FRY, J. C. (1982b). Temperature-compensated oscillations in respiration and cellular protein content in synchronous cultures of *Acanthamoeba castellanii*. *Proc. natn. Acad. Sci. U.S.A.* **79**, 3785–3788.

LLOYD, D., EDWARDS, S. W. & WILLIAMS, J. L. (1981). Oscillatory accumulation of total cell protein in synchronous cultures of *Candida utilis*. *FEMS Microbiol. Lett.* **12**, 295–298.

LLOYD, D., JOHN, L., EDWARDS, C. & CHAGLA, A. H. (1975). Synchronous cultures of microorganisms: Large-scale preparation by continuous-flow size selection. *J. gen. Microbiol.* **88**, 153–158.

LLOYD, D., PHILLIPS, C. A. & STATHAM, M. (1978). Oscillations of respiration, adenine nucleotide levels and heat evolution in synchronous cultures of *Tetrahymena pyriformis* ST prepared by continuous-flow selection. *J. gen. Microbiol.* **106**, 19–26.

LLOYD, D., POOLE, R. K. & EDWARDS, S. W. (1982a). The Cell Division Cycle: Temporal Organization and Control of Cellular Growth and Reproduction, 523pp. London: Academic Press.

VON LOH, D. & GOLENHOFEN, K. (1970). Temperature einfluesse auf die spontanaktivitat der isolierten *Taenia coli* vom meerschweinchen. *Pflügers Arch. ges.* **318**, 35–50.

MAJIMA, H. & GERWECK, L. E. (1983). Kinetics of thermotolerance decay in Chinese hamster ovary cells. Influence of the initial heat treatment and duration. *Cancer Res.* **43**, 2673–2677.

MALHOTRA, A., KRUUV, J. & LEPOCK, J. R. (1986). Sensitization of rat hepatocytes to hyperthermia by calcium. *J. Cell Physiol.* **128**, 279–284.

MARCUS, M., FAINSOD, A. & DIAMOND, G. (1985). The genetic analysis of mammalian cell-cycle mutants. *Ann. Rev. Genetics* **19**, 389–421.

MARQUES, N., EDWARDS, S. W., FRY, J. C., HALBERG, F. & LLOYD, D. (1987). Temperature-compensated ultradian variation in cellular protein content of *Acanthamoeba castellanii* revisited. In *Advances in Chronobiology* (ed. J. E. Pauly & L. E. Scheving). New York: Alan R. Liss. (in press).

MICHEL, U. & HARDELAND, R. (1985). On the chronobiology of *Tetrahymena*. III. Temperature compensation and temperature dependence in the ultradian oscillator of tyrosine amino transferase. *J. interdiscipl. Cycle Res.* **16**, 17–23.

NJUS, D., SULZMAN, F. M. & HASTINGS, J. W. (1974). Membrane model for the circadian clock. *Nature, Lond.* **248**, 116–120.

NURSE, P., THURIAUX, P. & NASMYTH, K. (1976). Genetic control of the cell division cycle in the fission yeast *Schizosaccharomyces pombe*. *Mol. Gen. Genet.* **146**, 167–178.

OVERGAARD, J. (ed). (1984). *Hyperthermic Oncology 1984*. London, Philadelphia: Taylor & Francis.

PAUL, J. (1986). Body temperature and the specific heat of water. *Nature, Lond.* **323**, 300.

PAVLIDIS, T. & KAUZMANN, W. (1969). Towards a quantitative biochemical model for circadian oscillators. *Archs Biochem. Biophys.* **132**, 338–348.

PETTIGREW, R. T., GALT, J. M., LUDGATE, C. M. & SMITH, A. N. (1974). Clinical effects of whole body hyperthermia in advanced malignancy. *Br. med. J.* **4**, 679–682.

PIRT, S. J. (1975). *Principles of Microbe and Cell Cultivation*. Oxford: Blackwell Scientific.

PITTENDRIGH, C. S. & BRUCE, V. G. (1957). *Rhythmic and Synthetic Processes of Growth*, p. 75. Princeton: Univ. Press.

POLANSHEK, M. (1977). Effects of heat shock and cycloheximide on growth and division of the fission yeast *Schizosaccharomyces pombe*. *J. Cell Sci.* **23**, 1–23.

RASMUSSEN, L. & ZEUTHEN, E. (1962). Cell division and protein synthesis in *Tetrahymena*, as studied with *p*-fluorophenylalanine. *C.r. Trav. Lab. Carlsberg* **32**, 333–358.

READ, R. A., FOX, M. H. & BEDFORD, J. S. (1983). The cell cycle dependence of thermotolerance. I. CHO cells heated to 42°C. *Radiat. Res.* **93**, 93–106.

READ, R. A., FOX, M. H. & BEDFORD, J. S. (1984). The cell cycle dependence of thermotolerance. II. CHO cells heated to 45°C. *Radiat. Res.* **98**, 491–505.

READEY, A. (1985). The photoentrainment of ultradian rhythms in 'Tetrahymena' occurs in permissive ranges defined by cell modal generation times. *Chronobiologia* **12**, 266.

RICE, G. C., GRAY, J. W., DEAN, P. N. & DEWEY, W. C. (1984a). Fluorescence-activated cell sorting analysis of the induction and expression of acute thermal tolerance within the cell cycle. *Cancer Res.* **44**, 2368–2376.

RICE, G. C., GRAY, J. W. & DEWEY, W. C. (1984b). Cell cycle progression and division of viable and non-viable Chinese hamster ovary cells following acute hyperthermia and their relationship to thermal tolerance delay. *Cancer Res.* **44**, 1802–1808.

ROBINSON, J. E., WIZENBURG, M. J. & McCREADY, W. A. (1974). Combined hyperthermia and radiation suggest an alternative to heavy particle therapy for reduced oxygen enhancement ratios. *Nature, Lond.* **251**, 521–522.

ROSBASH, M. & HALL, J. C. (1985). Biological clocks in *Drosophila*: finding the molecules that make them tick. *Cell* **43**, 3–4.

RUSCH, H. P., SACHSENMAIER, W., BEHRENS, K. & GRUTER, V. (1966). Synchronization of mitosis by the fusion of the plasmodia of *Physarum polycephalum*. *J. Cell Biol.* **31**, 204–209.

SACHSENMAIER, W., REMY, V. & PLATTNER-SCHOBEL, R. (1972). Initiation of synchronous mitosis in *Physarum polycephalum*. A model for the control of cell division in eukaryotes. *Expl Cell Res.* **73**, 41–48.

SCHNEIDERMAN, M. H., DEWEY, W. C. & HIGHFIELD, D. P. (1971). Inhibition of DNA synthesis in synchronized Chinese hamster cells treated in G_1 with cycloheximide. *Expl Cell Res.* **67**, 147–155.

SCHWEIGER, H. G. & SCHWEIGER, M. (1977). Circadian rhythms in unicellular organisms, an endeavour to explain the molecular mechanism. In *Int. Rev. Cytol.*, vol. 51 (ed. G. H. Bourne, J. F. Danielli & K. W. Jean), pp. 315–342. New York: Academic Press.

SELAWRY, O. S., GOLDSTEIN, M. N. & MCCORMICK, T. (1957). Hyperthermia in tissue cultured cells of malignant origin. *Cancer Res.* **17**, 785–791.

SEL'KOV, E. E. (1970). Two alternative, self-oscillating stationary states in thiol metabolism – two alternative types of cell division, normal and malignant ones. *Biophysika* **15**, 1065–1073.

STEVENSON, M. A., CALDERWOOD, S. K. & HAHN, G. M. (1986). Rapid increases in inositol trisphosphate and intracellular Ca^{2+} after heat shock. *Biochem. biophys. Res. Commun.* **137**, 826–833.

SUDBERY, P. E. & GRANT, W. D. (1976). The control of mitosis in *Physarum polycephalum. J. Cell Sci.* **22**, 59–65.

SWANN, M. M. (1957). The control of cell division: a review. I. General mechanisms. *Cancer Res.* **17**, 727–758.

SWEENEY, B. M. (1982). Interaction of the circadian cycle with the cell cycle in *Pyrocystis fusiformis. Plant Physiol.* **70**, 272–276.

SWEENEY, B. M. & HASTINGS, J. W. (1960). Effects of temperature upon diurnal rhythms. *Cold Spring Harbor Symp. quant. Biol.* **24**, 87–104.

THIEL, G., HARDELAND, R. & MICHEL, U. (1985). On the chronobiology of *Tetrahymena.* II. Further evidence for the persistence of ultradian rhythmicity in the absence of protein synthesis and growth. *J. interdiscipl. Cycle Res.* **16**, 11–16.

TOMASOVIC, S. P., HENLE, K. J. & DETHLEFSEN, L. A. (1979). Fractionation of combined heat and radiation in asynchronous CHO cells 2. The role of cell-cycle redistribution. *Radiat. Res.* **80**, 378–388.

TYSON, J. & SACHSENMAIER, W. (1978). Is nuclear division in *Physarum* controlled by a continuous limit cycle oscillator? *J. theor. Biol.* **73**, 723–728.

WIEGANT, F. A. C., TUYL., M. & LINNEMANS, W. A. M. (1985). Calmodulin inhibitors potentiate hypethermic cell killing. *Int. J. Hyperthermia* **1**, 157–169.

WOFFENDIN, C. & GRIFFITHS, A. J. (1985). Changes in respiratory activity and total protein during synchronous growth of *Dictyostelium discoideum, FEMS Microbiol. Lett.* **29**, 203–267.

ZEUTHEN, E. (1971). Synchrony in *Tetrahymena* by heat shocks spaced a normal cell generation apart. *Expl Cell Res.* **68**, 49–60.

ZEUTHEN, E. (1978). Induced reversal of order of cell division and DNA replication in Tetrahymena. *Expl Cell Res.* **116**, 39–46.

ZEUTHEN, E. & SCHERBAUM, O. H. (1954). Synchronous division in mass cultures of the ciliate protozoan *Tetrahymena pyriformis* as induced by temperature changes. *Colston Pap.* **7**, 141–155.

ZIELKE-TEMME, B. & HOPWOOD, L. (1982). Time-lapse cinematographic observations of heated G_1-phase Chinese hamster ovary cells. I. Division probabilities and generation times. *Radiat. Res.* **92**, 320–331.

Printed in Great Britain © *Society for Experimental Biology 1987*

CELLULAR HEAT INJURY: ARE MEMBRANES INVOLVED?

KEN BOWLER

Department of Zoology, University of Durham, Durham City, UK

Summary

Temperature has an all pervasive influence on cellular function. In consequence heat damage may occur at multiple sites. This is only one factor that makes the identification of the primary sites of heat perturbation difficult. It is also likely that cellular heat death is the consequence of a time-dependent accumulation of secondary damage that results from the primary lesion(s). It is important therefore to determine whether observed damage is a primary consequence of heating or a secondary 'knock-on' effect. The case will be considered for a role of cellular membranes as being a cellular site of the primary lesion in heat injury.

Introduction

It is well recognized that the sensitivity of animals to heat injury follows a hierarchical pattern, with the organism being more sensitive than the functioning of its organs and cells; macromolecules are found to be the least heat sensitive structures (Ushakov, 1964). It is for this reason that organism death is proposed to occur as a consequence of the functional loss of a homeostatic system (Lagerspetz, 1974; Kunnemann & Precht, 1979; Prosser & Nelson, 1981).

Heat injury is also experienced by bacteria, unicellular organisms, cells in culture and multicellular plants, at temperatures only just higher than the normal range. It is therefore pertinent to consider that the primary lesion in heat injury is the same in all cells. That is the same molecules, or molecular complexes, are perturbated by the heat stress, and this leads either directly to a loss in function at that site, or indirectly to a loss in function at a distant site. This is shown diagrammatically in Fig. 1, where it is suggested that heating perturbates function at a primary site(s) leading to an impairment of function at that site, which in turn affects the functioning at secondary sites causing their malfunction. This 'knock-on' effect of damage at one site causing malfunction at other sites leads to a cascade, and accumulation, of damage. At some point the accumulation of errors will cause an irreversible loss in function at some site(s), and this will result in cell death (Bowler, 1981).

Temperature has an all pervasive influence on cellular structures and this makes it difficult to identify the primary sites of lesion. Indeed at one time or another all the major cellular structures have been implicated as having a significant role in heat injury (see Roti Roti, 1982).

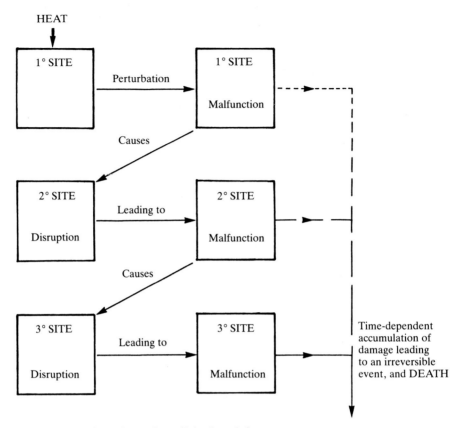

Fig. 1. A putative scheme for cellular heat injury.

The divers nature of possible cellular sites for the primary lesion is clearly a complicating factor in the study of heat injury. The failure to recognize that some observed damage may not result from the direct effect of heat (i.e. is secondary and tertiary damage) also compounds the issue. This may be particularly important if the nature of such damage is determined by the differentiated state of the cell. The competence of the cell to display such damage may depend upon its specialized biochemistry or structure. Heat damage to a muscle cell could appear quite different to that observed in a spermatocyte, even if the primary sites of the heat lesion were common.

Sound evidence exists to justify a distinction being made between primary damage and subsequent (secondary) damage to cells. Rofstad, Falkvoll & Oftedal (1984) report in human melanoma zenografts that the formation of abnormal mitotic figures and micronuclei reached their maxima in 1 or 2–3 days respectively after heating. Waters & Roti Roti (1982) also report that whilst heating caused a rapid suppression of DNA synthesis no strand breaks occurred in cells not replicating their DNA. A number of workers have described heat-induced strand breaks in DNA but this occurs only during incubation after the heating period (Waters & Henle, 1982; Wong & Dewey, 1982) and so may be a result of cell necrosis. Zielke-Temme & Hopwood (1982), have also proposed that the heat-induced death of CHO cells can best be explained as a direct effect (transitory) on membranes that causes secondary damage that is lethal.

An interesting mathematical model for cellular heat injury has recently been proposed by Jung (1986). From an analysis of survival curves he has derived a model which assumes cell killing to be a two-step process. First, heating causes non-lethal damage which is converted during continuation of the heat stress, into lethal damage. It is predicted that the production and conversion of the lethal lesions are random events and depend only on temperature. Cell death follows the conversion of one non-lethal into a lethal lesion.

In many respects the descriptive scheme shown in Fig. 1 resembles the model proposed by Jung (1986). It differs in that the primary lesion, which need not be or become irreversible, causes a cascade of secondary 'knock-on' effects that become irreversible with time. However, as Jung (1986) points out, the shoulder of survival curves may simply be the time-dependency of the conversion of non-lethal into lethal events.

Any theory proposed for cellular heat injury must also take into account the existing experimental evidence and so satisfy the following points.

  1. The shape of cell-killing curves.
  2. The action of agents that sensitize cells to heat (alcohols, local anaesthetics).
  3. The action of agents that protect cells from heat (polyols, D_2O).
  4. The sensitizing effect of anisotropic media.
  5. pH effects.
  6. Thermotolerance.
  7. 'Dietary' manipulation of thermal sensitivity.
  8. Thermal acclimation.
  9. Dependence of heat sensitivity on the cell cycle.

Cellular membranes – the site of the primary lesion?

Morphological studies

Increasing attention is focused on membranes as being a site for hyperthermic damage to cells. Some morphological evidence exists that membrane structure is disrupted by exposure to high lethal temperatures. Davison (1971) has shown that blowfly flight muscle mitochondria lose cristae and accumulate electron-dense particles following *in vivo* heat exposure. Fajardo, Egbert, Marmor & Hahn (1980) have also shown heat destruction of the plasma membrane and mitochondrial cristae in a tumour cell line. A discontinuous plasma membrane has also been observed by Schrek, Chandra, Molnar & Stefani (1980) in heated lymphocytes. A common observation in heat damaged cells is the loss of microvilli (Mulcahy, Gould, Hidvergi, Elson & Yatvin, 1981) and the formation of blebs (Willis, Findlay & McManus, 1976; Lionetti, Lin, Mattaliano, Hunt & Valeri, 1980; Mulcahy *et al.* 1981; Bass, Coakley, Moore & Tilley, 1982). Laudien & Bowler (unpublished) have found lifting of the plasma membrane from the underlying cytoplasm (see Fig. 2) a common form of damage in heated FHM-cells. Much of this evidence results from studies using E.M. techniques with the attendant possibility of artefact introduction. The report of Bass *et al.* (1982) however was carried out observing living CHO cells, which formed blebs during the heating procedure. What causes blebbing is not clear, but Coakley (1987), in this volume, argues that blebs may result from the breakdown of the cytoskeletal anchoring mechanism. Whether this is the result of the thermal inactivation of a membrane protein awaits further work.

Why might membranes be sensitive to temperature?

The currently held model of biological membranes is a modification of the fluid-mosaic model proposed by Singer & Nicholson (1972), with lipid molecules forming a fluid matrix in which proteins are dispersed. In this volume Lee & Chapman (1987) and Cossins (1987) both emphasize that a change in temperature will perturbate both the lipid and protein moieties that make up membranes.

Lee & Chapman (1987) reported that a change in temperature has two effects on membrane lipids. First, by gradually increasing the temperature the molecular motion of the lipids present will increase, this is simply a consequence of the increase in kinetic energy. The result will be a progressive increase in 'fluidity', in other words a decrease in the order of the lipid molecules. The second effect that a change in temperature might cause is a change in phase in the lipids from gel to liquid-crystalline. Below this transition temperature the fatty acyl chains of the phospholipids are packed in an ordered form, and 'melting' occurs because of the thermally-induced flexing of the acyl chains. Above the transition temperature the bulk of the lipids are liquid-crystalline and this state is considered to be essential for the function, and lateral mobility of integral membrane proteins (see Stubbs, 1983).

What is quite evident is that microorganisms (Sinensky, 1974) and poikilothermic animals (Hazel & Prosser, 1974; Cossins & Sinensky, 1984) respond to a change in environmental temperature by altering the degree of saturation of their membrane lipids. Cold acclimation results in an increase in unsaturation whereas warm acclimation results in more saturated membrane lipid fraction (Cossins, Friedlander & Prosser, 1977). This has been interpreted as an adaptive response to maintain 'fluidity' optimal for function. Sinensky (1974) has introduced the term 'homeoviscous adaptation' to describe this homeostatic mechanism. This idea, that organisms change the lipid composition of their cell membranes to compensate for direct effects of temperature on membrane physical properties, was an important step in appreciating that temperature has a powerful modulating influence on membrane structure.

Fig. 2. Electron micrographs of FHM-cells showing heat-induced morphological damage. A. Control cell. B. Cell given a 5 min heating at 41·5°C. Note the lifting of the plasma membrane from the underlying cytoplasm, the more numerous vesicles, and very densely staining mitochondria in the heated cell. (Prof. Dr H. Laudien, unpublished data.)

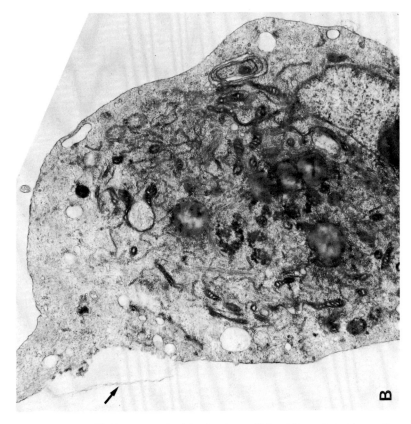

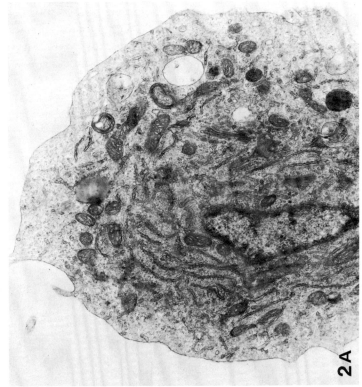

In membranes above their transition temperature it has been difficult to establish whether the *level* of membrane 'fluidity' is a factor in determining membrane function. As stressed by Cossins (1987) and Lee & Chapman (1987) in this volume, this arises not only because, in natural membranes the transition from gel to liquid-crystalline is not sharp, but also from the realization that the membrane is not physically homogeneous, but is composed of microdomains of differing physical properties (Stubbs, 1983).

However, implicit in the concept of homeoviscous adaptation is that for normal function, a cell requires its membranes to possess a specific level of 'fluidity'. Coupled to this is the notion that the functioning of membrane proteins might be regulated by, and be dependent on, membrane 'fluidity', (Cossins & Sinensky, 1984). If lipids are involved in cellular heat killing then clear evidence argues against this being as a consequence of a gel to liquid-crystalline phase change of the bulk membrane lipid (Lepock, Cheng, Al-Qysi & Kruuv, 1983), although such phase transitions do correlate with hypothermic cell death, impaired growth and cell division at low temperatures, in mammalian cells (Lepock *et al.* 1983; Kruuv, Glofcheski, Cheng, Campbell, Al-Qysi, Noland & Lepock, 1983), and in microorganisms (McElhany, 1985). However, it can not be excluded that a small, critical, fraction of membrane lipids are involved in gel to liquid-crystalline transitions at hyperthermic temperatures.

Hyperfluidity

An interesting extension of the 'fluidity' hypothesis suggests that hyperthermic temperatures have a deleterious hyperfluidizing effect on membrane lipids (Overath, Schairer & Stoffel, 1970; Yatvin, 1977). Dennis & Yatvin (1981) have produced a good correlation between membrane microviscosity (reciprocal of 'fluidity') and sensitivity to hyperthermia in an unsaturated fatty acid requiring mutant of *Escherichia coli*. The relationship in eukaryotic cells in much less clear. A good deal of indirect evidence, summarized in later sections, indicates that membrane 'fluidity' is significant in hyperthermic killing of mammalian cells, however direct evidence in missing. Lepock, Massicote-Nolan, Rule & Kruuv (1981) failed to show an increased heat sensitivity in V-79 cells with membranes fluidized by butylated hydroxytoluene. Yatvin, Vorphal, Gould & Lyte (1983) also found that increasing the microviscosity of ascites P388 and V79 cell membranes using cholesteryl hemisuccinate did not influence the response of the cells to heating.

The term 'fluidity', as commonly used in this field, is a poorly defined term (Lepock, 1982; Stubbs, 1983; Cossins, 1987) and includes a number of parameters of lipid motion as well as lipid order. Different probes measure these different parameters of fluidity, and this may well have contributed to the uncertainty (Lepock, 1982). Furthermore, Yatvin, Cree, Elson, Gipp, Tegmo & Vorpahl (1982) have questioned whether changes in 'fluidity' caused by chemical agents, such as butylated hydroxytoluene, is equivalent to that induced by diet or growth

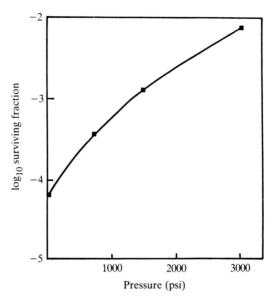

Fig. 3. The survival of HA-1 fibroblasts exposed to 44 °C for 90 min at pressures ranging up to 3000 p.s.i. Redrawn from Minton *et al.* (1980).

temperature manipulation. However in an interesting and unique study Minton, Stevenson, Kendig & Hahn (1980) report that high pressure inhibits hyperthermic cell killing, see Fig. 3. This can best be explained by a pressure reversal of the high temperature increase in membrane fluidity, but the authors did not discount the possibility of other sites being involved. Thus the importance of an increase in fluidity, *per se*, as a factor in hyperthermic cell killing is still an open question.

Membrane lipid composition and heat sensitivity

As previously discussed microorganisms (Sinensky, 1974) and poikilothermic animals cells (Hazel & Prosser, 1974) change their membrane lipid composition in response to a change in rearing temperature. In these organisms the principal changes occur to the fatty acids incorporated into the membrane phospholipid fraction. In mammalian cell membranes cholesterol has an important role (Lee & Chapman, 1987; in this volume). In brief cholesterol reduces the lipid order when added below the phase transition (gel phase) but increases the order when added to lipids in the liquid-crystalline phase (Stubbs, 1983). In consequence it broadens the phase change temperature range and reduces membrane 'fluidity' in cells at their growth temperatures. The closer packing of the lipid molecules reduces membrane permeability (Demel, van Kessel & van Deenen, 1972) and has been shown to modify enzyme activity (Madden, Chapman & Quinn, 1974). Sabine (1983) has recently proposed that mammalian cells have an optimal level of membrane cholesterol for function, at levels above and below that, membrane function will be impaired. Once again implicit in this is that some optimal level of membrane 'fluidity' exists for normal cell function. This argues that heating might move the cell to a non-optimal state, by 'fluidizing' the membrane lipids.

The reports by Cress & Gerner (1980) and Cress, Culver, Moon & Gerner (1982) that a good correlation exists between cell cholesterol content and cell thermal sensitivity is consistent with the reported stabilizing role for cholesterol in membranes. Using seven cell lines they found that, cells with a high cholesterol and phospholipid relative to protein content were more resistant to exposure at 43°C, (Fig. 4), but they were unable to show any dependency of heat sensitivity in fatty acid saturation of membrane lipids. Recently, Raaphorst, Vadasz, Azzam, Sargent, Borsa & Einspenner (1985) made a similar study using seven different cell lines from transformed C3H 10T1/2 cells. They describe that most of the transformed lines had comparable thermal sensitivities to the normal line, one was more heat sensitive and one more heat resistant. In no case did this detailed study find any correlation between heat sensitivity and membrane lipid composition. The same conclusion was reached by Konings & Ruifrok (1985), using 3 cell lines, who failed to find any correlation between cholesterol and phospholipid content and heat sensitivity of the cells. Anderson, Lunec & Cresswell (1985) reported that several indices of membrane composition of nine different cell lines correlated with a sensitivity of the cells to 44°C. They described positive correlations with cholesterol:cell; cholesterol:protein; protein:cell; phospholipid:cell and phospholipid:protein and heat sensitivity, but cholesterol:phospholipid molar ratios

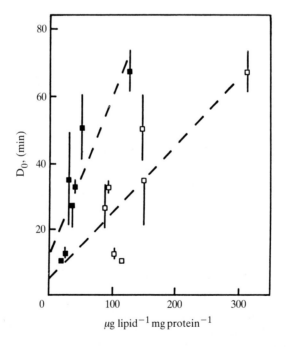

Fig. 4. The relationship between heat-sensitivity (D_o) as a function of either cholesterol (■) or phospholipid (□) content mg protein^{-1} for several different cell lines. Redrawn from Cress *et al.* (1982).

did not correlate. These workers considered their data was heavily biased by the inclusion of 3 very sensitive small celled lymphoid lines in the analysis. Evaluation of the data from the other six lines alone showed that the correlations between cell composition and heat sensitivity was weakened and only that of cholesterol and protein content per 10^6 cells were positive. They conclude that the levels of cholesterol and phospholipid cannot be used as reliable indicators of heat sensitivity.

A number of workers have found a positive correlation between growth temperature and cell heat resistance, and this adaptation response is marked in fish FHM cells (Bowler, Laudien & Laudien, 1981). Anderson, Minton, Li & Hahn (1981) have also shown that mammalian cells show this adaptation response to growth temperature. Cells grown at 32 °C were sensitized to 43 °C as compared with cells grown at 37 °C. Furthermore, in plateau phase cells a good correlation was obtained between membrane fluidity and culture temperature, with cells grown at lower temperatures displaying a higher membrane 'fluidity'. This was also found to correlate with changes in membrane cholesterol content. These workers suggest the changes in membrane lipid composition, in response to growth temperature, is an example of homeoviscous adaptation. Bates, LeGrimel-lec, Bates, Loufti & Mackillop (1985) confirmed the adaptive response to growth temperature in CHO cells. Cells grown to confluence at 37 °C and placed for 1 day at 40 °C were found to have lower membrane 'fluidity' across a range of temperatures as compared with cells maintained at 37 °C throughout. Once again this change in a membrane physical property correlated with an increased thermal resistance of the CHO cells at 43 °C or 45 °C. In a similar study, using rat embryo fibroblasts, Culver & Gerner (1982) found that the time course of the change in resistance to 43 °C of rat cells was dependent upon the direction of the shift in their culture temperature. In common with experiments from poikilothermic animals, they found the acclimation from 35° → 39 °C was completed in 8 h whereas acclimation from 39° → 35 °C to about 1 day. The time-course of these changes are shown in Fig. 5. Culver & Gerner also measured the time-course of the change in membrane cholesterol, phospholipid and protein during the temperature shift. The changes they found in these constituents were complex and did not correlate well with the change in cellular thermal sensitivity. Indeed only the generally lower levels of cholesterol in 35 °C cells as compared with 39 °C cells correlated with the change in sensitivity, see Fig. 5.

Gonzalez-Mendez, Minton & Hahn, (1982) have also reported that Chinese hamster ovary fibroblasts grown at 32°, 37°, 39° or 41 °C show a progressive increase in thermal resistance to exposure at 43 °C. These workers also found that growth at higher temperatures (39° & 41 °C) caused an increase in membrane cholesterol, and in membrane order, that correlated with the increase in thermal resistance. The conclusion drawn from these studies is that mammalian cells can make an adaptive response in cellular heat resistance to a temperature shift. This shift in heat resistance also correlates with changes in membrane cholesterol and

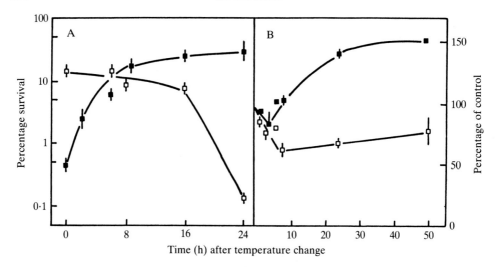

Fig. 5. Effect of growth temperature on survival and membrane composition of Rat-1 cells. (A) Time course of change in sensitivity to 43°C after a change in acclimation temperature 35°→39°C (■) or 39°→35°C (□). (B) Effect of a temperature change 35°→39°C (■) or 39°→35°C (□), on the cholesterol content ($\mu g/10^6$ cells) of Rat-1 cells. Modified after Culver & Gerner (1982).

'fluidity'. It is usually assumed in these studies that during growth at say 32° or 41°C there is no selection of resistant or sensitive lines of cells.

Yatvin and coworkers have shown that feeding mice, bearing ascites P388 tumour cells, on diets modified by incorporating saturated or unsaturated fatty acids, altered the *in vitro* sensitivity of the tumour cells to heat. The survival of mice receiving heated cells was longer when they received ascites cells from the diet enriched with unsaturated fatty acids (Hidvegi, Yatvin, Dennis & Hidvegi, 1980). In a subsequent morphological study it was reported that the pattern of heat damage was greater in the P388 cells from animals receiving the unsaturated fatty acid diet (Mulcahy *et al.* 1981). These studies have been repeated using solid, transplantable tumours (Yatvin, Abuirmeileh, Vorphal & Elson, 1983; Elegbede, Elson, Qureshi, Dennis & Yatvin, 1986) with substantially the same conclusion. Guffy, Rosenberger, Simon & Burns (1982) have demonstrated a similar shift in thermal sensitivity using dietary modification. In particular, a diet supplemented with the fatty acid 22:6 (docosahexaenoic acid) caused a significant increase in membrane 'fluidity' and in cell thermal sensitivity. This has been confirmed by Konings (1985) and Konings & Ruifrok (1985) using mouse fibroblast LM cells. The polyunsaturated fatty acid content of these cells was increased from 7 to 40 % by the addition of arachadonic acid (20:4) to the culture media. This caused an increase in membrane 'fluidity' and parallel increase in cell thermal sensitivity.

The important conclusion to be drawn from these various studies is that the modification of membrane lipid composition, and physical properties ('fluidity') by diet, or thermal adaptation, modifies cellular thermal sensitivity. There may well also be a relationship between membrane cholesterol content and thermal sensitivity in mammalian cells, but it may not be one of simple proportion. Indeed when the detailed responses are evaluated these various studies appear in conflict. In some, changes in membrane lipid saturation occur with no change in membrane cholesterol content, whilst others report no changes in membrane lipid saturation but significant modulation in cholesterol. In many of the studies the membrane fraction used has been poorly defined and characterized, and this may be the critical factor in evaluating such comparative work using different cell lines and diverse methodology. Those studies that have used whole cell extraction procedures for protein, cholesterol and phospholipid are particularly difficult to interpret in terms of membrane composition.

Anaesthetics, alcohols and heat sensitivity

The realization that membrane 'fluidity' might be a factor in hyperthermic cell killing has led to the use of membrane active agents such as alcohols and local anaesthetics in the potentiation of hyperthermic damage, see Yatvin *et al.* (1987). These molecules are thought to partition into the lipid matrix, disrupting lipid order (Paterson, Butler, Huana, Labelle, Smith & Schneider, 1972; Seeman, 1972) and so should act synergystically with heat in 'fluidizing' membranes.

Yatvin, Clifton & Dennis, (1979) demonstrated clearly that the simultaneous application of a local anaesthetic (lidocaine) and heat potentiated survival in tumour bearing mice. This result, with local anaesthetics, has been repeated for other tumour models (Hidvegi *et al.* 1980; Robbins, Dennis, Slattery, Lange & Yatvin, 1983; Barker, 1986) and these data also indicate a potentiation between heat and the anaesthetic in enhancing the host survival. Work using cells in culture also clearly shows a potentiation between heat and anaesthetic action in cell killing (Yatvin, Gipp, Rusy & Dennis, 1982). Yau (1979) also emphasized the membrane action of procaine, for he described that procaine caused blebbing in murine lymphoma cells, even at normal culture temperatures. Blebbing of the plasma membrane is indicative of damage and so it is not surprising that these cells were sensitized to heat exposure by procaine. More recently Konings (1985) has also shown that mouse fibroblast LM cells are sensitized to heat by procaine.

Li, Shui & Hahn (1980) have provided evidence for a similarity between ethanol and heat in cell killing. They concluded the same 'critical' targets are affected, but what is not clear is whether primary or secondary sites, if damaged, are implicated. Alcohols also have a potentiating effect on hyperthermic cell killing (Li & Hahn, 1978; Schrek & Stefani, 1981). Roti Roti (1982) discussed the correlation between heat-killing and the accumulation of proteins in the nucleus. Roti Roti & Wilson (1984) have recently shown that exposure of cells to alcohols (C_2 to C_5) and procaine without heating also causes an increase in nuclear protein in Hela Cells. They also confirm that these agents potentiate heat-induced cell killing, and with heat caused a larger increase in the nuclear accumulation of proteins. These workers concluded that the protein accumulation in the nucleus is a secondary event as procaine and alcohols are exerting their harmful effects at the membrane.

Implicit in this proposal is that heat must also be acting at the membrane. A similar conclusion was reached by Kampinga, Jorritsma & Konings (1985) to explain the sensitizing effect of procaine, on heat inactivation of nuclear DNA polymerase. Enzyme inactivation occurred when whole cells, but not isolated nuclei, were heated, thus the primary lesion occurred in cells at a site distant from the nucleus.

The problems of interpreting the effects of alcohols, and local anaesthetics, in potentiating heat sensitivity of cells is the lack of understanding of the site(s) and mode of action of these agents. There is much evidence to show that these molecules do partition into the membrane, particularly the liquid-crystalline phase (see Shinitzky, 1984). Ample evidence discussed elsewhere in this paper also demonstrates that these agents do increase membrane 'fluidity' (see also Yatvin et al. 1987). It is therefore tempting to speculate that their potentiating action with heat is an additive effect on 'fluidity'. There is however the possibility that these agents (alcohols and anaesthetics) are both exerting their effect on membrane proteins (McElhaney, 1985). Chan & Wang (1984) have provided evidence of an interaction between local anaesthetics and membrane proteins. Massicotte-Nolan, Glofcheski, Kruuv & Lepock (1981) have questioned the assumption that the enhancement of heat sensitivity by alcohols is a lipid mediated effect. They found the 'fluidizing' effect of a series of monohydric alcohols not to correlate well with their sensitization of cells to heat. They considered their evidence supported, more strongly, an effect on membrane proteins. The mechanism of potentiation of heat damage by these agents remains an open question, and clearly an action on both membrane order and membrane proteins should be considered likely.

Evidence from membrane permeability

If the damaging effect of heat on cells is by hyperfluidizing membrane lipids then it might be expected that hyperthermia would cause a breakdown in permeability. Animal cells contain a high $[K^+]_i$ and low $[Na^+]_i$ and $[Ca^{2+}]_i$ as compared with the extracellular fluids. Any impairment of the control of permeability at hyperthermic temperatures would allow the leak of K^+ from and Na^+ and Ca^{2+} into the cells. Evidence that this occurs is equivocal. Boonstra, Schamhart, de Laat & van Wijk (1984) have described an increased passive K^+ efflux at 42°C in hepatoma cells, but as this is accompanied by an increased active K^+ uptake the intracellular levels of K^+ remained unchanged. This agrees with the work of Vidair & Dewey (1986) who found that a heat dose that caused a 98% reproductive death of CHO cells had no effect on intracellular levels of Na^+, K^+ or Mg^{2+}. Yi, Chang, Tallen, Bayer & Ball (1983) and Ruifrock, Kanon & Konings (1985) however did find that heating lowered intracellular $[K^+]$. The latters workers report on interesting correlation between LM mouse fibroblast survival and $[K^+]_i$ at 16h following a dose at 44°C. As is shown in Fig. 6, this correlation held even when cells were heated in the presence of a sensitizer (procaine) or protector (erythritol) or even in thermotolerant cells.

The differences between these results are difficult to reconcile and are unlikely to result from differences in cell type used. Vidair & Dewey (1986) suggest that uncontrolled exchanges of ions might only occur in metabolically 'dead' cells, for their cells, whilst reproductively dead, were attached, and so metabolically functional.

The sensitizing action of anisotonic media might also imply membrane permeability is disrupted at hyperthermic temperatures (Raaphorst & Dewey, 1978; Henle & Dethlefson, 1979). Hypertonic media sensitize cells to hyperthermia more than do hypotonic solutions. It is not possible to implicate the plasma membrane as the only locus of this sensitizing effect. At normothermic temperatures hypotonic solutions initially cause cells to swell, they then shrink and K^+ are lost (Roti Roti & Rothstein, 1973). The cell may be able to establish a new osmotic equilibrium in a hypotonic medium even at hyperthermic temperatures through a loss of intracellular ions. In hypertonic media the cell loses water, shrinks in volume and remains under osmotic stress. A combination of hyperthermic and osmotic stress could be seen to produce sensitization. It is therefore pertinent that Yi *et al.* (1983) and Boonstra *et al.* (1984) both reported an increase in cell volume following hyperthermia alone. This would cause a fall in $[K^+]_i$ and Boonstra, Sybesma & van Wijk (1985) reported that heating cells in a high $[K^+]$ medium, elevated intracellular $[K^+]$ and this protected DNA synthesis, but not protein synthesis, from inhibition.

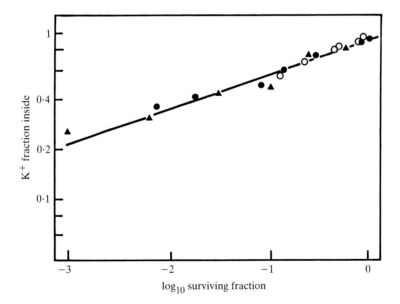

Fig. 6. Relation of the survival of mouse fibroblast LM cells and their residual potassium content measured 16h after hyperthermic treatment for various times up to 40 min at 44°C. The level of K^+ is related to the cells actually present at 16h. (●) Control cells; (▲) cells plus 2 mM procaine; (○) thermotolerant cells. Modified after Ruifrok *et al.* (1985).

Schanne, Kane, Young & Faber (1979) and Farber (1981) have proposed that changes in cell Ca^{2+} have a role in death caused by a number of toxic agencies. It was proposed that the toxic insult resulted in an increased Ca^{2+} flux, and a rise in $[Ca^{2+}]_i$ which converted potentially reversible to irreversible injury. However hyperthermia was not one of the harmful agents considered.

Intracellular calcium plays a key role in the control of many aspects of cellular activity. Subtle changes in free Ca^{2+} act as an intracellular messenger to activate a number of key enzymes (see Campbell, 1983). If heat caused an unregulated rise in $[Ca^{2+}]_i$ it is clear an uncontrolled train of cellular activities would be affected leading to a loss of cellular homeostasis. This would cause a cascade of damage as predicted in Fig. 1.

There are a number of reports that cells accumulate Ca^{2+} during hyperthermia (Anghileri, Escanye & Robert, 1985; Anghileri, Marcha, Escanye & Robert, 1985; Wiegant, Tuyl & Linnemans, 1985; Stevenson, Calderwood & Hahn, 1986) however whether the rise in cell Ca^{2+} is involved in cell killing is in dispute. Anghileri et al. (1985a,b) report an increased $^{45}Ca^{2+}$ influx and a parallel decrease in Ca^{2+} ATPase activity during hyperthermia. The Ca^{2+} intracellular homeostatic mechanisms would overload in this circumstance and Ca^{2+} induced toxicity would result (Farber, 1981). Vidair & Dewey (1986) found that CHO cells, immediately after heating at 45°C, showed no change in cellular $^{45}Ca^{2+}$. However during subsequent incubation at 37°C heated cells took up $^{45}Ca^{2+}$ in a heat dose-dependent manner. This net accumulation occurred because the active extrusion of intracellular Ca^{2+} was inhibited more strongly than was the passive influx. This is in partial agreement with Anghileri et al. (1985b). Vidair & Dewey (1986) further demonstrated that elevating intracellular $[Ca^{2+}]$ neither sensitized nor protected cells from heat, nor did the post-heat incubation at 37°C in different Ca^{2+} containing media influence cell survival. They concluded therefore that, although changes occur in cell calcium, as a consequence of heating, this played no role in hyperthermic cell death. Stevenson et al. (1986) found $^{45}Ca^{2+}$ influx increased markedly in CHO cells following heat shock at 45°C for 10 min, whereas $^{45}Ca^{2+}$ efflux was unaffected. Heat therefore caused a net inward movement of Ca^{2+}. The same authors also described an elevation of free Ca^{2+} during heat shock, and suggested this came mainly from intracellular storage sites. The rise in $[Ca^{2+}]_i$ caused by heat shock may be involved in the development of thermotolerance. However, it is not clear whether calcium is involved in cellular heat damage for discrepancies exist between these various studies, and in their interpretation, that are difficult to reconcile. It will not be resolved until measurements are made of changes in free cell calcium, rather than total cell calcium (which is largely in storage compartments), both during and immediately following a period of hyperthermia. Wiegant et al. (1985) in an interesting study, have shown that an important facet of the Ca^{2+} signalling system is impaired by heating. In both neuroblastoma and hepatoma cells calmodulin inhibiting drugs were found to enhance hyperthermic cell killing in a dose-dependent fashion. The inhibitors used prevent the formation of the activated Ca-calmodulin complex and so inhibit Ca^{2+}

mediated cellular activities requiring Ca-calmodulin for activation. The most likely role for Ca-calmodulin in cell survival during hyperthermia is the activation of the Ca^{2+}ATPase, responsible for active Ca^{2+} extrusion, and calcium sequestering mechanisms e.g. in mitochondria. This suggests that hyperthermia causes a rise in free cell Ca^{2+} and in absence of Ca-calmodulin, Ca^{2+} homeostasis fails and the cells die. An interesting corollary is that these workers found in the presence of calmodulin inhibitors, stress fibres did not breakdown in heated cells. This implies that the loss of stress fibres that normally occurs in heated cells (Coss, Dewey & Bamburg, 1982; Glass, de Witt & Cress, 1985) is not directly involved in cell killing by heat. It may however account for the formation of multinucleate cells found in the mammalian spermatogenic epithelium after periods of hyperthermia (Bowler, 1972).

Recent work by Malhotra, Kruuv & Lepock (1986) shows that the survival of hepatocytes at hyperthermic temperatures is modulated by extracellular calcium concentrations. The cells were sensitized to hyperthermia by both high calcium (10 mM) and the absence of extracellular calcium. In sensitizing conditions calcium influx precedes the loss of cell viability, in particular the ionophore A23187 is a potent sensitizer when extracellular Ca^{2+} is present. These data, when considered with those of Weigant *et al.* (1985), raise the question whether the Ca^{2+}ATPase is inactivated in hyperthermia.

Inactivation of membrane proteins and protein mediated function

A large number of observations implicate an alteration in the structure or function of a membrane protein in hyperthermic cell death. The plasma membrane mediated transport of Ca^{2+} *via* the Ca^{2+}ATPase and the co-transport of Na^+ and K^+ *via* the Na^+K^+ATPase are well characterized, and so it is not surprising these enzymes, and their associated ion fluxes, have been used to probe the protein perturbating effects of high temperature.

Burdon & Cutmore (1982) reported that the Na^+K^+ATPase activity was rapidly inactivated at 45°C, however this activity was partially restored by subsequent incubation at 37°C. Boonstra *et al.* (1984) found that incubation of hepatoma H35 cells at 45°C increased the active $^{86}Rb^+$ flux through the Na^+K^+ATPase which returned to normal levels on return to 37°C. Stevenson, Galey & Tobey (1983) also reported an increased K^+ transport in CHO cells at 43°C associated with the Na^+K^+ATPase.

In several recent papers rather more detailed investigations have been made of Na pump function at hyperthermic temperatures. Bates & Mackillop (1985) showed that ouabain-sensitive $^{86}Rb^+$ influx increased with temperature between 31° and 45°C, and only above 45°C did an irreversible inhibition of $^{86}Rb^+$ influx occur. However as temperatures rose above 37°C the affinity of Rb^+ for its transport site (K^+ site) progressively fell. Once again this decrease in affinity was reversible as long as the temperature did not rise above 45°C. However, the use of sigmoidal kinetics to analyse the affinity of Rb^+ would have been preferable to the

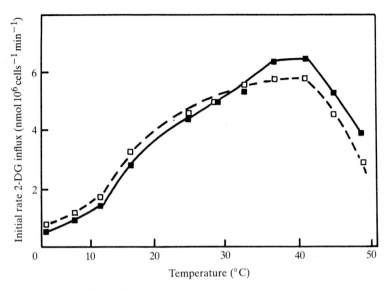

Fig. 7. Temperature dependence of permeability of control and thermally adapted CHO cells. Facilitated uptake of 5 mM 2-[³H]-deoxyglucose was measured for 2 min in control cells (—■—) and in cells thermally adapted at 40°C for 24 h (—□—). Redrawn from LeCavalier & Mackillop (1985).

method used (Robinson & Flashner, 1979). Anderson & Hahn (1985), in a more complete study, found that heating CHO cells at 45 °C inhibited ouabain-binding and $^{86}Rb^+$ uptake to the same extent, but ATPase activity was more resistant.

Kwock and coworkers have shown that Na^+-dependent amino acid transport is reversibly inactivated in a number of cell lines by hyperthermic treatment to the cells (Lin, Kwock, Hefter & Wallach, 1978; Kwock, Lin, Hefter & Wallach, 1978; Kwock, Lin & Hefter, 1985). This entry mechanism is believed to be a membrane protein mediated process, and its impairment would be expected to reduce the uptake of amino acids, which would contribute to the observed effect of heating on cell growth. Kwock et al. (1985) found that there was a 30 % increase in DTNB-sulphydryl group titration of surface proteins and this was associated with the reduction in Na^+-dependent amino acid uptake. This suggests that hyperthermic treatment caused, or permitted, a rearrangement to occur in membrane protein conformation.

LeCavalier & Mackillop (1985) showed that hyperthermic treatment of CHO cells impaired the facilitated entry of glucose. Whilst thermal adaptation at 40 °C increased survival at 43° and 45 °C no differences in the thermostability of the transport of glucose was observed (see Fig. 7). These workers also found that heating had little effect on the V_{max} or app.Km of this system. Once again, although it is possible to show hyperthermic impairment of a membrane protein function, it is not possible to relate this directly to cell killing.

Exposure of cells to hyperthermic temperatures altered the binding of growth factors (EGF). The affinity of the receptor for EGF was reduced after heating rat fibroblasts, but internalization was not found to be affected (Magun, 1981). The insulin receptor of Ha-1 cells is markedly more sensitive to 45°C than is the ATPase, however in contrast to the EGF data, the binding affinity is not affected, but the number of competent receptor proteins was reduced (Calderwood & Hahn, 1983). The concanavalin A binding glycoproteins of Ha-1 cells are heat resistant, even at 50°C (Stevenson, Calderwood & Hahn, 1983), a point that emphasizes that different proteins in the same cell membrane can be differentially sensitive to heat.

Davies, Rofstad & Lindmo (1985) have described that heating human FME cells caused qualitatively similar changes in three distinct surface antigens. Antigen expression was reduced after heating, and declined further to reach a minimum after about 1 day. Antigen expression however had recovered 1 week after heating. Davies *et al.* (1985) found the extent of loss of antigen expression was dose-dependent, with respect both to temperature and time, and indeed correlated with cell survival. The surface charge of mastocytoma cells was altered after heating at 42°C. Cell mobility in electrophoresis was decreased and this correlated with cell survival, Sato, Nakayama, Kojima, Nishimoto & Nakamura (1981) considered that heating altered the structure of membrane associated charge bearing groups. Lepock *et al.* (1983) has also shown that hyperthermic death of V79 cells correlated with transition in membrane proteins but not in membrane lipids. They found, using both intrinsic protein fluorescence, and a protein flurochrome energy transfer to paranaric acid, to probe transitions in structure, that both mitochondrial and plasma membrane proteins underwent irreversible transitions above 40°C. The strength of the signals received also implied a large fraction of the total protein was involved.

Polyols and simple sugars such as glycerol, erythritol, glucose and mannose, are known to protect cells from heat damage (Henle, Peck, Higashikubo, 1983; Henle, Monson, Moss & Nagle, 1984), however the mechanism of this protection is not understood. Henle *et al.* (1984) showed that this is not an osmotic effect simply resulting from entry of the molecules into the cell. Evidence was discussed suggesting that these protective molecules can act outside the cell, implying an action at the membrane, as well as on penetration. This protection maybe mediated through increased hydrophobic interactions (Back, Oakenfull & Smith, 1979) and is likely to be exerted on globular proteins.

A large number of studies also show D_2O protected cells from hyperthermic temperatures (McIver & Shrindhar, 1979; Li *et al.* 1980; Raaphorst & Azzam, 1982). The action of D_2O is so complex and widespread it is not possible to pinpoint the mechanism nor site of its protective action, but amongst many diverse effects D_2O caused structural changes in proteins, increased their thermostability, and modified the activity of enzymes (see Raaphorst & Azzam, 1982 for discussion).

The sensitization of cells to heat by polyamines reported by Fuller & Gerner (1982 & 1983) is also difficult to account for. These molecules occur naturally in cells at millimolar concentrations but, at much lower concentrations, outside the cell, cause sensitization to heat. Fuller & Gerner (1982) suggest they bind to membrane sites, probably proteins, but how this produces sensitization is not clear.

From a variety of *in vitro* thermal inactivation studies, Bowler, Duncan, Gladwell & Davison (1973) suggested cellular heat injury resulted from the thermal sensitivity of membrane lipoproteins. *In vivo* heating of blowflies to LD_{50} caused an impairment in mitochondrial function (Davison & Bowler, 1971; Bowler & Kashmeery, 1981), and three crayfish muscle membrane enzymes were found to be thermolabile in the range of temperatures lethal to the whole organism (see Fig. 8). As can be seen, organism heat death was modified by acclimation temperature, and it would be expected that enzyme thermostability would also depend on acclimation temperature. This was not found to be the case for the plasma membrane $Na^+K^+ATPase$ and $Mg^{2+}ATPase$ nor the sacroplasmic reticulum $Ca^{2+}ATPase$. What was clear however was that pyruvate kinase, a soluble enzyme from the muscle, was very much more thermostable.

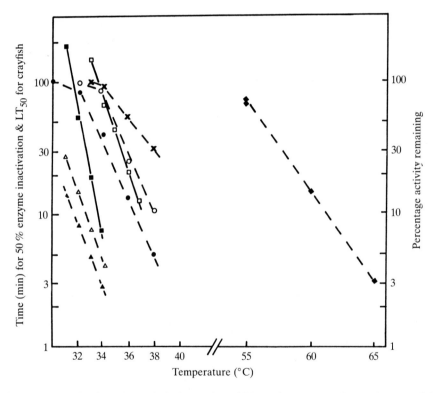

Fig. 8. Heat death points and thermal stability of enzymes from the crayfish *Austropotamobius pallipes*. LT_{50} death points for crayfish acclimated at 8°C (—■—) or 25°C (—□—). Time for 50% inactivation of sarcoplasmic reticulum $Ca^{2+}ATPase$ from 4°C (—▲—) or 25°C (—△—) acclimated animals at lethal temperatures shown. Time for 50% inactivation of muscle pyruvate kinase (—◆—). Activity remaining of muscle plasma membrane ATPase after a 10 min incubation at temperatures shown. $Mg^{2+}ATPase$ (—●—) 8°C acclimated (—○—) 25°C acclimated animals. $Na^+K^+ATPase$ (—×—) both 8° and 25°C acclimated animals. Redrawn with data from Cossins & Bowler (1976) and Bowler *et al.* (1973).

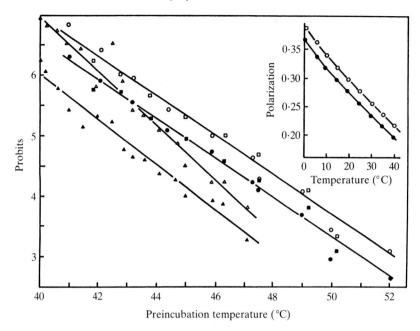

Fig. 9. The effect of acclimation on the thermal inactivation of goldfish brain synaptic membrane enzymes. Residual percentage activity is converted into probits and shown as a function of preincubation temperature, residual activity was measured at 30 °C. ACHEase: (—△—) 28 °C acclimated fish, (—▲—) 7 °C acclimated fish. $Na^+K^+ATPase$: (○, □) 28 °C acclimated fish, (●, ■) 7 °C acclimated fish. Inset, effect of temperature on polarization of DPH in synaptic membranes of 7 °C (●) and 28 °C (○) acclimated fish. Data from Cossins *et al.* (1981), and Bowler (unpublished).

Cossins, Bowler & Prosser (1981) however demonstrated a clear resistance acclimation effect on the thermal sensitivity of goldfish synaptic membrane $Na^+K^+ATPase$. As can be seen in Fig. 9 the enzyme from 28 °C acclimated fish was more thermally resistant than that from 6 °C acclimated fish. The thermal stability of synaptic membrane acetylcholinesterase was similarly dependent upon acclimation temperature. Fig. 9 also shows that warm acclimation caused an increase in the order parameter of the membrane lipids, which correlates well with the earlier report by Cossins (1977) that warm acclimation caused the incorporation of more saturated fatty acids into membrane phospholipids.

Both enzymes display a curvilinear time-course during isothermal inactivation, rather than the pseudo-first order decay kinetics expected (Joly, 1965). This is shown in Fig. 10 for the inactivation of the acetylcholinesterase at 43·5 °C. The final phase of the decay was linear at times greater than about 90 min and one explanation is that a sequential decay of a thermolabile into a thermostable species of the enzyme occurs.

$$\text{Thermolabile} \xrightarrow{K_1} \text{thermostable} \xrightarrow{K_2} \text{inactive species}$$

K_1 and K_2 are the first order decay constants, where $K_1 > K_2$. The curves have been fitted by a computer assisted least squares minimization procedure. The slope of the final decay process defines K_2, and the decay constant K_1, for the thermolabile species, can be derived by calculation, see Fig. 10.

The half-life for the thermolabile species shows a dependency on acclimation temperature for both the $Na^+K^+ATPase$ and the acetylcholinesterase and the values are shown in Table 1.

Thus the thermal inactivation of these two enzymes, from the same preparations, can be modified by thermal acclimation to about the same extent. This suggests a common causal relationship and this is likely to be the demonstrated acclimation-induced changes in membrane lipid saturation and lipid order.

In similar studies on the plasma membrane Mg^{2+} ATPase from liver and two transplantable rat tumours (MC7 & D23), the Mg^{2+} ATPase from the tumours was found to be considerably more thermolabile than the same enzyme from liver (Bowler *et al.* 1981; Barker, 1986). As is shown in Fig. 11 the thermal inactivation

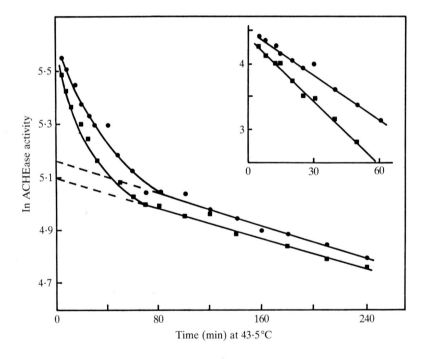

Fig. 10. Kinetics of the inactivation of goldfish brain Acetylcholinesterase at 43·5 °C. Observed decay for enzyme from 7° acclimated (—●—) and 28 °C acclimated (—■—) goldfish. Inset is shown the decay of the 'thermolabile' species of the enzyme from the two preparations, from which the decay constant K_1 is derived. Where activity (Z) of thermolabile species is obtained as

$$Z = \log_e ((\text{total activity}) - (e^x \cdot e^{-K_2 t}))$$

where x is the intercept of the extrapolated decay of the 'thermostable' species on the enzyme activity axis, K_2 is the decay constant of the 'thermostable' species, and t is the exposure time.

is sensitized in the presence of 1 mM tetracaine. What was also of significance was that the membrane lipid order and cholesterol content correlate with the thermal sensitivity of the three enzymes, this is shown in Fig. 12. The cholesterol:phospholipid (Molar ratios) and cholesterol:protein content both show a positive correlation with the inactivation temperature (50%). It is also the case that the

Table 1. *Mean values of the parameters describing the thermal decay of Acetylcholinesterase (at 43·5°C) and Na$^+$K$^+$ATPase (at 45·2°C) from brain synaptic membranes of goldfish acclimated at 6° or 28°C (*P = 0·05; **P = 0·001)*

Enzyme	°C	$K_1 \times 10^{-2}$ min^{-1}	First ½ life min	$K_2 \times 10^{-3}$ min^{-1}
ACHEase	6	4·24 ± 0·3*	16·76 ± 1·12*	1·97 ± 0·3
(n = 7)	28	3·05 ± 0·32*	23·95 ± 2·33*	2·28 ± 0·34
NaK ATPase	6	5·026 ± 0·38**	14·12 ± 1·09**	3·31 ± 0·37
(n = 5)	28	2·89 ± 0·12**	24·15 ± 1·04**	3·69 ± 0·78

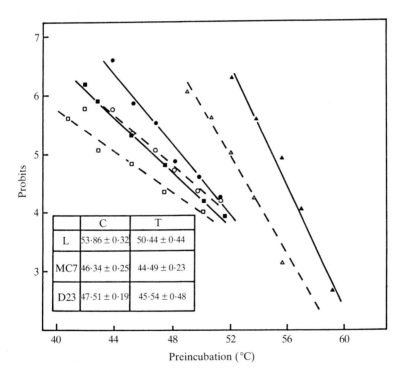

	C	T
L	53·86 ± 0·32	50·44 ± 0·44
MC7	46·34 ± 0·25	44·49 ± 0·23
D23	47·51 ± 0·19	45·54 ± 0·48

Fig. 11. Thermal inactivation of D23, MC7 and liver (L) plasma membrane Mg^{2+}ATPase activity. Residual activity was determined at 37°C after preincubation for 10 min at the temperature indicated in the presence of 1 mM tetracaine (T) or in the absence of tetracaine (C). Residual activity was expressed in the form of probits. Inset are the mean temperatures ± S.E. giving 50% inactivation (n = 5). Liver Mg^{2+}ATPase: (—▲—) control, (—△—) tetracaine. D23 Mg^{2+}ATPase: (—●—) control, (—○—) tetracaine. MC7 Mg^{2+}ATPase: (—■—) control, (—□—) tetracaine. Data from Barker (1986).

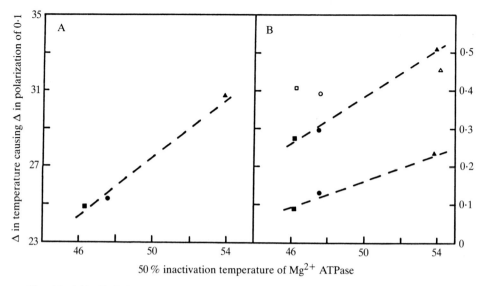

Fig. 12. (A) Relationship between membrane order parameter and thermal inactivation of Mg^{2+}ATPase for preparations from MC7 (■), D23 (●) and liver (▲) plasma membranes. (B) Relationship between membrane lipid composition and thermal inactivation of Mg^{2+}ATPase from MC7 (squares), D23 (circles) and liver (triangles) plasma membranes. Upper curve, Cholesterol:Phospholipid (Molar ratios). Lower curve, Cholesterol:Protein (μmole mg^{-1}). Open symbols, Cholesterol:Phospholipid (μmole mg^{-1}).

temperature rise needed to cause a change in polarization of 0·1 also correlates with the inactivation temperature of the enzyme. These limited data do point to the possible relationship between the thermal inactivation temperature of a membrane protein and the composition and physical state of the lipid matrix.

Conclusions

Membrane proteins appear to be more sensitive to thermal perturbation than do soluble proteins, and it is pertinent to consider why this should be. It is likely that the activity of integral membrane enzymes is just as sensitive to the microviscosity of their environment as are soluble proteins (Gavish & Werber, 1979). As Cossins *et al.* (1981) stress the anisotropic hydrophobic core of the membrane forms a relatively hindered, viscous, environment for enzyme functioning. In consequence, these enzymes may possess a relatively loose tertiary structure to permit the molecular flexibility essential for catalysis. This could also make these enzymes particularly sensitive to heat for such a structure could be susceptible to perturbation. A clear role would also exist for membrane lipids in this hypothesis. Hyperthermic temperatures would decrease lipid order and provide a less hindered environment for the proteins. This less viscous environment would facilitate a greater range of conformational movement, caused by the increased available kinetic energy, and lead to the adoption of configurations by proteins that are inactivating.

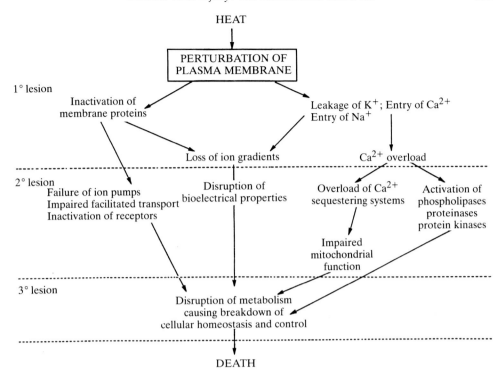

Fig. 13. Scheme to show possible sequences of events during the thermal death of an animal cell.

The inactivation of membrane proteins (enzymes, receptors, etc.) would lead to an impairment of function, and a loss of responsiveness by the compartment bounded by the membrane. Permeability and its control could also be affected. This would lead inevitably to a loss of cellular homeostasis and an irreversible loss of function, resulting in cell death. A scheme is shown in Fig. 13. Sensitizing agents could act by accelerating the primary events in membranes, by fluidizing or directly causing protein inactivation. Protecting agents could act by stabilizing protein structure against thermal perturbation.

In contrast to thermal adaptation, thermotolerance has been shown not to be associated directly with changes in membrane lipids (Gonzalez-Mendez *et al.* 1982; Lepock *et al.* 1981; Konings, 1985). This suggests that it is unlikely that the development of thermotolerance is related to a change in membrane composition. However, Stevenson *et al.* (1986) have recently provided evidence for membrane events being the proximal site triggering the heat shock response. It may well be that thermotolerance is associated with the protection of 'critical' intracellular sites that are the targets of the secondary and tertiary effects of hyperthermia.

References

ANDERSON, R. L. & HAHN, G. M. (1985). Differential effects of hyperthermia on the $Na^+K^+ATPase$ of Chinese hamster ovary cells. *Radiat. Res.* **102**, 314–323.

ANDERSON, R. L., LUNEC, J. & CRESSWELL, S. R. (1985). Cholesterol content and heat sensitivity of nine mammalian cell lines. *Int. J. Hyperthermia* **1**, 337–347.

ANDERSON, R. L., MINTON, K. W., LI, G. C. & HAHN, G. M. (1981). Temperature-induced homeoviscous adaptation in Chinese hamster ovary cells. *Biochim. biophys. Acta* **641**, 334–348.

ANGHILERI, L. J., ESCANYNE, M.-C. & ROBERT, C. M. J. (1985a). Calcium, calcium ATPase and cyclic AMP changes in Erhlich ascites cells submitted to hyperthermia. *Arch. Geschwuisttorsch* **55**, 171–175.

ANGHILERI, L. J., MARCHA, C., ESCANYE, M.-C. & ROBERT, C. M. J. (1985b). Effects of extracellular calcium on calcium transport during hyperthermia of tumour cells. *Eur. J. Cancer clin. Oncol.* **21**, 981–984.

BACK, J. F., OAKENFULL, D. & SMITH, M. B. (1979). Increased thermal stability of proteins in the presence of sugars and polyols. *Biochemistry* **18**, 5191–5196.

BARKER, C. J. (1986). The plasma membrane and cellular heat injury: a study on tumour cells. In *Temperature Relations in Animals and Man*. Biona Report 4 (ed. H. Laudien), pp. 67–76. Stuttgart: Gustav Fischer.

BASS, H., COAKLEY, W. T., MOORE, J. L. & TILLEY, D. (1982). Hyperthermia-induced changes in the morphology of CHO-K1 and their refractile inclusions. *J. thermal Biol.* **7**, 231–242.

BATES, D. A., LEGRIMELLEC, C., BATES, J. H. T., LOUFTI, A. & MACKILLOP, W. J. (1985). Effects of thermal adaptation at 40°C on membrane viscosity and the sodium-potassium pump in Chinese hamster ovary cells. *Cancer Res.* **45**, 4895–4899.

BATES, D. A. & MACKILLOP, W. J. (1985). The effect of hyperthermia on the sodium-potassium pump in Chinese hamster ovary cells. *Radiat. Res.* **103**, 441–451.

BOONSTRA, J., SCHAMHART, D. H. J., DE LAAT, S. W. & VAN WIJK, R. (1984). Analysis of K^+ and Na^+ transport and intracellular contents during and after heat shock and their role in protein synthesis in rat hepatoma cells. *Cancer Res.* **44**, 955–60.

BOONSTRA, J., SYBESMA, F. & VAN WIJK, R. (1985). Effects of external K^+ on protein and DNA synthesis during the after heat shock in rat hepatoma cells. *Int. J. Hyperthermia* **1**, 255–263.

BOWLER, K. (1972). The effects of repeated temperature stresses to testes of the laboratory rat. A histological study. *J. Reprod. Fertil.* **28**, 325–333.

BOWLER, K. (1981). Heat death and cellular heat injury. *J. thermal Biol.* **6**, 171–178.

BOWLER, K., DUNCAN, C. J., GLADWELL, R. T. & DAVISON, T. F. (1973). Cellular heat injury. *Comp. Biochem. Physiol.* **45**, 441–449.

BOWLER, K., LAUDIEN, H. & LAUDIEN, I. (1981). Cellular heat injury. *J. thermal Biol.* **8**, 426–430.

BOWLER, K. & KASHMEERY, A. M. S. (1981). Effects of *in vivo* heating of blowflies on oxidative capacity of flight muscle sarcosomes: A differential effect on glycerol 3 phosphate and pyruvate plus proline respiration. *J. thermal Biol.* **6**, 11–18.

BURDON, R. H. & CUTMORE, C. M. M. (1982). Human heat shock gene expression and the modulation of plasma membrane Na^+K^+ATPase activity. *FEBS lett.* **140**, 45–48.

CALDERWOOD, S. K. & HAHN, G. M. (1983). Thermal sensitivity and resistance of insulin-receptor binding. *Biochim. biophys. Acta* **756**, 1–8.

CAMPBELL, A. K. (1983). *Intracellular Calcium its Universal Role as Regulator*. Chichester: John Wiley & Sons.

CHAN, D. S. & WANG, H. H. (1984). Local anaesthetics can interact electrostatically with membrane proteins. *Biochim. biophys. Acta* **770**, 55–64.

COAKLEY, W. T. (1987). Hyperthermia effects on the cytoskeleton and on cell morphology. In S.E.B. Symposium XLI. *Temperature and Animal Cells* (ed. K. Bowler & B. J. Fuller), Cambridge: The Company of Biologists Limited.

COSS, R. A., DEWEY, W. C. & BAMBURG, J. R. (1982). Effects of hyperthermia on dividing Chinese hamster ovary cells and on microtubules *in vitro*. *Cancer Res.* **42**, 1059–1071.

COSSINS, A. R. (1977). Adaptation of biological membranes to temperature. The effect of temperature acclimation of goldfish upon the viscosity of synaptosomal membranes. *Biochim. biophys. Acta* **470**, 395–411.

COSSINS, A. R. & RAYNARD, R. S. (1987). Adaptive responses of animal cell membranes to temperature. This volume.

Cossins, A. R., Bowler, K. & Prosser, C. L. (1981). Homeoviscous adaptation and its effect upon membrane-bound proteins. *J. thermal Biol.* **6**, 183–187.

Cossins, A. R., Friedlander, M. J. & Prosser, C. L. (1977). Correlations between behavioural temperature adaptations of goldfish and the viscosity and fatty acid composition of their synaptic membranes. *J. comp. Physiol.* **120**, 109–121.

Cossins, A. R. & Sinensky, M. (1984). Adaptation of membranes to temperature, pressure and exogenous lipids. In *Physiology of Membrane Fluidity*, vol. II (ed. M. Shinitzky), pp. 1–20. Boca Raton, USA: CRC Press.

Cress, A. E., Culver, P. S., Moon, T. E. & Gerner, E. W. (1982). Correlation between amounts of cellular membrane components and sensitivity to hyperthermia in a variety of mammalian cell lines in culture. *Cancer Res.* **42**, 1716–1721.

Cress, A. E. & Gerner, E. W. (1980). Cholesterol levels inversely reflect the thermal sensitivity of mammalian cells in culture. *Nature, Lond.* **283**, 677–679.

Culver, P. S. & Gerner, E. W. (1982). Temperature acclimation and specific cellular components in the regulation of thermal sensitivity of mammalian cells. In *Cancer Therapy by Hyperthermia, Drugs and Radiation* (ed. L. Dethlefsen), pp. 99–101. Bethesda, USA: US Department of Health and Human Services.

Davies, D. de L., Rofstad, E. K. & Lindmo, T. (1985). Hyperthermia-induced changes in antigen expression in human FME melamoma cells. *Cancer Res.* **45**, 4109–4114.

Davison, T. F. (1971). Ultrastructural changes in sarcosomes after exposure of adult *Calliphora erythrocephala* to lethal and sub-lethal high temperatures. *J. cell. Physiol.* **78**, 49–58.

Davison, T. F. & Bowler, K. (1971). Changes in the functional efficiency of flight muscle sarcosomes during heat death of adult *Calliphora erythrocephala*. *J. cell. Physiol.* **78**, 37–48.

Demel, R. A., van Kessel, W. S. M. & van Deenen, L. L. M. (1972). The properties of polyunsaturated lecithins in monolayers and liposomes and the interactions of these lecithins with cholesterol. *Biochim. biophys. Acta* **266**, 26–40.

Dennis, W. H. & Yatvin, M. B. (1981). Correlation of hyperthermic sensitivity and membrane microviscosity in *E. coli* K1060. *Int. J. Radiat. Biol.* **39**, 265–271.

Elegbede, J. A., Elson, C. E., Qureshi, A., Dennis, W. H. & Yatvin, M. B. (1986). Increasing the thermal sensitivity of a mammary tumour (CA755) through dietary modification. *Eur. J. Cancer clin. Oncol.* **22**, 607–615.

Fajardo, L. F., Egbert, B., Marmor, J. & Hahn, G. M. (1980). Effects of hyperthermia in a malignant tumor. *Cancer* **45**, 613–623.

Farber, J. (1981). The role of calcium in cell death. *Life Sci.* **29**, 1289–1295.

Fuller, D. J. M. & Gerner, E. W. (1982). Polyamines: dual role in the modulation of cellular sensitivity to heat. *Radiat. Res.* **92**, 439–444.

Fuller, D. J. M. & Gerner, E. W. (1983). Sensitization to heat by the polyamines and their analogs. *Radiat. Res.* **95**, 124–129.

Gavish, B. & Werber, M. M. (1979). Viscosity-dependent structural fluctuations in enzyme catalysis. *Biochemistry* **18**, 1269–1275.

Glass, J. R., de Witt, R. G. & Cress, A. E. (1985). Rapid loss of stress fibres in Chinese hamster ovary cells after hyperthermia. *Cancer Res.* **45**, 258–262.

Gonzalez-Mendez, R., Minton, K. W. & Hahn, G. M. (1982). Lack of correlation between membranes lipid composition and thermotolerance in Chinese hamster ovary cells. *Biochim. biophys. Acta* **692**, 168–170.

Guffy, M. M., Rosenberger, J. A., Simon, I. & Burns, C. P. (1982). Effect of cellular fatty acid alteration on hyperthermic sensitivity in cultured L1210 murine leukemia cells. *Cancer Res.* **42**, 3625–3630.

Hazel, J. R. & Prosser, C. L. (1974). Molecular mechanisms of temperature compensation in poikilotherms. *Ann. rev. Physiol.* **54**, 620–677.

Heilbrunn, L. V. (1924). The colloid chemistry of protoplasma. IX. The best heat coagulation of protoplasma. *Am. J. Physiol.* **69**, 190–199.

Henle, K. J. & Dethlefsen, L. A. (1979). Sensitization to hyperthermia (45°C) of normal and thermotolerant CHO cells by anisotonic media. *Int. J. Radiat. Biol.* **36**, 387–397.

Henle, K. J., Monson, T. P., Moss, A. J. & Nagle, W. A. (1984). Protection against thermal cell death in Chinese hamster ovary cells by glucose, galactose, or mannose. *Cancer Res.* **44**, 5499–5504.

HENLE, K. J., PECK, J. W. & HIGASHIKUBO, R. (1983). Protection against heat-induced cell killing by polyols *in vitro*. *Cancer Res.* **43**, 1624–1627.

HIDVEGI, E. J., YATVIN, M. B., DENNIS, W. H. & HIDVEGI, E. (1980). Effect of altered membrane lipid composition and procaine on hyperthermic killing of ascites tumor cells. *Oncology* **37**, 360–363.

JOLY, M. (1965). *A physico-chemical approach to the denaturation of proteins*, pp. 350. New York: Academic Press.

JUNG, H. (1986). A generalised concept for cell killing by heat. *Radiat. Res.* **106**, 56–72.

KAMPINGA, H. H., JORRITSMA, J. B. M. & KONINGS, A. W. T. (1985). Heat-induced alterations in DNA polymerase activity of HeLa cells and of isolated nuclei. Relation to cell survival. *Int. J. Radiat. Biol.* **47**, 29–40.

KONINGS, A. (1985). Development of thermotolerance in mouse fibroblast LM cells with modified membranes and after procaine treatment. *Cancer Res.* **45**, 2016–2019.

KONINGS, A. W. T. & RUIFROK, A. C. C. (1985). Role of membrane lipids and membrane fluidity in thermosensitivity and thermotolerance of mammalian cells. *Radiat. Res.* **102**, 86–98.

KRUUV, J., GLOFCHESKI, D., CHENG, H. K., CAMPBELL, S., AL-QYSI, H., NOLAN, W. T. & LEPOCK, J. R. (1983). Factors influencing survival and growth of mammalian cells exposed to hyperthermia. I. Effects of temperature and membrane lipid perturbers. *J. Cell Physiol.* **115**, 179–185.

KUNNEMANN, H. & PRECHT, H. (1979). The influence of environmental temperature and salinity on animals. II. Resistance of poikilothermic animals to heat and cold. *Zoo. Anz.* **202**, 154–162.

KWOCK, L., LIN, P. S. & HEFTER, K. (1985). A comparison of the effects of hyperthermia on cell growth in human T and B lymphoid cells: relationship to alterations in plasma membrane transport properties. *Radiat. Res.* **101**, 197–206.

KWOCK, L., LIN, P. S., HEFTER, K. & WALLACH, D. F. H. (1978). Impairment of rat dependent amino acid transport in cultured human T cell line by hyperthermia and irradiation. *Cancer Res.* **38**, 83–87.

LAGERSPETZ, K. Y. H. (1974). Temperature acclimation and the nervous system. *Biol. Rev.* **49**, 477–514.

LeCAVALIER, D. & MACKILLOP, W. J. (1985). The effect on hyperthermia on glucose transport in normal and thermotolerant Chinese hamster ovary cells. *Cancer Lett.* **29**, 223–231.

LEE, D. C. & CHAPMAN, D. (1987). The effects of temperature on biological membranes and their models. In *Temperature and Animal Cells*, S.E.B. Symposium XLI (ed. K. Bowler & B. J. Fuller). Cambridge: The Company of Biologists Limited.

LEPOCK, J. R. (1982). Involvement of membranes in cellular responses to hyperthermia. *Radiat. Res.* **92**, 433–438.

LEPOCK, J. R., CHENG, H.-K., AL-QYSI, H. & KRUUV, J. (1983). Thermotropic lipid and protein transitions in Chinese hamster lung cell membranes; relationship to hyperthermic killing. *Can. J. Biochem. Cell Biol.* **61**, 421–427.

LEPOCK, J. R., MASSICOTE-NOLAN, P., RULE, G. S. & KRUUV, J. (1981). Lack of a correlation between hyperthermic cell killing, thermotolerance, and membrane lipid fluidity. *Radiat. Res.* **87**, 300–313.

LI, G. C. & HAHN, G. M. (1978). Ethanol-induced tolerance to heat and adriamycin. *Nature, Lond.* **274**, 694–701.

LI, G. C., SHIU, E. C. & HAHN, G. M. (1980). Similarities in cellular inactivation by hyperthermia or by ethanol. *Radiat. Res.* **82**, 257–268.

LIN, P. S., KWOCK, L., HEFTER, K. & WALLACH, D. F. H. (1978). Modification of rat thymocyte membrane properties by hyperthermia and ionising radiation. *Int. J. Radiat. Biol.* **33**, 371–382.

LIONETTI, F. J., LIN, P. S., MATTALIANO, R. J., HUNT, S. M. & VALERI, C. R. (1980). Temperature effects on shape and function of human granulocytes. *Expl. Haemat.* **81**, 3034–317.

MADDEN, T. D., CHAPMAN, D. & QUINN, P. J. (1974). Cholesterol modulates activity of calcium-dependent ATPase of the sarcoplasmic reticulum. *Nature, Lond.* **279**, 538–541.

MAGUN, B. E. (1981). Inhibition and recovery of macromolecular synthesis, membrane transport and lysosomal function following exposure of cultured cells to hyperthermia. *Radiat. Res.* **87**, 657–669.

MALHOTRA, A., KRUUV, J. & LEPOCK, J. R. (1986). Sensitisation of rat hepatocytes to hyperthermia by calcium. *J. cell Physiol.* **128**, 279–284.

MASSICOTTE-NOLAN, P., GLOFCHESKI, D. J., KRUUV, J. & LEPOCK, J. R. (1981). Relationship between hyperthermic cell killing and protein denaturation by alcohols. *Radiat. Res.* **87**, 284–299.

MCELHANEY, R. N. (1985). Membrane lipid fluidity, phase state, and membrane function in prokaryotic microorganisms. In *Membrane Fluidity in Biology*, vol. 4, pp. 147–208.

MCIVER, D. & SHRIDHAR, R. (1979). Deuterium oxide protects against hyperthermia cell death: evidence for a role of interfacial water. *Proc. 22nd A. Meet. Can. Fedn Biol. Soc.* **22**, 77.

MINTON, K. W., STEVENSON, M. A., KENDIG, J. & HAHN, G. M. (1980). Pressure inhibits thermal killing of Chinese Hamster ovary fibroblasts. *Nature, Lond.* **285**, 482–483.

MULCAHY, R. T., GOULD, M. N., HIDVERGI, E., ELSON, C. E. & YATVIN, M. B. (1981). Hyperthermia and surface morphology of P388 ascites tumour cells: Effects of membrane modifications. *Int. J. Rad. Biol.* **39**, 95–106.

OVERATH, P., SCHAIRER, H. U. & STOFFEL, W. (1970). Correlation of *in vivo* and *in vitro* phase transitions of membrane lipids in *Escherichia coli. Proc. natn. Acad. Sci. U.S.A.* **67**, 606–612.

PATERSON, S. J., BUTLER, K. W., HUANA, P., LABELLE, J., SMITH, I. C. P. & SCHNEIDER, H. (1972). The effect of alcohols on lipid bilayers: A spin label study. *Biochim. biophys. Acta* **266**, 597–602.

PROSSER, C. L. & NELSON, D. O. (1981). The role of nervous systems in temperature adaptation of poikilotherms. *A. Rev. Physiol.* **43**, 281–300.

RAAPHORST, G. P. & AZZAM, E. I. (1982). The effect of D$_2$O on thermal sensitivity and thermal tolerance in cultured Chinese hamster V79 cells. *J. thermal Biol.* **7**, 147–153.

RAAPHORST, G. P. & DEWEY, W. C. (1978). Enhancement of hyperthermic killing of cultured mammalian cells by treatment with anisotonic NaCl or medium solutions. *J. thermal. Biol.* **3**, 177–182.

RAAPHORST, G. P., VADASZ, J. A., AZZAM, E. I., SARGENT, M. D., BORSA, J. & EINSPENNER, M. (1985). Comparison of heat and/or radiation sensitivity and membrane composition of seven X ray-transformed C3H 10T1/2 cell lines and normal C3H 10T cells. *Cancer Res.* **45**, 5452–5456.

ROBBINS, H. I., DENNIS, W. H., SLATTERY, J. S., LANGE, T. A. & YATVIN, M. B. (1983). Systemic lidocaine enhancement of hyperthermia-induced tumour regression in transplantable murine tumour models. *Cancer Res.* **43**, 3287–3291.

ROBINSON, J. D. & FLASHNER, M. A. (1979). The (Na$^+$ + K$^+$)-activated ATPase enzymatic and transport properties. *Biochim. biophys. Acta* **549**, 145–176.

ROFSTAD, E. K., FALKVOLL, K. H. & OFTEDAL, P. (1984). Micronucleus formation in human melanoma xenografts following exposure to hyperthermia. *Radiat. environ. Biophys.* **23**, 51–60.

ROTI ROTI, J. L. (1982). Heat-induced cell death and radiosensitization: Molecular mechanisms. In *Proceedings of the Third International Symposium on Cancer Therapy by Hyperthermia, Drugs and Radiation* (ed. L. A. Dethlefsen), pp. 3–10. *National Cancer Insitute Monographs* **61**.

ROTI ROTI, L. W. & ROTHSTEIN, A. (1973). Adaptation of mouse leukemic cells L5178Y to anisotonic media. *Expl Cell Res.* **79**, 295–310.

ROTI ROTI, J. L. & WILSON, C. F. (1984). The effects of alcohols procaine and hyperthermia on the protein content of nuclei and chromatin. *Int. J. Radiat. Biol.* **46**, 25–33.

RUIFROK, A. C. C., KANON, B. & KONINGS, A. W. T. (1985). Correlation of colony forming ability of mammalian cells with potassium content after hyperthermia under different experimental conditions. *Radiat. Res.* **103**, 452–454.

SABINE, J. R. (1983). Membrane homeostasis: is there an optimal level of membrane cholesterol? *Bioscience Reports* **3**, 337–344.

SATO, C., NAKAYAMA, T., KOJIMA, K., NISHIMOTO, Y. & NAKAMURA, W. (1981). Effects of hyperthermia on cell surface charge and cell survival in mastocytoma cells. *Cancer Res.* **41**, 4107–4110.

SCHANNE, F., KANE, A. B., YOUNG, E. E. & FARBER, J. L. (1979). Calcium dependence of toxic cell death: a final common pathway. *Science* **206**, 700–702.

SCHREK, R., CHANDRA, S., MOLNAR, Z & STEFANI, S. S. (1980). Two types of interphase death of lymphocytes exposed to temperatures of 37–45 °C. *Radiat. Res.* **82**, 162–170.

SCHREK, R. & STEFANI, S. S. (1981). Effects of alcohol and hyperthermia on normal and leukemic lymphocytes. *Oncology* **38**, 69–71.

SEEMAN, P. (1972). The membrane actions of anaesthetics and tranquilizers. *Pharmacol. Rev.* **24**, 583–632.

SINENSKY, M. (1974). Homeoviscous adaptation – A homeostatic process that regulates the viscosity of membrane lipids in *Escherichia coli. Proc. natn. Acad. Sci. U.S.A.* **71**, 522–525.

SINGER, S. J. & NICHOLSON, G. L. (1972). The fluid mosaic model of the structure of cell membranes. *Science* **175**, 720–731.

SHINITZKY, M. (1984). *Physiology of Membrane Fluidity*, vol. II. Boca Raton, USA: CRC Press Inc.

STEVENSON, A. P., GALEY, W. R. & TOBEY, R. A. (1983). Hyperthermia-induced increase in potassium transport in Chinese hamster cells. *J. Cell Physiol.* **115**, 75–86.

STEVENSON, M. A., CALDERWOOD, S. K. & HAHN, G. M. (1983). Lectin binding and processing after heat in CHO HA-1 cells. *Radiat. Res.* **94**, 562.

STEVENSON, M. A., CALDERWOOD, S. K. & HAHN, G. M. (1986). Rapid increases in inositol triphosphate and intracellular calcium after heat shock. *Biochem. Biophys. Res. Commun.* **137**, 826–833.

STUBBS, C. D. (1983). Membrane fluidity: structure and dynamics of membrane lipids. In *Essays in Biochemistry*, vol. 19 (ed. P. N. Campbell & R. D. Marshall), pp. 1–39. London: Academic Press.

USHAKOV, B. P. (1964). Thermostability of cells and proteins of poikilotherms and its significance in speciation. *Physiol. Rev.* **44**, 518–560.

VIDAIR, C. A. & DEWEY, W. C. (1986). Evaluation of a role for intracellular Na^+, K^+, Ca^{2+}, and Mg^{2+} in hyperthermic killing. *Radiat. Res.* **105**, 187–200.

WATERS, R. L. & HENLE, K. J. (1982). DNA degradation in heated CHO cells. *Cancer Res.* **42**, 4427–4432.

WATERS, R. L. & ROTI ROTI, J. L. (1982). Hyperthermia and the cell nucleus. *Radiat. Res.* **92**, 458–462.

WIEGANT, F. A. C., TUYL, M. & LINNEMANNS, W. A. M. (1985). Calmodulin inhibitors potentiate hyperthermic cell killing. *Int. J. Hyperthermia* **1**, 157–169.

WILLIS, E. J., FINDLAY, J. M. & McMANUS, J. P. A. (1976). Effect of hyperthermia therapy on the liver. II. Morphological observations. *J. clin. Path.* **29**, 1–10.

WONG, R. S. L. & DEWEY, W. C. (1982). Molecular studies on the hyperthermia inhibition of DNA synthesis in Chinese hamster ovary cells. *Radiat. Res.* **92**, 370–395.

YATVIN, M. B. (1977). The influence of membrane lipid composition and procaine on hyperthermic death of cells. *Int. J. Radiat. Biol.* **32**, 513–522.

YATVIN, M. B., ABUIRMEILEH, N. M., VORPAHL, J. W. & ELSON, C. E. (1983). Biological optimization of hyperthermia: modification of tumor membrane lipids. *Eur. J. Cancer clin. Oncol.* **19**, 657–663.

YATVIN, M. B., CLIFTON, K. H. & DENNIS, W. H. (1979). Hyperthermia and local anaesthetics: Potentiation of survival of tumour bearing mice. *Science* **205**, 195–196.

YATVIN, M. B., CREE, T. C., ELSON, C. E., GIPP, J. J., TEGMO, I.-M. & VORPAHL, J. W. (1982a). Probing the relationship of membrane 'fluidity' to heat killing of cells. *Radiat. Res.* **89**, 644–646.

YATVIN, M. B., DENNIS, W. H., ELEGBEDE, J. A. & ELSON, C. E. (1987). Sensitivity of tumour cells to heat and ways of modifying the response. In S.E.B. Symposium XLI, *Temperature and Animal Cells* (ed. K. Bowler & B. J. Fuller). Cambridge: The Company of Biologists Limited.

YATVIN, M. B., GIPP, J. J., RUSY, B. F. & DENNIS, W. H. (1982b). Correlation of bacterial hyperthermic survival with anaesthetic potency. *Int. J. Radiat. Biol.* **42**, 141–149.

YATVIN, M. B., VORPAHL, J. W., GOULD, M. N. & LYTE, M. (1983). The effects of membrane modification and hyperthermia on the survival of P-388 and V-79 cells. *Eur. J. Cancer clin. Oncol.* **19**, 1247–1253.

YAU, T. M. (1979). Procaine-mediated modification of membranes and of the response to X-irradiation and hyperthermia in mammalian cells. *Radiat. Res.* **80**, 523–541.

YI, P. N., CHANG, C. S., TALLEN, M., BAYER, W. & BALL, S. (1983). Hyperthermia-induced intracellular ionic level changes in tumor cells. *Radiat. Res.* **93**, 534–544.

ZIELKE-TEMME, B. & HOPWOOD, L. (1982). Time lapse cinematographic observations of heated G1-Phase Chinese Hamster ovary cells. *Radiat. Res.* **92**, 320–321.

Printed in Great Britain © Society for Experimental Biology 1987   187

HYPERTHERMIA EFFECTS ON THE CYTOSKELETON AND ON CELL MORPHOLOGY

W. T. COAKLEY

Department of Microbiology, University College, Cardiff, UK

Summary

Human erythrocyte ghost membranes undergo five thermal transitions at temperatures between 50 and 75°C. Spontaneous fragmentation of whole cells occurs at 50°C, a transition temperature which has been associated with denaturation of the cytoskeletal protein spectrin. Haemolysis occurs at 65°C and microvesiculation of the resulting ghost membrane is seen at temperatures in excess of 70°C. The cell fragmentation develops through spatially periodic growth of surface waves on the erythrocyte membrane. The interfacial instability associated with the surface wave growth arises from thermal impairment of the stabilizing function of spectrin. Interfacial instability is also associated with the beading pattern which arises when long processes drawn mechanically from erythrocytes are heated. Similar beading of cell processes is a feature of many cytoskeleton-weakening agents acting on nonerythroid cells.

The complexities of the cytoskeletons of eucaryotic cells including structure, composition and interaction of cytoskeletal microfilaments, microtubules and intermediate filaments, both with each other and with the cell membrane, are outlined. Attention is drawn to the importance of the function of proteins which interact with the cytoskeletal elements and to the influence of calcium concentration on those proteins. Actin monomers are denatured (and are no longer polymerizable) at temperatures a few degrees above the growth temperature of the cell source of the actins. Actin in the filament form requires much higher denaturation temperatures. This greater thermal lability of actin monomers would be expected to result (because of treadmilling in microfilaments) in a gradual depolymerization of the filaments. Depolymerization of microtubules occurs at temperatures close to the cell growth temperature and may be dependent on a thermal effect on microtubule-associated proteins.

The response of spread interphase mammalian cells to temperatures around 43°C includes central retraction of membrane, loss of microvilli, concentration of organelles in a juxtanuclear position, rounding up of the cell, retention of contact with the substratum by processes which are sometimes beaded and blebbing of the cell membrane. The morphological effects of heat are compared here with those of cytochalasin, colcemid and a number of morphology modifying agents. Blebbing of membrane is a fairly general response of cells to stress. Proteins in blebs diffuse as if released from a lateral constraint. Moderate heating has been shown to cause cortical microfilament separation from the plasma membrane. There is increasing evidence that cell death correlates with the occurrence of advanced stages of cell blebbing at least for cells in G1 stage of the cell cycle. A major contrast between thermal effects on erythrocytes and on cultured mammalian cells is that erythrocyte fragmentation occurs immediately the denaturation temperature of spectrin is reached while morphological change in cultured cells is a gradual process. The implications of this difference are discussed.

Introduction

Because of the relative lack of complexity of the shape of the human erythrocyte and the wealth of information available on the structure of its cytoskeleton heat-induced morphological changes have been well studied in that system and have also been correlated with changes in the membrane-cytoskeleton complex. These morphological changes in red blood cells will be treated in some detail below both for their own sakes and for the insights they provide to the general case of morphological change in heated animal cells. The structure of the cytoskeleton of nucleated cells will be discussed. Heat-induced shape changes in animal cell morphology will be described and compared with those seen when cells are exposed to a variety of other shape-modifying agents. The extent to which a relationship between morphological change and cell viability has been established for heated cultured cell lines will be examined.

Thermal transitions in human erythrocyte membrane ghosts

The membrane complex which forms the interfacial region (Gibbs, 1928; Coakley & Deeley, 1980) between the cell suspending phase and the erythrocyte cytoplasm consists of a glycocalyx, a membrane bilayer and a tangential cytoskeleton. The erythrocyte glycocalyx consists of part of the transmembrane protein band 3 and of Glycophorins A, B and C (Bennett, 1985). The cytoplasmic aspect of the anion-transporting transmembrane protein, band 3, is linked, by interaction with ankyrin (a monomeric phosphoprotein of molecular weight 215 000), to spectrin, the major protein of the membrane skeleton (Bennett, 1985). Spectrin is a flexible rod-shaped molecule about 100 nm in length composed of two parallel α and β polypeptide chains of molecular weight 260 000 and 225 000 respectively. Spectrin heterodimers self-associate at one end of the molecule to form tetramers of 200 nm length. As well as its binding site for ankyrin, spectrin can bind by lateral association to F-actin (Bennett, 1985). This binding is promoted by band 4.1 protein (band 4.1 in turn has the ability to bind to Glycophorin A (Anderson & Lovrien, 1984)). Erythrocyte actin has all the properties of other cellular actins. It has been visualized in the erythrocyte membrane skeletons as oligomeric short filaments rather than as extended filaments. Fig. 1 (adapted from Bennett, 1985) is a schematic diagram of a proposed model of the organization of the erythrocyte skeleton.

Differential scanning calorimetry studies show five thermal transitions during heating of erythrocyte ghosts (Davio & Low, 1982). The A transition (49·5°C) has been shown to involve the partial denaturation of spectrin (Brands, Erickson, Lysko, Schwartz & Taverna, 1977). The B1 transition (56°C) may involve bands 2.1, 4.1, and 4.2 but the nature of the endotherm is unclear. The B2 transition (62°C at pH 7·4) has been identified with the thermal unfolding of the cytoplasmic domain of band 3 (Appell & Low, 1982). The C transition corresponds to the denaturation temperature of the transmembrane portion of band 3 protein (Davio & Low, 1982).

Thermal effects on whole human erythrocytes

It has been known since the last century that human erythrocytes fragment on heating to temperatures around 50°C. Schultze's (1865) microscopic examination of red blood cells on a heated stage at 52°C revealed that, initially, superficial indentations on the rim of the cells deepened and rapidly led to the production of spherical fragments which usually remained in contact with each other as if on a fine stem. Through a variety of different fragmentation pathways there developed gradually more and more regularly spherical formations until after 15–30 min the samples contained only small spheres of a range of sizes (see Coakley, Bater & Deeley (1978) for scanning electron micrographs). Ham, Shen, Fleming & Castle (1948) list early studies on the effects of heat on erythrocytes *in vivo* and *in vitro*. In their own work they found that morphological changes in heated human erythrocytes preceded increases in osmotic and mechanical fragilities and final haemolysis of the cells. Subdivision of the cells occurred without haemolysis. When cells are heated rapidly to temperatures in excess of the thermal fragmentation temperature 50% haemolysis is observed at a temperature of 65°C

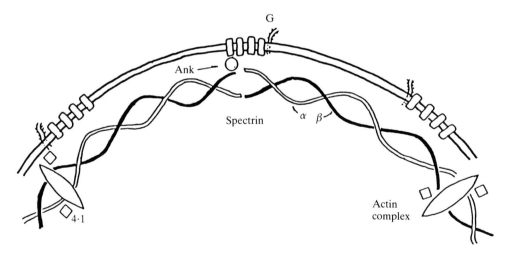

Fig. 1. Schematic representation of the organization of the erythrocyte membrane. The transbilayer anion channel protein Band 3 is linked, by interaction with the protein ankyrin (ank) to the cytoskeletal protein spectrin. A further linkage is shown from a glycophorin (G) through Band 4.1 (□) to the actin oligomer. (Adapted from Bennett, 1985).

(Coakley, Nwafor & Deeley, 1983). Microvesiculation of the ghost membrane occurs at temperatures in excess of 70°C (Coakley, Bater, Crum & Deeley, 1979). Ponder (1949) described the morphological alterations he observed, by light microscopy, in red blood cells at temperatures between 48 and 52°C. He concluded that the molecular pattern on which the discoidal shape depended had undergone an irreversible modification.

The temperatures at which thermal transitions occur in ghost membranes and the temperatures at which observable (light microscopy) changes in erythrocytes occur are shown in Table 1. The principal morphological changes occur close to the temperature for transitions A, C and D, respectively.

In addition to the occurrence of a thermal transition at the fragmentation temperature of erythrocytes changes in the deformability of the red cell at temperatures just below 50°C (Rakow & Hochmuth, 1975; Williamson, Shanahan & Hochmuth, 1975; Rakow, Simchon, Sung & Chien, 1981) have been attributed to heat-induced changes in spectrin. It has also been shown that the phosphorylation of spectrin in human erythrocyte membranes is slightly decreased at 46°C and totally inhibited at 50°C (Mohandas, Greenquist & Shohet, 1978). It is now widely accepted that the thermal fragmentation of erythrocytes is associated with the thermal denaturation of spectrin.

Cinemicroscopy of the process of fragmentation of erythrocytes as they were heated through the denaturation temperature of spectrin at rates of $0.1–1 \, \mathrm{K \, s^{-1}}$ showed that many of the products and fragmentation pathways previously reported (Schultze 1865; Ponder, 1949) had arisen from hydrodynamic stress resulting from temperature-gradient induced flow in the cell suspending phase (Coakley, Bater, Crum & Deeley, 1979). It has been shown that when cells are protected from such flow (by heating the cells in rectangular microslides) only two fundamental forms of fragmentation occur (Coakley & Deeley, 1980; Deeley & Coakley, 1983). In one case a wave grows on the rim of the heated erythrocyte and vesicles pinch from the crest of the growing surface wave while in the second case no wave grows on the cell rim but membrane internalization can occur at the cell dimple.

The cinemicrographs of Fig. 2 show examples of the mode of fragmentation where surface waves grew in the radial direction on the rims of erythrocytes as they

Table 1. *Erythrocyte ghost membrane thermal transitions and changes in whole cells*

Thermal transition	Temperature (°C)	Whole cell change	Temperature (°C)
A	49·5	Fragmentation	50
B_1	55		
B_2	62		
C	66	Haemolysis	65
D	77	Microvesiculation	>70

Ghost thermal transition temperatures (Brandts *et al.* 1977; Krishnan & Brandts, 1979; Davio & Low, 1982). Whole cell changes (Coakley *et al.* 1979, 1983).

were heated in microslides through the denaturation temperature of spectrin (Coakley & Deeley, 1980). The sequence shown for each of these cells lasted for about 1 s. In the second sequence the ellipse-like shape of the cell just below the fragmentation temperature is replaced by a profile with five growing wave crests. The 11th frame of the sequence shows the axis of symmetry about which the disturbances grow, implying that the phenomenon is a property of the whole surface. Wave growth on the cell rim continues to a stage where beads form simultaneously from the growing wave crests at opposite ends of the cell. About 3 % of cells heated in physiological saline did not fragment by wave growth on the cell rim. Rather the cell rounded up, tending to enclose the dimple region. This

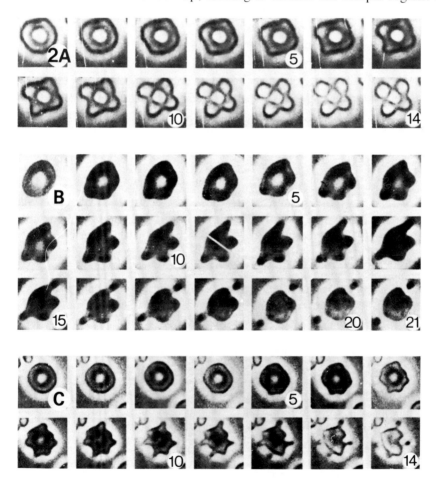

Fig. 2. Cinemicrographs of erythrocytes illustrating surface wave growth as the cells were heated (under a variety of experimental conditions) through the denaturation temperature of spectrin; (A) Successive frames (16 frames s^{-1}) of a cell in isotonic low ionic strength buffer at a heating rate of 1·0°K s^{-1}. (B) Successive frames (16 frames s^{-1}) of a cell in 154 mmol l^{-1} buffered saline at a heating rate of 0·33°K s^{-1}. The line drawn on frame 11 shows the axis of symmetry of the wave development; (C) Alternate frames (50 frames s^{-1}) of cells in 154 mmol l^{-1} saline at a heating rate of 1·0°K s^{-1} (Coakley & Deeley, 1980).

dimple could either remain as an intact, almost totally internalized surface or could develop a wave at the rim of the folded dimple and send saline-containing vesicles into the cell (Deeley & Coakley, 1983).

The involvement of the growth of spatially periodic surface waves in the fragmentation process suggested that the biconcave disc shape was no longer a stable shape when the contribution of the elasticity of the cytoskeleton to the properties of the interfacial region membrane complex was impaired. In making this point it is useful to recall that an interface is stable if, following a small disturbance it returns to its initial form. Disturbances (perturbations) can originate outside the system, as with various forms of environmental vibration, or inside, when random thermal motion leads to small fluctuations in interfacial composition or shape (i.e. there is no requirement that an observer provide a driving force). If a sine disturbance of any wavelength may grow then the system is unstable. The growth of a spatially periodic sine disturbance does not imply that the interface is vibrating as it would for a sinusoidal disturbance periodic with time. Rather, the amplitude of the spatially periodic disturbance grows exponentially with time. The unstable interface may reach a new steady state with the interface deformed but intact or it may break up to form droplets or particles (Miller, 1978; Dimitrov & Jain, 1984; Gallez & Coakley, 1987).

It has been shown that erythrocyte crenation (involving a morphological change from the discoid form to a spherical surface with regularly spaced spicules all over its surface) can easily be brought about by simple procedures such as a change of pH (Glaser, 1982) or even by contact with a glass surface. In order to establish that the regular wave growth on the rim of heated cells (Fig. 2) did not arise from any peculiar property of the organization of the erythrocyte cytoskeleton when the cell is in the discoid form a system was developed in which cylindrical processes were drawn by flow stress from erythrocytes at the moment of spectrin denaturation (Coakley et al. 1978; Crum, Coakley & Deeley, 1979). The departure from the discoid to the cylindrical form was of interest because the involvement of interfacial instability in the breakup of cylinders of liquid into lines of droplets is well understood (Meister & Scheele, 1967).

Fig. 3 shows a cell which is swept off a glass substratum by flow. A ring of membrane remains attached to the glass surface. Contact is maintained between the cell body and the membrane ring by means of a membrane tether (Blackshear et al. 1971). Fig. 3 shows that the ring of membrane breaks up into regularly spaced spherical bodies by growth of a surface wave in a time of the order of 0·3 s. The tether also consists of a row of beads. It is known that cylinders of viscoelastic liquids with a sufficiently high elastic component become unstable not by breaking into a line of regular discrete droplets but by changing into an irregularly beaded thread (Huebner & Chu, 1971). Irregularly beaded threads have been observed (Deeley, Crum & Coakley, 1979) in membrane tethers produced when erythrocytes were stressed before the denaturation of spectrin was complete (Fig. 4). There was a temperature-time window for the production of uniformly beaded tethers (Table 2). The lower limits were set by the need to denature spectrin while the upper limit arose from a stiffening of the membrane as the denatured spectrin cross-linked to give an elastic rigidity to the cell (Deeley *et al.* 1979).

Beading of long cylindrical cell processes when the cytoskeleton is weakened appears to be a general phenomenon. Such beading has been observed in erythrocytes stressed rapidly at room temperature (Hochmuth *et al.* 1973), in the axopodia of the protozoan *Actinosphaerium nucleofilum* when the axopodial microtubular system disintegrated *in vivo* under high pressure (Tilney *et al.* 1966), in well-extended pseudopodia of *Amoeba proteus* following a pressure-induced shift in the cytoplasmic sol-gel equilibrium (Marsland, 1964) and in brush border microvilli exposed to calcium in the range necessary to activate the actin severing protein, villin (Burgess & Prum, 1982).

Analysis of the fragmentation of heated erythrocytes has provided quantitative information on the extent to which different morphogens can influence the bending properties of membranes. The thermal fragmentation patterns of control erythrocytes in physiological saline or of cells exposed to morphogens while in

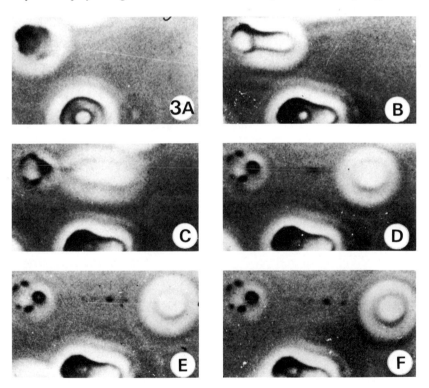

Fig. 3. Cinemicrographs of vesicle formation on a membrane ring and membrane cylinder drawn from an erythrocyte by flow stress applied at the denaturation temperature of spectrin. The frames are sequential with a time interval of 63 ms between frames: (A) Stress has just commenced, (B) the cells are distorted by flow, (C) the upper cell is pulled from the glass substratum and rapidly pulls out a tether; note that the membrane ring which remains attached to the glass has begun to break up, (D) the ring continues to break up (E,F) beads are now discernible on the tether (Crum *et al.* 1979).

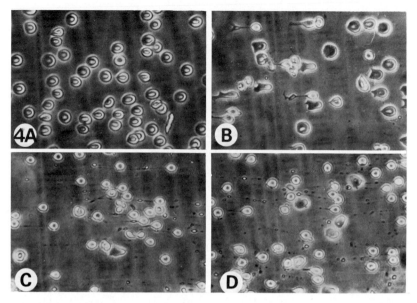

Fig. 4. Photomicrographs showing different categories of response of erythrocytes to flow stress at elevated temperatures: (A) Category 1; shows no tether formation (B) Category 2; unbeaded tethers on some cells (C) Category 3; beading is present but it is not well defined and is irregular (D) Category 4 predominance of long regularly beaded tethers (Deeley *et al.* 1979).

saline, were quantified by calculating the average (W) of the number of waves which grew on the cell rims. W decreased when cell surface charge was depleted (Coakley & Deeley, 1980 and Doulah, Coakley & Tilley, 1984) or when the diffusion potential was increased (Doulah *et al.* 1984), when divalent cations were in the suspending phase (Coakley & Doulah, 1984) or in the presence of cationic drugs (Nwafor & Coakley, 1985; 1986). The value of W increased in the presence of anionic drugs (Nwafor & Coakley, 1985; 1986).

To summarize the information obtained from the erythrocyte system:

(1) The temperature for morphological change is the denaturation temperature of the main cytoskeletal protein.

Table 2. *Different categories of response of cells to a stress pulse after incubation at various times and temperatures. The numbers refer to the different categories as illustrated in Fig. 4 (Deeley* et al. *1979)*

Temperature (°C)	Time (s)							
	30	60	90	120	180	300	600	900
48	1	1	1	1	1	1	1–2	1–2
49	1	1	1	1	1–2	1–2	1–2	1–2
50	1	1	1–2	2	2	2	2–3	2–3
51	3	3	3–4	4	4	2	2	1–2
52	3	3–4	3–4	4	4	3	2	1–2
53	3	3	3–4	4	4	3–4	3	1
55	3	3	3	3	2–3	1–2	1	1

(2) This denaturation occurs at a lower temperature than that required for the permeability change which results in haemolysis.

(3) The morphological change is instantaneous when the heating rate is rapid. At low heating rates or on continued heating at the lower limit of the denaturation temperature range renaturation of denatured cytoskeletal protein can result in a cytoskeleton which retains some elastic properties. Morphological changes at the denaturation temperature are then much less dramatic and, for instance, a low amplitude wavy profile can develop without resulting in cell fragmentation.

(4) Cell processes bead in a spatially regular way if the elastic properties of the cytoskeleton are seriously impaired. Less regular beading implies that the cytoskeleton is behaving like a viscoelastic system. The presence of irregular beading of processes may be taken to imply impairment of the cytoskeletal function.

(5) Thermally stressed cells require careful handling to avoid modification of morphological change by hydrodynamic flow prior to cell examination.

The cytoskeleton of eucaryotic cells

Composition and structure

While spectrin is found, to some small degree, in the cytoskeleton of most mammalian cells, the main cytoskeletal proteins in any one cell are actin, tubulin and one or more intermediate filament (IF) monomers (e.g. vimentin in cultured fibroblasts). These proteins are organized in the three major classes of cytoskeletal fibres: 7nm diameter actin microfilaments, 24nm diameter microtubules and 7–15nm diameter intermediate filaments. These fibres, with their associated proteins, are involved in cell shape change and in various cell movements. Both actin filaments and microtubule filaments are formed by polymerization of protein subunits G actin and tubulin, respectively. In contrast, there do not appear to be soluble pools of IF proteins. Nor are intermediate filaments known to undergo reversible cycles of polymerization and depolymerization although they are polymers of monomer units. Myosin is found in small amounts in all non-muscle cells.

Tubulin monomers have a temperature optimum for polymerization at 37°C (Olmsted & Borisy, 1973). Polymerization requires GTP and is inhibited by calcium (Timasheff & Grisham, 1980). During polymerization the rate of addition of subunits to one end of a tubule is several times greater than at the other end. Depolymerization of dimers also occurs at different rates at the two ends. Thus at steady state the microtubules function as treadmills. Colchicine depolymerizes microtubules through its ability to bind to tubulin dimers. A dimer with a bound colchicine molecule can add to the end of a microtubule. The presence of the colchicine inhibits the addition of further dimers. Colchicine does not affect the disaggregation process, therefore depolymerization occurs. Although the kinetic situation is quite complex actin microfilaments are also believed to go through a treadmilling process (Pollard & Cooper, 1986). Cytochalasins specifically block the polymerization of actin filaments by binding to an end of a growing filament. Cytochalasin does not affect depolymerization.

In many non-muscle cells an abundance of actin microfilaments is found in the cortex. Adhesion of cells to a substrate induces conversion of much of this random assortment of actin microfilaments into stress fibres (Vasiliev, 1985), which are very long microfilament bundles. Stress fibres are generally located in the region of cytoplasm just inside the plasma membrane and many appear to insert an end into the plasma membrane at a point of cell-substrate attachment. Stress fibres have been shown to contain myosin, α-actinin and tropomyosin intermittently along their length and also the regulatory protein myosin light chain kinase (Darnell, Lodish & Baltimore, 1986).

In non-mitotic cultured fibroblasts microtubules form a complex cytoplasmic network that criss-crosses the cell. The microtubule free ends radiate towards the plasma membrane from microtubule organizing centres (Brinkley, 1985) near the cell nucleus. Intermediate filaments often terminate at the nuclear membrane and at the cell periphery (Steinert & Parry, 1985). Spectrin is involved in the membrane-IF binding site (Mangeat & Burridge, 1984). IFs in axons appear to be connected to the intracellular lattice of microtubules (Hirokawa, 1982). Treatment of fibroblasts with colchicine, an inhibitor of the polymerization of tubulin, causes a complete dissolution of microtubules over a period of several hours while vimentin remains in fibres which clump in bundles near the nucleus (Hynes & Destree, 1978). Because IFs are very insoluble and are permanent features of the cytoplasmic architecture it is often assumed that they play a crucial role in the regulation of cell shape. However, this may not be the case, at least in cultured fibroblasts. Injection of a monoclonal antibody specific to the IF protein causes a rapid collapse of IFs to a perinuclear region. This disruption occurs without any apparent alteration in the microfilament or microtubule systems and without any obvious effect on cell morphology, mitosis or motility (Klymkowsky, Miller & Lane, 1983).

The complexity of the assembly and function of the actin system in cells (where the number of distinct regulatory proteins may exceed ten for many cells) has recently been reviewed by Pollard & Cooper (1986). These proteins include the following. (i) Actin monomer binding proteins which can sequester monomers in a non-polymerizable complex. (ii) Capping proteins which bind to one end of actin filaments and influence subunit reactions there. These proteins, like cytochalasins, inhibit the addition of actin monomers to growing barbed ends of actin filaments. Capping proteins accelerate spontaneous polymerization of actin monomers resulting in numerous short filaments with capped barbed ends. Moreover addition of capping proteins to solutions of long filaments very rapidly results in capped short filaments (filament severing). Gelsolin, a capping protein found in many vertebtrate cells, requires calcium for activity and is therefore a possible agent for actin filament breakup when calcium homeostasis is impaired. (iii) Low molecular weight severing proteins which bind actin monomers and sever actin filaments. (iv) Proteins that bind to the sides of actin filaments. One group binds to only one filament at a time. A second divalent group cross-links filaments while 'bundlers' bind filaments into tight parallel arrays. Some bundling proteins are inhibited by calcium. The cross-linking proteins include spectrin analogues and calcium-inhibitable alpha-actinin. (v) Proteins that bind actin and actin-binding proteins to membranes. Most of these proteins bind the sides of actin filaments to membranes. Actin filaments appear to be attached to the membrane in more than one way as only part of the membrane-associated actin is removed from membrane fractions by non-ionic detergents (Taylor *et al.* 1976). Godman & Miranda (1978) suggested that the different actin-binding mechanisms may be required for cytoskeletal (coplanar with the membrane) and for contractile functions of actin. Vinculin, a $130 \times 10^3 M_r$ protein localized at attachment plaques in fibroblasts where actin filament bundles meet the membrane, has, historically, received a lot of attention as an actin/membrane binding protein (Pollard & Cooper, 1986). Alpha-actinin which, with vinculin, interacts with the ends of actin filaments in muscle cells is also found in non-muscle cells. Bennett (1985) has pointed out that many cells have proteins like spectrin, band 4.1 and membrane-associated ankyrin (Fig. 1) and has reviewed the evidence that spectrin and ankyrin may be one of the means of mediating association of actin, microtubules and possibly intermediate filaments to the plasma membrane. Spectrin in nucleated cells may not have the membrane stabilizing property it displays in erythrocytes (Mangeat & Burridge, 1984). Actin-binding protein, which serves (as does spectrin) as a substrate for a calcium-stimulated protease, links actin to platelet membranes (Fox, 1985).

Temperature effects on some cytoskeletal proteins

Comparative studies of the structural stabilities of skeletal muscle actin from vertebrates adapted to living at different temperatures show that (for vertebrates living at atmospheric pressure) the heat stabilities of the actin monomers increased with rising average body temperature (Swezey & Sobero, 1982). The heat stabilities of the G-actins were determined from an assay, at 38 °C of DNase 1 inhibition by polymerizable actin. Since thermal denaturation of rabbit and of chicken (body temperatures 37 °C and 39 °C resp.) actin occurred at 38 °C these studies of thermal denaturation are of limited guidance to the situation which might arise for monomers heated *in vivo*. The major contributor to the stabilization free energy of actin filaments (as a particular example of the general situation for multisubunit protein systems) is thought to be hydrophobic interactions whereas polar and/or charged group interactions establish the specificity of protein–protein binding (Swezey & Sobero, 1982). The extreme sensitivity of actin filament assembly to temperature, pressure and solutes has been attributed to the effects of these variables on hydrophobic interactions and on protein tertiary structure. Some loss of alpha helical structure can occur at temperatures slightly above the normal temperature of a vertebrate. Physical studies of actin conformation (Lehrer & Kerwar, 1972) show thermal denaturation of G-actin on exposure to temperatures in excess of 40°C. F-actin required temperatures of 55°C before denaturation occurred. This important observation suggests a way in which monomer, released in the normal way by treadmilling of a filament, would join the monomer pool which was undergoing denaturation and would thus not be available for repolymerization to maintain the filament.

In low ionic strength medium sea urchin sperm microtubules depolymerize rapidly at 40°C (Stephens, 1970). The author concluded that the thermal depolymerization was an effect on equilibrium rather than simple melting since the components do not become soluble at single discrete temperatures. Elevated temperatures have been shown to disrupt the cytoplasmic microtubule organization *in vivo* (Rieder & Bajer, 1977) in a manner which is relevant to elevated temperature effects on cell shape (Turi *et al.* 1981). Microtubules isolated from calf (body temperature about 38·5°C) will disassemble at 41°C and will denature at 43°C. In a viscometric study of crude brain extracts from a variety of vertebrates it was concluded that the (reversible) disassembly temperatures of microtubule proteins were within the range of fever temperature of these species. Temperatures lethal to the animals cause microtubule protein denaturation (irreversible disassembly of microtubules (Turi, Lu & Lin, 1981)). Purified microtubule protein was not as sensitive to 41°C as crude extracts. It was suggested that the elevated temperatures may promote the binding of depolymerization factor to the microtubules in the crude extracts. The temperature stability of a number of microtubule-associated proteins has been determined (Timasheff & Grisham, 1980).

Two major helix-coil transitions in the myosin rod appear to stem from melting in the short S-2 and light meromyosin domains. The midpoints of these transitions are near 46° and 53°C respectively (Tsong, Himmelfarb & Harrington, 1983) while a melt in the LMM-HMM hinge region appears to occur at around 41°C.

Morphological effects on heated mammalian cultured cell lines

Most cultured mammalian cell lines appear to be able to survive exposures to temperatures of 40°C for times of the order of 3 h. Loss of cell viability is detected over such time intervals at incubation temperatures about 42°C.

The morphological changes which occur in Chinese hamster ovary (CHO) cells heated in microcapillaries in a water bath at 43°C for times up to 3h have been described by Bass, Coakley, Moore & Tilley (1982). These authors identified the following stages of morphological change:

Stage A (Control cells): Interphase cells were relatively flat and covered a large area of the substratum. The upper cell surface bore numerous microvilli. The distribution of organelles in the cell cytoplasm was uniform (Fig. 5A).

Stage B (30 min): The cytoplasm had separated into (i) a clear cytoplasm and (ii) a more granular cytoplasm containing the nucleus and cell organelles near the centre of the cell. The nucleus contained dark patches of condensed chromatin. There was some suggestion of ruffling at the cell margins (Fig. 5B).

Stage C (30–60 min): The margin of most cells had retracted, leaving some attachments to the substratum. The attachments were longer and narrower than in controls and some were beaded. Beaded fibres were seen in the vicinity of the cells. The dense cytoplasm and organelles were concentrated around the perinuclear region. Some cells showed a network extending from the cell margin towards the nucleus (Fig. 5D). Beading of cell processes was observed (Fig. 6A). A few cells had blebs on the cell surface.

Stage D (1–2 h): The blebs were larger and appeared to be fluid-filled and bound by a membrane. Cell organelles were excluded from the bleb and they seemed to

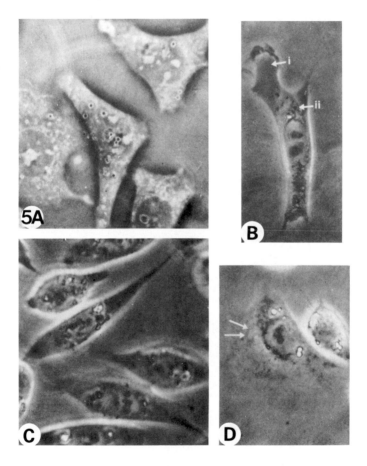

Fig. 5. CHO-K1 cells grown in glass microcapillaries at 37°C and heated at 43°C for various times. (A) Time zero – the cells have a large area in contact with the substratum. (B) 30 min – the cytoplasm has separated into two types: (i) clear cytoplasm and (ii) a more granular cytoplasm. (C) 1 h – the dense cytoplasm and organelles were concentrated around the perinuclear region. The margin of most cells had retracted leaving some attachments to the substratum. (D) 1 h – a network appears to extend from the cell margin towards the nucleus (arrows) (Bass *et al.* 1982).

be kept within the dense cytoplasm. After 2 h at 43°C the majority of the cells were rounded and there was usually more than one bleb per cell. 90 % of cells bore blebs after 2 h compared with 10 % after 1 h (Fig. 6B).

Stage E (2–3 h): Few cells had lysed in the microcapillary system, almost all of the intact cells were circular refractile bodies. Most of the cells had lost their blebs.

At 42°C no blebs were seen until the elevated temperature had been maintained for 150 min. The small amount of lysis observed in the microcapillary system at 43°C suggested that some of the lysis observed in earlier work (Bass *et al.* 1978) in

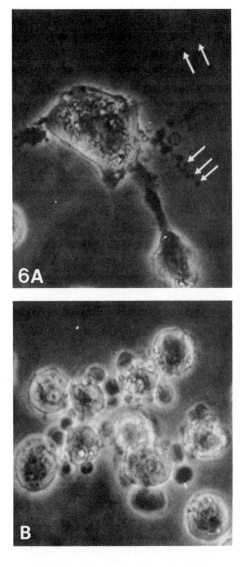

Fig. 6. CHO-K1 cells as in Fig. 5 (A) 1 h – several beaded fibres are seen in the vicinity of the cells (B) 2 h – the majority of the cells were rounded and some cells had multiple blebs (Bass *et al.* 1982).

cells heated while attached to coverslips resulted from stress applied to the cells during the experimental and cell-fixing procedures.

Light and scanning electron microscopy have shown that juxtanuclear organelle condensation, bleb formation and loss of microvilli (Lin, Lui & Tsai, 1978) are striking features of heated cells (Bass, Moore & Coakley, 1978; Borrelli, Wong & Dewey, 1986). The observed changes will be discussed under two general headings (i) effects on the cytoskeleton organization such as membrane retraction, loss of microvilli, rounding up of the cell and (ii) membrane bleb formation.

Heat induced cell retraction

Bass *et al.* 1982 pointed out that the morphological effects of heat were essentially the reverse of the shape changes observed when cells spread on a glass substratum (Wang & Goldman, 1978). Cell spreading involves the coordinated interaction of microfilaments, microtubules and intermediate filaments (Vasiliev, 1985).

Studies of the numerous effects, which have parallels in thermal effects on morphology, brought about in animal cells by exposure to cytochalasins contributed significantly to the early developments in the understanding of the organization of the cell cytoskeleton. These effects were reviewed comprehensively at the end of that period by Godman & Miranda (1978). Cytochalasins can directly and distinctly affect at least two different cellular sites: (i) the membrane, because of the hydrophobic nature of cytochalasins and (ii) the microfilament assemblies (Tannenbaum, 1978). The morphological changes wrought by the cytochalasins suggest that they occur in successive stages, of which the first and most important is a sustained general contraction.

The early effects of cytochalasin are dose dependent and, in the range of efficient concentration ($0 \cdot 25 - 0 \cdot 75 \, \mu g \, ml^{-1}$ cytochalasin D) become visible within 1–5 min (Godman & Miranda, 1978). These early effects are: (i) inhibition of undulations ruffling movements of the cell surface: (ii) withdrawal of microvilli, (iii) unequivocal alteration and redistribution of the membrane-associated microfilament apparatus with accompanying rearrangements at the cell surface (iv) zeiosis and its variants (with some cells only, not, for instance, with lymphocytes): (v) Cell retraction; because adhesion is not disturbed many kinds of cells develop a dendriform appearance as the lamellar cytoplasm between attachment sites is withdrawn centrally. In the fully retracted state most of the cytoplasm is withdrawn. Retraction processes radiate from subapical parts of the rounded cell body. (vi) Often eventual flattening or arborization of the cell occurs. These manifestations are cell-type-dependent; one or other of them assume greater prominence in the different cell types and each individual effect is dose dependent. The time required to achieve a maximal stable effect varies widely (10 min to 48 h). At moderate concentrations of drug the effects are all readily reversible (and are independent of new protein synthesis (Croop & Holtzer, 1975)) on withdrawal of the drug. High concentrations or long exposure times can compromise viability (Godman & Miranda, 1978).

Cell retraction is a general response to physiological, morbid or pharmacologic stimuli including mitotic rounding, exposure to cold or to high hydrostatic pressure, trypsinization, actinomycin D, cationic local anaesthetics (Godman & Miranda, 1978) and hypoxia (Lemasters *et al.* 1987). In each of the above examples cell rounding has been thought to involve disassembly of the membrane-related microtubule apparatus (often arguably related to augmented intracellular Ca^{2+} levels) and also of most microfilament cables. In each case adherence to substrate is either lost or weakened. The superficial appearance that results is the same: rounded cell body from which straight radial spikes, retraction fibres, emerge at subapical levels of the cell body and attach to the substrate at their tips. This appearance and the supposed sequence of events differ in detail from cell retraction induced by cytochalasins, in which microtubules, IFs (until late in the course of exposure to cytochalasin) and sites of adherence remain quite intact (Godman & Miranda, 1978).

When cytochalasin-retracted fibroblasts are subsequently exposed to colcemid while still in the presence of cytochalasin the cylinder-like connections of the rounded cell to the substratum break up into beaded forms (Croop & Holtzer, 1975). This observation, because of the type of beaded fibre effect (as in Fig. 4) shown in Fig. 7A, tends to place thermal damage with the cell retracting systems mentioned above which have more complete cytoskeleton disorganizing effects than cytochalasin. Exposure to colchicine alone causes a spread cell to round up somewhat (Croop & Holtzer, 1975; Mangeat & Burridge, 1984) but not to retract as in cytochalasin.

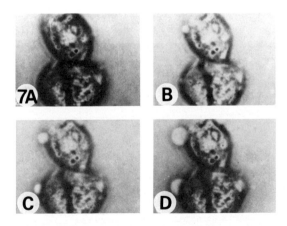

Fig. 7. Selected frames from a cinemicroscopy record of bleb formation on cells grown in glass microcapillaries at 37°C and heated at 57°C on a microscope stage for (A) 180 s, (B) 200 s, (C) 215 s and (D) 230 s. (Bass *et al.* 1982).

Bleb formation on heated cells

Bleb formation has received attention from a number of workers. Lin, Wallach & Tsai (1973) and Lin *et al.* (1978) showed that relatively short exposures of cultured lymphoid cells to temperatures of 45–50°C resulted in a loss of microvilli and a cobblestone plasma membrane morphology. Membrane blebs have been observed on the surfaces of heated biliary epithelial cells (Wills, Findlay & McManus, 1976), on CHO cells (Bass *et al.* 1978, 1982; Borrelli *et al.* 1986) and on heated human granulocytes (Lionetti, Lin, Mattaliano, Hunt & Valeri, 1980).

Bass *et al.* (1982) used cine-photomicroscopy to study the development of blebs on individual cells. In order to accumulate data on a large number of cells in a reasonable time the cine-microscopy was carried out at temperatures in the range 48–75°C. The same principal cellular changes – e.g. retraction, rounding – as were recorded by photomicroscopy at 43°C were seen at the higher temperatures. The stages of bleb formation at the higher temperatures were similar to those observed at 43°C but they appeared sooner.

The granular cytoplasm of cells heated rapidly to 57°C and maintained at that temperature for 60 s was concentrated in the perinuclear region. After 180 s bulges appeared at the margin of the cells. Each bulge enlarged to form a bleb filled with an apparently clear fluid – no organelles were visible (Fig. 7). When more than one bleb grew on a single cell the growth began at the same time at various points on the cell periphery and the blebs grew at the same rate (Bass *et al.* 1982). This observation of synchronous growth of blebs on individual cells contrasts with the wide variation in degree of blebbing seen by Borrelli *et al.* (1986) on a population of synchronous G1 phase CHO cells heated in suspension. The contrast implies that once an individual cell suffered the lesion which initiated a bleb then blebbing at more than one point on that cell was very likely. The blebs observed at 57°C often burst as if the membrane could not expand further (Bass *et al.* 1982). Adjacent blebs on the cell surface sometimes merged to form a larger bleb which remained attached to the cell.

Bass *et al.* (1982) pointed out that the percentage (90%) of cells showing blebs after heating for 2 h in microcapillaries at 43°C was the same as the number of cells losing their viability in separate experiments in culture bottles exposed to the same heating regime. Borrelli *et al.* (1986) examined the suitability of bleb formation as a marker to identify cell death immediately after heat treatment without the delay involved in viability studies. The ability to separate doomed cells by parameters other than colony formation might provide insight not only into the mechanism of hyperthermia-induced cell death, but might also provide a means of separating (for experimental studies) viable from non-viable cells within a population of heated cells. Experiments on blebbing and viability were carried out on a synchronous culture of CHO cells (Borrelli *et al.* 1986). Immediately after heating in suspension at 45·5°C an aliquot of cells was taken from the flask and individual cells were scored for both blebbing and for subsequent growth.

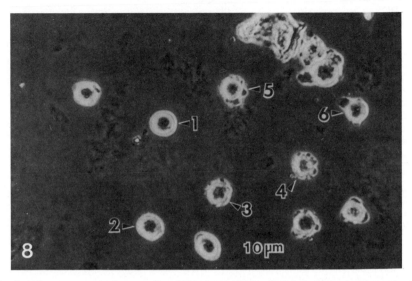

Fig. 8. Phase contrast micrographs of G_1 phase CHO cells heated for 10 min at 45·5 °C. Some cells retained shapes indistinguishable from control cells (state 1) while others exhibited various degrees of membrane surface blebbing (states 2–6) (Borrelli *et al.* 1986).

A scoring system (the six states of which are illustrated in Fig. 8) was established based on the diameter of the largest blebs on each cell relative to the diameter of the cell itself. Within 1 h of heating many of the smaller blebs were reabsorbed while the larger blebs were released. The percentage of cells scored as states 1,2,3 was equal to the percentage of cells surviving each heat treatment. Only for heating durations such that blebs were released during heating and consequently not scored did the agreement fail.

The correlation between membrane blebbing and cell death held only for cells heated in the G_1 phase of the cell cycle. Many state 1–3 cells heated in S phase failed to form colonies. This result was taken (Borrelli *et al.* 1986) to be consistent with the observation that S phase cells are more sensitive than G_1 cells to a given hyperthermic exposure (where cell death has been attributed to effects involving nucleic acids, Westra & Dewey, 1971). The observation of Miranda, Godman & Tanenbaum, 1974) that cytochalasin-induced cell retraction is less in S phase (where viability is most sensitive to heat) than in G_1 phase further suggests that the state of integrity of microfilaments does not play a large part in the death of cells heated in S phase. Care must be exercised in interpreting the significance of colony formation results for cells heated in suspension (Borrelli *et al.* 1986) in case the reduced colony count represents an impairment of the ability of the cell to spread rather than the development of a lethal lesion (which would also kill cells which were spread before heating).

Transmission electron microscopy of the heated cells showed that the blebs did not contain mitochondria or endoplasmic reticulum. Heated cells with low blebbing scores exhibited morphological alterations at the level of electron microscopy resolution to a lesser extent than cells with high scores. Further work is in progress to determine if there is a one-to-one relationship between scoring state and the extent of intracellular morphological change.

Bass *et al.* (1982) showed that the activation energy of bleb formation in the temperature range 43–75 °C was $164 \, kJ \, mol^{-1}$ which was significantly lower than the activation energy of $596 \, kJ \, mol^{-1}$ reported for cell lethality by Sapareto, Hopwood, Dewey, Mundundi & Gray (1978). Borrelli *et al.* (1986) draw attention to the fact that the viability activation energy for cell death decreases with increasing temperature (Landry & Marceau, 1978) and suggest that the bleb/cell-death correlation may hold over the temperature range of 42–75 °C.

Bleb development in general

The term 'bleb' has been broadly used in the literature to describe rounded protuberances of the cell membrane. Two types of bleb, potocytotic and zeiotic, are recognized. Potocytotic blebs contain clear fluid and their contents are separated from the cell nucleus and organelles by a distinct boundary line. Zeiotic blebs begin as herniation of cytoplasm and endoplasmic reticulum through the cell cortex. Occasionally endoplasmic organelles enter the protuberances (Bass *et al.* 1982). The observations of Bass *et al.* (1982) and of Borrelli *et al.* (1986) confirm that heat-induced blebs are of the potocytotic type.

Potocytotic blebs arise under a variety of conditions (Bass, 1980), e.g. when tissue explants are exposed to hypertonic or hypotonic solutions (Hogue, 1919) or to ether vapour (Rosenfeld, 1932). Many cell types, with the exception of squamous cells, developed potocytotic blebs when incubated in physiological saline and buffered Ringer's solution containing glucose (Zollinger, 1948). Fluid-filled blebs can be induced *in vivo* and *in vitro* by a variety of agents which have an affinity for sulphhydryl groups (Belkin & Hardy, 1961; Scott, Perkins, Szchunke, Hoerl & Maercklein, 1979). Depletion of cellular ATP has been associated both with inhibition (Scott *et al.* 1979) and with appearance of membrane blebs (Lemasters, Di Guiseppi, Niemenen & Herman, 1987). Borek & Fenoglio (1976) reported blebbing following X-irradiation and Lemasters *et al.* (1987) observed *in vivo* blebbing in liver tissue under hypoxic conditions.

It has been shown that the membrane of potocytotic blebs is derived from and has the same characteristics as the plasma membrane, e.g. the membrane of the bleb has intramembranous particles with a density similar to that present on intact cells and a valid enzyme marker for fibroblast plasma membrane is found on the bleb membrane (Scott *et al.* 1979). Freeze-fracture electron microscopy revealed no change in membrane ultrastructure of heated blebs (Borrelli *et al.* 1986) except on about 15 % of the blebs that contained bald patches devoid of membrane. The diffusion coefficient of integral proteins on muscle cell blebs is greater than that on normal membrane and is close to the value predicted for hydrodynamic drag in the lipid membrane (Tank, Wu & Webb, 1982). The enhanced molecular diffusibility was attributed to release from unidentified natural constraints that is induced in some way by detachment of the cell membrane. Coss *et al.* (1979) observed a detachment of cortical microfilaments from bleb membranes of cells heated to 41·5°C.

Exposure of a rat carcinoma cell line to a carbamoylating nitrosourea inhibits glutathione reductase and depletes thiols. The most prevalent drug-induced lesions in interphase cells include membrane blebbing, development of irregular crescent shaped nuclei and loss of plasma membrane filopodia (Tew, Kyle, Johnson & Wang, 1985). The morphological changes correlate with cytotoxic responses of the cell and would be consistent with drug induced inhibition of glutathione reductase. Supplementing the isolation medium for lung epithelial cells by the constituent amino acids of glutathione reduced surface membrane blebbing compared with unsupplemented controls (Mettler, Yano, Kikkawa & Ivasauskas, 1984). The addition of free -SH groups in the form of $0·002–0·004$ mol l^{-1} L-cysteine inhibited or retarded chemically induced blebbing (Belkin & Hardy, 1961). However, Bass (1980) found that L-cysteine had no effect on heat-induced blebbing. She suggested that heat might produce the same injurious end point as the compounds binding -SH groups even though -SH groups do not seem to be directly involved in the heat effect. Moore, Thor, Moore, Nelson, Moldeus & Orrenius (1985) reported that acetaminophen and its major toxic metabolite (NaPQI) in hepatocytes isolated from pretreated rats caused an elevation of cytosolic Ca^{2+}, which was accompanied by plasma membrane blebbing. This disruption of Ca^{2+} homeostasis always preceded cell death. NaPQI also produced a rapid depletion of both cytosolic and mitochondrial reduced glutathione as well as a loss of protein-bound-SH groups. The authors suggest that the toxic effects arise from a disruption of Ca^{2+} homeostasis secondary to the depletion of soluble and protein-bound thiols and that the mechanism may be of general applicability to a variety of hepatotoxins.

Lemasters *et al.* (1987) tested, in cyanide and iodoacetate treated rat hepatocytes, the hypothesis that a rise in cytosolic free calcium is the stimulus for bleb formation and is the final common pathway to irreversible cell injury. They found, unexpectedly, that cytosolic free calcium did not change during bleb formation or before loss of cell viability. Cell death was precipitated by a sudden breakdown of the plasma membrane permeability barrier, possibly caused by rupture of a cell surface bleb. The bleb formation was attributed in this case to interference with microfilament and microtubule assembly by the 'chemical hypoxia' induced depletion of ATP.

Conclusions

Morphological changes in cultured cell lines differ from those in erythrocytes in a number of important respects. For example erythrocyte fragmentation arises from denaturation of a specific protein, spectrin, on short exposure to temperatures much lower than those necessary to bring about the permeability change required to induce haemolysis. The rapid (within 1 s) fragmentation of the erythrocyte at a threshold temperature contrasts with the situation for heated cultured cell lines where the shape changes, even when cells are heated rapidly to temperatures as high as 57°C, can require over 200 s before bleb development begins. Bleb growth and swelling therefore involve more than the simple denaturation of the proteins linking the membrane to the cytoskeleton. The relatively long time delay to cell death, the degree of similarity between thermal shape changes and those induced by cytochalasin and the observation of microtubule depolymerization in heated cells argue that depolymerization of the cytoskeleton fibres occurs as a consequence of hyperthermia. The evidence that thermal denaturation of actin monomers and a thermal effect on microtubule-associated proteins could bring about a breakdown of the treadmilling process has been reviewed above.

The influence of calcium concentration on actin binding and actin severing proteins and on proteases which degrade actin binding proteins highlights the importance of calcium regulation for the correct functioning of the cytoskeleton. The techniques are now available (Lemasters *et al.* 1987) which allow specific questions to be answered and will help decide between different hypotheses of thermal damage to cells. There is increasing evidence that advanced stages of membrane blebbing correlate with cell death at least in particular stages of the cell cycle.

Mangeat & Burridge (1984) have separated, for platelets, the concept of the membrane skeleton and the cytoskeleton in a eucaryotic cell. The apparent absence of any sudden effect of heat on morphology during heating may reflect the dominance of the multicomponent cytoskeleton on the stability of the eucaryotic cell. On the other hand as further information on the relative cytoskeletal protein compositions of differentiated cells becomes available (for instance membranes from lung and lens have large amounts of spectrin (Bennett, 1985)) it may be worthwhile to extend the very small number of cell types whose morphological response to temperature rise has been determined in order to establish if heating conditions exist which would lead to cell death through fragmentation on heating past a specific protein melting temperature.

The author is indebted to many colleagues with whom he collaborated on different phases of some of the work described above. He is particularly grateful to Dr J. O. T. Deeley in respect to the erythrocyte work and to Dr H. Bass and Dr J. L. Moore in relation to the heating of cultured cells. Parts of the work were supported by the M.R.C., The Wellcome Trust and by Tenovus.

References

ANDERSON, R. A. & LOVRIEN, R. E. (1984). Glycophorin is linked by band 4.1 protein to the human erythrocyte membrane skeleton. *Nature, Lond.* **307**, 655–658.

APPEL, F. C. & LOW, P. S. (1982). Evaluation of structural interdependence of membrane spanning and cytoplasmic domains of Band 3. *Biochemistry* **21**, 2151–2157.

BASS, H. (1980). The effects of hyperthermia of mammalian cells. Ph.D. Thesis, University of Wales, UK.

BASS, H., COAKLEY, W. T., MOORE, J. L. & TILLEY, D. (1982). Hyperthermia-induced changes in the morphology of CHO-K1 and their refractile inclusions. *J. therm. Biol.* **7**, 231–242.

BASS, H., MOORE, J. L. & COAKLEY, W. T. (1978). Lethality in mammalian cells due to hyperthermia under oxic and hypoxic conditions. *Int. J. Radiat. Biol.* **33**, 57–67.

BELKIN, M. & HARDY, W. G. (1961). Relation between water permeability and integrity of sulphydryl groups in malignant normal cells. *J. Biophys. Biochem. Cytol.* **9**, 733–745.

BENNETT, V. (1985). The membrane skeleton of human erythrocytes and its implications for more complex cells. *Ann. Rev. Biochem.* **54**, 273–304.

BLACKSHEAR, P. L. JR, FORSTOM, R. J., DORMAN, F. D. & VOSS, G. O. (1971). Effects of flow on cells near walls. *Fed. Proc.* **30**, 1600–1611.

BOREK, C. & FENOGLIO, C. M. (1976). Scanning electron microscopy of surface features of hamster embryo cells transformed *in vitro* by X-irradiation. *Cancer Res.* **36**, 1325–1334.

BORRELLI, M. J., WONG, R. L. L. & DEWEY, W. C. (1986). A direct correlation between hyperthermia-induced membrane blebbing and survival in synchronous G_1 CHO cells. *J. cell. Physiol.* **126**, 181–190.

BRANDTS, J. F., ERICKSON, L., LYSKO, K., SCHWARTZ, A. T. & TAVERNA, R. D. (1977). Calorimetric studies of the structural transitions of the human erythrocyte membrane. The involvement of spectrin in the A transition. *Biochemistry* **16**, 3450–3454.

BRINKLEY, B. R. (1985). Microtubule organising centers. *Ann. Rev. Cell Biol.* **1**, 145–172.

BURGESS, D. R. & PRUM, B. E. (1982). Reevaluation of brush border motility: Calcium induces core filament solation and microvillar vesiculation. *J. Cell Biol.* **94**, 97–107.

COAKLEY, W. T., BATER, A. J., CRUM, L. A. & DEELEY, J. O. T. (1979). Morphological changes, haemolysis and microvesiculation of heated human erythrocytes. *J. therm. Biol.* **4**, 85–93.

COAKLEY, W. T., BATER, A. J. & DEELEY, J. O. T. (1978). Vesicle production on heated and stressed erythrocytes. *Biochim. Biophys. Acta* **512**, 318–330.

COAKLEY, W. T. & DEELEY, J. O. T. (1980). Effects of ionic strength, serum protein and surface charge on membrane movement and vesicle production in heated erythrocytes. *Biochim. Biophys. Acta* **602**, 355–375.

COAKLEY, W. T. & DOULAH, F. A. (1984). Divalent cation effects on membrane bending in heated erythrocytes. *J. biol. Phys.* **12**, 85–92.

COAKLEY, W. T., NWAFOR, A. & DEELEY, J. O. T. (1983). Tetracaine modifies the fragmentation mode of the heated human erythrocyte and can induce heated cell fusion. *Biochim. Biophys. Acta* **727**, 303–312.

COSS, R. A., DEWEY, W. C. & BAMBURG, J. R. (1979). Effects of hyperthermia (41·5°C) on Chinese hamster ovary cells analysed in mitosis. *Cancer Res.* **39**, 1911–1918.

CROOP, J. & HOLTZER, H. (1975). Response of myogenic and fibrogenic cells to cytochalasin B and to colcemid. *J. Cell Biol.* **65**, 271–285.

CRUM, L. A., COAKLEY, W. T. & DEELEY, J. O. T. (1979). Instability development in heated human erythrocytes. *Biochim. Biophys. Acta* **554**, 76–89.

DARNELL, J., LUDISH, H. & BALTIMORE, D. (1986). *Molecular Cell Biology*. New York: Scientific American Books.

DAVIO, S. R. & LOW, P. S. (1982). Characterisation of the calorimetric C transition of the human erythrocyte membrane. *Biochemistry* **21**, 3585–3593.

DEELEY, J. O. T. & COAKLEY, W. T. (1983). Interfacial instability and membrane internalisation in human erythrocytes heated in the presence of serum albumin. *Biochim. Biophys. Acta* **554**, 90–101.

DEELEY, J. O. T., CRUM, L. A. & COAKLEY, W. T. (1979). The influence of temperature and incubation time on deformability of human erythrocytes. *Biochim. Biophys. Acta* **554**, 90–101.

DIMITROV, D. S. & JAIN, R. K. (1984). Membrane stability. *Biochim. Biophys. Acta* **779**, 437–468.

DOULAH, F. A., COAKLEY, W. T. & TILLEY, D. (1984). Intrinsic electric fields and membrane bending. *J. Biol. Phys.* **12**, 44–51.

FOX, J. E. B. (1985). Identification of actin-binding protein as the protein linking the membrane skeleton to glycoproteins on platelet plasma membranes. *J. biol. Chem.* **260**, 11970–11977.

GALLEZ, D. & COAKLEY, W. T. (1987). Interfacial instability at cell membranes. *Progress in Biophysics & mol. Biol.* In press.

GIBBS, J. W. (1928). *The Collected Works of J. Willard Gibbs*, vol. 1, 2nd edn, pp. 55–349. New York: Longmans Green.

GLASER, R. (1982). Echinocyte formation induced by potential changes of human red blood cells. *J. Membrane Biol.* **66**, 79–85.

GODMAN, G. C. & MIRANDA, A. F. (1978). Cellular contractility and the visible effects of cytochalasin. In *Cytochalasins: Biochemical & Cell Biological Aspects* (ed. S. W. Tanenbaum), pp. 277–430. Amsterdam: North Holland.

HAM, T. H., SHEN, S. C., FLEMING, E. M. & CASTLE, W. B. (1948). Studies on the destruction of red blood cells. *Blood* **3**, 373–403.

HIROKAWA, N. (1982). Cross-linker system between neurofilaments, microtubules and membranous organelles in frog axons revealed by the quick-freeze deep etching method. *J. Cell Biol.* **94**, 129–142.

HOCHMUTH, R. M., MOHANDAS, N. & BLACKSHEAR, P. L. (1973). Measurement of the elastic modulus for red cell membrane using a fluid mechanical technique. *Biophys. J.* **13**, 747–762.

HOGUE, M. J. (1919). The effects of hypotonic and hypertonic solutions on fibroblasts of the embryonic chick heart *in vitro*. *J. exp. Med.* **30**, 617–648.

HUEBNER, A. L. & CHU, H. N. (1971). Instability and breakup of charged jets. *J. Fluid Mech.* **49**, 361–372.

HYNES, R. O. & DESTREE, A. (1978). 10 nm filaments in normal and transformed cells. *Cell* **13**, 151–163.

KLYMKOWSKY, M. W., MILLER, R. H. & LANE, E. B. (1983). Morphology behaviour and interaction of cultured epithelial cells after the antibody-induced disruption of keratin filament organisation. *J. Cell Biol.* **96**, 494–509.

KRISHNAN, K. S. & BRANDTS, J. F. (1979). Interaction of phenothiazine and lower aliphatic-alcohols with erythrocyte membranes – Scanning calorimetric study. *Molec. Pharmacol.* **16**, 181–188.

LANDRY, J. & MARCEAU, N. (1978). Rate-limiting events in hyperthermic cell killing. *Radiat. Res.* **75**, 573–585.

LEHRER, S. S. & KERWAR, G. (1972). Physical studies of actin conformation. *Biochemistry* **11**, 1211–1217.

LEMASTERS, J. J., DI GUISEPPI, J., NIEMENEN, A.-L. & HERMAN, B. (1987). Blebbing, free Ca^{2+} and mitochondrial membrane potential preceding cell death in hepatocytes. *Nature, Lond.* **325**, 78–80.

LIN, P. S., LUI, P. & TSAI, S. (1978). Heat-induced ultrastructural injuries in lymphoid cell. *Expl molec. Path.* **29**, 281–290.

LIN, P. S., WALLACH, D. F. J. & TSAI, S. (1973). Temperature-induced variations in the surface topology of cultured lymphocytes as revealed by scanning electron microscopy. *Proc. natn. Acad. Sci. U.S.A.* **70**, 2492–2496.

LIONETTI, F. J., LIN, P. S., MATTALIANO, R. J., HUNT, S. H. & VALERI, C. R. (1980). Temperature effects on shape and function of human granulocytes. *Expl Hemat.* **81**, 304–317.

MANGEAT, P. H. & BURRIDGE, K. (1984). Immunoprecipitation of non-erythrocyte spectrin within live cells following microinjection of specific antibodies. Relation to cytoskeletal structure. *J. Cell Biol.* **98**, 1363–1377.

MARSLAND, D. A. (1964). *Primitive Motile Systems in Cell Biology* (ed. R. D. Allen & N. Kamiga), pp. 173–185. New York: Academic Press.

MEISTER, B. J. & SCHEELE, G. F. (1967). Generalised solution of the Tomotika stability analysis for a cylindrical jet. *A.I.Ch.E. Journal* **13**, 682–688.

METTLER, N. R., YANO, S., KIKKAWA, Y. & IVASAUKAS, F. (1984). Type II epithelial cells of the lung. VII. The effect of ascorbic acid and glutathione. *Lab. Invest.* **51**, 441–448.

MILLER, C. A. (1978). Stability of interfaces. In *Surface and Colloid Science*, vol. 10 (ed. E. Matijevic), pp. 227–293. New York: Plenum Press.

MIRANDA, A., GODMAN, G. & TANENBAUM, S. W. (1974). Action of cytochalasin D on cells of established lines. II. Cortex and microfilaments. *J. Cell Biol.* **62**, 406–423.

MOHANDAS, N., GREENQUIST, A. C. & SHOHET, S. B. (1978). Effects of heat and metabolic depletion on erythrocyte deformability, spectrin extractability and phosphorylation. In *The Red Cell*, vol. 21 (ed. G. J. Brewer), pp. 453–472. Progress in Clinical and Biological Research. New York: Alan R. Liss Inc.

MOORE, M., THOR, H., MOORE, G., NELSON, S., MOLDEUS, P. & ORRHENIUS, S. (1985). The toxicity of acetaminophen and N-acetyl-p-benzoquione imine in isolated hepatocytes is associated with thiol depletion and increased cytosolic Ca^{2+}. *J. biol. Chem.* **260**, 13035–13040.

NWAFOR, A. & COAKLEY, W. T. (1985). Drug-induced shape change in erythrocytes correlates with membrane potential change and is independent of glycocalyx charge. *Biochem. Pharmacol.* **34**, 3329–3336.

NWAFOR, A. & COAKLEY, W. T. (1986). Charge independent effects of drugs on erythrocyte morphology. *Biochem. Pharmacol.* **35**, 953–957.

OLMSTED, J. B. & BORISY, G. G. (1973). Characterisation of microtubule assembly in porcine brain extracts by viscometry. *Biochemistry* **12**, 4282–4289.

POLLARD, T. D. & COOPER, J. A. (1986). Actin and actin-binding proteins. A critical evaluation of mechanisms and functions. *Ann. Rev. Biochem.* **55**, 987–1035.

PONDER, E. (1949). Shape and shape transformations of heated human red cells. *J. exp. Biol.* **26**, 35–45.

RAKOW, A. I. & HOCHMUTH, R. M. (1975). Thermal transition in the human erythrocyte membrane: effect on elasticity. *Biorheology* **12**, 1–3.

RAKOW, A., SIMCHON, S., SUNG, L. A. & CHIEN, S. (1981). Aggregation of red cells with membrane altered by heat treatment. *Biorheology* **18**, 3–8.

RIEDER, C. & BAJER, A. S. (1977). Heat-induced reversible hexagonal packing of spindle microtubules. *J. Cell Biol.* **74**, 717–725.

ROSENFELD, M. (1932). The action of ether on cells in mitosis. *Arch. Exp. ZellForsch.* **12**, 570–586.

SAPARETO, S. A., HOPWOOD, L. E., DEWEY, W. C., RAJU, M. R. & GRAY, J. W. (1978). Effects of hyperthermia on survival and progression of Chinese hamster ovary cells. *Cancer Res.* **38**, 393–400.

SCHULTZE, M. (1865). Einheizbarer objecttish und seine verwendung bei untersuchungendes blutes. *Arch. Mikrosk. Anat. Entwiech.* **1**, 1–42.

SCOTT, R. E., PERKINS, R. G., SZCHUNKE, M. A., HOERL, B. H. & MAERCKLEIN, P. B. (1979). Plasma membrane vesiculation in 3T3 AMP SV3T3 cells. *J. Cell Sci.* **35**, 229–243.

STEINERT, P. M. & PARRY, D. A. D. (1985). Intermediate filaments. *Ann. Rev. Cell Biol.* **1**, 41–65

STEPHENS, R. E. (1970). Thermal fractionation of outer fibre doublet microtubules into A- and B-subfibre components: A- and B-tubulin. *J. mol. Biol.* **47**, 353–363.

SWEZEY, R. R. & SOMERO, G. N. (1982). Polymerisation thermodynamics and structural stabilities of skeletal muscle actins from vertebrates adapted to different temperatures and hydrostatic pressures. *Biochemistry* **21**, 4496–4503.

TANK, D. W., WU, E. S. & WEBB, W. W. (1982). Enhanced molecular diffusibility: muscle membrane blebs, release of lateral constraints. *J. Cell Biol.* **92**, 207–212.

TANNENBAUM, J. (1978). Approaches to the molecular biology of cytochalasin action: In *Cytochalasins: Biochemical and Cell Biological Aspects* (ed. S. W. Tanenbaum), pp. 521–529. Amsterdam: North Holland.

TAYLOR, D., WILLIAMS, V. & CRAWFORD, N. (1976). Platelet membrane action: solubility and binding studies with ^{125}I-labelled actin. *Biochem. Soc. Trans.* **4**, 156–160.

TEW, K. D., KYLE, G., JOHNSON, A. & WANG, A. L. (1985). Carbamoylation of glutathione reductase and changes in cellular and chromosome morphology in a rat cell line resistant to nitrogen mustards but collaterally sensitive to nitrosoureas. *Cancer Res.* **45**, 2326–2333.

TILNEY, L. G., HIRAMOTO, Y. & MARSLAND, D. (1966). Studies on the microtubules in Helizoa. III. A pressure analysis of the role of the structures in the formation and maintenance of the axopodia of *Actinosphaerium nucleofilum*. *J. Cell Biol.* **29**, 77–95.

TIMASHEFF, S. N. & GRISHAM, L. M. (1980). *In vitro* assembly of cytoplasmic microtubules. *Ann. Rev. Biochem.* **49**, 565–591.

TSONG, T. Y., HIMMELFARB, S. & HARRINGTON, S. F. (1983). Stability and melting kinetics of structural domains in the myosin rod. *J. mol. Biol.* **136**, 431–450.

TURI, A., LU, R. C. & LIN, P-S. (1981). Effect of heat on the microtubule and its relationship to body temperatures. *Biochem. Biophys. Res. Comm.* **100**, 584–590.

VASILIEV, J. M. (1985). Spreading of non-transformed and transformed cells. *Biochim. Biophys. Acta* **780**, 21–65.

WANG, E. & GOLDMAN, R. D. (1978). Functions of cytoplasmic fibres in intracellular movements in BHK-21 cells. *J. Cell Biol.* **79**, 708–726.

WESTRA, A. & DEWEY, W. C. (1971). Variation in sensitivity to heat shock during the cell cycle of Chinese hamster cells *in vitro*. *Int. J. Rad. Biol.* **19**, 467–477.

WILLIAMSON, J. R., SHANAHAN, M. O. & HOCHMUTH, R. M. (1975). The influence of temperature on red cell deformability. *Blood* **46**, 611–624.

WILLS, E. J., FINDLAY, J. M. & McMANUS, J. P. A. (1976). Effect of hyperthermia therapy on the liver. II. Morphological observations. *J. clin. Path.* **29**, 1–10.

ZOLLINGER, H. U. (1948). Cytologic studies with the phase microscope. *Ann. J. Path.* **24**, 545–567.

Printed in Great Britain © *Society for Experimental Biology 1987*

ROLE OF ENERGY IN CELLULAR RESPONSES TO HEAT

STUART K. CALDERWOOD

Joint Center for Radiation Therapy, Harvard Medical School,
50 Binney Street, Boston, MA 02115, USA

Summary

We have examined the effect of heat on energy-generating processes and on parameters of bioenergetic status in animal cells. Heat inactivates several processes involved in uptake and metabolism of nutrients. In particular, insulin-stimulated hexose transport in HA-1 fibroblasts and electron transport in blowfly sarcosomes (Bowler, 1981) exhibit thermal sensitivities that reflect the vulnerability to heat of the whole cell or organism. These heat-induced lesions in energy production are not, however, reflected by parameters of energy status in most cells studied. In HA-1 fibroblasts, for instance, over 99 % of cells are killed by 45°C heat before a decrease is observed in any parameter of energy status. A general role for energy in cellular responses to heat thus seems unlikely, although the thermal responses of tissues *in vivo* may differ from those of cells *in vitro*.

Introduction

Most cells and tissues grow and divide only within a limited temperature range (Bowler, 1981; Hahn, 1982). Damage and death occur, especially in mammalian cells, at temperatures a few °C above normal. Many species do, however, possess homeostatic mechanisms that permit them to acclimatise to temperature shifts (Bowler, 1981) and to acquire resistance to abrupt heat shocks (Hahn, 1982). As biological responses to heat are similar over a surprisingly broad range of species, it seems possible that one or a few common underlying lesions may mediate them. The agent(s) involved might be expected to be minatory in nature; pleiotropic on transient heat stimulation, toxic after prolonged exposure. In the present report, we have investigated energy production as a possible cellular target for heat and the adenylates, the universal units of energy exchange, as biochemical mediators of thermal effects. Energy production seems a logical candidate as a common target in heat injury due to the trans-species similarity in the biochemical pathways of energy metabolism. In addition it has been shown repeatedly that mitochondrial function is highly temperature-sensitive (Westermark, 1927; Morris & King, 1962; Christiansen & Kvamme, 1969; Mondovi, 1976; Bowler, 1981; Puranam *et al.* 1984).

In this report I have discussed the effects of heat on energy production and adenylate levels in mammalian cells. The data are considered under 3 main headings; transport of energy substrates, metabolism and adenylates.

Materials and methods

Cells and culture conditions

Chinese hamster ovary (HA-1) fibroblasts were grown in Eagle's minimal essential medium supplemented with 15% foetal calf serum, penicillin and streptomycin. The cultures were maintained in a humidified incubator with a mixture of 95% air and 5% CO_2 and routinely checked for mycoplasma. Confluent monolayers of cells, fed daily with fresh medium, were used for experiment on Day 3 after reaching confluence. At this time, cultures contained $(2 \cdot 4 – 2 \cdot 6) \times 10^7$ cells.

Heat treatment

Monolayers of cells on plastic Petri dishes were heated in a hot-water bath located inside an incubator with an atmosphere maintained at a ratio of 95% air and 5% CO_2. This gas mixture maintained the pH of the culture medium between 7·2 and 7·4 in all the heating conditions (pH was monitored before and after heating). Water-bath temperature was controlled to within ±0·1°C. The time required to reach a steady temperature was about 3 min and is included in the heating times reported later. Immediately preceding and following each heat treatment, the cultures were overlaid with 4 ml of fresh medium.

Cell survival

The techniques for measurement of cell survival used in the present study have been described in detail previously (Calderwood & Hahn, 1983b). In brief, at the conclusion of treatment, cells were trypsinized and counted by haemocytometer and then diluted to yield approximately 100–200 colonies per 60-mm Petri dish. Trypsinization has been shown previously to have minimal effect on survival in heated or unheated culture. After 10 days incubation at 37°C colonies were fixed, stained and counted; surviving fraction was determined by comparison with untreated control cultures. All experiments were performed on a minimum of two occasions.

[125]I-labelled insulin

Insulin was labelled with ^{125}I and purified as described previously (Calderwood & Hahn, 1983b).

Insulin-binding assay

Binding studies were carried out on confluent monolayers of HA-1 cells $(2 \cdot 5 \times 10^7$ cells per dish).

Cells were incubated for 150 min on ice (0–1°C) in 2 ml phosphate-buffered saline + 0·1 % albumin + 0·1 mmol^{-1} bacitracin containing 20 ng ^{125}I-labelled insulin. Under these conditions, binding was confined almost exclusively to the cell surface, and internalization and degradation of the label was minimal. At the end of the incubation period, monolayers were washed four times in 5 ml ice-cold phosphate-buffered saline + 0·1 % albumin (pH 7·4). Cells were then digested in 2·0 ml 0·5 mol l^{-1} NaOH (overnight at room temperature) and the viscous digest transferred to a plastic counting vial. The dish was washed with 1 ml NaOH, and this washing combined with the 2 ml digest was counted in a Beckman γ-spectrometer. All assays were carried out in replicate dishes, at least three for each data point. All data were expressed as specific binding; this parameter is obtained after subtraction of the nonspecific fraction from total insulin binding. Nonspecific binding was determined as the amount of ^{125}I-labelled insulin uptake in the presence of a large excess (10 μg ml^{-1}) of unlabelled insulin. This concentration appeared to give an accurate estimate of nonspecific binding; higher concentrations (100 μg ml^{-1}) led to an underestimation of nonspecific binding. Binding of ^{125}I-labelled insulin to cell-free culture dishes was both minimal (100–200 cts min^{-1}) and nonspecific; binding to the dishes was thus accounted for by the nonspecific binding control.

The insulin-receptor binding equilibrium was studied at a range of receptor occupancies using the Scatchard plot. Binding was carried out with insulin concentrations ranging from 0·02 to 2000 ppm. For each concentration, it was necessary to do a separate control for nonspecific binding. Data were plotted as the ratio of bound insulin to free insulin.

Control experiments were also carried out to ensure that the heating and binding assay conditions used did not lead to loss of cells from the monolayers (and thereby underestimation of the degree of insulin binding). Control experiments employing the most severe heating conditions (160 min 40°C; 80 min 44°C; 40 min 45°C) indicated no such loss. Cell loss was, however, routinely checked for in each experiment by measuring the protein content of the NaOH digest used for ^{125}I counting. Protein concentrations were measured by the method of Lowry *et al.* (1951).

Hexose transport assay

Cells were incubated in (^{3}H)-2-deoxyglucose ((^{3}H)2-DOG) or (^{3}H)3-*O*-methyl glucose (^{3}H-OMG), and uptake of the sugar was monitored from 0 to 60 s (for measurement of initial rate of hexose transport) or from 0 to 30 min to measure steady-state uptake; incubations were carried out in Dulbecco's phosphate-buffered saline (PBS) with 2-DOG or 3-OMG at an activity of 2·5 μCi ml^{-1} and a temperature of 25°C. After incubation, cells were rapidly washed 5 times in ice-cold PBS. Monolayers were solubilized in 1·5 ml 0·1 % sodium dodecyl sulphate (120 min) and ^{3}H uptake measured by radioassay. Non carrier-mediated transport was measured using phloretin. Previous work (Gay & Hilf, 1980) indicated essentially complete inhibition of carrier-mediated transport by $400 \mu mol \, l^{-1}$ phloretin and this was confirmed for the present cell line (S. K. Calderwood, unpublished data).

Respiration and glycolysis

Respiration and glycolysis were measured as O_2 utilization and CO_2 production using classical Warburg manometry (Dickson & Calderwood, 1979). Respiration was assayed in Krebs-Ringer phosphate buffer, pH 7·4 with 16 % KOH in the centre well. For respiration, the buffer contained $0·013 \, mol \, l^{-1}$ sodium succinate. Anaerobic glycolysis was measured in Krebs-Ringer bicarbonate – phosphate buffer pH 7·4 and a gas phase of 95 % N_2:5 % CO_2 in the presence of $2 \, g \, l^{-1}$ glucose (Mondovi *et al.* 1969). Results were expressed as $\mu l \, O_2$ consumed (respiration) or $\mu l \, CO_2$ produced (glycolysis) per 10 mg tumour slices or 10^7 cells per hour.

Measurement of adenylate energy charge

Cultures were incubated with medium containing $0·1 \, \mu mol \, l^{-1}$ (^3H) adenine at an activity of $1 \, \mu Ci \, ml^{-1}$ ^3H for 24 h prior to experiment. The medium was then changed and labelled cells subjected to experimental conditions. Acid extracts for thin layer chromatography (TLC) analysis were then prepared as follows.

Medium was rapidly decanted, removing the last drops by Pasteur pipette. Cultures were placed immediately on ice, in a small volume ($200 \mu l$) of ice-cold $0·4 \, mol \, l^{-1}$ perchloric acid, and the cells scraped off with a rubber policeman. After 15 min on ice, samples were collected and centrifuged at $11\,000 \, g$ (4°C). The supernatant was removed to a clean tube and neutralized with $3 \mu l$ $8·0 \, mol \, l^{-1}$ KOH and $9 \mu l$ $2·0 \, mol \, l^{-1}$ KOH/$0·5 \, mol \, l^{-1}$ $KHCO_3$. Small aliquots ($5 \mu l$) of $0·4 \, mol \, l^{-1}$ KOH were then added until neutrality. Samples were centrifuged ($11\,000 \, g$) at 4°C and stored at -20°C until assayed (within 5 days).

Samples were separated by two-dimensional thin layer chromatography on polyethyleneimine anion exchange cellulose (Pedersen & Catterall, 1979). A $10 \mu l$ sample was applied to one corner of a marked-out plate along with a $2 \mu l$ aliquot of adenylate standard solution ($10 \, mmol \, l^{-1}$ AMP, ADP, ATP). Solutions were added in small drops from the tip of a $10 \mu l$ micropipette and drops dried when absorbed to present excessive spreading. The plate was washed in methanol (60 s) and dried thoroughly. The sample was then eluted in dimension I in the following solutions, $0·5 \, mol \, l^{-1}$ formic acid/$0·5 \, mol \, l^{-1}$ sodium formate (5 min), $2·0 \, mol \, l^{-1}$ formic acid/$2·0 \, mol \, l^{-1}$ sodium formate (15 min), $4·0 \, mol \, l^{-1}$ formic acid/$4·0 \, mol \, l^{-1}$ sodium formate (40 min). The plate was then dried and viewed under a short-wave mineral light. The three spots visible (due to addition of adenylate standards) were arranged (in order of distance travelled from the origin) AMP, ADP, ATP. The spots were covered in transparent tape, cut out of the plates and transferred to scintillation vials. Samples were counted for 10 min on a Beckman scintillation spectrometer. Adenylate energy charge was then calculated, $EC = (cts\,min^{-1}\,ATP) + 1/2\,(cts\,min^{-1}\,ADP)/(cts\,min^{-1}\,ATP) + cts\,min^{-1}\,ADP + (cts\,min^{-1}\,AMP)$.

In some of the later experiments adenylate concentrations were determined by a high performance liquid chromatography (HPLC) technique. The latter has been described previously (Bump *et al.* 1984). In brief, cells were deproteinized by perchloric acid extraction. The perchloric acid was then extracted into freon as the triethylamine ion pair, and the adenylates were separated by HPLC and determined by absorbance at 254 nm. Both methods gave similar values for EC under comparable conditions.

Measurement of [ATP] by luciferase

Cells were rinsed twice in phosphate-buffered saline and then deproteinized in $1\cdot0$ ml ice-cold $0\cdot4\,mol\,l^{-1}$ perchloric acid (PCA) prior to luciferase assay (Seitz & Neary, 1975). For the assay, $50\,\mu l$ of PCA extract were added to $1\cdot0$ ml arsenate buffer ($0\cdot1\,mol\,l^{-1}$ sodium arsenate, $40\,mmol\,l^{-1}$ magnesium sulphate (pH 7·9) $1\cdot0$ ml Dulbecco's phosphate-buffered saline (pH 7·4) and $1\cdot0$ ml distilled H_2O) in a glass scintillation vial (Addanki *et al.* 1966; Stanley & Williams, 1969). Then $50\,\mu l$ of luciferase was added to the samples and the vials rapidly shaken, transferred to a scintillation counter, and counted for $0\cdot1$ min. The counter was prepared for the assay by switching off the coincidence gate for the photomultiplier tubes as suggested by Stanley & Williams (1969) for employing a scintillation counter to measure luminescence. The instrument was calibrated using a standard curve with [ATP] ranging from 10^{-11} to 10^{-8} moles/assay tube.

This method allowed us to determine absolute concentrations of adenylates which was difficult using the radioisotopic method. The two techniques (EC and luciferase) indicated similar overall changes in ATP concentration in response to perturbation and thus the values they indicate are comparable.

Measurement of inorganic phosphate

Cell monolayers were washed once in $4\cdot0$ ml of phosphate-free, 20 mm Hepes-buffered saline (pH 7·4) at $0-1\,°C$, then rapidly deproteinized in $1\cdot0$ ml of $0\cdot6\,mol\,l^{-1}$ perchloric acid. Acid extracts were neutralized with $1\cdot0\,mol\,l^{-1}$ NaOH and buffered to pH 7·0 with tris. Inorganic phosphate was then assayed (Anderson & Davis, 1982). Protein assays (Lowry *et al.* 1951) were carried out on the pellets remaining after PCA precipitation. Phosphate determined as nmoles/10^6 cells or nmoles/μg^{-1} protein correlated well over a range of treatments.

Results and discussion

Effects of heat on uptake of energy substrates

Most cell types derive energy from substrates obtained from the extracellular environment. One likely effect of heat might thus be to inhibit influx of such substrates. The effect of heat shock on the initial velocity (V_c) of hexose transport in HA-1 cells is shown in Fig. 1. Hexose transport appeared to be markedly temperature resistant, with 50 % inhibition (I_{50}) requiring heating times at 45 °C and 47 °C far in excess of those required for cell killing. There was, however, a relatively small, heat-sensitive component of V_c which was rapidly inactivated at 43 °C, 45 °C and 47 °C. This heterogeneity in heat sensitivity of hexose transport might reflect the presence of two classes of hexose transport system. The existence of two separate hexose transport systems is indicated in these cells by kinetic analysis (S. K. Calderwood, unpublished data); Eadie-Hofstee analysis of 2-deoxyglucose transport indicated two kinetically distinct transport systems, one with high capacity and low affinity for the sugar ($Km = 5.8$ mmol l^{-1};

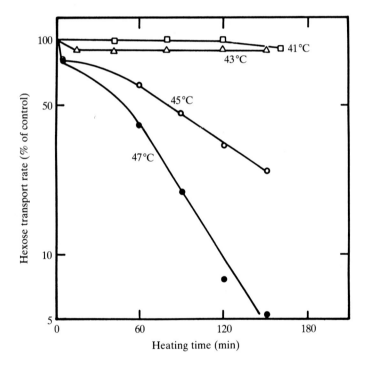

Fig. 1. Effect of heat shock at 41 °C to 47 °C on the initial velocity of ^{3}H-2-deoxyglucose influx. Cells in monolayer were heated in full medium. They were then washed 3× with 5 ml Dulbecco's phosphate-buffered saline, left for 30 min at 37 °C and then ^{3}H-2-DOG uptake measured at 25 °C. Uptake was measured from 15–120 s in 0.4 mmol l^{-1} ^{3}H-2-deoxyglucose (2.5 μCi ml^{-1}), in phosphate-buffered saline. Incubations were stopped by rapid washing in ice-cold saline containing 5 mmol l^{-1} D-glucose. Results are from duplicate experiments. Data are expressed as % of control velocity.

$V_{max} = 15$ nmol min^{-1} 10^{-6} cells) and one low-capacity, high-affinity component ($Km = 0.28$ mmol l^{-1}; $V_{max} = 1.0$ nmol min^{-1} 10^{-6} cells). Existence of multiple hexose transport systems has been observed previously in mammalian cells (reviewed by D'Amore & Lo, 1986).

For many mammalian cells, hexose transport capacity (V_{max}) is inadequate at physiological glucose concentrations (5 mmol l^{-1}). They require the binding of agonists such as insulin in order to accumulate glucose (Gliemann & Rees, 1983). Fig. 2 shows the effect of insulin on V_c in HA-1 cells. Insulin pretreatment stimulates ^{3}H-2-dOG uptake mainly at the higher concentration (5.0 mmol l^{-1}), indicating that the hormone acts mainly by increasing V_{max} for transport as observed in other cell types (reviewed by Gliemann & Rees, 1983). The hormone is known to induce the recruitment of hexose transport systems to the plasma membrane. The effect of heat on steady-state accumulation of 2-DOG, both basal and insulin-stimulated, is indicated in Fig. 3. Heating (45°C) had only a minor

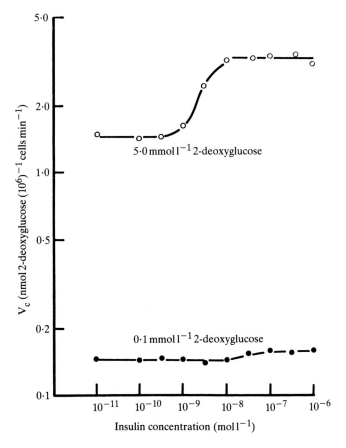

Fig. 2. Insulin dependence of ^{3}H-2-deoxyglucose uptake in HA-1 fibroblasts. Cells were incubated in Dulbecco's phosphate-buffered saline at the insulin concentrations shown for 60 min prior to the addition of 2.5 μCi ml^{-1} ^{3}H-2-deoxyglucose at concentrations of 0.1 mmol l^{-1} and 5.0 mmol l^{-1}. Uptake was measured over 60 s at 25°C. Data shown are means of triplicate assays. Experiments were repeated 4 times.

effect on basal transport but in contrast rapidly abolished insulin-stimulated hexose transport (Fig. 3).

We have reported previously that heat-induced inhibition of insulin-receptor binding closely parallels the kinetics of the heat killing of cells and that development of thermotolerance is paralleled by an acquired heat resistance in insulin receptors (Calderwood & Hahn, 1983*a,b*). This is indicated in the Scatchard curve in Fig. 4; heat (80 min at 43 °C) has little effect on binding affinity but causes a loss in receptor number from 12 940 to 6500 binding sites/cell. Arrhenius analysis of receptor inactivation (42·5–47 °C) indicates an activation energy for receptor loss of 149 Kcal/mol, suggesting thermal denaturation of receptors (Johnson *et al.* 1974). This effect seems to be fairly specific to insulin as heat did not cause loss of epidermal growth factor receptors (Magun & Fennie, 1981) or lectin acceptor proteins (Stevenson *et al.* 1983*b*) under similar heating conditions. We are aware of few additional data on the effect of heat shock on nutrient transport in mammalian cells. However, Kwock *et al.* (1985) showed that Na^+-dependent amino acid transport was inhibited by heat in lymphoid cell lines and that the inhibition was paralleled by growth delay and reduction of ^3H-thymidine incorporation into DNA. Resumption in growth coincided with recovery in amino acid transport (Kwock *et al.* 1985).

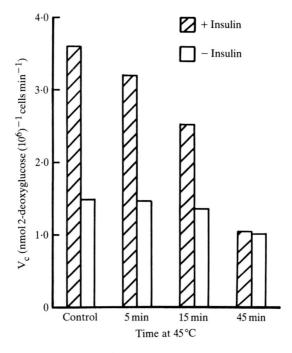

Fig. 3. Effect of heat (45 °C) on ^3H-2-deoxyglucose influx in the presence (▨) or absence (□) of 10^{-7} m insulin. Cells were preheated at 45 °C (as in Fig. 1), incubated with or without insulin for 30 min, prior to addition of ^3H-2-deoxyglucose (2·5 μCi; 5·0 mmol l^{-1}). Uptake was then assayed over 90 s at 25 °C. Data are means of triplicate assays; experiments were performed twice.

Respiration and glycolysis

It has been known for a considerable period of time that temperature shock inhibits oxygen consumption in many animal cells and tissues (Westermark, 1927; Morris & King, 1962; Kristiansen & Kvamme, 1969; Mondovi, 1976; Dickson & Calderwood, 1979; Bowler, 1981). Fig. 5 shows inhibition of O_2 consumption in rat Yoshida sarcoma cells at 42°C, but no effect of 40°C heating. A number of studies have been carried out on respiratory processes in isolated mitochondria. Christiansen & Kvamme (1969) showed inhibition of electron transport, uncoupling of oxidative phosphorylation and loss of respiratory control in rat mitochondria from brain, liver and Ehrlich carcinoma cells heated at temperatures between 41°C and 45°C. In blowfly mitochondria, Bowler (1981) also observed heat-

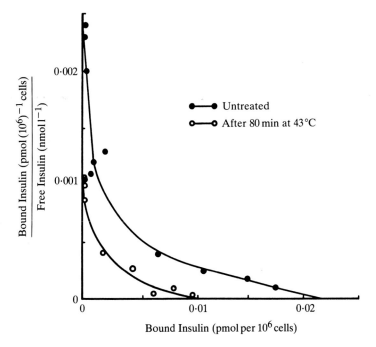

Fig. 4. (After Calderwood & Hahn, 1983*a*.) Scatchard curves of total insulin binding to unheated cells (●) and immediately after 80 min at 43°C (○). The binding assay was carried out at 0°C for 150 min in phosphate-buffered saline + 0·1 % bovine serum albumin + 1 mmol l^{-1} bacitracin (pH 7·4). (Similar curves were obtained if binding was carried out at 37 or 4°C in the absence of bacitracin.) Affinity constants and receptor numbers derived from the curves shown above are tabulated below. K_e, affinity constant of empty receptor; K_f, affinity constant of filled receptor.

	Derived binding parameters	
	Unheated cells	80 min at 43°C
Binding capacity (sites/cell)	12 940	6500
$\bar{K}_e$ (mol l^{-1})	1×10^9	$8·23 \times 10^8$
$\bar{K}_f$ (mol l^{-1})	$2·60 \times 10^7$	$2·59 \times 10^7$

induced loss of respiratory control. He found a close correlation between the inhibition of respiration and heat killing of the organism. A particularly heat-sensitive site in electron transport appears to be located close to site II in the respiratory chain; several studies suggest a temperature-sensitive site in the vicinity of coenzyme Q and site II (Morris & King, 1962; Christiansen & Kvamme, 1969; Bowler, 1981). Puranam et al. (1984) showed also that kidney mitochondria from rats subjected to mild hyperthermia showed a decrease in state 3 oxidation which could be repaired by the addition of cytochrome c. No loss of respiratory control was observed under these mild conditions.

Little information is available on mitochondrial morphology after heat. Mito-chondria have been shown to undergo a redistribution to a perinuclear site in heated rat embryo fibroblasts (Welsh & Suhan, 1985; D. Chase, personal communication). This redistribution may be secondary to collapse of intermediate filaments towards the nucleus in heated cells (Welsh et al. 1985; Welsh & Suhan, 1985). A number of ultrastructural changes have been observed in cells and tissues after heat. They include: (1) the appearance of electron-dense granules in the mitochondrial matrix of BHK fibroblasts (M. Borelli & W. C. Dewey, personal communication); (2) the assumption of a condensed morphology immediately

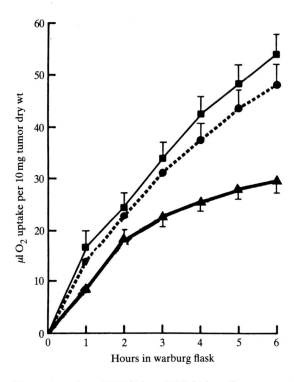

Fig. 5. Effect of hyperthermia at 40°C (■) or 42°C (▲) on O_2 consumption in Yoshida rat sarcoma compared to untreated controls (●). Cells were heated in vivo and O_2 utilization measured in vitro at 38°C. Points are means ± s.D. from 7 separate experiments (after Dickson & Calderwood, 1979).

after heat in mouse mammary carcinoma and BHK cells (Overgaard & Overgaard, 1976; Borelli & Dewey); (3) eventual mitochondrial swelling (Overgaard & Overgaard, 1976; Welsh & Suhan, 1985; M. Borrelli & W. C. Dewey, personal communication).

The biochemical and ultrastructural studies suggest complex changes in heated mitochondria. At least two types of effect may be involved. Inhibition of electron transport indicates direct damage, particularly in the region of site II in the electron transport chain. Loss of respiratory control and uncoupling may indicate permeability changes in the inner mitochondrial membrane and partial collapse of the proton electrochemical gradient (Hackenbrock, 1981). Mitochondrial condensation has been linked to increased oxidative phosphorylation (Hackenbrock *et al.* 1971). Thus during heat or immediately afterwards oxidative phosphorylation may be stimulated, perhaps by increased ATP demand. The studies of Mirtsch *et al.* (1984), indicating heat-induced increases in the specific activity of ^{32}P-ATP in the presence of ^{32}P-orthophosphate, would tend to support this suggestion. Mitochondrial swelling seems a relatively late event and may be secondary to decline in electron transport and to membrane damage (Overgaard & Overgaard, 1976). The appearance of granules in heated mitochondria (M. Borelli & W. C. Dewey, in preparation) may indicate calcium deposition (Shen & Jennings, 1972; Campbell, 1983). This suggests indirect effects of heat on mitochondria, possibly mediated through Ca^{2+}. A number of studies indicate a large increase in cellular Ca^{2+} influx and concentration after heat (Stevenson *et al.* 1986; Drummond *et al.* 1986; Vidair & Dewey, 1986). The majority of the Ca^{2+} influx in heated HA-1 fibroblasts is blocked by uncouplers (2,4-DNPH, CCP) and inhibitors of electron transport such as rotenone and antimycin A (M. A. Stevenson, S. K. Calderwood & G. M. Hahn, in preparation). A role for mitochondrial Ca^{2+} uptake in cellular responses to heat is indicated by the studies of O'Byrne-Ring *et al.* (1981) who showed that in Ts mutants of *Drosophila* Ca^{2+} did not stimulate electron transport. Thus mitochondrial accumulation of Ca^{2+} may be a homeostatic response in cells subjected to heat shock-induced Ca^{2+} loading. The overall role of respiration as a site for thermal killing is not clear. Inhibition of O_2 consumption was observed in many studies of normal and malignant mammalian cells, although cell viability was not measured in most studies (reviewed by Mondovi, 1976). Landry *et al.* (1985) found that chick embryo fibroblast cells selected for respiration deficiency by treatment with ethidium bromide were of similar heat sensitivity to cells with a full respiratory complement. However, in blowfly muscle, an intensely respiring tissue, thermal inhibition of respiration correlates very well with viability of the organism (Bowler, 1981). It seems therefore that there is insufficient evidence to assess the general role of respiration as a process limiting the heat resistance of tissues. The studies of Christiansen & Kvamme (1969) indicated major differences in the heat sensitivity of respiratory processes between different tissues of the same animal. In accordance with many subsequent studies, they observed that respiration in malignant cells was more heat-sensitive than in normal cells (Westermark, 1927; Christiansen & Kvamme, 1969; Mondovi, 1969). One factor which may make respiration in many malignant cells heat-sensitive is the abnormally high mitochondrial membrane potential observed in a wide range of carcinomas by Chen & co-workers (Davis *et al.* 1985; Nadakavudakaren *et al.* 1985). High membrane potentials were shown to correlate with heat sensitivity in a range of cell types (Ushakov, 1981) with the ionic imbalance perhaps rendering cells susceptible to perturbation by heat.

A number of studies have indicated that mitochondrial damage may be involved in the induction of heat shock gene expression in *Drosophila* (Leenders *et al.* 1974). This has not, however, been confirmed in other species and mitochondrial damage is not thought to constitute a major signal for the heat shock response (Lanks, 1986).

Many investigations have been carried out on glycolysis in normal and malignant cells; the overall conclusion was that glycolysis is not a temperature-sensitive process in most cells (reviewed by Mondovi, 1976).

Adenylate energy charge, phosphorylation potential and ATP concentration

There is evidence that two adenylate ratios, adenylate energy charge (Atkinson, 1977) and phosphorylation state or potential (Slater, 1979) may carry out the function of coupling between energy production and utilization. Energy charge (EC), an indicator of the degree of phosphorylation of the ATP-ADP-AMP system, is calculated from the following ratio, $EC = [ATP] + 1/2 \ [ADP]/ATP + [ADP] + [AMP]$.

Cells appear to maintain EC at a value of approximately $0·85$ (Atkinson, 1977; Pradet & Raymond, 1983). Decreases in EC led to activation of ATP generating sequences due to the response of their regulatory enzymes to relative concentrations of ATP, ADP and AMP (Matsumoto *et al.* 1979). Increases in EC led to the reverse changes (Helgerson *et al.* 1983). Phosphorylation potential is calculated from the ratio, $[ATP]/[ADP] \times [Pi]$. The ratio is related to the mass action constant for ADP phosphorylation. The $[ATP]/[ADP] \times [Pi]$ ratio is maintained at approximately $1–2 \times 10^3 \ mol \, l^{-1}$ in a number of cell types (Wilson *et al.* 1974; Erecinska, 1977). The two indices differ in that AMP is a primary regulator in EC control of metabolism (Ball & Atkinson, 1975; Matsumoto *et al.* 1979) while the use of $ATP/[ADP] \times [Pi]$ as an energy index emphasizes the role of inorganic phosphate (Yushok, 1971; Erecinska *et al.* 1977; Sauer, 1978). In order to assess whether heat might kill cells through an inhibition of energy-producing capacity and thus lead to death by starvation, we compared the effects of both starvation and heat on parameters of energy status.

Energy levels were investigated under starvation conditions (Fig. 6). In cells incubated in nutrient-free buffer [ATP] declined rapidly ($T1/2 = 108 \ min$). Despite this decline in [ATP], no corresponding increase in the levels of the other adenylates occurred until 5 h when [AMP] increased tenfold. [ADP] remained relatively constant. Due to the failure of AMP and ADP to accumulate in the first 3 h of starvation, EC remained relatively high in this period despite the progressive

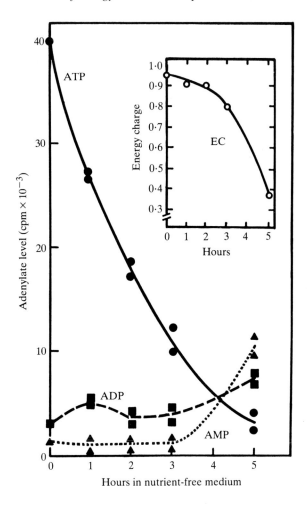

Fig. 6. Levels of ATP, ADP and AMP in HA-1 CHO cells incubated in nutrient-free medium. Inset in the figure is the EC value calculated for the adenylate levels at each time point. Each data point in the main figure is the mean of a triplicate assay carried out on the PCA extract from one cell culture; variation in each case was below 8% of the mean value. Inorganic phosphate was also determined in cultures treated as above. Using these [Pi] values, the [ATP]/[ADP] × [Pi] ratio was determined. Pi values shown below are means ± s.D. of duplicate assays on five replicate cultures. Experiments were done twice and yielded reproducible values. (After Calderwood *et al.* 1985.)

Time in nutrient-free buffer (h)	Pi (nmoles $(10^6)^{-1}$ cells)	$\dfrac{[ATP]}{[ADP] \times [Pi]}$ (mol l^{-1})
0	5·34 (±0·36)	2250
1	6·23 (±0·60)	790
2	6·81 (±0·99)	860
3	7·39 (±0·52)	330
5	11·56 (±1·23)	60

fall in [ATP]. Inorganic phosphate levels increased gradually until 3 h and then more rapidly to approximately double control values at 5 h. The [ATP]/[ADP] × [Pi] ratio calculated from these values declined with similar kinetics to those of [ATP] decrease.

The effect of starvation on cell viability and proliferation is shown in Fig. 7. In nutrient-deprived cells an initial resistant period (0–2 h) over which no killing occurred preceded a phase of rapid exponential cell killing. The shape of the curve was similar to that showing EC decline in starved cells (inset, Fig. 6); this curve indicated also an initial resistant portion (0–3 h) and a portion of rapid decline. Cell killing did not seem to be related to the ATP level or [ATP]/[ADP] × [Pi] ratio which declined progressively after cells were placed in glucose-free medium (Fig. 6) in contrast to the biphasic cell survival curve (Fig. 7). EC thus may be an important parameter in cells under starvation conditions.

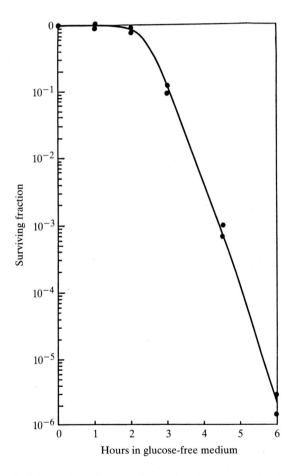

Fig. 7. Cell survival in nutrient-free medium. Cells in cultures incubated for longer than 3 h without glucose tended to detach from the monolayers. In these cases, the medium overlying each monolayer was spun at 1000 g and the cell pellet assayed for survival. (After Calderwood et al. 1985.)

The effect of 45°C on EC is shown in Fig. 8. There was a threshold period of approximately 40 min 45°C^{-1} before any effect was observed on EC; EC thereafter declined in a progressive and linear manner to minimal values by 180 min at 45°C. This decline in EC was mediated by a rapid fall in ATP concentration after 30–40 min. AMP levels rose to three to four times the control value by 60 min and

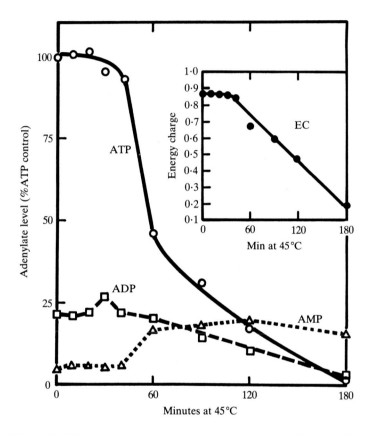

Fig. 8. Effect of 45°C on adenylate concentrations and energy charge in HA-1 cells. Each point is derived from a duplicate assay. The experiment was repeated twice using the ^{3}H-adenine/TLC method and HPLC (Bump *et al.* 1984) to measure adenylate levels. The two methods gave similar results. Pi was determined in separate cultures heated at 45°C. Pi values shown below are means ± S.D. from four replicate cultures, each assayed in duplicate. Experiments were done three times and yielded reproducible values. (After Calderwood *et al.* 1985.)

Time at 45°C (min)	Pi (nmoles $(10^6)^{-1}$ cells)	$\dfrac{[ATP]}{[ADP] \times [Pi]}$ (mol l^{-1})
0	5·54 (±0·75)	1628
15	5·83 (±0·52)	1412
30	6·57 (±0·16)	1422
60	6·18 (±0·52)	669
120	5·88 (±0·07)	598
180	6·18 (±0·19)	54

remained at this elevated level. ADP levels declined progressively after 60 min. Heat had no marked effect on Pi levels, which remained constant from 0 until 180 min at 45°C. Decrease in [ATP]/[ADP] × [Pi] ratio after heating paralleled the decline in [ATP] and EC.

As may be seen from Fig. 9, there was no significant correlation between EC and cell survival in cells heated at 45°C. Even at survival levels of 10^{-6}, no significant decrease in EC occurred. This finding was consistent when energy was measured immediately (Fig. 9) or 24 h after heating (data not shown). Cells killed by nutrient deprivation alone, in contrast, demonstrated an excellent correlation between EC and cell survival (data from Figs 6, 7).

The starvation studies (Figs 6, 7) indicate that regulation of EC is closely correlated with cell survival in CHO fibroblasts under conditions of prolonged deprivation of reducible substrates. Such a relationship has been observed previously in procaryotes (Swedes *et al.* 1975, 1979). The correlation between EC and clonogenic capacity in HA-1 cells indicates a possible role for EC in cell

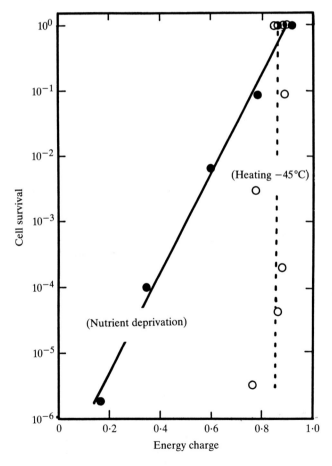

Fig. 9. Correlation between EC and cell survival in heated (Fig. 8) and glucose-deprived (Fig. 6) cells. (After Calderwood *et al.* 1985.)

viability in energy-stressed conditions. A likely mechanism is loss of volume control due to inability to operate energy-linked solute fluxes. Maintenance of volume in animal cells is thought to involve the active efflux of Na^+ ions (Macknight & Leaf, 1980).

The changes in EC induced by heating at 45°C indicate two sharply divided phases. In the initial period (30–40 min) no decline in energy status is seen; this may indicate that cells are protected from heat-induced lesions by the redundant capacity of their energy-producing pathways (Nagle *et al.* 1980; Soderberg *et al.* 1980). The decline in ATP and EC after 30–40 min may indicate inactivation of multiple energy-producing pathways and/or may reflect uncontrolled ATP utilization. We have demonstrated in previous studies that EC and cell survival could be reduced by a combination of metabolic inhibitors (2-deoxyglucose and rotenone; Bump *et al.* 1984). After 30–40 min, decline in EC appeared to be resisted by another mechanism; AMP and ADP did not accumulate, indicating AMP catabolism as in starved HA-1 cells (Fig. 6). Whether the transition at 40 min 45°C^{-1} reflects decreased energy production or increased demand is not clear from the present study. However, the recent work of Mirtsch *et al.* (1984) indicates that although ATP levels decrease in heat-stressed melanoma cells, ^{3}H-adenine incorporation into ATP increases, indicating increased ATP synthesis despite a decline in absolute levels. This may implicate increased ATP demand as the major cause of EC decline in heated cells. One likely source of increased energy demand could be altered flux of major ions across heat-damaged membranes necessitating increased energy utilization for active ion efflux. We have shown previously that 30–40 min at 45°C induces a large, reversible decrease in concanavalin-A receptor number in HA-1 cells, probably indicating major alterations in plasma membrane structure (Stevenson *et al.* 1983*b*). Heat-induced changes in membrane protein structure are also implicated by the physico-chemical studies of Lepock (1982). In addition, heat appears to uncouple Rb^+ transport from ATP hydrolysis in this cell type after 40 min at 45°C (Anderson & Hahn, 1987); such uncoupling would cause a major increase in ATP demand and might be involved in the heat-induced decline in cellular energy status observed in the present study. A number of other reports indicate effects of heat on ion flux and membrane potential in eucaryotic cells which might be due to ion leaks or alterations in the efficiency of active transport (Mikkelsen *et al.* 1981; Yi, 1979; Bowler, 1981; Stevenson *et al.* 1983*a*; Boonstra *et al.* 1984). Thus, decline in energy status of heated cells due to increased energy demand for maintenance of cellular ion homeostasis may be indicated.

The thermotropic changes in energy status observed in the present cell line, although drastic, do not appear to mediate thermal cell killing (Figs 8, 9). Similar observations were made by Lunec & Cresswell (1983) who found that [ATP] in Ehrlich ascites tumour cells was not decreased by heating for periods of up to 60 min at 44°C. These workers found that the L5178YS lymphoma line was much more prone to heat-induced changes in [ATP], although even in these cells [ATP] remained at above 50 % control levels at a cell survival value of 10^{-3}. Additionally

Jankowsky *et al.* (1981) observed no major alteration in EC during thermal adaptation of *Anguilla anguilla*.

Although energy seems precluded as a major factor in thermal cell killing in a nutritionally adequate micro-environment, it may have a role under sub-optimal conditions (Lilly, 1984; Calderwood *et al.* 1985). Cells become extremely heat-sensitive in nutrient-deprived media and when subjected to metabolic inhibitors (Calderwood *et al.* 1985; Nagle *et al.* 1980; Haveman & Hahn, 1981; Hahn, 1982; Calderwood & Dickson, 1983). This may imply that under energy-deprived conditions thermal killing is shifted to an alternative, energy-linked pathway, or that nutrient deprivation directly activates a step involved in the pathway of thermal killing observed under nutritionally adequate conditions. Sensitization by nutrient deprivation may be of significance for tumour cells *in vivo* which exist in a milieu which is deficient in nutrients (Gullino, 1975; Calderwood & Dickson, 1983).

Overall conclusions

The data thus indicate that cells contain discrete molecules and processes involved in bioenergy production that reflect the temperature sensitivity of the whole cell or organism. It may be of significance that these sites (insulin receptor, site II on the respiratory chain) are located in membranes. Much evidence suggests a membrane site for heat killing in most cells (Bowler, 1981; Cossins *et al.* 1981; Hahn, 1982; Lepock, 1982; Calderwood & Hahn, 1983a,b). In most studies of cultured cells, however, decline in parameters of energy status appears to be a relatively late phenomenon only observed after prolonged heat stress.

I thank my co-workers Drs J. A. Dickson, G. M. Hahn, M. A. Stevenson, E. A. Bump, I. Van Kersen and E. Shiu for permitting me to quote both the published and unpublished work which we carried out together. I would particularly like to thank Dr Hahn with whom much of this work was carried out. I also thank Drs M. Borelli and D. Chase for helpful discussions on mitochondrial ultrastructure.

References

ADDANKI, S., SOTOS, J. F. & REARICK, P. D. (1966). Rapid determination of picomole quantities of ATP with a liquid scintillation counter. *Analyt. Biochem.* **14**, 261–264.

ANDERSON, R. L. & DAVIS, S. (1982). An organic phosphorous assay which avoids the use of hazardous perchloric acid. *Clin. Chim. Acta* **121**, 111–116.

ANDERSON, R. L. & HAHN, G. M. (1987). Differential effects of hyperthermia on the Na^+K^+ ATPase of Chinese hamster ovary cells. *Radiat. Research* (in press).

ATKINSON, D. E. (1977). *Cellular Energy Metabolism and its Regulation.* New York: Academic Press.

BALL, W. J. JR & ATKINSON, D. E. (1975). Adenylate energy charge in Saccharomyces Cerevisiae during starvation. *J. Bacteriol.* **121**, 975–982.

BOONSTRA, T., SCHAMHART, D. H. J., DELAAT, S. W. & VAN WIJK, R. (1984). Analysis of K^+ and Na^+ transport and intracellular contents during and after heat shock and their role in protein synthesis in rat hepatoma cells. *Cancer Res.* **44**, 955–960.

BOWLER, K. (1981). Heat death and cellular heat injury. *J. therm. Biol.* **6**, 171–179.

BUMP, E. A., CALDERWOOD, S. K., SAWYER, J. & BROWN, J. M. (1984). Role of the adenylate energy charge in the response of Chinese Hamster Ovary Cells to radiation. *Int. J. Radiat. Oncol. Biol. Phys.* **10**, 1411–1414.

CALDERWOOD, S. K. & DICKSON, J. A. (1983). pH and tumor response to hyperthermia. *Adv. Radiat. Biol.* **19**, 135–186.

CALDERWOOD, S. K. & HAHN, G. M. (1983a). Thermal sensitivity and resistance of insulin-receptor binding. *Biochim. biophys. Acta* **756**, 1–8.

CALDERWOOD, S. K. & HAHN, G. M. (1983b). Thermal sensitivity and resistance of insulin-receptor binding in thermotolerant cells. *Biochim. biophys. Acta* **734** 76–82.

CALDERWOOD, S. K., BUMP, E. A., STEVENSON, M. A., VAN KERSEN, I. & HAHN, G. M. (1985). Investigation of adenylate energy charge, phosphorylation potential and ATP concentration in cells stressed with starvation and heat. *J. cell. Physiol.* **124**, 261–268.

CAMPBELL, A. C. (1983). *Intracellular Calcium, its Universal Role as a Regulator*, pp. 393–425. Chichester, New York: Wiley.

CHRISTIANSEN, E. N. & KVAMME, E. (1969). Effects of thermal treatment on mitochondria of brain, liver and ascites cells. *Acta physiol. scand.* **76**, 472–484.

COSSINS, A. R., BOWLER, K. & PROSSER, C. L. (1981). Homeoviscous adaptation and its effect upon membrane-bound proteins. *J. therm. Biol.* **6**, 183–188.

DAVIS, S., WEISS, M. J., WONG, J. R., LAMPIDIS, T. J. & CHEN, L. B. (1985). Mitochondrial and plasma membrane potentials cause unusual accumulation and retention of Rhodamine 123 by human breast adenocarcinoma-derived MCF-7 cells. *J. biol. Chem.* **260**, 13 844–13 850.

D'AMORE, T. & LO, T. C. Y. (1986). Hexose transport in L6 rat myoblasts. I. Rate limiting step, kinetic properties and evidence for two systems. *J. cell. Physiol.* **127**, 95–106.

DICKSON, J. A. & CALDERWOOD, S. K. (1979). Effects of hyperglycemia and hyperthermia on the pH, glycolysis and respiration of the Yoshida sarcoma in vivo. *J. natn. Cancer Inst.* **63**, 1371–1381.

DRUMMOND, I. A. S., McCLURE, S. A., POENIE, M., TSIEN, R. Y. & STEINHARDT, R. A. (1986). Large changes in intracellular pH and calcium observed during heat shock are not responsible for the induction of heat shock proteins in *Drosophila melanogaster*. *Mol. Cell Biol.* **6**, 1767–1775.

ERECINSKA, M., STUBBS, M., MIYATA, Y., DITRE, C. M. & WILSON, D. F. (1977). Regulation of cellular metabolism by phosphate. *Biochim. biophys. Acta* **462**, 20–35.

GAY, R. G. & HILF, R. (1980). Density-dependent and adaptive regulation of glucose transport in primary cell cultures of the R3230AC rat mammary adenocarcinoma. *J. cell. Physiol.* **102**, 155–174.

GLIEMANN, J. & REES, W. D. (1983). The insulin sensitive hexose transport system in adipocytes. *Current Topics in Membranes and Transport* **18**, 339–377.

GULLINO, P. M. (1975). Extracellular compartments of solid tumors. In *Cancer, A Comprehensive Treatise*, vol. 3 (ed. F. Becker), pp. 327–350. New York, London: Plenum Press.

HACKENBROCK, C. R., REHN, T. G., WEINBACH, E. C. & LEMASTERS, T. L. (1971). Oxidative phosphorylation and ultrastructural transormation in mitochondria in the intact ascites tumor cell. *J. Cell Biol.* **51**, 123–137.

HACKENBROCK, C. R. (1981). Lateral diffusion and electron transport in the mitochondrial inner membrane. *Trends Biochem. Sci.* **6**, 151–156.

HAHN, G. M. (1982). *Hyperthermia and Cancer*. New York: Plenum Press.

HAMMERSTEDT, R. H. & LARDY, H. A. (1983). The effect of substrate cycling on the ATP yield of sperm glycolysis. *J. biol. Chem.* **258**, 8759–8768.

HAVEMAN, J. & HAHN, G. M. (1981). The role of energy in hyperthermia induced mammalian cell inactivation. A study on the effects of glucose starvation and an uncoupler of oxidative phosphorylation. *J. cell. Physiol.* **108**, 237–241.

HELGERSON, S. L., REQUADT, C. & STOECKENIUS, W. (1983). *Halobacterium halobium* photophosphorylation: illumination dependent increase in the adenylate energy charge and phosphorylation potential. *Biochemistry, N.Y.* **22**, 5746–5763.

JANKOWSKY, H. D., HOTOPP, W. & VSIANSKY, P. L. (1981). Effects of assay and accumulation temperatures on incorporation of amino acids into protein of isolated hepatocytes from the European eel, *Anguilla anguilla* L. *J. therm. Biol.* **6**, 201–208.

JOHNSON, F. H., EYRING, H. & STOVER, B. J. (1974). *The Theory of Rate Processes in Biology and Medicine*. New York, London: Wiley.

KWOCK, L., LIN, P. S. & HEFTER, K. (1985). A comparison of the effects of hyperthermia on cell growth in human T and B lymphoid cells: relationship to alterations in plasma membrane transport properties. *Radiat. Res.* **101**, 197–206.

LANDRY, J., CHRETIEN, P., DE MUYS, T. N. & MORAIS, R. (1985). Induction of thermotolerance and heat shock protein synthesis in normal and respiration deficient chick embryo fibroblasts. *Cancer Res.* **45**, 2240–2247.

LANKS, K. W. (1986). Modulations of the eucaryotic heat shock response. *Expl Cell Res.* **165**, 1–10.

LEENDERS, H. J., KEMP, A., KONING, J. F. J. G. & ROSING, J. (1974). Changes in cellular ATP, ADP and AMP levels following treatment affecting cellular respiration and the activity of certain nuclear genes in *Drosophila* salivary glands. *Expl Cell Res.* **86**, 25–30.

LEPOCK, J. R. (1982). Involvement of membranes in cellular responses to hyperthermia. *Radiat. Res.* **92**, 433–438.

LILLY, M. B., NG, T. C., EVANOCHKO, W. T., KATUOLI, C. R., KUMAR, N. G., GLOGAVISH, G. A., DURENT, T. R., HIRAMOTO, R., GHANTA, V. & GLICKSON, J. D. (1984). Loss of high energy phosphate following hyperthermia demonstrated in vivo by ^{31}P NMR spectroscopy. *Cancer Res.* **441**, 633–638.

LOWRY, O. H., ROSEBROUGH, N. J., FARR, A. L. & RANDALL, R. J. (1951). Protein measurement with the folin phenol reagent. *J. biol. Chem.* **193**, 265–275.

LUNEC, T. & CRESSWELL, S. R. (1983). Heat-induced thermotolerance expressed in the energy metabolism of mammalian cells. *Radiat. Res.* **93**, 588–597.

MACKNIGHT, A. D. C. & LEAF, A. (1980). Regulation of cellular volume. In *Membrane Physiology* (ed. T. E. Andreoli *et al.*), pp. 315–335. New York, London: Plenum Press.

MAGUN, B. E. & FENNIE, C. W. (1981). Effects of hyperthermia on binding of epidermal growth factor. *Radiat. Res.* **86**, 133–146.

MATSUMOTO, S. S. & RAIVIO, K. O. & SEEGMILLER, J. E. (1979). Adenine nucleotide degradation during energy depletion in human lymphoblasts. *J. biol. Chem.* **254**, 8956–8962.

MIKKELSEN, R. & KOCH, B. (1981). Thermosensitivity of the membrane potential of normal and SV40 transformed hamster lymphocytes. *Cancer Res.* **41**, 209–215.

MIRTSCH, S., STREFFER, C., VAN BENNINGEN, D. & REBMAN, A. (1984). ATP metabolism in human melanoma cells after treatment with hyperthermia (420C). *Proc. 4th Int. Symp. on Hyperthermic Oncology* (ed. J. Overgaard), pp. 19–22. London: Taylor and Francis.

MONDOVI, B. (1976). Biochemical and ultrastructural lesions. In *Proc. 1st Int. Symp. on Cancer Therapy by Hyperthermia Radiation* (ed. M. J. Wizenberg & J. E. Robinson), pp. 3–15. Chicago: American College of Radiology.

MORRIS, R. O. & KING, T. E. (1962). Thermal denaturation of the heart muscle preparation with respect to its capacity for DPNH oxidation. *Biochemistry, N.Y.* **1**, 1017–1024.

NADAKAVUKAREN, K. K., NADAKAVUKAREN, J. J. & CHEN, L. B. (1985). Increased Rhodamine 123 uptake by carcinoma cells. *Cancer Res.* **45**, 6093–6099.

NAGLE, W. A., MOSS, A. J., ROBERTS, H. G. & BAKER, M. O. (1980). Effects of 5-thio-D-glucose on cellular adenosine triphosphate levels and deoxyribonucleic acid rejoining in hypoxic and aerobic Chinese hamster cells. *Radiol.* **137**, 203–211.

O'BYRNE-RING, N., BEHAN, A. & DUKE, E. J. (1981). Mitochondrial calcium flux in temperature sensitive mutants of *Drosophila*. *J. therm. Biol.* **6**, 195–200.

OVERGAARD, K. & OVERGAARD, J. (1976). Pathology of heat damage studies on the histopathology in tumor tissue exposed to in vivo hyperthermia. *Procs Int. Symp. on Cancer Therapy by Hyperthermia and Radiation* (ed. J. E. Robinson *et al.*), pp. 115–127. Chicago: American College of Radiology.

PEDERSEN, P. L. & CATTERALL, W. A. (1979). The use of thin layer chromatography on poly (ethyleneimine) cellulose to facilitate assays of ATP-ADP exchange, ATP-Pi exchange, adenylate kinase and nucleoside diphosphokinase activity. *Methods in Enzymol.* **55**, 283–289.

PRADET, A. & RAYMOND, P. (1983). Adenine nucleotide ratios and adenylate energy charge in energy metabolism. *A. Rev. Plant Physiol.* **34**, 199–224.

PURANAM, R. S., SHIVASWAMY, C. K., KURUP, C. K. R. & RAMASARMA, P. (1984). Oxidations in kidney mitochondria of heat-exposed rats: regulation by cytochrome *c*. *J. Bioenergetics and Biomembranes* **16**, 421–431.

SAUER, L. A. (1978). Control of adenosine monophosphate catabolism in Mouse ascites tumor cells. *Cancer Res.* **38**, 1057–1063.

SEITZ, W. R. & NEARY, M. P. (1975). Recent advances in bioluminescence and chemiluminescence assay. *Meths biochem. Anal.* **23**, 161–188.

SHEN, A. C. & JENNINGS, R. B. (1972). Myocardial calcium and magnesium in acute ischaemic injury. *Am. J. Pathol.* **67**, 417–452.

SLATER, E. C. (1979). Measurement and importance of phosphorylation potential: calculation of free energy of hydrolysis in cells. *Methods Enzymol.* **55**, 235–245.

SODENBERG, K., NISSINEN, E., BAKAY, B. & SCHEFFLER, F. E. (1980). The energy charge in wild type and respiration-deficient Chinese hamster cell mutants. *J. cell. Physiol.* **103**, 169–172.

STANLEY, P. E. & WILLIAMS, S. G. (1969). Use of the liquid scintillation spectrometer for determining adenosine triphosphate by the luciferase enzyme. *Analyt. Biochem.* **29**, 381–392.

STEVENSON, M. A., CALDERWOOD, S. K. & HAHN, G. M. (1983). Lectin binding and processing after heat in CHO-HA-1 cells. *Radiat. Res.* **94**, 562a.

STEVENSON, M. A., CALDERWOOD, S. K. & HAHN, G. M. (1986). Rapid increases in inositol trisphosphate and intracellular Ca^{2+} after heat shock. *Biochem. Biophys. Res. Commun.* **137**, 826–833.

STEVENSON, A. P., GALEY, W. R. & TOBEY, R. A. (1983a). Hyperthermia induced increase in potassium transport in Chinese hamster cells. *J. cell. Physiol.* **115**, 75–86.

SWEDES, J. S., DIAL, M. E. & MCLAUGHLIN, J. D. (1979). Regulation of protein synthesis during energy limitation of Saccharomyces Cerevisiae. *J. Bacteriol.* **138**, 162–170.

SWEDES, J. S., SEDO, R. J. & ATKINSON, D. E. (1975). Relation of growth and protein synthesis to the adenylate energy charge in an adenine-requiring mutant of *Escherichia coli*. *J. biol. Chem.* **250**, 6930–6938.

USHAKOV, V. (1981). Non-thermal control of thermal stability. *J. therm. Biol.* **6**, 290–215.

VIDAIR, C. A. & DEWEY, W. C. (1986). Evaluation of a role for intracellular Na^+, K^+, Ca^{2+} and Mg^{2+} in hyperthermic cell killing. *Radiat. Res.* **105**, 187–200.

WELCH, W. J., FERAMESCO, J. R. & BLOSE, S. H. (1985). The mammalian stress response and the cytoskeleton: alterations in intermediate filaments. In *Intermediate filaments, Ann. N.Y. Acad. Sci.* **455**, 57–67.

WELSH, W. J. & SUHAN, J. P. (1985). Morphological study of the mammalian stress response: characterization of changes in cytoplasmic organelles, cytoskeleton, and nucleoli, and appearance of intranuclear actin filaments in rat fibroblasts after heat shock treatment. *J. Cell Biol.* **101**, 1198–1211.

WESTERMARK, N. (1927). The effect of heat upon rat tumors. *Skand. Arch. Physiol.* **52**, 257–302.

WILSON, D. F., STUBBS, M., OSHINO, M. & ERECINSKA, M. (1974). Thermodynamic relationships between the mitochondrial oxidation-reduction reactions and cellular ATP levels in ascites tumor cells and perfused rat liver. *Biochemistry, N.Y.* **13**, 5305–5311.

YI, P. N. (1979). Cellular ion content change during and after hyperthermia. *Biochem. biophys. Res. Commun.* **91**, 177–182.

YUSHOK, W. D. (1971). Control of adenine nucleotide metabolism of ascites tumor cells. *J. biol. Chem.* **246**, 1607–1617.

Printed in Great Britain © *Society for Experimental Biology 1987*

SENSITIVITY OF TUMOUR CELLS TO HEAT AND WAYS OF MODIFYING THE RESPONSE

MILTON B. YATVIN[1], *WARREN H. DENNIS*[2],
J. ABIODUN ELEGBEDE[3] *and CHARLES E. ELSON*[4]

Departments of Human Oncology[1], Physiology[2] and Nutritional Sciences[4],
University of Wisconsin-Madison and Department of Biochemistry[3],
Ahmadu Bello University, Zaria, Nigeria

Summary

In our view, the initial effect of hyperthermia on cells is the disorganization of the membrane lipid. Such disorganization alters the membrane's biophysical properties leading to passive changes in transmembrane permeability, shifts in surface charge, and altered stereoorganization of macromolecules associated with the membrane. For example, the passive permeability changes could account for the observed increase in the association of non-histone proteins to chromatin. Surface charge changes resulting from relative changes in the phospholipids could shift the concentration and types of membrane-bound proteins. Such events could initiate hyperthermic cell death.

Membranes of tumour cells are characterized by elevated cholesterol. Such differences in cholesterol concentration with their attendant shift of biophysical characteristics could explain the variation in heat sensitivity between cell lines and within the cell cycle. Further support for lipids being the initial target comes from our studies demonstrating enhanced thermosensitivity when anaesthetics are present. Thermosensitivity of solid tumours is not further influenced by lidocaine in host animals fed diets enriched in linoleic acid, a diet which markedly modifies fatty acid, phospholipid patterns and cholesterol concentration of cellular membranes. One should recognize that global measures of change, such as the ratio of unsaturated to saturated fatty acids, in unsaturated fatty acid index, or percentage of unsaturated fatty acids, may not accurately reflect changes in fatty acid patterns which are related to changes in thermosensitivity. For example, we now recognize that double bond location with respect to the headgroup must be considered as well as the relative content of unsaturated fatty acids in the membrane.

Further studies bearing on the role of diet and anaesthetics on cell killing and on metastatic spread are needed. An increased understanding of the relationship of membrane biophysics and biochemistry correlated with how cells respond to heat could aid in elucidating the mechanisms of cell death. Such knowledge could provide a more rational basis for cancer therapy.

The association between pyrexia and tumour regression was noted as early as 1866 by Busch who observed neoplasm remission in patients afflicted with severe erysipelas (Busch, 1866). Over the past 80 years clinicians have on occasion treated tumours by heat alone or, more recently, in combination with radiotherapy (Dickson, 1979; Jensen, 1903). Coley treated patients with inoperable and advanced malignant disease first by deliberate infection with erysipelas and later by inoculation of mixed filtrates from haemolytic streptococci and *Bacillus prodigiosus* (Coley, 1893; Nauts *et al.* 1953). The toxin injection induced a pyrexia of 40°C of 4–6 h duration. The roles played by heat, the bacterial products and the immune system response in tumour regression were never defined (Reimann & Nishimura, 1949; Dickson, 1977).

Westermark's report (1927) of the destructive effects of heat on animal tumours attracted little attention since the orientation of the medical community in 1927 was to the new disciplines of radio- and chemotherapy. Advances in ultrasound, microwave and radio-frequency technologies, over the past decade have contributed to the renewed interest in heat as a therapeutic modality (Fajardo *et al.* 1980).

Survival studies of normal and neoplastic cells in culture revealed the latter to be more sensitive to the lethal effects of hyperthermia (Auersperg, 1966; Levine & Robbins, 1970). Bhuyan (1979) argues that the value of these studies is limited by a lack of quantitation in the measurement of cell death (survival was based on subjective assessment of the rate of growth), the use of doubtful criteria to evaluate cell death (e.g. loss of cell membrane permeability), and the poor choice of cells (non-growing fibroblasts *versus* epithelial carcinoma; spleen cells *versus* Erlich ascites) used for the comparisons. The relationship has been studied in a more rigorous and quantitative manner (Love *et al.* 1970; Giovanella *et al.* 1973). Chen & Heidelberger (1969) compared mouse prostate cells 'transformed' by a carcinogenic hydrocarbon *in vitro* with normal prostate cells. Giovanella *et al.* (1976) compared human colon carcinoma cells with interstitial epithelial cells and melanoma cells with melanocytes. In each study, tumour cells were more thermosensitive than normal cells based on their decreased cloning efficiency in culture and decreased ability to produce tumours upon inoculation into an appropriate recipient.

As yet it remains unexplained why some tumour cells or subpopulations within a tumour are apparently more heat sensitive than others or why certain tumour cells are more sensitive to heat than normal tissues. One explanation may be that in some tumours heating leads to impaired blood flow resulting in the tumour becoming hotter than normal tissues because of reduced cooling by the blood. However, the viable regions of tumours in general have greater blood flows than most normal tissues (Jirtle, 1981). Targets proposed to explain the thermolethal effects of hyperthermic insult on neoplastic cells/tissues include the nucleic acids (Corey *et al.* 1977; Lunec *et al.* 1981; Mills & Meyn, 1981), proteins (Tomasovic *et al.* 1979; Warters & Roti Roti, 1981) and membranes (Bowler *et al.* 1973; Cossins & Prosser, 1978; Anderson *et al.* 1981; Yatvin, 1977).

Nucleic acids

Several observations point to hyperthermic actions on the nucleus. Chinese hamster ovary (CHO) (Bhuyan *et al.* 1977; Sapareto *et al.* 1978; Westra & Dewey, 1971), L1210 (Bhuyan *et al.* 1977), HeLa (Palzer & Heidelberg, 1973) and EMT-6 (Leith *et al.* 1979; Marmor *et al.* 1977) cells are most sensitive to hyperthermia when in S phase or in mitosis. Westra & Dewey (1971) reported that heat during S phase produces chromosomal aberrations in mitotic cells. These aberrations do not appear to be related to heat-initiated breaks in DNA though heat does cause single stranded regions to remain for a considerable period of time in newly replicated DNA. The repair of high (Corry *et al.* 1977; Lett & Clark, 1978) and low (Mills & Meyn, 1981) dose radiation-induced DNA strand breaks is inhibited by hyperthermia. DNA strand breakage following a hyperthermic treatment (43°C for 15 min) in L1210 cells *in vitro* has been reported (Bowden & Kasumic, 1981).

Heat affects many activities and structures in the cell. The protein mass of chromatin increases after hyperthermia (Tomasovic *et al.* 1978) rendering DNA less accessible to repair enzymes (Jorritsma & Konings, 1983) and to attack by micrococcal nuclease (Warters & Roti Roti, 1981). The rate of DNA synthesis and the synthetic activities of polymerases alpha and beta are reduced in CHO cells (Mondovi *et al.* 1969; Henle & Leeper, 1979) after exposure to hyperthermia. Heat inhibits the initiation and elongation of replicons thereby limiting the initiation of cellular repair processes and also increases the non-specific association of non-histone proteins with DNA (Tomasovic *et al.* 1978). In cell-free systems, HeLa cell DNA polymerase and DNase were not more sensitive to heat inactivation (Weniger *et al.* 1979). These observations support the hypothesis that heat-induced changes in chromatin structure inhibit the cells' repair capacity (Jorristma & Konings, 1983).

Heat affects at least one step in the metabolism of RNA. Inhibition of RNA synthesis and a reversible transformation of granular to fibrillar ribonucleoprotein in the nucleolus were evident in Syrian hamster cells after exposure to 42–45°C (Simard & Bernhard, 1967). Mondovi *et al.* (1969) also demonstrated the heat-mediated depression of ^{3}H-uridine incorporation into the RNA of minimal deviation hepatoma 5123, Novikoff hepatoma, and osteosarcoma whereas the incorporation of labelled RNA, DNA and protein precursors in regenerating rat livers was elevated.

Protein

The protein denaturation hypothesis that heat acts directly on protein structure rather than on its environmental milieu is frequently cited in explanation of cell hyperthermic death. The principal support for this hypothesis is the high activation energy calculated from pseudo-Arrhenius plots of the logarithm of the reciprocal of cell inactivation rate *versus* the reciprocal of absolute temperature. The values obtained are in the range found for heat inactivation of enzymes (Bhuyan, 1979).

Recently, attention has been drawn to the 'heat shock proteins'. The heat shock response was first observed in the fruitfly (*Drosophila*) some 20 years ago (Ritossa, 1962). Within the past five years this response has been shown additionally in plants and animals. In cells or tissues exposed to elevated growth temperatures, there is a rapid synthesis and accumulation of a small number of polypeptides and a concomitant decrease in the production of normal cellular proteins. Temperatures above 30 °C induce a heat shock response in the fruitfly and soybean, species which normally live at ambient temperatures in the vicinity of 25 °C. Temperatures from 42–45 °C induce this response experimentally in warm-blooded animals (Marx, 1983). Exposure to certain drugs (Hightower, 1980), heavy metals (Levison *et al.* 1980), various amino acid analogues (Kelley & Schlesinger, 1978) and fever (Marx, 1983) evoke the synthesis of similar stress proteins (Welch *et al.* 1982).

The specific function of heat shock proteins, if any, is not yet clear. The heat shock proteins may confer a degree of protection to the cell upon subsequent stress situations; such protection might be contingent upon prior synthesis of the stress proteins (Welch *et al.* 1982). Lindquist *et al.* (1982) proposed that heat shock proteins bind to and protect the chromatin during exposure to high temperatures. These proteins may act as bookmarks marking the site of transcription prior to heat shock so that transcription may resume at the site when the temperature is lowered (Marx, 1983; Lindquist *et al.* 1982).

In our opinion, the cellular membranes are most likely the initial target of hyperthermia. The hyperthermia-mediated perturbation of the membrane initiated a cascade of effects which ultimately may be manifested in nuclear and protein functional aberrations leading to cell death. Our hypothesis is attractive in a clinical sense, in that of all the cellular components, the membrane lipids are most amenable to modification. Hence, the possibility of the development of adjuvant approaches exists (Fig. 1).

Membranes

A possible link between growth temperature and the stability of cellular membranes was first alluded to in the works of Heilbrunn (1924) and Belehradek (1931). Though the presence of lipids in membranes had not been confirmed at the time, they postulated that organisms adapt to changes in temperature by altering their plasma membrane lipid composition. Resistance to heat was, they proposed, related to the melting temperature of the lipids (Heilbrun, 1924; Belehradek, 1931). Later *in vitro* studies showed that mycoplasma (Huang *et al.* 1974), bacteria (Sinensky, 1974; Esser & Souza, 1974), yeast (Arthur & Watson, 1976), higher plants (Simon, 1974) and animal cells (Ferguson *et al.* 1975; Johnston & Roots, 1964; Cossins & Prosser, 1978) respond to changes in environmental temperatures by altering the degree of saturation of the fatty acyl chains of their cellular membrane phospholipids. Decreasing the growth temperature tends to decrease the degree of saturation of the fatty acyl chains. Thus, the increasing proportion of unsaturated fatty acids tends to maintain the fluidity of the membrane as the temperature falls and the reverse occurs with increased temperature (Anderson *et al.* 1981).

Films of cellular lipids display an ordered set of structural transitions as temperature is increased, passing from gel to laminar to hexagonal arrangements (Luzzati & Husson, 1962). This last change occurs over a narrow range of temperatures slightly above the normal temperature of cells. Although the lipid structure in these model experiments differs significantly from that found in cell membranes, the concept of a change occurring abruptly with increasing temperature is attractive. If a similar reorganization of the membrane lipids over a small temperature range occurs, one could predict modified permeability and altered relationships of lipids with cell macromolecules. Bowler *et al.* (1973) and Bowler (1981) have suggested that heat injury is secondary to the stability of lipoprotein complexes that affects enzymatic activity dependent on integrity of the cell membrane.

Sinensky (1974) observed that the phospholipids of *Escherichia coli* grown at 10°C contained a disproportionate quantity of unsaturated fatty acids whereas

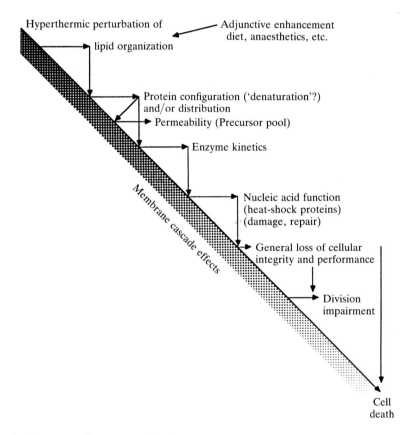

Fig. 1. A proposed sequence of cellular events occurring after exposure to heat and their modification by anaesthetics and dietary manipulations.

those lipids in cells grown at elevated temperatures contained a disproportionate quantity of saturated fatty acids. The viscosities of the membrane lipids, measured at growth temperatures did not vary. When determined at 37 °C, the lipids derived from the cells grown at 10 °C were the least viscous.

The fatty acid auxotroph *E. coli* K1060, requires unsaturated fatty acids for growth and by varying the unsaturated fatty acids in the growth medium, membrane composition can be markedly altered. On exposure to mild hyperthermia, the sensitivity of these *E. coli* cells increased in proportion with increased unsaturation index of the cellular lipids (Yatvin, 1977). The above and similar observations led us to the proposal that the killing effect of hyperthermia is at the level of membrane disorganization characterized by a decrease in membrane lipid order (Yatvin, 1977; Dennis & Yatvin, 1981).

One may reasonably presume that heat induced alterations in physical properties of the membrane lipid matrix lead in turn to changes in the protein-lipid interactions described by Bowler *et al.* (1973). Such disruptions could modify the energetics of the cell and other membrane processes associated with cell homeostasis. When such disruptions are large enough or exist for a prolonged period of time, the cell loses its ability to reproduce (Yatvin, 1977). Treatment of mammalian cells with anaesthetics, which are membrane-active agents (Yatvin *et al.* 1979), or with alcohol (Li *et al.* 1980) increase cell killing following hyperthermic treatment *in vivo* and *in vitro*. Surface membrane alterations in mammalian cells have also been reported (Lin *et al.* 1973; Bass *et al.* 1978; Mulcahy *et al.* 1981) and indirectly implicate membrane damage in the pathogenesis of cell inactivation following hyperthermia (Fig. 1).

A hyperthermic therapeutic enhancement could be realized if tumour cell membranes were either naturally more fluid or could be made so (Yatvin, 1977). Others (Burns *et al.* 1979) have altered the membrane lipid composition and, thereby, the membrane microviscosity of murine L1210 cells by diet modification. Likewise, we have modified the membrane lipid composition (Hidvegi *et al.* 1980) and microviscosity of murine P388 ascites cells by feeding the host mice diets high in either saturated or unsaturated fatty acids. Yau (1979) in an extensive study of procaine-mediated response of mammalian cells to both hyperthermia and radiation found that procaine could reversibly influence cellular morphology. In agreement with our findings, Yau (1979) suggested that procaine may exert its modifying influence on radiation and heat-induced effects *via* membrane-mediated mechanisms.

We studied surface alterations of heated P388 cells by scanning electron microscopy (Mulcahy *et al.* 1981). Cell killing in similarly treated cells was correlated to membrane fatty acid composition. Cells high in saturated fatty acids responded differently to hyperthermic treatment *in vitro* than cells high in unsaturated fatty acids (UFA). The morphological response of highly saturated cells was modified by procaine and resembled heated cells obtained from animals fed a diet high in UFA. A cell's response to hyperthermia is related to its membrane fluidity at the time of treatment (Dennis & Yatvin, 1981; Mulcahy *et al.* 1981). These observations are discussed in greater detail later in this review.

The contemporary model of membrane structure is that proposed by Singer & Nicolson (1972). Based on consideration of the thermodynamics of macromolecules, properties of the proteins and lipids of functional membranes and other experimental evidence, these authors concluded that a 'fluid mosaic model' of the gross organization and structure of the proteins and lipids of biological membranes is consistent with restrictions imposed by thermodynamics. They also suggested that the proteins in this model are a heterogeneous set of globular molecules, each arranged in an amphipathic structure. The ionic and polar groups protrude from the membrane into the aqueous phase and the non-polar groups are largely buried in the hydrophobic interior of the membrane (Singer & Nicolson, 1972; Hopkins, 1978).

Membrane composition

The major constituents of membranes are lipids, proteins and polysaccharides (Simon, 1974; Chapman, 1967; Singer & Nicolson, 1972). Generally, 50 % or more of the weight of isolated plasma membrane consists of protein (Nystrom, 1973; Rosenberg & Guidotti, 1969) which can further be separated into several different fractions. Some of the proteins located in the membrane provide pores through which water and other small molecules penetrate (Simon, 1974), other proteins serve structural roles conferring mechanical stability and others possess catalytic functions (Nystrom, 1973).

Lipids, considered to be more important than protein as structural components of membranes, consist largely (90 %+) of cholesterol and phospholipids with lesser quantities of neutral fat (Brossa *et al.* 1980; Williams & Chapman, 1970). These lipids encompass up to seven classes of phospholipids (PL), their lyso-derivatives and neutral glycerolipids, all of which exhibit varying patterns of fatty acids primarily of 16 or 18 carbons in saturated bonds (SFA) or with *cis* unsaturated bonds at (MUFA) or distal to (PUFA) the 9-position of the chain. Fatty acids of chain lengths to 22 carbons are also present; these acids may contain as many as 6 *cis* bonds. Throughout nature, PL containing a SFA at the 1-position and a $\Delta 9$ *cis* MUFA or PUFA at the 2-position appear in the bilayer of the membrane (Tattrie *et al.* 1968; Innis & Clandinin, 1981). The polar headgroup of the PL may be zwitterionic (e.g. phosphatidylcholine (PC)) or acidic (e.g. phosphatidylethanolamine (PE), phosphatidylinositol, phosphatidylserine). Assembly of the polar headgroups in an orderly bilayer is facilitated by the nature of PC wherein the positive charge of the quaternary nitrogen is shielded by the methyl groups. The phosphate has its charge distributed between two oxygens, each at the corner of the phosphate tetrahedron. This distributed negative charge enhances the coupling of two cations to the single anionic phosphate or the divalent cation may attract two anionic phosphates. The interlocking ionic charges between PC and an acidic PL gives rigidity; two PC, each having cationic and anionic charges would not interlock and hence would decrease rigidity. In this way ionic interactions between headgroups contribute stability to the bilayer.

The *cis* bond at the Δ9-position of the UFA gives order between the Δ9-carbon and the membrane surface; below the Δ9-carbon, the 120° angle of the bond introduces disorder towards the centre of the bilayer. The ubiquitous *cis* bond at the Δ9-position also allows van der Waal's forces to form at this level sharply rigidifying the proximal region of the chain. Introduction of a *cis* bond distal to the 9-position orients the remainder of the chain parallel to the surface of the bilayer. Thus, fatty acids extending into the bilayer from each side can interact with the formation of van der Waal's forces; additional *cis* bonds, however, disrupt this layering. Within the bilayers the van der Waal's forces allow lateral fluidity or motion. A protein embedded in the rigid portion of the bilayer (the paraffin barrier) might have vertical mobility (bobbing motion) which would be expressed as the unsaturation of the disordered layer (allowing the membrane to thin) increases. The interlocking headgroups of the PL hold their fatty acid chains at a fixed height perpendicular to the plane of the membrane. The double bonds at the Δ9-positions of the fatty acids thus are held at the same depth relative to the ionic sheet only if the UFA of adjacent molecules are esterified at the same position, the 2-position of the PL. Random positioning would not allow such order (Williams & Chapman, 1970; Haines, 1979).

Since cellular PL are present primarily in the membrane fractions, the total PL content of a cell serves as an approximate measure of its membrane content. The subcellular membrane PL composition varies from organelle to organelle (Simon, 1974). A single membrane type also exhibits species to species variation in sterol, PL and fatty acid compositions (Nystrom, 1973). PL are relatively insoluble in water and exist in several forms when in the solid state. When heated, PL do not undergo direct transformation from crystalline to liquid; rather they pass through an intermediate mesomorphic or liquid-crystalline state. A change in physical characteristics of PL due to dehydration or to temperature change modifies the architecture of membranes. The loss of PL molecules or the modification of their hydrocarbon chains which affects their packing by lowering the extent of hydrophobic bonding also affects the architecture of membranes and influences membrane permeability (Simon, 1974).

However, membrane fluidity is influenced not only by the degree of saturation of fatty acyl chains but by other factors such as fatty acyl chain length, PL headgroup composition and membrane cholesterol content (Anderson *et al.* 1981). Cholesterol is an amphipathic molecule which exhibits dual roles with regard to membrane fluidity. In artificial membranes, cholesterol interacts with the membrane PL thereby reducing their areas, suppresses the motion of acyl chains and decreases the fluidity of the membrane (Poznansky *et al.* 1973). Cholesterol reduces the thermal motion of acyl chains at temperatures above the PL phase transition when membrane becomes highly fluid, whereas below this temperature, cholesterol increases acyl chain mobility by disrupting the van der Waal's forces. Thus, cholesterol appears to introduce an intermediate degree of fluidity in bilayers. Since the plasma membrane of mammalian cells contains equal molar amounts of cholesterol and PL, its influence on membrane fluidity must be considered. In addition, cholesterol may further influence neoplastic cells as neoplastic hepatocytes have higher plasma membrane cholesterol-phospholipid ratios (Van Hoven & Emmelot, 1972).

Fluidity is usually assessed either by measurement of the degree of rotational mobility of 'fluorescent polarization' or of 'electron spin resonance' probes embedded within the lipid bilayers (Yatvin *et al.* 1982*a*). The more fluid a membrane, the greater the rotational mobility of the probe. Lepock (1982) points out that with increasing temperature the rate of lipid motion is increased while the lipid order is decreased. The fluorescent probe diphenylhexatriene primarily monitors order while other probes are more sensitive to rate of motion. Each technique, Lepock argues, measures only one aspect of what constitutes fluidity. He suggests that a change in 'lipid order' is a more precisely defined mechanism for the effect of lipid composition than is the concept of fluidity. Yatvin (1977) and others (Dennis & Yatvin, 1981; Yatvin *et al.* 1981) reported that diet-(media-) and temperature-induced alterations in the unsaturated fatty acid of mammalian cell and *E. coli* membrane lipids cause significant changes in fluidity at a fixed temperature as measured by fluorescent probes and fatty acid spin labels. According to these investigators, the measured 'fluidity' positively correlated with sensitivity of the cells to hyperthermia. The argument (Lepock *et al.* 1981) seems to be with the definition of the term 'fluidity'. But as Yatvin *et al.* (1982*a*) pointed out, 'membrane fluidity' should be viewed as a functional concept that may or may not correlate with physical properties observed by every probe technique.

Differences in fatty acid and sterol composition of cell membranes could provide a rational explanation for the considerable variation in heat sensitivity between cell lines as well as the cell-cycle sensitivity (Dewey *et al.* 1977). A striking relationship has been observed between thermal sensitivity and cholesterol content of six different cell lines, with no significant pattern of saturated to unsaturated fatty acid ratio found (Cress & Gerner, 1980). Furthermore, Hahn *et al.* (1977) have reported data indicating a synergism between a polyene antibiotic which acts by removing cholesterol from the membrane and hyperthermia.

From the foregoing, it is obvious that cell membranes have an acknowledged role in determining sensitivity of cells to hyperthermic cell killing and that an understanding of the relation of membrane biochemistry and response to heat could provide a rational basis for cancer therapy by hyperthermia. Thus, membrane characteristics of normal and neoplastic cells become an important consideration in assessing the likelihood of a cell being damaged by a given hyperthermic insult (Martinez-Palomo, 1979). Support for this contention is forthcoming from the analysis of different types of malignant neoplastic tissue which has revealed a variety of cellular modifications, none of which is common to all tumours studied. However, among the features studied, the surface of the cell

frequently displayed significant structural alterations. The most commonly observed alterations were: (1) increased thickness of the carbohydrate containing
surface coat; (2) relative absence of cell contact; (3) increased mobility of
carbohydrate surface receptors; (4) diminished number of subplasmalemmal
microfilaments and; (5) increase in intermembranous particles. In our view, such
differences could greatly influence cellular response to heat and account for some
of the differential responses to heat of tumours.

*Membrane characteristics of metastasizing and non-metastasizing tumours and
normal cells*

Kim (1979) developed paired lines of mammary carcinomas which differ in their
ability to metastasize to distant sites. These metastasizing and non-metastasizing
carcinomas display interesting differences in their biochemical and plasma
membrane properties. The most outstanding characteristics, however, of the
metastasizing rat tumour cells is that they seem to have extremely unstable plasma
membrane structures. This causes them to constantly shed incompletely assembled membrane constituents with concomitant renewal of the loss. Such
membrane instability may permit the leakage of various degradative lysosomal
enzymes ordinarily meant for the physiological cell renewal or homeostatic control
mechanisms. Such enzymes may then help in weakening and breaking down cell to
cell bonds in a tumour cell mass as well as in clearing the path for dispersion of
freed cells. The locomotive properties of these cells is probably a manifestation of
plasma membrane instability. The dissociated plasma membrane constituents
include glycoprotein and glycolipid-antigens, surface marker enzymes, glycocyl-
transferases, glycosidases, and some degradative enzymes.

In this vein, Chernavskii *et al.* (1982) have also proposed that normal and
malignant cells differ in the physical organization of their plasma membranes.
Namely, they propose that there is a viscoelastic mechanical frame on the surface
of normal cells, whereas malignant cells have no integral closed frame although
the major components (protein, glycoproteins, glycolipids, etc.) of both are
present. Gallez (1984) put forth a model of cell membranes after malignant
transformation which suggests that transformed cells have stable surface characteristics. The milieu in which tumour cells grow affect the characteristics of the
plasma membrane. Van Blitterswijk *et al.* (1985) found that fluidity of mouse
lymphoma cells as measured by fluorescence polarization using diphenylhexatriene and (free) cholesterol/phospholipid molar ratios of whole cells varied with
site of implantation. They suggest that the observed differences in membrane
fluidity between distinct subsets of tumour cells may be relevant to the sensitivity
of such cells to immune attack or to drugs.

Recently, Barz *et al.* (1985) studied cellular and extracellular plasma membrane
vesicles from a non-metastasizing lymphoma and its metastasizing variant. The
tissues differed in their lipid composition, indicating a more fluid state of plasma
membranes derived from the highly metastatic tumour line. Likewise the

metastasizing cells shed much greater amounts of their plasma membrane in the form of extracellular membrane vesicles. These could be isolated from ascites fluid and they differed markedly from the plasma membranes from which they were derived.

It is possible that metastatic foci may, as a result of the membrane characteristics that allowed them to metastasize initially, be more responsive than normal tissues to killing by whole body hyperthermia. Factors such as age, size and site of the metastatic lesions may also greatly modify the response and must be carefully evaluated. The possibility of selectively killing cells seeded to the lung from tumours such as osteogenic sarcoma and Ewing's sarcoma by regional or whole body hyperthermia is an exciting possibility that should be investigated. It holds the possibility of being less damaging than other forms of elective therapeutic control (radiation in combination with chemotherapy) against the ever present danger of metastatic spread of such malignant tumours to the lungs.

Whether metastasizing rat carcinomas are analogous to metastases of human breast cancer remains to be established. Nevertheless, based on our previous work implicating the cell membrane as an important target in hyperthermic cell death, and Kim's finding that plasma membranes of metastasizing tumour cells are unstable and severely compromised, we felt that it was not unreasonable to expect a greater heat sensitivity in metastasizing rat carcinomas compared to their non-metastasizing counterparts.

Effect of heat on various tumour cell types

Tomasovic and co-workers have demonstrated that heterogenous cell lines cloned from a single rat tumour display different heat sensitivities and different thermotolerance development. The tumour also differed in metastatic potential but there was no correlation with thermal tolerance (Tomasovic *et al.* 1983; Tomasovic *et al.* 1984; Welch *et al.* 1984). Rofstad & Brustad (1985) report heterogeneity in heat sensitivity and development of thermotolerance of cell lines derived from a single human melanoma xenograft.

Therefore, we investigated the response to water bath heating of legs implanted with paired Kim tumours. Generally, the response of legs of all animals heated in 43·5°C or 43·0°C water heat for 1 h was similar. Within 12 h of heating there was prominent oedema. The oedema gradually subsided, and except for some minor scaling on the foot, the legs appeared normal. Heating at 42·5°C for 60 min produced only minor swelling. Tumour response to the heat treatment was evaluated by estimation of the time required for regrowth to treatment size. The data in Table 1 illustrates the marked differences in delays for the metastatic and non-metastatic tumours.

The non-metastatic tumours had extremely short average delays. In some treated animals there was no delay in growth. At the other extreme, regression of tumour and even complete disappearance was observed in a few animals 2 or 3

Table 1. *Response of metastatic (TMT-081) and non-metastatic (MT-100) tumours to water bath heating*

	Treatment	Regrowth delay* (days)	Cures†
Non-metastatic tumours			
MT-W9B	43·5°C, 60 min	1·78 ± 0·84	0
MT-100	43·5°C, 60 min	3·10 ± 2·56	0
Metastic tumours			
SMT-2A	43·5°C, 60 min	10·95 ± 3·18	0
TMT-081	43·5°C, 60 min	19·9 ± 2·63	12 of 14

* Time after heating required for tumour to return to treatment size.
† No visible sign of tumour at treatment site 21 days after treatment.

days after treatment. However, reappearance and/or regrowth of the tumours occurred with 5 or 6 days, and cures were not observed.

On the other hand, the response of the metastatic tumour to an identical heating (43·5°C for 60 min) was strikingly different. Extensive shrinkage and in many cases total disappearance of the tumour was seen in both the SMT-2A and TMT-081 tumours within 3 days of treatment. For the SMT-2A the regrowth delay was 10·95 days but no local cures were obtained. More impressive results were obtained with the TMT-081 tumour. In two separate experiments 12 of 14 animals treated showed local cures for at least 21 days. For both tumours, the reappearance of tumours in all cases was preceded by the appearance of metastatic lesions in the leg above the treatment site. Thus, it cannot be ruled out that the reemergence of a tumour at the treatment site was due to metastatic spread from above.

In studies in which metastatic tumours were heated at lower temperatures or for shorter times at 43·5°C, the TMT-081 tumour is quite sensitive to moderate heat treatments. Treatments at 43·0°C for 60, 43·5°C for 30 and 42·5°C for 60 min resulted in lag times that were comparable to those at 43·5°C for 60 min. Although no cures resulted, local regrowth was again preceded by the appearance of metastatic lesions. The response of the SMT-2A tumour to either a shorter heating at 43·5°C or a lower temperature (42·5°C) for 60 min was not significantly different from controls. Thus, for these pairs of tumours at least, the metastatic mammary tumours are markedly more sensitive to heat than the non-metastatic counterparts.

In order to rule out the possibility that the response of the tumours was due to differences in tissue temperature, intra-tumour temperature measurements were made. The results shown in Table 2 indicate that for the TMT-081 and MT-100 tumours no significant temperature differences exist. In both tumours, an equilibrium tumour temperature was attained in 3–5 min and remained nearly constant throughout the duration of heating.

Clearly, the two mammary adenocarcinomas are distinct with regard to their fatty acid composition (Table 3). About twice as much arachidonic acid is present in the non-metastasizing as compared to the metastasizing tumour (Yatvin *et al.*

1982*b*). On the other hand, the metastasizing tumour contains less linoleic acid and displays a higher ratio of palmitic to stearic acid than the non-metastasizing one. The ratio of PL to protein in metastasizing tumours is twice that of its non-metastasizing counterpart. One may postulate that the metastasizing tumour cell attempts to increase the microviscosity of the membrane by reducing its content of arachidonic acid (homeoviscous adaptation). In the SMT-2A, it may be needed to compensate for the loss of proteins from the membrane.

Preliminary studies using fluorescence polarization to assess the microviscosity of these tumour cells have not indicated large differences. Perhaps the relatively greater amount of longer chain saturated fatty acids and the decrease in arachidonic acid may have compensated for the loss of proteins. Other explanations for the greater heat sensitivity of the metastatic tumour are conceivable and must be tested in the future. For example, blood flow in the metastasizing tumour was found to be only about one-half of that in the non-metastasizing tumour (Jirtle, 1981) thus, the greater response to hyperthermia of the metastasizing tumour might be explained on the basis of a smaller heat transfer from the tumour. The lack of any significant difference in intra-tumour temperature (Table 3) argues against such an explanation. Whatever the reasons may be for the differences in sensitivity to hyperthermic treatment between metastasizing and non-metastasizing tumours, they seem to open potentially valuable avenues of approach for the understanding of mechanisms of metastatic behaviour and for the treatment of metastatic lesions.

The influence of hyperthermia on metastatic spread has been of concern to some. Yerushalmi (1976) has suggested that hyperthermia may cause metastatic spread from the primary tumour if treatment is inadequate. This conclusion has been challenged by Kim & Hahn (1979), who found that when they used the Dunn osteogenic sarcoma and amputation techniques, mild to moderate heating (40·5–42·5 °C) alone or in combination with radiation did not increase metastatic spread.

Modification of cellular lipids

Since the report of a possible correlation between cellular membrane composition and thermosensitivity, attention has focused on means of modifying cellular

Table 2. *Temperatures of metastatic (TMT-081) and non-metastatic (MT-100) tumours heated in a water bath*

Tumour	Water bath temperature, °C	Tumour temperature, °C*
TMT-081 (5)	43·5 ± 0·05	42·89 ± 0·15
MT-100 (7)	43·5 ± 0·05	42·85 ± 0·21

* Average temperature of tumour for 60 min heating. Both tumours reached equilibrium in 3–5 min.

Table 3.

Tumour	Phospholipid/ protein	16:0	16:1	Fatty acid – Area % 18:0	18:1	18:2	20:4
SMT-2A	0·332 ±0·108	20·23 ±1·33	4·60 ±0·98	18·45 ±2·09	31·50 ±1·86	14·95 ±3·05	10·26 ±1·44
MT-W9B	0·146 ±0·026	24·11 ±1·07	5·19 ±0·58	14·17 ±0·43	26·94 ±1·91	8·96 ±0·82	20·63 ±2·82

membrane lipids of tumours to render them more sensitive to hyperthermic insult. A number of these studies involved the use of bacteria and/or fungi, particularly the K1060 auxotroph of *E. coli*, which is defective both in its ability to synthesize and to degrade unsaturated fatty acids. It has been demonstrated (Sinensky, 1974; Yatvin, 1976, 1977; Yatvin *et al.* 1980*a,b*) that by varying the fatty acid composition of the medium, the membrane fatty acid composition of these organisms could be altered. Similar responses in the fatty acid composition of mammalian cells grown in tissue culture have been reported. Such modified cells include human lymphoid cells (George *et al.* 1983), CHO (Rintoul & Simoni, 1977) and human endothelial cells (Spector *et al.* 1980).

There are reports also of *in vivo* modification of the cellular composition of mammalian tissues by varying the lipid composition of the diets (Montfoort *et al.* 1971). Hidvegi *et al.* (1980) modified the membrane lipid composition of murine P388 ascites cells by feeding hosts diets high in either saturated or unsaturated fatty acids. The diet modified cells were heated *in vitro* and injected into recipient mice and their survival determined. Exposing the cells to 43·5°C significantly increased survival of mice receiving the ascites cells grown in animals fed the highly unsaturated fatty acid diet. The hyperthermic killing effect was potentiated by the addition of 1 mmol l^{-1} procaine. In a related study, Mulcahy *et al.* (1981) studied the quantitative distribution of cell surface alterations of heated P388 ascites tumour cells as determined by scanning electron microscopy. Cells harvested from host animals maintained on a standard rodent chow diet or one high in saturated fatty acids responded differently to identical hyperthermic treatment *in vitro* than cells obtained from animals on a highly unsaturated fat diet. Erhlich ascites cells grown in animals fed either highly saturated or unsaturated fatty acid-containing diets also respond differently to hyperthermia *in vitro* (Ehrnstrom & Harms-Ringdahl, pers. comm.). These same workers using ESR probes found the membranes from cells from mice fed the highly unsaturated diet were less well organized than those cells from animals fed the more saturated diet. The morphological response of cells from chow animals fed a diet high in saturated fatty acids was modified by addition of procaine to the incubation medium, a membrane-active drug. The pattern of response observed after these cells were heated in the presence of procaine resembled that seen following heat treatment of ascites cells obtained from animals fed the diet high in unsaturated fatty acids.

These data are consistent with the hypothesis that a cell's response to hyperthermic insult is related to its membrane fluidity at the time of treatment.

Alterations in surface morphology produced in response to hyperthermia are different for cells with differing membrane fluidities (Figs 2–7; Mulcahy *et al.* 1981). Further, heating ascites cells from control animals in the presence of procaine produces a surface pattern which is nearly identical to that observed with ascites cells from animals fed a diet high in unsaturated fat. Since in these experiments we were unable to correlate specific morphological membrane alterations with cell lethality or irreversible damage, we cannot state with certainty that ascites cells with more fluid membranes were damaged more than ascites cells with rigid membranes. However, the significant amount of cell lysis, a terminal form of damage, observed for heat-treated cells from unsaturated hosts or for control cells heated in the presence of procaine, suggests that cells with fluid membranes are more sensitive to hyperthermia than cells with less fluid membranes.

Bass *et al.* (1978) studied cell surface morphology of *in vitro* cultured Chinese hamster cells following heating at 43 °C for 3 h. The surface of those cells was characterized by a reduction in the number of microvilli, bleb and debris formation, and decreased adherence to the substratum.

Fig. 2. Scanning electromicrograph of P388 ascites cells removed from a mouse fed the saturated fatty acid diet. The cells were fixed immediately after removal from the peritoneum. Ascites cells fixed immediately after removal from mice in the other dietary groups had the same surface morphology. ×2600. From Mulcahy *et al.* (1981). *Int. J. Radiat. Biol.* **39**, 95–106. With permission.

Fig. 3. Ascites cells from standard chow fed-animals after incubation at 43 °C for 3 h.
×3000. From Mulcahy *et al.* (1981). *Int. J. Radiat. Biol.* **39**, 95–106. With permission.
Also Figs 4–7.

Recently, Yatvin *et al.* (1983) and Burns *et al.* (1983) reported selective modification of mouse tissue lipids by dietary manipulations. Both groups fed diets
containing 16 % fat, either high (safflower oil/sunflower oil) or low (beef
tallow/coconut oil) in linoleic acid. Yatvin *et al.* (1983) studied the modification of
membrane lipids from liver and solid tumour (mammary adenocarcinomas CA755
and MtGB grown in the hind legs of BDF1 and C3H mice) while Burns' group
examined changes in the plasma membrane fatty acid composition of liver, heart,
spleen, bone marrow and thymus and liver of DBA/2 mice.

Yatvin *et al.* (1983) observed that the safflower oil diet elevated the diene and
tetraene content but lowered the hexaene content of liver membrane PL whereas
in the membrane of the solid tumour (CA755) the PL contained higher
proportions of tetraene (20:4) and hexaene (22:6). While host liver responded to
the high linoleate diet by reducing synthesis of hexaene, tumours growing in these
animals were incapable of manifesting any control.

Burns *et al.* (1983) reported that tissue PL of animals fed coconut oil diets were
enriched in monoenoics whereas those fed the sunflower diet were enriched in
polyenoics. They also observed differences in the extent and type of modification
in the different tissues and between whole liver and liver plasma membrane. For
example, the fatty acid composition of purified liver plasma membrane PL was
higher in monoenoics and lower in polyenoics content compared to the liver
phospholipid extract. The authors concluded that dietary modification of the fatty

acid composition of normal tissues differs in extent and type according to the tissue, and that liver plasma membrane PL composition is different from that of total liver PL.

In normal tissue the perturbation of membrane lipids by diet apparently achieves a plateau, whereas, in tumour tissue seemingly there is no plateau. If such is true, prolonged feeding could result in a tumour membrane with a lipid composition that will enhance thermal killing to a greater extent in tumour tissue than normal tissue.

The influence of membrane components in cellular response to hyperthermic insult is still under investigation, particularly in *in vivo* systems. *In vitro*, however, there is a large body of evidence linking the different membrane components to thermosensitivity. An increase in the degree of unsaturation of fatty acyl chains of membrane lipids enhances the thermal sensitivity of cultured cells.

Influence of adjuvant therapy

We have extended our observations of adjuvant therapy to a mammary adenocarcinoma growing in the legs of mice. When tumours are treated by heating in combination with an infiltration of local anaesthetics, greater tumour cell killing is obtained as indicated by a prolongation of animal survival (Yatvin *et al.* 1979). Tumours of approximately 4×4 mm were infiltrated with 50 μl of 4 % lidocaine

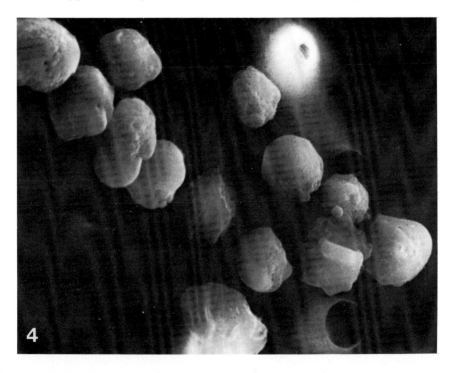

Fig. 4. P388 cells from hosts maintained on an unsaturated fatty acid diet. These cells were incubated for 3 h at 43 °C. ×2600.

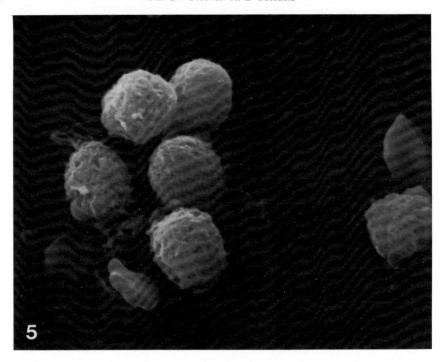

Fig. 5. The alterations apparent after 3 h incubation at 43 °C in these P388 ascites cells from mice maintained on a diet high in saturated fatty acids are very similar to those observed when ascites cells from standard chow-fed animals are similarly treated (Fig. 3). ×2600.

(2 mg/mouse) and the leg containing the tumour heated for 1 h in a well-regulated water bath. Heat alone (43·5 °C) results in increased survival. Lidocaine alone has no effect on survival. There is a significant interaction of the drug with heating at 43·5 °C for 1 h. Five of 31 mice whose tumours were treated with lidocaine and 43·5 °C for 1 h showed complete remission of the tumour (Fig. 8). Such 'cures' are unusual in this tumour.

The survival data from three experiments were combined and analysed using a Cox survival model extended to include covariates as well as right-censored data. Mouse survival post-treatment as a function of time are shown in Fig. 9 based on the Cox model. In one experiment bupivacaine at a dose of 0·375 mg/mouse yields similar results (Fig. 10).

The increase in doubling time of the tumour volume (Fig. 11) is another measure of the effect of these treatments. Again, analysis shows that the significant treatments are heating at 43·5 °C and the interaction of the heating at 43·5 °C with local anaesthetic infiltration. Using a different local anaesthetic, tetracaine, in combination with heating at 44 °C, Bowler (1987) reported that the effectiveness of tetracaine in combination with heat as a 'cure' varied inversely with the heat sensitivity of two rat tumour lines. The greatest effect was seen in the more heat-sensitive D23 hepatoma as compared with the less heat-sensitive MC7 sarcoma (Barker, 1986).

Because of its clinical implications, systemic lidocaine was tested in the above model system and in a murine fibrosarcoma tumour model. An equivalent supraadditive, tumour-inhibitory effect of heat and lidocaine was obtained with both systemically and intratumour-administered lidocaine. The serum levels of lidocaine necessary to achieve tumour regression were within the therapeutic range for the control of arrhythmia in humans. Several treatment schedules, varying the mode of drug delivery, were evaluated (Robins *et al.* 1983).

The ability of the local anaesthetic to prolong survival following hyperthermia most likely is due to its membrane disordering property. Hyperthermia alone and hyperthermia with lidocaine cause changes in the fine structure of the CA755 tumour cell as well as the breakdown of the tumour vasculature (Clark *et al.* 1983). The first structural change, observed immediately after termination of hyperthermia of 43·5°C for 1 h, is the vesiculation of the Golgi apparatus. Other structural changes occur later but with variable times of onset. The changes appear to be unrelated to the presence of lidocaine. Vascular breakdown results in haemorrhaging within the tumour, and its onset and intensity appear to vary directly with the size of the tumour. Breakdown of the tumour cell plasmalemma and degenerative changes of the cytoplasm and nucleoplasma are seen more frequently in large tumours and in the interior of small tumours at any given time after the end of hyperthermia. The vesiculation of the Golgi persists in treated cells for as long as 30 h. This modification may represent an intensification in the function of

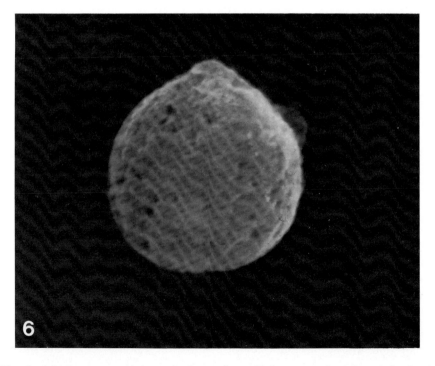

Fig. 6. A higher power micrograph of an ascites cell, from a standard chow fed animal, after exposure at 43°C for 3 h. ×9000.

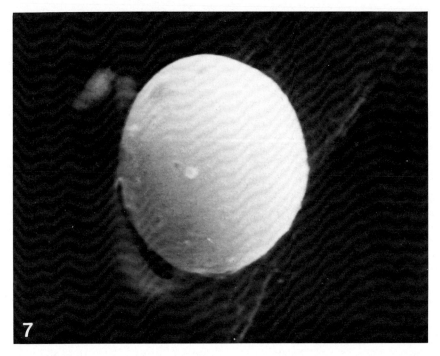

Fig. 7. A P388 ascites cell, harvested from a standard chow-fed animal, and then heated as in Fig. 6 with the addition of $1\,\mathrm{mmol\,l^{-1}}$ procaine added to the incubation medium. ×9000.

the Golgi apparatus; however, it closely corresponds to that found in a variety of other cells treated with a class of compounds, including lidocaine, that specifically inhibits the function of the Golgi apparatus. Unlike that of hyperthermia, the effect of these compounds is rapidly reversible. Since the Golgi apparatus probably is crucial in repairing any deleterious effects of hyperthermia, any impairment of its normal function would place most treated tumour cells in a difficult position. The rate of tumour destruction may ultimately depend on the breakdown of the tumour vasculature following hyperthermia and lidocaine.

Using a spleen colony assay to evaluate fractional survival of AKR leukaemia and normal bone marrow cells after *in vitro* heat exposure Robins *et al.* (1984) found that neoplastic cells were inherently more sensitive to thermal killing than normal syngeneic stem cells. The differential effect of hyperthermia on AKR murine leukaemia and AKR bone marrow cells could be further enhanced by the addition of lidocaine (Fig. 12) or thiopental to the incubation mixtures during heating. A role for membrane fluidity in the sensitivity of cells to hyperthermia both in *in vivo* systems is thus suggested.

As noted, the goals of specifically modifying tumour cells by diet and the ability to target anaesthetics to only tumour cells still remain to be accomplished. Injectable local anaesthetics readily leave the site of injection. Recently, sustained local release of anaesthetic agents has been made possible by microencapsulation

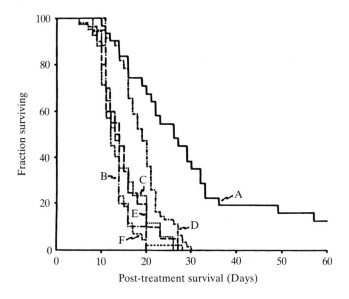

Fig. 8. Fraction of tumour-bearing mice surviving as a function of days elapsed after treatment for (A to C) groups receiving lidocaine and (D to F) groups receiving either saline injection or no injection. In groups A and D, the tumour-bearing legs were heated at 43·5 °C for 1 h; in B and E, they were heated at 42·0 °C for 1 h; and in C and F, they were not heated. Numbers of animals were (A) 31, (B) 10, (C) 17, (D) 51, (E) 20, and (F) 42. From Yatvin *et al.* (1979). *Science* **205**, 195–196. With permission.

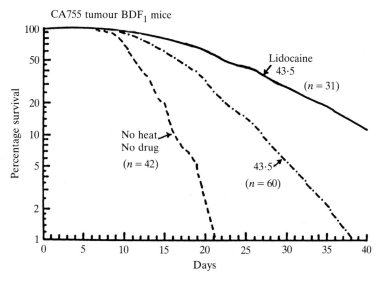

Fig. 9. Survival data of BDF1 mice with a CA755 adenocarcinoma is shown as a function of time for untreated mice, mice treated by heating at 43·5 °C for 1 h and for mice treated by the combination of lidocaine infiltration of the tumour with heating at 43·5 °C for 1 h. The curves were calculated from a Cox survival model fit to the data. From Yatvin *et al.* (1980*a*). *Molecular Mechanisms of Anesthesia*, vol. 2, pp. 495–499. With permission.

of the drug in a biodegradable polymer (Williams *et al.* 1984). Likewise, methoxy-fluorane, an inhalation anaesthetic characterized by slow onset and high tissue affinity, has been prepared as an injectable, tissue compatible stable formulation by sonication in the presence of lecithin. Methoxyfluorane's low vapour pressure and H_2O solubility result in its tight binding and slow release by tissue with a $10\times$ increase in duration of local anaesthesia (Haynes, 1984). The use of sustained

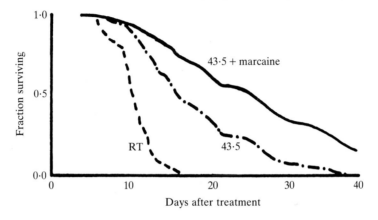

Fig. 10. Data as shown in Fig. 9, with the exception that the drug infiltrated was bupivacaine. From Yatvin *et al.* (1980*a*). *Molecular Mechanisms of Anesthesia*, vol. 2, pp. 495–499. With permission.

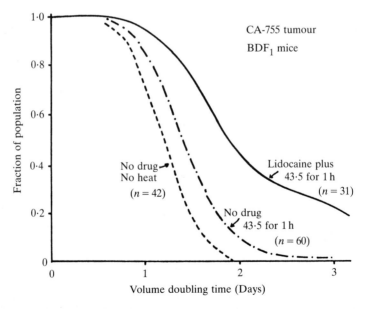

Fig. 11. The fraction of the treated population whose tumour volume doubling time is equal to or exceeds the value shown on the abscissa is shown for three treatments. From Yatvin *et al.* (1980*a*). *Molecular Mechanisms of Anesthesia*, vol. 2, pp. 495–499. With permission.

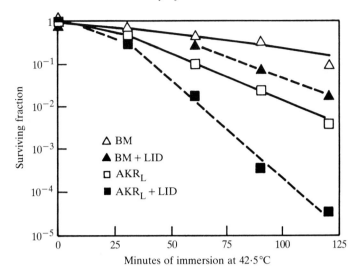

Fig. 12. Effect of lidocaine on survival during 42·5°C hyperthermia. Surviving fraction is estimated as the ratio of spleen colony counts at each time point to the count at 0 time, the time of immersion into the bath, with no drug present. This chart presents surviving fraction for normal bone marrow (BM) and for AKR leukaemia (AKR_L) during heating *in vitro* at 42·5°C. Closed symbols represent the addition of lidocaine (LID) to cell suspensions. The size of the symbols used is greater than ±one relative standard error of the colony count. Redrawn from Fig. 2. Robins *et al.* (1984).

release anaesthetics has obvious clinical implications. If they can provide greater local control of the primary tumour and large metastases while sparing normal tissue during whole body hyperthermia, this theoretically would enable a patient to deal more effectively with disease at non-injectable sites.

As a part of their dietary modification studies discussed earlier, Yatvin *et al.* (1983) investigated the influence of dietary lipids on the thermosensitivity of mouse CA755 mammary tumours. The tumours were exposed to hyperthermia (43·5 ± 0·1°C) when tumour volume indices (TVIs) ranged between 80 and 200 (Fig. 13). Where indicated, lidocaine was infused into the tumour prior to heat treatment. Thermal sensitivity of the linoleic acid-enriched tumours was enhanced over the controls. Lidocaine enhanced the thermal sensitivity of the control but not of the linoleic acid-enriched tumours (Fig. 14; Yatvin *et al.* 1983).

The influence of dietary lipids on thermosensitivity can also be demonstrated on both a more thermo-resistant mammary adenocarcinoma (MTgB) as well as a more heat sensitive one (Fig. 15).

Cells or organisms show adaptive responses to dietary PUFA. When fed in the diet, high levels of linoleic acid increase membrane fatty acid unsaturation index (UI) and produce changes in the fatty acid mean chain length (MCL). In addition to shifts in fatty acid composition, there also occur shifts in the membrane cholesterol concentration, the polar headgroup composition (PE/PC) of the PL and in Ca^{2+}-dependent phospholipase A_2 activity. The diet-mediated increase in

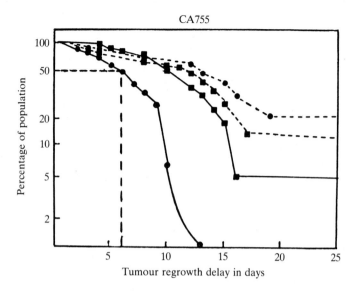

Fig. 13. The influences of the dietary regimen. Control (●——●), pre-fed safflower oil, safflower oil diet + hyperthermia (●---●), combined safflower oil diet + hyperthermia-lidocaine (■---■), and chow diet + hyperthermia-lidocaine (■——■) on the tumour regrowth delay response of CA755 tumours. No tumour regrowth occurred in 3 of 15 (●---●), 2 of 17 (■---■) and 1 of 18 (■---■) experimental hosts. From Yatvin *et al.* (1983). *Eur. J. Cancer clin. Oncol.* With permission.

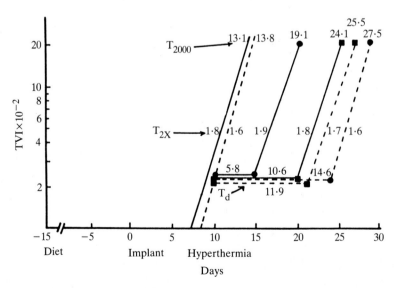

Fig. 14. The influences of dietary regimen (control ——, pre-fed safflower oil ---), hyperthermia (●) and combined hyperthermia-lidocaine modalities (■) on CA755 tumour growth and regrowth. From Yatvin *et al.* (1983). *Eur. J. Cancer clin. Oncol.* With permission.

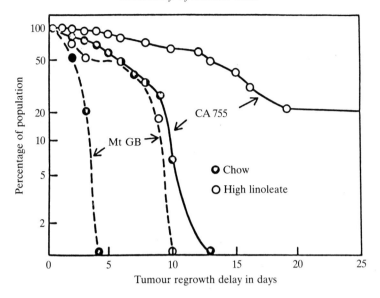

Fig. 15. A comparison of heat sensitivity of two mammary adenocarcinoma growing in legs of mice fed diets high in saturated or unsaturated fatty acids.

membrane UI is countered by an increase in the rigidity imparted by the change in the concentrations of the zwitterionic and acidic PL (Innis & Clandinin, 1981).

In normal cells, the adaptive process is rapid. Whether the changes in PL polar groups are due to *in situ* polar group turnover, *in situ* polar group conversion, microsomal and mitochondrial syntheses of specific PL or insertion and extraction exchanges mediated by the specific PL-transfer proteins is not known. A transfer protein specific for cardiolipin has not been identified. This lipid, which may be involved in Ca^{2+} transport, is apparently resistant to diet-mediated changes. An inverse relationship exists between diet-mediated UI of membrane fatty acids and the membrane cholesterol concentration (Haeffner & Privett, 1975).

Membranes of tumour cells generally have a content (2–3 fold) of cholesterol higher than that of the host tissue (Brossa *et al.* 1980). Tumour membranes contain higher proportions of the acidic- and lyso-PL and a lesser proportion of the zwitterionic PL. Furthermore, cholesterol synthesis in tumour cells is not sensitive to feedback regulation (Sabine, 1975).

Exposure to hyperthermia imposes a certain amount of stress on a tissue. Normal cells apparently have the capacity to modify their membrane composition to resist hyperthermic stress. Tumour cells exist with a more rigid membrane and have less reserve capacity to maintain a constant plasma membrane microviscosity. We postulate that a high PUFA dietary treatment will expend part of the tumour's capacity to maintain a constant plasma membrane viscosity. On exposure to hyperthermic stress, the already compromised membrane's remaining capacity to respond will be overwhelmed. Thus, the physical change effected by heating cannot be corrected thereby enhancing hyperthermic therapeutic ratio.

Studies were designed to test this hypothesis that a diet-mediated increase in the polyunsaturated fatty acid content of CA755 tumour membranes is accompanied by major shifts in the PL headgroup pattern, in saturated fatty acids, and in increases in the cholesterol and PL phosphorous concentrations. Materials isolated from barley (Inh) inhibit cholesterol biosynthesis (Qureshi et al. 1982; Burger et al. 1984). Feeding the Inh prevents the compensatory increase in cholesterol (Table 4) and, as postulated, markedly enhanced thermosensitivity (Fig. 16). Inbar & Shinitzky (1974) and Alderson & Green (1975) have demonstrated that by enriching the cell surface membranes of host mice with cholesterol, the malignancy of lymphoma cells or of Erlich ascites carcinoma cells can be inhibited. Cress & Gerner (1980) observed an inverse relationship between the cholesterol level of cellular membrane and the thermal sensitivity of a number of mammalian cell lines in cultures.

Dietary manipulation of tumour cell membrane to enhance the tumour's response to hyperthermia is possible. It is therefore not inconceivable that manipulation of dietary lipids and anaesthetics could improve therapeutic ratios in situations where hyperthermia is employed as a modality in the treatment of cancer (Fig. 14).

Table 4. *Effect of dietary fat on the biochemical parameters in the livers of normal and host mice and transplanted CA755*

	BT	BTI	SO	SOI
Liver proteins (mg g^{-1})		(10% Fat)		
Normal:	220·6 ± 22·4	224·2 ± 27·4	218·6 ± 19·4	214·8 ± 37·4
Host:	200·8 ± 17·8	209·8 ± 14·5	207·9 ± 17·2	197·7 ± 10·7
Liver cholesterol (nmol mg^{-1} protein)				
Normal:	32·2 ± 9·4	27·8 ± 4·2	28·0 ± 9·2	28·4 ± 4·9
Host:	34·0 ± 5·8	29·5 ± 3·0	27·8 ± 4·6	29·1 ± 2·7
CA755 protein (mg g^{-1})	126·0 ± 7·5	123·0 ± 6·9	121·4 ± 15·2	126·1 ± 16·4
CA755 cholesterol (mg g^{-1}):	0·89 ± 0·1	0·93 ± 0·2	1·06 ± 0·3	0·77 ± 0·2
nmol mg protein^{-1}:	18·2 ± 1·8	19·8 ± 5·1	20·9 ± 10·7	15·8 ± 6·1
Host liver membrane		(16% Fat)		
Protein (mg g^{-1}):	38·2 ± 6·3	59·2 ± 2·3	38·0 ± 9·2	65·5 ± 20·5
Cholesterol (nmol mg^{-1} protein):	37·0 ± 14·2	35·6 ± 1·3	25·0 ± 10·4	24·4 ± 1·7
CA755 membrane				
Protein (mg g^{-1} tumour):	10·8 ± 0·1	15·0 ± 0·9	9·8 ± 2·8	19·0 ± 0·7
Cholesterol (nmol mg^{-1} protein):	69·4 ± 8·1	71·6 ± 16·2	162·2 ± 58·6	76·5 ± 25·7

BT = Beef tallow diet.
BTI = Beef tallow diet plus cholesterol inhibitor.
SO = Safflower oil diet.
SOI = Safflower oil diet plus cholesterol inhibitor.
Elegbede et al. 1986. *Eur. J. Cancer clin. Oncol.* **22**, 607–615. With permission.

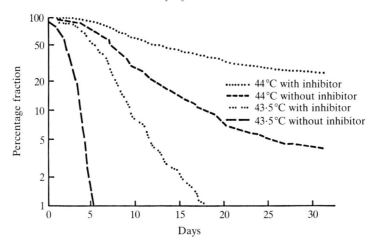

Fig. 16. The influence of the inhibitor of cholesterol biosynthesis of tumour regrowth delay. From Elegbede *et al.* (1986). *Eur. J. Cancer clin. Oncol.* **22**, 607–615. With permission.

In summary it is our view that the effects of hyperthermia on cells is initiated by disorganizing the membrane lipid. Such disorganization alters the membrane's biophysical properties leading to passive changes in trans-membrane permeability, shifts in surface charge, and altered stereoorganization of macromolecules (proteins) associated with the membrane (Bowler *et al.* 1973; Yatvin, 1977; Bowler, 1981). The passive permeability could account for the increase in the association of non-histone proteins to chromatin (Warters & Roti Roti, 1981). Thus, the mileau (lipids) of membrane-bound proteins could be the critical target which subsequently initiates the changes in protein reported by Lepock *et al.* (1981, 1983).

Membranes of tumour cells are characterized by elevated cholesterol. Differences in cholesterol concentration could explain the variation in heat sensitivity between cell lines and within the cell cycle (Leith *et al.* 1977; Thrall *et al.* 1976). Hahn *et al.* (1977) have reported a synergism between a polyene antibiotic whose action is on cholesterol and hyperthermia. A relationship also exists between thermal sensitivity and cholesterol content of six cell lines (Cress & Gerner, 1980). Support for lipids being the initial target comes from our studies in which anaesthetic agents enhance thermosensitivity (Yatvin, 1977; Yau, 1979). Conceivably, the hydrophobic areas of the membrane bound protein could be a site of anaesthetic action. Thermosensitivity of solid tumours is not further influenced by lidocaine in host animals fed diets enriched in linoleic acid, a diet which markedly modifies fatty acid and PL patterns and cholesterol concentration of cellular membranes (Yatvin *et al.* 1983). This lack of effect appears to rule out a meaningful lidocaine-protein interaction in heat killing. However, global measures of change, such as the ratio of unsaturated to saturated fatty acids or in UI or % UFA, may not reflect accurately changes in fatty acid patterns that are related to changes in thermosensitivity. For example, we now recognize that double bond location with respect to the headgroup must be considered.

Anaesthetics may affect membrane 'fluidity' leading to shedding of membrane components (Kim, 1979; Poste & Fidler, 1980). Therefore, one must evaluate the influence of UFA rich diets and/or the use of local anaesthetics in combination with hyperthermia on metastatic spread. Anaesthetics increase post-operative metastases of 3LL Lewis Lung carcinoma and B16 melanoma (Shapiro *et al.* 1981) possibly by a suppressive action on the immune system or by an action on tumour cell membranes. Further studies bearing on the role of diet and anaesthetics on metastatic spread are needed. An increased understanding of the relationship of membrane biophysics and biochemistry and how cells respond to heat could aid in elucidating the mechanisms of cell death. Such knowledge could provide a more rational basis for cancer therapy.

Supported by National Cancer Institute grants R01-CA24872 and P30-CA14520.

References

ALDERSON, C. E. & GREEN, C. (1975). Membrane cholesterol content and malignancy of Ehrlich ascites-carcinoma cells. *Biochem. Soc. Trans.* **3**, 1009–1011.

ANDERSON, R. L., MINTON, K. W., LI, G. C. & HAHN, G. M. (1981). Temperature-induced homeoviscous adaptation of Chinese hamster ovary cells. *Biochim. biophys. Acta* **641**, 334–348.

ARTHUR, H. & WATSON, K. (1976). Thermal adaptation in yeast: Growth temperatures, membrane lipid, and cytochrome composition of psychrophilic, mesophilic and thermophilic yeasts. *J. Bacteriol.* **128**, 56–68.

AUERSPERG, N. (1966). Differential heat sensitivity of cells in tissue culture. *Nature, Lond.* **209**, 415–416.

BARKER, C. J. (1986). The plasma membrane and cellular heat injury: A study on tumour cells. In *Temperature Relations in Animals and Man*, Biowa report 4 (ed. H. Laudien), pp. 229. Stuttgart: Gustav Fischer.

BARZ, D., GOPPELT, M., SZAMEL, M., SHIRRMACHER, V. & RESCH, K. (1985). Characterization of cellular and extracellular plasma membrane vesicles from a non-metastasizing lymphoma (Eb) and its metastasizing variant (ESb). *Biochim. biophys. Acta* **814**, 77–84.

BASS, H., MOORE, J. A. & COAKLEY, W. T. (1978). Lethality in mammalian cells due to hyperthermia under oxic and hypoxic conditions. *Int. J. Radiat. Biol.* **33**, 57–62.

BELEHRADEK, J. (1931). Le mecanisme physico-chimique de L'adaptation thermique. *Protoplasma* **12**, 406–434.

BHUYAN, B. K. (1979). Kinetics of cell kill by hyperthermia. *Cancer Res.* **39**, 2277–2284.

BHUYAN, B. K., DAY, K. J., EDGERTON, C. E. & OGUNBASE, O. (1977). Sensitivity of different cell lines and of different phases in the cell cycle to hyperthermia. *Cancer Res.* **37**, 3780–3784.

BOWDEN, G. T. & KASUNIC, M. D. (1981). Hyperthermic potentiation of the effects of a clinically significant x-ray dose on cell survival, DNA damage, and DNA repair. *Radiat. Res.* **87**, 109–120.

BOWLER, K. (1981). Heat death and cellular heat injury. *J. Therm. Biol.* **6**, 171.

BOWLER, K. (1987). Cellular heat injury: are membranes involved? *Symp. Soc. exp. Biol.* **41**, 000–000.

BOWLER, K., DUNCAN, D. J., GLADWELL, R. T. & DAVIDSON, T. F. (1973). Cellular heat injury. *Comp. Biochem. Physiol.* **45**, 441–449.

BROSSA, O., GARCIA, R., CAMUTO, R. A., GANTERO, B. & FEU, F. E. (1980). Cholesterol and phospholipid contents of the inner and outer membranes of highly purified mitochondria from

Yoshida hepatoma AH-130. *I.R.C.S. Med. Sci., Alimentary System; Biochem; Cancer; Cell and Memb. Biol; Path.* **8**, 73–74.

BURNS, C. D., LUTTENEGGER, D. G., DUDLEY, D. T., BUETTNER, G. R. & SPECTOR, A. A. (1979). Effect of modification of plasma membrane fatty acid composition on fluidity and methotrexate transport in L1210 murine leukemia cells. *Cancer Res.* **39**, 1726–1730.

BURNS, C. P., ROSENBERGER, J. A. & LUTTENEGGER, D. G. (1983). Selectivity in modification of the fatty acid composition of normal mouse tissues and membranes *in vivo. Ann. Nutr. Metab.* **27**, 268–277.

BURGER, W. C., QURESHI, A. A., DIN, Z. Z., ABUIRMEILEH, N. & ELSON, C. E. (1984). Suppression of cholesterol biosynthesis by constituents of barley kernel. *Atherosclerosis* **51**, 75–87.

BUSCH, W. (1866). Uber den Einfluss welchen heftigere Erysiplen zuweilen auf organisierte Neubildungen ausuben. *Verh. Naturh. Preuss. Rhein Westphal.* **23**, 28–30.

CHAPMAN, D. (1967). The effect of heat on membranes and membrane constituents. In *Thermobiology* (ed. A. H. Rose), pp. 122–217. New York: Academic Press.

CHEN, T. T. & HEIDELBERGER, C. (1969). Quantitative studies on the malignant transformation of mouse prostate cells by carcinogenic hydrocarbons *in vitro. Int. J. Cancer* **4**, 166–178.

CHERNAVSKII, D. S., COLEZHAEV, A. A. & VOLKOV, E. I. (1982). What are the differences between normal and tumour cell surface? Academy of Sciences of the U.S.S.R., P. H. Lebedev Physical Institute, Creprint N 94.

CLARK, A. W., ROBINS, I. H., VORPAHL, J. W. & YATVIN, M. B. (1983). Structural changes in murine cancer associated with hyperthermia and lidocaine. *Cancer Res.* **43**, 1716–1723.

COLEY, W. B. (1893). The treatment of malignant tumors by repeated inoculation of erysipelas. With a report of 10 original cases. *Am. J. med. Sci.* **105**, 487–511.

CORRY, P. M., ROBINSON, S. & GETZ, S. (1977). Hyperthermic effects on DNA repair mechanisms. *Radiology* **123**, 475–482.

COSSINS, A. R. & PROSSER, C. L. (1978). Evolutionary adaptation of membranes to temperature. *Proc. natn. Acad. Sci. U.S.A.* **75**, 2040–2043.

CRESS, A. E. & GERNER, E. W. (1980). Cholesterol levels inversely reflect the thermal sensitivity of mammalian cells in culture. *Nature, Lond.* **283**, 677–679.

DENNIS, W. H. & YATVIN, M. B. (1981). Correlation of hyperthermic sensitivity and membrane microviscosity in *E. coli* K1060. *Int. J. Radiat. Biol.* **39**, 265–271.

DEWEY, W. C., HOPWOOD, L. E., SAPARETO, S. A. & GERWECK, L. E. (1977). Cellular responses to combinations of hyperthermia and radiation. *Radiology* **123**, 463–474.

DICKSON, J. A. (1977). The effects of hyperthermia in animal tumor systems. *Rec. Res. Cancer Res.* **59**, 43–111.

DICKSON, J. A. (1979). Hyperthermia in the treatment of cancer. *Lancet* **i**, 202–205.

ELEGBEDE, A. J., ELSON, C. E., QURESHI, A., DENNIS, W. H. & YATVIN, M. B. (1986). Increasing the thermosensitivity of a mammary tumor (CA755) through dietary modification. *Eur. J. Cancer clin. Oncol.* **22**, 607–615.

ESSER, A. F. & SOUZA, K. A. (1974). Correlation between thermal death and membrane fluidity in *Bacillus stearothermophilus. Proc. natn. Acad. Sci. U.S.A.* **71**, 4111–4115.

FAJARDO, L. F., EGBERT, B., MARMOR, J. & HAHN, G. M. (1980). Effects of hyperthermia in a malignant tumor. *Cancer* **45**, 613–623.

FERGUSON, K. A., GLASER, M., BAYER, W. H. & VAGELOS, P. R. (1975). Alteration of fatty acid composition of LM cells by lipid supplementation and temperature. *Biochemistry* **14**, 146–151.

GALLEZ, D. (1984). Cell membranes after malignant transformation, Part I and Part II. *J. theor. Biol.* **111**, 323–354.

GEORGE, A. M., LUNEC, J. & CRAMP, W. A. (1983). Effect of membrane fatty acid changes on the radiation sensitivity of human lymphoid cells. *Int. J. Radiat. Biol.* **43**, 363–378.

GIOVANELLA, B. C., MORGAN, A. C., STEHLIN, J. S. & WILLIAMS, L. J. (1973). Selective lethal effect of supranormal temperatures on mouse sarcoma cells. *Cancer Res.* **33**, 2568–2578.

GIOVANELLA, B. C., STEHLIN, J. S. & MORGAN, A. C. (1976). Selective lethal effect of supranormal temperatures on human neoplastic cells. *Cancer Res.* **36**, 3944–3950.

HAEFFNER, E. W. & PRIVETT, O. S. (1975). Influence of dietary fatty acids on membrane properties and enzyme activities of liver mitochondria of normal and hypothesectomized rats. *Lipids* **10**, 75–81.

HAHN, G. M., LI, G. C. & SHIU, E. (1977). Interaction of amphotericin B and 43°C hyperthermia. *Cancer Res.* **37**, 761–764.

HAINES, T. H. (1979). A proposal on the function of unsaturated fatty acids and ionic lipids: The role of potential composition in biological membranes. *J. theoret. Biol.* **80**, 307–323.

HAYNES, D. (1984). Pharmacologist, Abst., Sept. p. 220. Thirty-hour local anesthesia produced by Duracaine: A methoxyfluorane microdroplet preparation.

HEILBRUNN, L. V. (1924). The colloid chemistry of protoplasm IV. The heat coagulation of protoplasm. *Am. J. Physiol.* **69**, 190–199.

HENLE, K. J. & LEEPER, D. B. (1979). Effects of hyperthermia (45°C) on macromolecular synthesis in Chinese hamster ovary cells. *Cancer Res.* **39**, 2665–2674.

HIDVEGI, E. J., YATVIN, M. B., DENNIS, W. H. & HIDVEGI, E. (1980). Effect of altered membrane lipid composition and procaine on hyperthermic killing of ascites tumor cells. *Oncology* **37**, 360–363.

HIGHTOWER, L. E. (1980). Cultured animal cells exposed to amino acid analogues or puromycin rapidly synthesize several polypeptides. *J. cell. Physiol.* **102**, 407–427.

HOPKINS, C. R. (1978). In *Structure and Function of Cells*, pp. 57–104. London, Philadelphia, Toronto: W. B. Saunders Company, Ltd.

HUANG, L., LORCH, S. K., SMITH, G. G. & HAUG, A. (1974). Control of membrane lipid fluidity in *Acholeplasma laidlawii*. *FEBS Lett.* **43**, 1–5.

INBAR, M. & SHINITZKY, M. (1974). Increase of cholesterol levels in the surface membrane of lymphoma cells and its inhibitory effect on ascites tumor development. *Proc. natn. Acad. Sci. U.S.A.* **71**, 2128–2130.

INNIS, S. M. & CLANDININ, M. T. (1981). Mitochondrial membrane polar headgroup composition is influenced by diet fat. *Biochem. J.* **193**, 155–167.

JENSEN, C. O. (1903). Experimentelle untersuchungen uber kreb bei mausen. *Zbl Bakt.* I-Abt. orig. **34**, 28–122.

JIRTLE, R. L. (1981). Blood flow to lymphatic metastases in conscious rats. *Eur. J. Cancer* **17**, 53–60.

JOHNSTON, P. V. & ROOTS, B. I. (1964). Brain lipid fatty acids and temperature acclimation. *Comp. Biochem. Physiol.* **11**, 303–309.

JORRITSMA, J. B. M. & KONINGS, A. W. T. (1983). Inhibition of repair of radiation-induced strand breaks by hyperthermia, and its relationship to cell survival after hyperthermia alone. *Int. J. Radiat. Biol.* **43**, 505–516.

KELLEY, P. M. & SCHLESINGER, M. J. (1978). The effect of amino acid analogues and heat shock on gene expression in chicken embryo fibroblasts. *Cell* **15**, 1277–1286.

KIM, J. H. & HAHN, E. W. (1979). Clinical and biological studies of localized hyperthermia. *Cancer Res.* **39**, 2258–2261.

KIM, U. (1979). Factors influencing metastasis of breast cancer. In *Breast Cancer*, vol. 3 (ed. W. L. McGuire), pp. 1–49. Place: Plenum Publishing Corp.

LEITH, J. T., MILLER, R. C., GERNER, E. W. & BOONE, M. L. M. (1977). Hyperthermic potentiation: Biological aspects and applications to radiation therapy. *Cancer* **39**, 766–779.

LEPOCK, J. R. (1982). Involvement of membrane cellular responses to hyperthermia. *Radiat. Res.* **92**, 433–438.

LEPOCK, J. R., MASSICOTTE-NOLAN, P., RULE, G. F. & KRUUV, J. (1981). Lack of correlation between hyperthermic cell killing, thermotolerance and membrane lipid fluidity. *Radiat. Res.* **87**, 300–313.

LEPOCK, J. R., CHENG, K.-H., AL-QYSI, H. & KRUUV, J. (1983). Thermotropic lipid and protein transitions in Chinese hamster lung cell membranes: Relationship to hyperthermic cell killing. *Can. J. Biochem. Cell Biol.* **61**, 421–427.

LETT, L. & CLARK, E. P. (1978). In *Cancer Therapy: By Hyperthermia and Radiation* (ed. C. Streffer). Munich: Urban and Swarzenberg.

LEVINE, E. M. & ROBBINS, E. B. (1970). Differential temperature sensitivity of normal and cancer cells in culture. *J. cell. Physiol.* **76**, 373–379.

LEVINSON, W., OPPERMANN, H. & JACKSON, J. (1980). Transition series metals and sulfhydryl reagents induce the synthesis of four proteins in eukaryotic cells. *Biochim. biophys. Acta* **606**, 170–180.

LI, G. C., SHIU, E. C. & HAHN, G. M. (1980). Similarities in cellular inactivation by hyperthermia or by ethanol. *Radiat. Res.* **2**, 257–268.

LIN, P. S., WALLACH, D. F. H. & TSA, S. (1973). Temperature-induced variations in the surface topology of cultured lymphocytes as revealed by scanning electron microscopy. *PNAS* **70**, 2494–2496.

LINDQUIST, S., DIDOMENICO, B., BUGAISKY, G., KURTZ, S., PETKO, L. & SONODA, S. (1982). Regulation of the heat-shock response in *Drosophila* and yeast. In *Heat Shock: From Bacteria to Man* (ed. M. J. Schlesinger, M. Ashburner & A. Tissieres), pp. 167–175. Cold Spring Harbor Laboratory, Cold Spring Harbor, New York, U.S.A.

LOVE, R., SORIANO, R. Z. & WALSH, R. J. (1970). Effect of hyperthermia on normal and neoplastic cells *in vitro*. *Cancer Res.* **30**, 1525–1533.

LUNEC, J., HESSLEWOOD, I. P., PARKER, R. & LEAPER, S. (1981). Hyperthermic enhancement of radiation cell killing in HeLa S3 cells and its effect on the production and repair of DNA strand breaks. *Radiat. Res.* **85**, 116–125.

LUZZATI, V. & HUSSON, F. (1962). The structure of the liquid-crystalline phases of lipid-water systems. *J. Cell Biol.* **12**, 207–219.

MARMOR, J. B., HAHN, N. & HAHN, G. M. (1977). Tumor cure and cell survival after localized radiofrequency heating. *Cancer Res.* **37**, 879–883.

MARTINEZ-PALMOMO, A. (1979). The nature of neoplastic cell membranes. *Exper. molec. Path.* **31**, 219–235.

MARX, J. L. (1983). Surviving heat shock and other stresses: Heat shock genes and their protein products help to protect cells against damage induced by stress and also aid studies of gene control. *Science* **221**, 251–253.

MILLS, H. D. & MEYN, R. E. (1981). Effects of hyperthermia on repair of radiation-induced DNA strand breaks. *Radiat. Res.* **87**, 314–328.

MONDOVI, B., AGRO, A. F., ROTILIO, G., STROM, R., MORICCA, G. & FANELLI, A. R. (1969). The biochemical mechanism of selective heat sensitivity of cancer cells. II. Studies on nucleic acids and protein synthesis. *Eur. J. Cancer* **5**, 137–146.

MONTFOORT, A., VAN GOLDE, L. M. G. & VAN DEENEN, L. L. M. (1971). Molecular species of lecithins from various animal tissues. *Biochim. biophys. Acta* **231**, 335–342.

MULCAHY, R. T., GOULD, M. N., HIDVEGI, E. G., ELSON, C. E. & YATVIN, M. B. (1981). Hyperthermia and surface morphology of P388 ascites tumor cells: Effects of membrane modifications. *Int. J. Radiat. Biol.* **39**, 95–106.

NAUTS, H. C., FOWLER, G. A. & BOGATKO, F. H. (1953). A review of the influence of bacterial infection and of bacterial products (Coley's toxins) on malignant tumors in man. *Acta med. Scand.* **276**, 1–103.

NYSTROM, R. A. (1973). In *Membrane Physiology*, pp. 6–62. New Jersey: Prentice-Hall, Inc.

PALZER, R. J. & HEIDELBERGER, C. (1973). Influence of drugs and synchrony on the hyperthermic killing of HeLa cells. *Cancer Res.* **33**, 422–427.

POSTE, G. & FIDLER, I. J. (1980). The pathogenesis of cancer metastasis. *Nature, Lond.* **283**, 139.

POZNANSKY, M., KIRKWOOD, D. & SOLOMON, A. K. (1973). Modulation of red cells (K^+) transport by membrane lipids. *Biochim. biophys. Acta* **330**, 351–355.

QURESHI, A. A., BURGER, W. C., ELSON, C. E. & BENEVENGA, N. J. (1982). Effects of cereals and culture filtrate of *Trichoderma viride* on lipid metabolism of swine. *Lipids* **17**, 924–934.

REIMANN, S. P. & NISHIMURA, E. T. (1949). Attempts at the chemotherapy of cancer. *J. Mich. med. Soc.* **48**, 454–484.

RINTOUL, D. A. & SIMONI, R. D. (1977). Incorporation of a naturally occurring fluorescent fatty acid into lipids of cultured mammalian cells. *J. biol. Chem.* **252**, 7916–7918.

RITOSSA, F. (1962). A new puffing pattern induced by temperature shock and DNP in *Drosophila*. *Experientia* **18**, 571–573.

ROBINS, I. H., DENNIS, W. H., MARTIN, P. A., SONDEL, P. M., YATVIN, M. B. & STEEVES, R. A. (1984). Potentiation of differential hyperthermia sensitivity of AKR leukemia and normal bone marrow cells by lidocaine and thiopental. *Cancer Res.* **54**, 2831–2835.

ROBINS, I. H., DENNIS, W. H., SLATTERY, J. S., LANGE, T. A. & YATVIN, M. B. (1983). Systemic lidocaine enhancement of hyperthermia-induced tumor regression in transplantable murine tumor models. *Cancer Res.* **43**, 3187–3191.

ROFSTAD, E. K. & BRUSTAD, T. (1985). Heterogeneity in heat sensitivity and development of thermotolerance of cloned cell lines derived from a single human melanoma xenograft. *Int. J. Hyperthermia* **1**, 85–96.

ROSENBERG, S. A. & GUIDOTTI, G. (1969). Fractionation of the protein components of human erythrocyte membranes. *J. biol. Chem.* **244**, 5118–5124.

SABINE, J. R. (1975). Defective control of lipid biosynthesis in cancerous and precancerous liver. *Progr. Biochem. Pharmacol.* **10**, 269–307.

SAPARETO, S. A., HOPWOOD, L. E., DEWEY, W. C., RAJU, M. R. & GRAY, J. W. (1978). Effects of hyperthermia on survival and progression of Chinese hamster ovary cells. *Cancer Res.* **38**, 393–400.

SCHRECK, R. (1966). Sensitivity of normal and leukemic lymphocytes and leukemic myeloblasts to heat. *J. natn. Cancer Inst.* **37**, 649–654.

SHAPIRO, J., JERSKY, J., KATSAV, S., FELDMAN, M. & SEGAL, S. (1981). Anesthetic drugs accelerate the progression of post-operative metastases of mouse tumors. *J. clin. Invest.* **68**, 678–685.

SIMARD, R. & BERNHARD, W. (1967). A heat-sensitive cellular function located in the nucleolus. *J. Cell Biol.* **34**, 61–76.

SIMON, E. W. (1974). Phospholipids and plant membrane permeability. *New Phytol.* **73**, 377–420.

SINENSKY, M. (1974). Homeoviscous adaptation – A homeostatic process that regulates the viscosity of membrane lipids in Escherichia coli. *Proc. natn. Acad. Sci. U.S.A.* **71**, 522–525.

SINGER, S. J. & NICOLSON, G. L. (1972). The fluid mosaic model of the structure of cell membranes. *Science* **175**, 720–731.

SPECTOR, A. A., HOAK, J. C., FRY, G. L., DENNING, G. M., STOLL, L. L. & SMITH, J. B. (1980). Effect of fatty acid modification on prostacyclin production by cultured human endothelial cells. *J. clin. Invest.* **65**, 1003–1012.

TATTRIE, N. H., BENNETT, J. R. & CYR, R. (1968). Maximum and minimum values for lecithin classes from various biological sources. *Can. J. Biochem.* **46**, 819–824.

THRALL, D. E., GERWECK, L. E., GILLETTE, E. L. & DEWEY, W. C. (1976). Response of cells *in vitro* and tissues *in vivo* to hyperthermia and x-irradiation. *Adv. Radiat. Biol.* **6**, 211–226.

TOMASOVIC, S. P., HENLE, J. K., DETHLEFSEN, L. A. (1979). Fractionation of combined heat and radiation in asynchronous CHO cells. II. The role of cell-cycle redistribution. *Radiat. Res.* **80**, 378–388.

TOMASOVIC, S. P., ROSENBLATT, P. L. & HEITZMAN, D. (1983). Heterogeneity in induced thermal resistance of rat tumor cell lines. *Int. J. Radiat. Oncol. Biol. Phys.* **9**, 1675–1681.

TOMASOVIC, S. P., ROSENBLATT, P. L., JOHNSTON, D. A., KUANG, T. & LEE, P. S. Y. (1984). Heterogeneity in induced heat resistance and its relation to synthesis of stress proteins in rat tumor cell clones. *Cancer Res.* **44**, 5850–5865.

TOMASOVIC, S. P., TURNER, G. N. & DEWEY, W. C. (1978). Effects of hyperthermia on non-histone proteins isolated with DNA. *Radiat. Res.* **73**, 535–552.

VAN BLITTERSEWIJK, W. J., HILKMAN, H. & HENGEVELD, T. (1984). Differences in membrane lipid composition and fluidity of transplanted GRSL lymphoma cells, depending on their site of growth in the mouse. *Biochim. biophys. Acta* **778**, 521–529.

VAN HOVEN, R. P. & EMMELOT, P. (1972). Studies on plasma membranes. XVIII. Lipid class composition of plasma membrane isolated from rat mouse liver and hepatoma. *J. Memb. Biol.* **9**, 105–124.

WARTERS, R. L. & ROTI ROTI, J. L. (1981). The effect of hyperthermia on replicating chromatin. *Radiat. Res.* **88**, 69–78.

WELCH, D. R., EVANS, D. P., TOMASOVIC, S. P., MILAS, L. & NICOLSON, G. L. (1984). Multiple phenotypic divergence of mammary adenocarcinoma cell clones. II. Sensitivity to radiation, hyperthermia and FUdR. *Clin. expl Metastasis* **2**, 357–371.

WELCH, W. J., GARRELS, J. I. & FERAMISCO, J. R. (1982). The mammalian stress proteins. In *Heat Shock: From Bacteria to Man* (ed. M. J. Schlesinger, M. Ashburner & A. Tissieres), pp. 257–266. Cold Spring Harbor Laboratory, Cold Spring Harbor, New York, U.S.A.

WENIGER, P., WAWRA, E. & DOLEJS, I. (1979). Die wirkung van hyperthermie auf DNA-reparatur vorgange. (The action of hyperthermia on DNA repair.) *Radiat. Environ. Biophys.* **16**, 135–141.

WESTERMARK, N. (1927). The effect of heat upon rat tumors. *Skand. Arch. Physiol.* **52**, 257–322.

WESTRA, A. & DEWEY, W. C. (1971). Variation in sensitivity to heat shock during the cell cycle of Chinese hamster cells *in vitro*. *Int. J. Radiat. Biol.* **19**, 467–477.

WILLIAMS, D. L., NUWAYSER, E. S., CREEDEN, D. E. & GAY, M. H. (1984). *Proc. 11th International Symposium on Controlled Release of Bioactive Materials*, July 23–24, Ft. Lauderdale, Florida, U.S.A.

WILLIAMS, R. M. & CHAPMAN, D. (1970). Phospholipids, liquid crystals and cell membranes. *Prog. Chem. Fats* **11**, 3–79.

YATVIN, M. B. (1976). Evidence that survival of gamma-irradiated *Escherichia coli* is influenced by membrane fluidity. *Int. J. Radiat. Biol.* **30**, 571–575.

YATVIN, M. B. (1977). The influence of membrane lipid composition and procaine on hyperthermic death of cells. *Int. J. Radiat. Biol.* **32**, 513–521.

YATVIN, M. B., ABUIRMEILEH, N. M., VORPAHL, J. W. & ELSON, C. E. (1983). Biological optimization of hyperthermia: Modification of tumor membrane lipids. *Eur. J. Cancer clin. Oncol.* **19**, 657–663.

YATVIN, M. B., CLIFTON, K. H. & DENNIS, W. H. (1979). Hyperthermia and local anesthetics: Potentiation of survival in tumor-bearing mice. *Science* **205**, 195–196.

YATVIN, M. B., CREE, T. C., ELSON, C. E., GIPP, J. J., TEGMO, I.-M. & VORPAHL, J. W. (1982*a*). Probing the relationship of membrane 'fluidity' to heat killing of cells. *Radiat. Res.* **89**, 644–646.

YATVIN, M. B., SCHMITZ, B. J., RUSY, B. F. & DENNIS, W. H. (1980*a*). Local anesthetic alteration of cell survival after hyperthermia or irradiation. In *Molecular Mechanisms of Anesthesia*, vol. 2 (ed. B. R. Fink), pp. 495–499. New York: Raven Press.

YATVIN, M. B., SCHMITZ, B. J. & DENNIS, W. H. (1980*b*). Radiation killing of *E. coli* K1060: Role of membrane fluidity, hyperthermia and local anesthetics. *Int. J. Radiat. Biol.* **37**, 513–519.

YATVIN, M. B., VORPAHL, J. W. & ELSON, C. E. (1981). Influence of dietary lipid on membrane composition, microviscosity, heat sensitivity, and response to lidocaine of two mouse mammary adenocarcinomas. *Radiat. Res. Soc.*, Minneapolis.

YATVIN, M. B., VORPAHL, J. & KIM, U. (1982*b*). Differential response to heat of metastatic and non-metastatic rat mammary tumors. In *Advances in Experimental Medicine and Biology*, vol. 157 (ed. H. Bicher & D. Bruley), pp. 177–184. New York: Plenum Press.

YAU, T. M. (1979). Procaine-mediated modification of membranes and of the response to irradiation and hyperthermia in mammalian cells. *Radiat. Res.* **80**, 523–541.

YERUSHALMI, A. (1976). Influence of metastatic spread of whole body or local tumor hyperthermia. *Eur. J. Cancer* **12**, 455–463.

Printed in Great Britain © Society for Experimental Biology 1987

THERMOTOLERANCE AND THE HEAT SHOCK PROTEINS

ROY H. BURDON

Department of Bioscience & Biotechnology, Todd Centre, University of Strathclyde, Glasgow G4 0NR, UK

Summary

Mammalian cells can dramatically increase their tolerance to thermal damage after prior heat conditioning. The thermal history, the heat fractionation interval and the recovery conditions, all modify significantly the degree of thermotolerance exhibited.

Several lines of evidence have suggested that perhaps that shock proteins (hsps) provide the protective mechanism. For example by following the synthesis and degradation of heat shock proteins during development and decay of thermotolerance, strong circumstantial evidence has been obtained in certain cases that hsps are involved in the acquisition, maintenance and decay of thermotolerance. The levels of certain heat shock proteins, particularly the class at 68–70 kDa, can also correlate with thermotolerance. On the other hand these correlations do not always hold. Moreover cycloheximide treatment during the heat shock does not appear to block the development of thermotolerance. In addition depletion of medium Ca^{2+} which also inhibits hsp induction can produce thermotolerance. However it should be emphasized that whilst hsp synthesis is elevated after heat shock, it is clear that there are low level of hsps, always present in unheated cells, which may be sufficient to confer tolerance.

Other data now show that thermotolerance measured in terms of cells survival is closely paralled by thermal resistance of *total* protein synthesis. Moreover it is possible to demonstrate the development of thermotolerance in the *total* protein synthetic activity of Hela cells, either held continuously at 42°C or treated briefly at 45°C and returned to normal growth temperature. This development of tolerance in the *total* protein synthetic activity is nonetheless reduced by actinomycin D and a role for the nucleolus is suggested.

The properties of hsps that might be of significance in the generation of thermotolerance are examined. For example there appears to be a specific role of hsp 70 in aiding repair of heat damaged nucleoli. Another heat shock protein appears to be ubiquitin which is likely to have a role in the degradation of the abnormal proteins which are postulated to be involved in the transcriptional activation of the hsp genes themselves.

Thermotolerance

Mammalian and other cells can dramatically increase their tolerance to thermal damage after prior heat conditioning (Gerner *et al.* 1976; Henle & Dethlefsen,

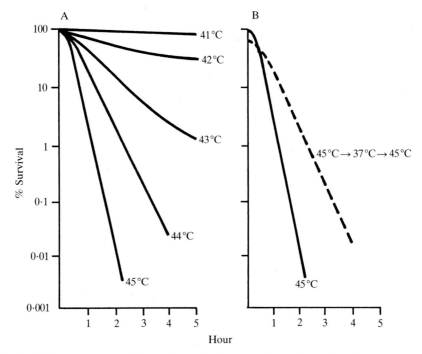

Fig. 1. Schematic mammalian cell survival curves illustrating the phenomenon of thermotolerance. (A) Illustrates the normal survival curve at various supraoptimal temperatures. (B) Illustrates the normal survival at 45°C (——) and the survival curve with reduced slope that is obtained if the cells are pretreated by heating at 45°C for 20 min followed by recovery at 37°C for 3 h, before return to 45°C (– – –).

1978). The mechanism of this cellular thermotolerance is still unclear but it is apparent that the thermal history, the heat fractionation interval and the recovery conditions, all modify significantly the degree of thermotolerance. There are two main forms of thermotolerance encountered in mammalian cells (Henle *et al.* 1978; Lepock & Kruuv, 1980). During *continuous* heat treatment, at less than approximately 42·5°C, heat survival curves show a biphasic response, the cells becoming more heat resistant after a few hours of hyperthermia (see Fig. 1(A)). An alternative form results from an acute heat treatment at temperatures approximately between 43°C and 46°C (see Fig. 1(B)). Under these conditions tolerance does not actually develop. Instead it needs time at 37°C to express itself. In both these situations however the thermotolerance is transient, decaying over a period of time depending on a number of factors. Since thermotolerance is stable at 4°C, it may be that an enzymatic catabolic process is involved in the decay process.

For the development of thermotolerance it has been suggested that at least two processes are required (Li & Hahn, 1984), one occurring over the entire temperature range ('*trigger*') and one needing time at the lower range of temperatures ('*development*'). Thus for the first type of thermotolerance both 'trigger' and 'development' would occur simultaneously at 42·5°C or below.

Heat shock proteins and thermotolerance

The possibility that heat shock proteins (hsps) (see Burdon, R. H., this volume) might be involved in the development of thermotolerance is suggested from a number of experimental approaches. For example mutant analysis in *Dictyostelium* has indicated the involvement of small hsps (Loomis & Wheeler, 1982) and mutatations specifically in the hsp 70 gene of yeasts were found to confer temperature sensitive growth in yeast (Craig & Jacobson, 1984). That hsps are connected with the acquisition, maintenance and decay of thermotolerance in mammalian cells was indicated from experiments of Landry *et al.* (1982) in which the synthesis and degradation of heat shock proteins was followed during development and decay of thermotolerance. In addition the levels of certain hsps, particularly the hsp 70 class, in murine tumours (Li & Mak, 1985) murine embryos (Muller *et al.* 1985), *Xenopus* embryos (Heikkila *et al.* 1985) and Chinese hamster fibroblasts (Li & Werb, 1982) have been reported to correlate with thermotolerance.

On the other hand Omar & Lanks (1984) compared normal and SV40-transformed mouse embryo cells. The transformed cells had higher basal levels and consistently synthesised the major hsps at a higher rate, both at physiological temperature and after exposure to heat shock (43–45°C). Parallel determination of cell survival showed that the transformed cells were nevertheless *more* susceptible to killing by hyperthermia than their normal counterparts. Thus it appears that the higher intrinsic resistance of the normal cells to killing by heat is not directly related to basal hsp levels, nor to the degree to which synthesis of these proteins is induced following hyperthermia. Heterogeneity in induced heat resistance and its relation to hsps is also observed in rat tumour cell clones (Tomasovic *et al.* 1984).

A complication in all these studies is that certain cell lines have a higher intrinsic heat resistance than other cell lines. Such resistance is genetically inherited and those cells are referred to as 'heat-resistant' as distinct from the normal 'heat-sensitive' cell lines. Thermotolerance can however be induced in *both* 'heat-resistant' and 'heat-sensitive' cell lines following hyperthermic treatment. Whilst the thermotolerance *induced* in the 'heat-sensitive' cells is quantitatively greater than that *induced* in the 'heat-resistant' cells, the final level of heat resistance attained after the induction process is nevertheless highest in the 'heat-resistant' cell lines. Thus some sort of overlap may exist between mechanisms of thermotolerance induction and genetically inherited heat resistance. Whilst heat sensitive and heat resistant melanoma cells have equal amount of hsps (Anderson *et al.* 1984), Chinese hamster cells show elevated levels of 70 kDa hsps associated with both inherited heat resistance and with induced thermotolerance (Laszlo & Li, 1985). Thus despite the complexities of overlap between two possibly distinct phenomena, the correlation studies provide a strong basis for careful examination of the role of hsps in the development of thermotolerance.

A cautionary note is nevertheless expressed by Hall (1983) and Widerlitz *et al.* (1986) whose experiments indicate that thermotolerance development in yeasts and in rat fibroblasts does not require protein synthesis as judged from use of cycloheximide as inhibitor. Recent experiments with yeast involving the use of anti-sense RNA to inhibit hsp 26 synthesis suggests that that particular hsp is not required for growth at high temperatures or for thermotolerance (Petko & Lindquist, 1986). These data would appear to argue against a role for at least certain hsps in thermotolerance but it should be remembered that other hsps (e.g. hsps 70s) are present at lower concentrations even in unheated cells. Another situation where hsp induction is blocked yet mammalian cells show heat tolerance is where the medium is depleted of Ca^{2+} (Lamarche *et al.* 1985).

Thermotolerance in the protein synthetic system

A somewhat different standpoint is expressed by Schamhart *et al* (1984*a,b*) who found that thermotolerance measured in terms of cell survival was paralleled by the thermal resistance of *total* cellular protein synthesis. Indeed the thermosensitivity of survival is reflected by the rate of recovery of protein synthesis rather than by the heat induced initial changes in protein synthesis. This of course does not rule out a role for hsps as they could be involved in the recovery of cellular protein synthesis and the development of thermal resistance in the protein synthetic apparatus. In this connection we have examined the heat induced thermotolerance in the Hela cell protein synthetic system and explored the possible involvement of hsps.

As can be seen in Fig. 2 when Hela cell cultures are subject to hyperthermia an immediate effect is a marked inhibition of protein synthesis. Nevertheless when the hyperthermia is continuous at temperatures less than 45·2°C, we find that protein synthesis recovers slowly to reach levels higher than in untreated cells. In short after the initial inhibition it becomes *intrinsically* thermotolerant to temperatures of 42°C or less. In situations where acute hyperthermia is applied (43°–46°C) again there is an immediate inhibition of protein synthesis but if the treatment is brief (10–15 min) total cellular protein synthesis recovers, again to levels higher than in unheated cells if the temperature is returned to 37°C (Fig. 3). However by 24 h after the acute hyperthermia the levels of protein synthesis are back to about normal. An important feature of the recovery is that if a second acute hyperthermic treatment is carried out say after 2 h, the protein synthetic system is now quite thermotolerant. This thermotolerance however is transient and is absent in the cells allowed to recover for 24 h. The maximum tolerance is observed at around 2 h of recovery which correlates well with the maximum of induced hsp synthesis (see Slater *et al.* 1981). This raises the possibility of hsp involvement.

A first approach was to ascertain the effects of cycloheximide on the process. As can be seen from Fig. 4 use of this protein synthesis inhibitor raises certain problems. For example the protein synthetic system in Hela cells actually becomes more   thermotolerant   in   cells   treated   wtih   cycloheximide.

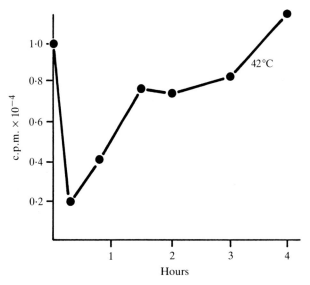

Fig. 2. Effect of continuous hyperthermia at 42°C on ^{35}S-methionine incorporation into Hela cell protein.

Monolayer cultures of Hela cells were set up as described for Fig. 3. Hyperthermia was at 42°C for various periods and ^{35}S-methionine incorporation was determined subsequently at 42°C for 30 min.

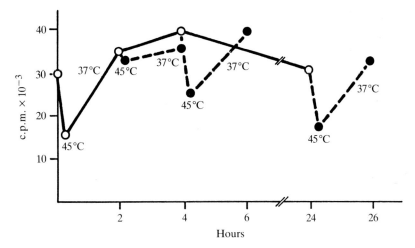

Fig. 3. Effects of repeated hyperthermic treatment at 45°C on ^{35}S-methionine incorporation into Hela cell protein.

Duplicate monolayer cultures of Hela cells (0.5×10^6 cells per 2·5 cm dish in Eagles MEM supplemented with 10% calf serum) were allowed to grow for 24 h at 37°C before being subject to hyperthermia. Cultures were initially treated at 45°C for 10 min then returned to 37° for various time (O——O). Certain cultures however were taken at various times during recovery and subject to a further treatment at 45°C for 10 min followed by 2 h recovery at 37°C (●–––●). All determinations of subsequent ability to incorporate ^{35}S-methionine were carried out at 37°C for 30 min as described by Burdon *et al.* (1982).

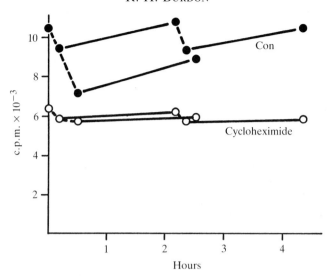

Fig. 4. Effect of cycloheximide on the ability of Hela cells to incorporate ^{35}S-methionine into protein following repeated hyperthermia.

Triplicate Hela cell monolayer cultures were set up as described in Fig. 2. After 24 h, to one set (○) was added cycloheximide at 20 μg ml^{-1} whereas there were no addition to the set of control cultures (●).

After 15 min each set was then subject to 45 °C for 10 min (– – –), recovery at 37 °C (——) followed by second treatment at 45 °C (– – –) and a second period of recovery at 37 °C (——).

After the various treatments the cultures were washed three times with fresh medium and then their ability to incorporate ^{35}S-methionine at 37 °C was determined by the procedures of Burdon et al. (1982).

Moreover the cycloheximide does not appear to impair recovery either from a first or second hyperthermic treatment. On the other hand when the range of proteins synthesised under these conditions was examined by SDS–PAGE, it was clear that whatever thermotolerance and recovery was achieved in the presence of cyclohexi-mide, there was no heat-induced increase in hsp synthesis. As such data could be simply a manifestation of the particular mode of action of cycloheximide, the alternative use of puromycin was explored. Whilst this did lead to a lack of recovery of protein synthesis and thermotolerance after acute brief hyperthermia (Fig. 5(A)), a problem with puromycin was the exposure of cells to this drug over a prolonged period at 37 °C at 200 μg ml^{-1} leads to a gradual reduction in overall protein synthetic capacity (Fig. 5(B)). For these reasons the data obtained with cycloheximide, or puromycin, are difficult to interpret with regard to the role of induced hsp synthesis and induced thermotolerance in protein synthesis.

Perhaps more helpful are the data obtained using the RNA synthesis inhibitor actinomycin D. This drug, however, was not without its problems. Exposure of Hela cells to levels of this drug sufficient to block all cellular RNA synthesis (e.g. 2 μg ml^{-1}) also caused a reduction in cellular protein synthetic capacity on prolonged exposure (e.g. after 4 h protein synthesis was reduced to at least (70 %). However at low concentrations (0·05 μg ml^{-1}) this was not such a significant

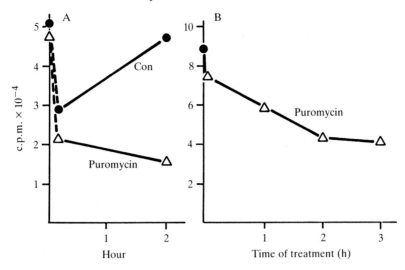

Fig. 5. Effect of puromycin on the ability of Hela cells to incorporate ^{35}S-methionine into protein following hyperthermia. (A) As for Fig. 3, triplicate cultures were set up and allowed to grow for 24 h. To one set was added puromycin at $200\,\mu g\,ml^{-1}$ ($\triangle$) whereas no puromycin was added to control cells ($\bullet$). Each set was then subject to hyperthermia at 45 °C for 10 min (– – –) followed by recovery at 37 °C (——). (B) Cells were simply incubated with puromycin at $200\,\mu g\,ml^{-1}$ for various times.

After treatment cell in (A) and (B) were washed 3 times with fresh medium and then assessed for their ability to incorporate ^{35}S-methionine into protein at 37 °C over 30 min (Burdon *et al.* 1982).

problem, at least over the time period involved, as the loss was much less (10–15 %). When the ability of Hela cells to recover their protein synthetic activity after hyperthermia was assessed after low actinomycin D treatment it was found to be partly impaired (Fig. 6). In addition such actinomycin D treatment resulted in less thermotolerance being developed in the system. The actinomycin D treated cells were less able to withstand a second hyperthermic treatment (Fig. 6), and also recovered poorly from it.

The nucleolus and thermotolerance in protein synthesis

Since low levels of actinomycin D are known to preferentially impair nucleolar RNA synthesis (Perry, 1967) rather than mRNA and tRNA synthesis, it is possible that ribosomal RNA synthesis is important for *both* recovery and thermotolerance development in Hela cell protein synthesis. Already (Burdon, R. H., this volume) it has been argued that ribosomal RNA from the nucleolus may be required in the 'rescue' of heat damaged cytoplasmic ribosomes.

When the synthesis of RNA, sensitive to low levels of actinomycin D, is examined after acute hyperthermia and recovery, it is clear that although it is quite heat sensitive, it nevertheless recovers to become impressively heat resistant (Fig. 7). Under conditions of *continuous* hyperthermia at 42 °C there is an initial depression in low actinomycin D sensitive RNA synthesis, but the synthesis

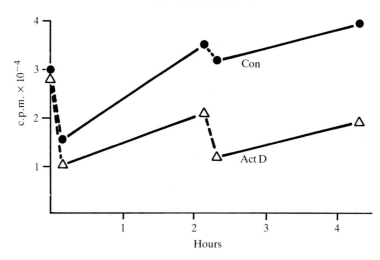

Fig. 6. The effect of low levels of actinomycin D on the ability of Hela cells to incorporate ^{35}S-methionine into protein following hyperthermia.
    The procedure adopted was exactly that in Fig. 4, except that actinomycin D $(0.05\,\mu g\,ml^{-1})$ was used in place of cycloheximide.

recovers to become intrinsically heat tolerant (Fig. 8). Thus it appears that the processes sensitive to low actinomycin D are not only important for the recovery and thermotolerance of protein synthesis, but themselves recover and become thermotolerant. This raises the question of whether any hsp synthesised during the recovery process (Slater *et al.* 1981; Burdon *et al.* 1982) might influence the RNA synthetic patterns of the nucleolus. Although the mammalian hsp 110 is a normal nucleolar component (Subjeck *et al.* 1983) hsp 70 type heat shock proteins are known to migrate from cytoplasm to nucleoli after heat shock (Welch & Feramisco, 1984; Pelham, 1984). There is also evidence that hsp 70 aids in the repair of heat damaged nucleoli, a suggestion being that this is achieved by binding to exposed hydrophobic surfaces (Pelham, 1984). Recently a model has been proposed in which repeated binding and ATP-driven release of hsp 70 help to solubilise hydrophobic precipitates of abnormal denatured proteins (Lewis & Pelham, 1985).
    Whilst the migration of hsps to the nucleolus might be responsible for the acquired thermal resistance of low actinomycin D-sensitive RNA synthesis the actual induction of new hsp synthesis may not be necessary as the intracellular level of constitutively synthesised hsps in unheated cells is significant. On the other hand we find that conditions that stimulate the induction of hsp synthesis in Hela cells, bring about a change in the sensitivity of RNA synthesis to cycloheximide. This inhibitor of protein synthesis causes a selective decrease in the activities of RNA polymerases I and III. This appears to be due to the rapid depletion of transcription factors that are required for initiation by RNA polymerases I and III (Gokal *et al.* 1986). Amongst these is the ribosomal DNA initiation factor TFIC. Treatment of Hela cells with cycloheximide will reduce total RNA synthesis but, if the cells are first subject to hyperthermic protocols known to induce hsp synthesis,

this reduction is no longer observable (Table 1). Clearly the possibility of interaction of hsps with nucleolar components involved in ribosomal DNA transcription needs investigation and may be important both for protein synthesis recovery and for thermotolerance.

Non-nucleolar hsps and thermotolerance

Whilst the above speculation centres around a role for hsps in the nucleolus, hsps are of course to be found in other cellular locations (see Burdon, R. H., this volume). A major portion of hsps in fact remains cytoplasmic. Hsp 90 is entirely cytoplasmic and in avian cells, whilst it has been reported to assist with the transit of the oncogene product pp60[src] to the plasma membrane, it also forms a part of the 8S steroid receptor complex (Schuh *et al.* 1985). Other hsps are known to associate with cytoskeleton elements possibly at their connections to the plasma membrane (Hughes & August, 1982; La Thangue, 1984). It is conceivable that the cytoskeleton associated hsps are involved in the repair of known cytoskeleton

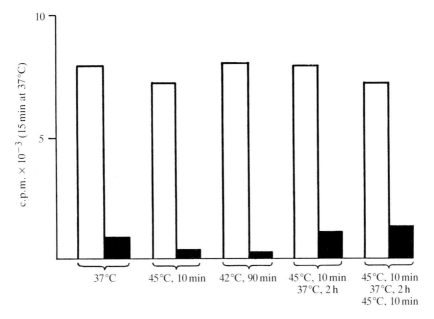

Fig. 7. The effects of various heating protocols on ^{3}H-uridine incorporation into RNA by Hela cells.

Duplicate monolayer cultures (see Fig. 2) were treated to various hyperthermic protocols in the presence or absence of low levels of actinomycin D ($0.05\,\mu g\,ml^{-1}$) as indicated. They were then labelled with ^{3}H-uridine ($2\,\mu C\,plate^{-1}$, $41\,Ci\,mmol^{-1}$) for 15 min at 37°C. To determine the level of ^{3}H-radioactivity in RNA, the cultures were washed three times with ice-cold 5% trichloroacetic acid and the cell monolayer dissolved in $880\,\mu l$, $0.4\,M$ NaOH, neutralised with $20\,\mu l$ glacial acetic acid and the ^{3}H-determined by scintillation spectrometry.

The solid histograms represent ^{3}H-incorporation into RNA that was sensitive to the low levels of actinomycin D used, whereas the open histograms are a measure of the incorporation that is insensitive to the actinomycin D concentrations used.

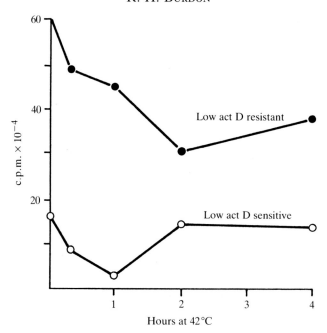

Fig. 8. The effect of continuous hyperthermia at 42 °C and ^{3}H-uridine incorporation into Hela cell RNA.

The experimental were conducted as described for Fig. 7 except that the hyperthermia was continuous at 42 °C. Low actinomycin D sensitive ^{3}H incorporation (○); actinomycin D resistant incorporation (●).

Table 1. *Effect of cycloheximide on Hela cell RNA synthesis*

Pretreatment	Dry treatment (37 °C)	^{3}H-uridine incorporation (cpm per 30 min × 10⁻⁴)
none	none	21·9
	cycloheximide (20 µg ml⁻¹ 2 h)	20·1
45 °C 10 min then 37 °C 2 h	none	21·2
	cycloheximide (20 µg ml⁻¹ 2 h)	23·3
42 °C 2 h	none	21·2
	cycloheximide (20 µg ml⁻¹ 2 h)	26·5

Triplicate Hela cell monolayer cultures were set up as for Fig. 3. After growth at 37° for 24 h, they were divided into three groups. One served as a control whereas the other two were subject to various hyperthermic pretreatments. Each group was further subdivided into two, one of the sub-groups of each main group being treated at 37° with 20 µg ml⁻¹ cycloheximide as indicated. The ability of all cultures to incorporate ^{3}H-uridine was assessed by addition of 2 µC ^{3}H-uridine per culture and incubation at 37 °C for 30 min. ^{3}H-radioactivity in RNA was determined by the procedure described in Fig. 7.

damage following hyperthermia and possible development of thermotolerance. Hsps might be able to confer thermal resistance to cytoskeletal elements to prevent cells rounding-up. Cytoskeletal associations may also have implication for plasma membrane systems, albeit indirectly. For example the integrity of the cytoskeleton may affect the heat sensitivity of NaK-ATPase activity (Burdon & Cutmore, 1982). This enzyme activity, which appears quite stable in isolated plasma membranes when they are heated *in vitro*, is nonetheless demonstrably heat sensitive in heated intact cells and hsp induction appears to be involved in its repair and development of thermotolerance in its activity following hyperthermia (Burdon *et al.* 1984).

Another hsp that may have considerable relevance is ubiquitin. Transcription of the gene for this small 8 kDa hsp was found to be actively induced by hyperthermia in avian fibroblasts (Bond & Schlesinger, 1985). An increase in ubiquitin levels might well serve to protect cells from the deleterious effects of an accumulation of damaged proteins.

A final comment of the subject of possible mechanism of thermotolerance development relates to Fig. 2. The recovery of *total* protein synthesis is to levels around 100–200 % of that in unheated cells. This could explain the two-fold increase in such enzymes as superoxide dismutase reported to follow acute brief hyperthermia (Loven *et al.* 1985). Such an enzyme, as already pointed out (Burdon, R. H. this volume), could be of considerable relevance to the phenomenon of thermotolerance, especially if its overall level were to increase following heat shock. Indeed a positive correlation between superoxide dismutase levels and thermotolerance development has recently reported (Loven *et al.* 1985).

Normally, treatment of Hela cells with 0·5 mM diethyldithiocarbamate (DDC), an inhibitor of superoxide dismutase, will cause a progressive loss in cellular protein synthetic activity (up to 80 % loss after 4 h at 37°C). Surprisingly, treatment with this inhibitor does not block recovery of protein synthetic activity after hyperthermia (see Burdon, R. H., this volume). Thus levels of superoxide dismutase may increase sufficiently following hyperthermia to overcome the effects of DDC at 0·5 mM. However it remains to be seem whether hyperthermia increases superoxide dismutase gene transcription.

Reduction of thermotolerance

Whilst the involvement of hsps in thermotolerance still awaits final analysis, it is clear that both cellular thermotolerance and protein synthesis thermotolerance are not inhibited by cycloheximide. This, as has been argued, does not actually rule out a role for hsps. On the other hand is should be emphasized that there are situations where heat induced hsp synthesis is blocked, such is in media depleted Ca^{2+}, yet thermotolerance is observed. Conversely treatment of mammalian cells with 0·3 mM benzaldehyde will inhibit the development of thermotolerance (Mizuno *et al.* 1984). We find that benzaldehyde however will not block hsp

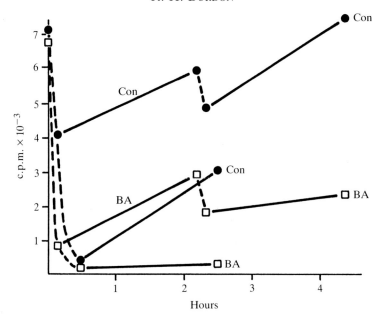

Fig. 9. The effect of benzaldehyde on the ability of Hela cells to incorporate ^{35}S-methionine into protein following hyperthermia.
    The procedure adopted was exactly as described in Fig. 4 except that 0·3 mM benzaldehyde was used in place of cycloheximide.

induction by heat, but it will inhibit recovery of protein synthesis after hyperthermia (Fig. 9). Moreover it will reduce the development of thermotolerance in the protein synthetic system is judged by its effects on a second hyperthermic treatment of benzaldehyde treated cells. Development of drugs related to benzaldehyde may be clinical use where tolerance can be an impediment to the use of hyperthermia as a modality in cancer therapy.

Other roles for heat shock proteins

If a role for hsps in thermotolerance is still *sub-judice*, what are other possible functions? Their impressive evolutionary conservation suggests primary biological importance. Recent observations indicate them to be constitutively synthesized at specific times in the normal development of insect, frog and mouse embryos (see Bienz, 1985). The mechanism for this specific developmental control of hsp expression is not known. There may of course be specific periods of stress during embryogenesis, but there may be more subtle controls of hsp gene expression. For instance it is known that the expression of hsp 70 can be induced by the oncogene products of the E1A gene of the DNA tumour virus adenovirus 5 (Nevins, 1982), as well as by the rearranged c-*myc* mouse oncogene (Kingston *et al.* 1984). Such oncogenes are implicated in the immortalisation of primary cells in culture and it may be that hsp 70 has a critical pole to play in animal cell growth.

Thus the continued study of temperature- and stress-related adaptive mechanisms may yield considerable insights relevant to fundamental biological phenomena.

Relationship of thermotolerance to other stress resistance

As a postscript it should be pointed out that whilst heat shock will induce tolerance to further heat stress is also leads to increased tolerance to other types of stress. For example the application of heat stress to yeasts greatly increases their subsequent tolerance to ethanol (Watson & Cavicchioli, 1983). Heat shock of hamster cells also increases their resistance to ethanol (Li *et al.* 1980) and conversely, exposure to ethanol, also an hsp inducer, increases the resistance of hamster cells to both heat shock and drugs such as adriamycin (Li & Hahn, 1978). Clearly the range of stress protection that is possible remains to be fully explored, however as already mentioned (Burdon, R. H., this volume) *prior* heat treatment does not protect Hela cells from the effects of cold exposure.

The work is partly supported by a grant from the Cancer Research Fund and from the Scottish Home & Health Department.

References

ANDERSON, R. C., TAO, T.-W. & HAHN, G. M. (1984). Cholesterol: phospholipid ratios decrease in heat resistant variants of B16 melanoma cells. In *Hyperthermic Oncology* 1984, vol. 1 (ed. J. Overgaard), pp. 123–126. London: Taylor & Francis.

BIENZ, M. (1985). Transient and developmental activation of heat-shock genes. *Trends biochem. Soc.* **10**, 157–161.

BOND, U. & SCHLESINGER, M. J. (1985). Ubiquitin is a heat shock protein in chicken embryo fibroblasts. *Molec. Cell Biol.* **5**, 949–956.

BURDON, R. H. & CUTMORE, C. M. M. (1982). Human heat shock gene expression and the modulation of plasma membrane Na, K-ATPase activity. *FEBS Letts.* **140**, 45–48.

BURDON, R. H., KERR, S. M., CUTMORE, C. M. M., MUNRO, J. & GILL, V. (1984). Hyperthermia, Na,K-ATPase and lactic acid production in some human tumour cells. *Br. J. Cancer* **49**, 437–445.

BURDON, R. H., SLATER, A., MCMAHON, M. & CATO, A. C. B. (1982). Hyperthermia and the heat shock proteins of Hela cells. *Br. J. Cancer* **45**, 953–963.

CRAIG, E. A. & JACOBSEN, K. (1984). Mutations of the heat inducible 70 kilodalton genes of yeast confer temperature sensitive growth. *Cell* **38**, 841–849.

GERNER, E. W., BOONE, R., CONNOR, W. G., HICKS, J. A. & BOONE, M. L. M. (1976). A transient thermotolerant survival response produced by single thermal doses in Hela cells. *Cancer Res.* **36**, 1035–1040.

GOKAL, P. K., CAVANAUGH, A. H. & THOMPSON, E. A. (1986). The effects of cycloheximide upon transcription of rRNA, 5SRNA and tRNA genes. *J. biol. Chem.* **361**, 2536–2541.

HALL, B. G. (1983). Yeast thermotolerance does not require protein synthesis. *J. Bact.* **156**, 1363–1365.

HEIKKILA, J. J., KLOC, M., BURY, J., SCHULTZ, G. A. & BROWDER, L. W. (1985). Acquisition of the heat shock response and thermotolerance during early development of *Xenopus laevis*. *Devl Biol.* **107**, 483–489.

HENLE, K. J. & DETHLEFSEN, L. A. (1978). Heat fractionation and thermotolerance: a review. *Cancer Res.* **38**, 1843–1878.

Henle, K. J., Karamuz, J. E. & Leeper, D. B. (1978). Induction of thermotolerance in Chinese Hamster Ovary cells by high (45°) or low (45°) hyperthermia. *Cancer Res.* **38**, 570–574.

Hughes, E. N. & August, J. T. (1982). Coprecipitation of heat shock protein with a cell surface glycoprotein. *Proc. natn. Acad. Sci. U.S.A.* **79**, 2305–2309.

Kingston, R. E., Baldwin, A. S. & Sharp, P. A. (1984). Regulation of heat shock 70 gene expression by c-myc *Nature, Lond.* **312**, 280–282.

Lamarche, S., Chretien, P. & Landry, J. (1985). Inhibition of the heat shock response and synthesis of glucose regulated proteins in Ca^{2+}-deprived rat hepatoma cells. *Biochem. biophys. Res. Commun.* **131**, 868–876.

Landry, J., Berneir, D., Chretien, P., Nichole, L. M., Tanguay, R. M. & Marceau, N. (1982). Synthesis and degradation of heat shock proteins during development and decay of thermotolerance. *Cancer Res.* **42**, 2457–2461.

La Thangue, N. B. (1984). A major heat shock protein defined by a monoclonal antibody. *EMBO J.* **3**, 1871–1879.

Lazlo, A. & Li, G. C. (1985). Heat-resistant variant of Chinese hamster fibroblasts altered in expression of heat shock protein. *Proc. natn. Acad. Sci. U.S.A.* **82**, 8029–8033.

Lepock, J. R. & Kruuv, J. (1980). Thermotolerance as a possible cause of the critical temperature at 43°C in mammalian cells. *Cancer Res.* **40**, 4485–4488.

Lewis, M. J. & Pelham, H. R. B. (1985). Involvement of ATP in the nuclear and nucleolar functions of the 70kD heat shock protein *EMBO J.* **4**, 3137–3143.

Li, G. C. & Hahn, G. M. (1978). Ethanol induced tolerance to heat and to adriamycin. *Nature, Lond.* **274**, 699–701.

Li, G. C. & Hahn, G. M. (1984). Mechanisms of thermotolerance. In *Hyperthermic Oncology* 1984, vol. 2 (ed. J. Overgaard), p. 231–237. London: Taylor & Francis.

Li, G. C. & Mak, J. Y. (1985). Induction of heat shock protein synthesis in murine tumours during the development of thermotolerance. *Cancer Res.* **45**, 3816–3827.

Li, G. C. & Werb, Z. (1982). Correlation between synthesis of heat shock proteins and development of thermotolerance in Chinese hamster fibroblasts. *Proc. natn. Acad. Sci. U.S.A.* **79**, 3218–3222.

Li, G. C. Shiu, E. C. & Hahn, G. M. (1980). Similarities in cellular inactivation by hyperthermia or by ethanol. *Radiat. Res.* **82**, 257–263.

Loomis, W. F. & Wheeler, S. A. (1982). Chromatin-associated heat shock proteins in *Dictyostelium*. *Devl. Biol.* **90**, 412–418.

Loven, D. P., Leeper, D. B. & Oberley, L. W. (1985). Superoxide dismutase levels in Chinese hamster ovary cells and ovarian carcinoma cells after hyperthermia and exposure to cycloheximide. *Cancer Res.* **45**, 3029–3033.

Mizuno, S., Ishida, A. & Miwa, N. (1984). Enhancement of hyperthermic cytotoxicity and prevention of thermotolerance induction by benzaldehyde in SV40-transformed cells. In *Hyperthermic Oncology* 1984, vol. 1 (ed. J. Overgaard), pp. 99–102. London: Taylor & Francis.

Muller, W. U., Li, G. C. & Goldstein, L. S. (1985). Heat does not induce synthesis of heat shock proteins or thermotolerance in the earliest stage of mouse embryo development. *Int. J. Hyperthermia* **1**, 97–102.

Nevins, J. R. (1982). Induction of the synthesis of a 70,000 dalton mammalian heat shock protein by adenovirus E1A gene product. *Cell* **29**, 913–919.

Omar, R. A. & Lanks, K. W. (1984). Heat shock protein synthesis and cell survival in clones of normal and simian virus 40-transformed mouse embryo cells. *Cancer Res.* **44**, 3976–3982.

Pelham, H. R. B. (1984). Hsp 70 accelerates the recovery of nucleolar morphology after heat shock. *EMBO J.* **3**, 3095–3100.

Perry, R. P. (1967). The nucleolus and the synthesis of ribosomes. *Prog. Nuclei Acad. Res. molec. Biol.* **6**, 219–248.

Petko, L. & Lindquist, S. (1986). Hsp 26 is not required for growth at high temperatures, nor for thermotolerance, spore development or germination, *Cell* **45**, 885–894.

Schamhart, D. H. J., Berendson, V., van Rijn, J. & van Wijk, R. (1984a). Comparative studies on heat sensitivity of several rat hepatoma cell lines and hepatocytes in primary culture. *Cancer Res.* **44**, 4507–4516.

SCHAMHART, D. H. J., VAN WALRAVEN, H. S., WIEGANT, F. A. C., LINNEMANS, W. A. M., VAN RIJN, J., VAN DEN BERG, J. & VAN WIJK, R. (1984b). Thermotolerance in cultured hepatoma cells: cell viability, cell morphology, protein synthesis and heat shock proteins. *Radiat. Res.* **98**, 82–95.

SCHUH, S., YONEMOTO, W., BRUGGE, J., BAUER, V. J., RIEHL, R. M., SULLIVAN, W. P. & TOFT, D. O. (1985). A 90,000 dalton protein common to both steriod receptors and the Rous Sarcoma virus transforming protein pp60$^{\text{v-src}}$ *J. biol. Chem.* **260**, 14292–14296.

SLATER, A., CATO, A. C. B., SILLAR, G. M., KIOUSSIS, J. & BURDON, R. H. (1981). The pattern of protein synthesis induced by heat shock of Hela cells. *Eur. J. Biochem.* **117**, 341–346.

SUBJECK, J. R., SHYY, T., SHEN, J. & JOHNSON, R. J. (1983). Association between the mammalian 110,000 dalton heat-shock protein and nucleoli. *J. Cell Biol.* **97**, 1389–1395.

TOMASOVIC, S. P., ROSENBLATT, P. C., JOHNSTON, D. A., TANG, K. & LEE, P. S. Y. (1984). Heterogeneity in induced heat resistance and its relation to synthesis of stress proteins in rat tumour cell clones. *Cancer Res.* **44**, 5850–5856.

WATSON, K. & CAVICCHIOLI, R. (1983). Acquisition of ethanol tolerance in yeast cells by heat shock. *Biotech. Letts* **5**, 683–688.

WELCH, W. J. & FERAMISCO, J. R. (1984). Nuclear and nucleolar localization of the 72,000-dalton heat shock protein in heat-shocked mammalian cells. *J. biol. Chem.* **259**, 4501–4513.

WIDERLITZ, R. B., MAGUN, B. E. & GERNER, E. W. (1986). Effects of cycloheximide on thermotolerance expression, heat shock protein synthesis, and heat shock protein mRNA accumulation in rat fibroblasts. *Molec. Cell Biol.* **6**, 1088–1094.

Printed in Great Britain © *Society for Experimental Biology 1987*

COLD TOLERANCE IN MAMMALIAN CELLS

JOHN S. WILLIS

Department of Physiology and Biophysics, University of Illinois, 524 Burrill Hall,
407 South Goodwin, Urbana, Illinois 61801, USA

Summary

As whole organisms, most mammals have a poor tolerance for hypothermia. But their cells may have a capacity for a far wider cold tolerance, which may be expressed in peripheral tissues, sporadically in core tissue and in cultured cells. Against this background the cold resistance of cells of deep hibernators may be seen as the extreme of a continuum and is complicated by the consideration that the voluntary hypothermia of hibernation is probably in most cases a metabolic adaptation to forestall starvation. Similarly, cold resistance of peripheral tissues may in diving animals be confounded by the need to be adapted to hypoxia as well. Hence, attempts to analyse cold resistance by comparisons of absolute rates of arbitrarily chosen reactions may be misleading. A more useful approach is analysis of maintenance of *balance*: balance between ATP synthesis and utilization, balance between macromolecule synthesis and degradation and balance between pumps and leaks.

Cation pumps and leaks constitute a major component of energy utilization and are central to other cell functions, even during minimal metabolism. Hence, the maintenance of ion gradients is a central issue in understanding adaptation not only to hypothermia but also to starvation and hypothermia. Of the three hypometabolic states, hypothermia has been best studied in this regard. In most cases, passive permeability is more reduced at low temperature in cold-tolerant cells than in cold-sensitive ones. In some cases there is also a difference in Na-K pump activity and perhaps in ATP dependent Ca-pump activity.

Pump activities and probably the maintenance of minimal leak require ongoing metabolism. The question of whether, in cold-sensitive cells, energy supplies are adequate at low temperature was once the focus of this field, but has been ignored for a decade without having been fully resolved. There are many instances of less temperature sensitivity of specific metabolic activities (mitochondrial respiration, etc.) in hibernators than in non-hibernators, without any verification of whether this is essential for survival at low body temperature. Certainly, robust pumping has been found in some failing cold-sensitive cells at low temperature, suggesting no shortage of ATP in these cases, but in other cases the issue may be a more complex one than just that of ATP availability.

For the pump, assuming ATP is adequate in both sensitive and insensitive cells, analysis of differences in tolerance requires an unravelling of the mechanism of the pump itself, whereas for passive permeability, with our present state of knowledge, the problem is one mainly of identification of which of several specific parallel pathways are involved. Comparing ground squirrel (cold-tolerant hibernator) and guinea pig (cold-sensitive non-hibernator), the difference which accounts for higher passive permeability of K at low temperature in the Ca-sensitive K channel and for Na the corresponding difference in Na-permeation is in the Na-H exchange mechanism. Na-K-Cl cotransport, a passive carrier mechanism which may function as a first line of defence against swelling disappears in cold-sensitive cells at 5°C but continues to function in ground squirrel cells. The simple (pore-mediated?) leak of Na and K at low temperature does not seem to be different among species or to correlate with cold-sensitivity. Causes of failure in cold-sensitive mammalian cells at low temperature appear to be diverse among cell types and species. Tolerance, therefore, may depend more upon the mechanisms of interactions between and continued integration of metabolism, pumps and leaks than upon adaptation of any one of these components.

Introduction

Prevalence of cold tolerance

High body temperature, at least in a terrestrial environment, is a prerequisite for thermal regulation because it permits control of heat loss and, therefore, variable heat production (Burton & Endholm, 1955). Maintenance of high body temperature is costly not only because it obligates higher energy expenditure but also increased oxygen acquisition and water loss. Since voluntary, reversible cooling of tissues offers the possibility of conserving energy (Morrison, 1960; French, 1986), oxygen (Bullard, David & Nichols, 1960) and water (Fisher & Manery, 1967), it is not surprising to find numerous occurrences of it among birds and mammals.

Hence, the notion of 'hypothermia' as an abnormal state implying life-threatening blockage of cell function, is very parochial. Most cells of most species of animals experience and are tolerant to a wide range of temperatures. Even among mammals and birds, significant cold tolerance may be more the rule than the exception. Only the cores of adult mammals and birds are so rigidly stenothermic that their cells can afford to lose this natural tolerance. Mammalian cold tolerance is commonly found in two rather different situations: in peripheral tissues whose temperature is allowed to vary (Chatfield, Lyman & Irving, 1953; Miller & Irving, 1963) and in cells generally of species that allow variations in core temperature, the champions in this category being the long term deep hibernators like ground squirrels. Even in the core tissues of pristinely homeothermic mammals, tolerant cells can sometimes be found. Tolerance may also be influenced or induced by developmental stage, by prior experience of cold, by seasonality and by diet (Willis, 1967, 1978; Fuller, Marley & Green, 1985).

Table 1. *Persistence of cold sensitivity or cold resistance with varying degrees of isolation from the organism*

Degree of isolation	Example	Criterion	Species difference maintained?	If not, why not?
Intact organism	Brain	Ion gradients	Yes	
Freshly excised tissue or organ	Heart	Rhythmicity	Yes	
	Kidney	Ion gradients	Yes	
	Erythrocytes	Ion gradients	Yes	
	Peripheral nerves	Excitability	Yes	
	Brain	Ion gradients	No	HS
	Aortic smooth muscle	Ion gradients	Yes	
	Liver	Ion gradients	No	HS
Stored tissues	Kidney	Ion gradients	Yes	
	Kidney	Gluconeogenesis	Yes	
	Kidney	Protein synthesis	Yes	
	Brain	Metabolism	Yes	
Primary cell culture	Heart	Rhythmicity	No	NHR
	Kidney	Ion gradients	Yes	
	Liver	Ion gradients	Yes	
	Embryonic fibroblasts	Ion gradients, dye exclusion, proliferation	Yes	
Early passaged cultures ('secondaries')	Embryonic fibroblasts	Ion gradients, dye exclusion	Variable	Some HS
Continuous cell lines (many passages)	HeLa, mouse, L cells, others	Dye exclusion, ion gradients, proliferation	No	NHR

Symbols: NS = hibernator cells become cold sensitive; NHR = cells from non-hibernator are cold resistant. Most cases are cited in Willis, 1977. Liver cases are from Willis *et al.* 1981; Gluconeogenesis in kidney is based on Fuller *et al.* 1986. Aortic smooth muscle is based on Kamm *et al.* 1979*a*.

Isolation of cells from the organism may also determine tolerance (Table 1). There are varying degrees of such isolation – freshly excised tissues (cell, tissue and whole organ preparations), stored organs or tissues, primary cultures, established cultures, transformed cultures. In most cases freshly excised tissues from species that experience low body temperature (hibernators) exhibit greater tolerance than those from rigidly normothermic species, and this provides the basis for the analytical work discussed below. As the separation of the cells from the body increases, however, the differences between species tend to diminish, largely because of the emergence of wider tolerance of low temperature in long term cultures from originally temperature-sensitive sources (for review see Willis, 1967). Such observations, along with cold tolerance of neonatal and peripheral tissues, suggest that the capacity for wide temperature tolerance persists in the genome of even conventionally cold-sensitive birds and mammals, and that it is only the expression of this capacity that is altered.

Criteria of tolerance

Criteria for hypothermic survival of cells may be divided conveniently into differentiated function (nerve conduction, heart beating, transepithelial transport, secretion of specific products), generalized or metabolic measures of cellular integrity (maintenance of ion gradients, dye exclusion, cell swelling, retention of proteins, protein synthesis, gluconeogenesis), and measures of cell growth and proliferation (cell division, colony formation, incorporation of nucleic acid precursors).

These diverse measures may, of course, give different answers about the retention or even the nature of cold tolerance. Thus, retention of species-related differences in cold tolerance of regulation of Na/K gradient may persist longer than cardiac rhythmicity (Table 1), and cold sensitivity of cell division is more pronounced at high sub-optimal temperatures than at low (Nelson, Kruuv, Koch & Frey, 1971), making the exposure dependent upon phase of the cell cycle (e.g. static cultures being more cold-resistant than those in log phase (Kruuv, Glofcheski & Lepock, 1985)).

By any of these criteria one must also distinguish between changes occurring while the cell is actually cold and those that occur subsequent to rewarming. Both aspects are clearly important for both practical and theoretical considerations, but conclusions may be affected depending upon which is chosen. Thus, in many tissues maintenance of Na and K gradients may be of paramount importance to avoid primary fatal consequences of the organism while it is still cold (hypo-kalemia, blockage of nerve and muscle conduction), whereas maintenance of Ca and H ion gradients might be more crucial to the avoidance of long term damage to cells following rewarming (Hume, 1986).

Strategies of investigation

From the foregoing it seems clear that it may not be meaningful to try to analyse the 'adaptation' of cold tolerant cells. Tolerance being arguably the ground state, it is necessary at the same time to understand the sensitivity to cold found in 'typical' mammals. Two strategies have emerged for doing this. The first is to study cold-sensitive cells unilaterally and to try by manipulation of independent variables (such as conditions of incubation) to confer cold-tolerance upon them. This approach is best exemplified by studies of tissue, organ and cell storage. It relies primarily upon criteria of survival based upon recovery from hypothermia. While this approach has provided much useful information, it is essentially pragmatic in nature, and, consequently, its limitation is that seldom are the effects of temperature *per se* studied in isolation from the other attendant factors of storage or incubation.

The other 'strategy' is to compare specific aspects of cell function in cold-sensitive and cold-tolerant cells. This approach is best characterized by work on hibernator cold adaptation, but it could equally well be applied to cold tolerance of peripheral tissues. It usually relies upon criteria of survival and function assessed when the cells are actually cold. The two strategies could be combined with profit, but this has seldom been attempted (Willis, Foster & Behrends, 1975; Green, Marley & Fuller, 1985).

The comparison between tolerant and sensitive states can be made between two species or within one species between cold-adapted and non-adapted states. In either case it is necessary to establish (by criteria such as listed above) the difference in cold tolerance of the compared forms. This is obvious in making comparisons between species and usually has been done. Many studies of seasonal hibernators, however, have dealt with detailed cell mechanisms compared in winter and summer preparations without the difference in temperature sensitivity having been established for the two states, but merely, and unwarrantedly, assumed.

For example, findings of increased dependence on sarcoplasmic reticular Ca stores (Kondo & Shibata, 1984) and increased capacity for Ca uptake (Belke, Pehowich & Wang, 1986) in cardiac muscle during hibernation can be related to increased myocardial contractility at low temperature during hibernation (Zhou, Dryden & Wang, 1986). On the other hand, even though chemically driven Ca channels are more cold resistant in ground squirrel intestinal smooth muscle and become more dominant during hibernation (Wolowyk, Li & Wang, 1986), we know nothing of the difference in capacity for activity at low temperature of this muscle between the two species or of any change with hibernation. Besides the possibility of cold adaptation, seasonal changes also influence nutritional and reproductive state and the level of activity. In hibernation some tissues are active (heart, diaphragm), others not. In ground squirrels hibernation is a period of inanition and such changes as seen in the intestine with hibernation or with cooling might be related more to insuring inactivity than the opposite.

Given the complex web of reactions in a cell, how can one hope to make sense of the failure that occurs in sensitive cells? Early work by the 'comparative' approach laid heavy emphasis on reaction rates (mitochondrial respiration, glycolysis, muscle contraction, etc.), invariably expressed as Arrhenius functions (for review, see Willis, 1982). This was perhaps useful in illustrating the 'superior' cold tolerance of hibernator tissues, but has contributed relatively little in terms of untangling the 'web'. A major theme of the work of my mentor, C. P. Lyman, was that a stunning and crucial difference between hibernation and forced hypothermia is that, in the former, homeostasis is maintained in the whole organism for many systems (thermoregulation, acid-base balance, blood pressure). The extension of this realization to the cellular level, for me, means that one should focus on the *balance* of critical reactions: synthesis and utilization of ATP, pumping and leakage of ions, synthesis and breakdown of proteins or other macromolecules, and not upon the absolute rates of the constituent reactions, considered in isolation.

The discussion that follows focuses on the 'comparative' approach outlined above, with its emphasis on acute survival of profound low body temperature (i.e.

approaching 0°C). Since death of cold-sensitive forms during hypothermia is thought to be the immediate result of failure of plasma membrane function (fibrillation, nerve blockage, cell swelling), and since so many of the criteria of tolerance of isolated cells also hinge upon plasma membrane integrity, our concern will be confined largely to the causes of the breakdown of gradients in cold-sensitive forms.

Membrane failure in the cold

Overview

A paradigm has been suggested for the analysis of differences between hibernator and non-hibernator cells with respect to Na and K regulation (Willis, Fang & Foster, 1972) which can be paraphrased and generalized as follows: (1) Is the cytoplasmic/extracellular gradient for the ion lost at low temperature (*in vivo* and *in vitro*)? (2) If 'yes', (a) is the metabolically dependent pump reduced (compared with a cold-tolerant form) or (b) is the leak for the ion elevated (again, compared with a cold-tolerant form)? (3) If 2(a) is 'yes', can the inhibition be attributed to failure of metabolism (lack of sufficient ATP)? (4) If 3 is 'no' or 2(b) is 'yes', what are the differences in the transport mechanisms themselves?

Maintenance of gradients

At the first level of our dichotomous paradigm, differences in retention of K ion at low temperature between tissues of hibernators and non-hibernators have been described for numerous tissues: skeletal muscle, aortic smooth muscle, kidney cortex (both as slices and cultured cells), smooth muscle, red blood cells (for review, see Willis, Ellory & Cossins, 1981). For liver, there are no differences in incubated slices between hamsters and guinea pig, even though K is unchanged in the hibernating animal (Becker, Willis & Sobolewski, 1982). There is, however, a perfect retention of K at 5°C in one-day cultured hamster hepatocytes, whereas guinea pig cells lose K. It is also notable that though there have been numerous studies of cold-tolerant *vs* cold-sensitive cardiac muscle rhythmicity and excitability (Willis, 1978, 1979), the difference has never yet been shown to be due to ion regulation.

The cellular correlates of maintained ion gradients, such as membrane potential, excitability, synaptic transmission and cell volume regulation have also long since been shown to be more intact at low temperature in selected tissues of hibernators than those of non-hibernators.

Pump vs leak

Attempts to analyse the relative contribution of pumps and leaks in intact cells, using net ion movements or flux measurements, have only been made in a few cell types. Comparing unidirectional isotopic fluxes in red blood cells of one species of hibernator (ground squirrel) with one species of non-hibernator (guinea pig), Kimzey & Willis (1971) showed that both for Na ion and for K ion both pumps and leaks were different (leaks slower, pumps faster in the hibernator). This is probably the most complete analysis of this type to date. The conclusion regarding pump (measured as ouabain-sensitive K influx) was confirmed in seven species of hibernator and nine species of non-hibernator, with the interestingly exceptional observation that Syrian hamsters (which hibernate in three-day bouts) showed a 'cold-sensitive' pump and moles (not known to practise temperature drops) possessed a 'cold-resistant' pump (Willis, Ellory & Wolowyk, 1980).

Both pump and leak were thought to be involved in differences in retention and net uptake of K at low temperature of kidney slices (rat, guinea pig *vs* hamster, ground squirrel). Later studies based on K fluxes in cultured kidney cells (hamster and ground squirrel *vs* guinea pig) confirmed the difference in temperature sensitivity of passive permeability to K, but also showed that differences in temperature sensitivity of the pump were not great in this tissue for these species (i.e. pumping was not so drastically reduced in guinea pig cells as had been thought, Zeidler & Willis, 1976). The retention of K by hamster hepatocytes at 5°C is almost unaffected by ouabain, implying a total dependence upon reduced passive permeability. A more active Na/K pump at low temperature was inferred from flux studies in aortic strips (smooth muscle) of ground squirrel compared with rat (Kamm *et al.* 1979).

(Studies of Na-K ATPase have produced conflicting results [for review see Charnock, 1978; Charnock & Simonson, 1978; Willis *et al.* 1981]. Comparison of temperature effect on intact pump with broken membrane Na-K ATPase [Willis, Ellory & Becker, 1978] suggests that the isolated enzyme may not provide a reliable index of cold sensitivity. One advantage of Na-K ATPase measurements over many flux measurements is that they are typically made under 'V_{max}' conditions, whereas most flux determinations are usually done at near steady-state ionic concentrations, so that Na ion is typically far below that required for V_{max}. If in cold-sensitive cells Na concentration is increasing, the comparison with a truly steady-state cold-tolerant cell would unduly favour the cold-sensitive cells.)

Thus, depending upon species and tissue, either pumping or passive pathways, or both, may account for the difference in maintenance of gradients between tolerant and sensitive cells.

ATP vs *pump*

Older attempts to analyse differences between hibernators and non-hibernators at the cellular level dealt largely with cellular metabolism (for review, see Willis, 1982). Except for studies of mitochondrial lipids, this emphasis has died out and a consensus emerged that ATP *per se* was not necessarily the limiting factor for survival at low temperature. In part, this may have been due to studies of stored red cells which show only gradual decline in ATP levels, even though loss of ion gradients is rapid and progressive (Wood & Beutler, 1967; for review see Willis *et al.* 1981). Also, the early evidence with Na-K ATPase showing less temperature sensitivity in hibernating hibernators than awake hibernators or rats (Bowler & Duncan, 1969; Willis & Li, 1969), although this was later disputed (Charnock, 1978), suggested that an ATP hypothesis was not necessary. Finally, it was thought that the affinity of the Na/K pump for ATP was sufficiently high (half-activation about 0·2 mM) to provide a safety margin against any but the most extreme depletion of ATP (van Rossum, 1972). Certainly, the sanguine view that ATP is probably not a problem has not been based upon any extensive or modern data about comparative differences in phosphorylated nucleotides as a function of temperature.

It is time that this issue be more intensively investigated. Dog kidney and liver perfused at low temperature maintains better gradients and survives storage better when ATP levels are kept high by the addition of adenosine to the perfusion medium (Southard, Rice & Belzer, 1985). Chemical estimates of ATP cell concentrations do not necessarily reflect concentration of energy available, since this also depends upon the concentrations of inorganic phosphate and ADP (Burlington, Meininger & Thurston, 1976), since temperature might change binding of Mg to ATP to form the natural substrate for Na-K ATPase, and since variation in pH may alter the enthalpy of hydrolysis (Dawson, Gadian & Wilkie, 1980; Dawson & Wilkie, 1984). Of course, availability of total high energy phosphates (2,3-diphosphoglycerate in erythrocytes, creatine phosphate in muscle) needs to be considered as well as rate of turnover of ATP. Finally, recent studies (Soltoff & Mandel, 1984; Tessitore, Sakhrani & Massry, 1986) have shown that, at least in kidney cells, the affinity of the Na-K pump for ATP is much lower than previously imagined. It is also relevant in this regard that the effect of temperature on the kinetics of the Na-K ATPase have not been determined.

Na-K pump

Attempts to analyse at a mechanistic level the operation of the Na-K pump in an intact cell membrane at low and high temperatures have only been made in red blood cells, where there is known to be a difference between hibernators and non-hibernators (for a fuller account see Willis, 1986). There have, however, been numerous analytical studies of Na-K ATPase with temperature including some on enzyme from hibernator sources (for review, see Willis *et al.* 1981).

The result of all this effort has not been such as to point to a clear explanation of cold-block of the pump in cold-sensitive forms nor the differences in cold-tolerant forms. In general, the results with intact red cells do not accord well with work on broken membrane preparations, the latter of which predict blocks at specific points in the overall reaction of the pump (for a recent example, see Kaplan & Kenney, 1985). Activation of the pump by cytoplasmic Na and by extracellular K is altered by temperature, but the shifts in sensitivity to the two ions are parallel in cold-tolerant and cold-sensitive cells (Ellory & Willis, 1982). Reactions involving only a part of the pump or the pump operating under an unusual mode either are not interestingly different between hibernators and non-hibernators or are too difficult to measure at the very low temperatures (5°C) of interest. There *are* differences in the mechanics of the pump at 37°C between ground squirrel and guinea pig, but it is not yet known if these differences can be related to cold adaptation (Willis & Ellory, 1985*a*,*b*).

Passive K permeability

The analysis of Na and K passive permeation must be carried out separately for the two ions, because for each ion there is a variety of parallel pathways through the plasma membrane. The specific pathways present depend not only upon the ion but also upon the species and type of cell. For human (and rodent) red blood cells the pathways for K permeation are shown in Fig. 1.

The recognition of so many permeability mechanisms is a relatively recent development in cell physiology and is still proceeding apace. It is largely based upon the use of selective inhibitors such as loop diuretics (bumetanide, furosemide) which block (among other things) the so-called Na-K-Cl cotransport mechanism, and amiloride which inhibits the Na-H exchange. Unfortunately, none of these inhibitors is as selective for specific mechanisms as, say, ouabain is thought to be for the Na-K pump. Nevertheless, by use of these inhibitors singly and in combination, it is possible to dissect the contribution of each path to total unidirectional flux of radioactive K or Na isotope into or out of the cell.

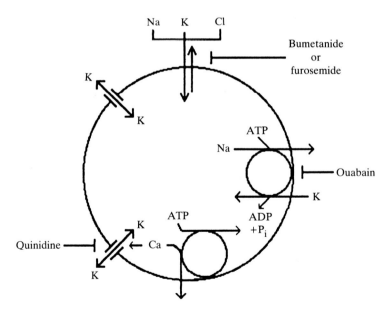

Fig. 1. Pathways of K permeation in red blood cells, based on findings in human and rodent red blood cells. Circles represent ATP-dependent pumps; matched arrows, carrier-mediated mechanisms; double-headed arrows, channels; blunted lines, inhibitors.

Application of this method to the effects of low temperature in cold-sensitive and cold-tolerant cells has been successfully undertaken so far only in red blood cells.

As shown in Fig. 1, if there is no Ca ion in the medium (and hence none in the cytoplasm), treatment of rodent or human red blood cells with the combination of ouabain and bumetanide, leaves only a residual leak remaining. Under these conditions influx of K is a linear function of concentration over a wide range, as though it might represent permeation through a simple hole or pore. One might expect diffusion through a simple pore to be little affected by cooling, but when Stewart, Ellory & Klein (1981) tested this point they obtained a surprising result: an actual *increase* in influx with cooling between 12 °C and 0 °C. (They found the same for Na permeation, measured as Na influx. The same phenomenon had been observed previously by Wieth, 1970, who substituted nitrate ion for chloride to eliminate the cotransport system.) When this experiment was repeated amongst a selection of rodents (hibernators and non-hibernators), no 'paradoxical' rise was observed at low temperatures (Hall & Willis, 1986). Instead, K influx declined steeply at higher temperatures (37 to about 18 °C) and then declined very gradually with cooling down to 0 °C, much as one might expect of diffusion through a pore. There were no apparent differences between cold tolerant and cold sensitive forms; in particular there was no difference between red blood cells of ground squirrel and guinea pig.

The last observation was surprising because it was in these two species that a marked difference in temperature sensitivity of passive (ouabain-insensitive) K influx had previously been described (Kimzey & Willis, 1971). However, in those earlier studies, unlike the recent ones, Ca ion had been included in the medium. When the ouabain-and-bumetanide K influx was compared in ground squirrel and guinea pig cells in medium containing Ca ion, there was no effect in ground squirrel cells, but in guinea pig cells there was a rise in influx at lower temperatures, very like the 'paradoxical' effect (Hall & Willis, 1984). Ca activation of K permeability is a well known phenomenon, originally discovered by Gardos in red blood cells. One of its features is that it can be blocked by quinine; in guinea pig cells the Ca-induced 'paradoxical' effect was blocked by quinine.

The question then was whether the difference between the species might be due to absence of the Gardos channel in ground squirrel or to its insensitivity to Ca at low temperature. Subsequent experiments showed that when the cell's Ca regulation was by-passed by using a Ca ionophore to make the membrane highly permeable to Ca, or by depleting them of ATP to stop the ATP-dependent Ca pump, ground squirrel cells also exhibited a Ca-activated increase in K permeability. There was no large difference in sensitivity or response of the system to Ca in the two species as a function of temperature (Hall & Willis, 1984).

The inference from these experiments was that the mechanisms of Ca regulation were failing in the guinea pig cells, making them vulnerable to increased K leakiness due to the activation of this special channel. Ground squirrel red blood cells presumably retain their Ca gradient at low temperature and are able thereby to preserve a low permeability to K ion.

Na-K-Cl cotransport

The cotransport system found in red blood cells is probably a widespread mechanism. Like other carrier-mediated passive permeability systems it should not be regarded as a mere 'leak', but rather it probably has the effect of mediating sudden changes in pump activity or in volume of the cell (Brugnara, Canessa, Cusi & Tosteson, 1986; Canessa, Brugnara, Cusi & Tosteson, 1986; Kregenow, 1981). One role may be to allow for regulatory volume increase following sudden osmotic shrinkage. Conceivably, it might also function to compensate for sudden rise in cytoplasmic Na at the expense of dumping some K. Thus, it should be regarded as a back-up to the normal, basal Na-K pump operation.

In this light it is of interest that in the comparative studies described above between ground squirrel and guinea pig red blood cells, cotransport was still detectable at 5°C in ground squirrel cells but had completely disappeared in guinea pig cells. A similar observation has been made in cultured kidney cells of the same two species (Becker, 1986). This pathway is not always present in red blood cells. Among those species whose red cells possess it, there is not a very good correlation between its temperature sensitivity (expressed as activity at 5°C relative to that at 37°C) and the ability of the species to hibernate (Hall & Willis, 1987).

Passive Na permeability

The routes for movement of Na ion through the plasma membrane of the red blood cell partially overlap and are partially distinct from those for K (Fig. 2). In addition to cotransport and the residual permeation pathway there is also a Na-H exchange mechanism which is inhibited by the drug amiloride and stimulated by cytoplasmic Ca (Escobales & Canessa, 1985a,b). Na-Ca exchange, which is a feature of many kinds of cells, including red blood cells of carnivores (Parker, 1977), has not been reported in rodent and human red blood cells.

The disparity in decline of Na influx in ground squirrel red blood cells relative to guinea pig cells (i.e. steeper in ground squirrel cells) is especially large in the range from 37°C to 20°C (Fig. 3). In fact, there is virtually no decrease in Na influx over this temperature range in guinea pig cells at all. When influx is measured in the presence of amiloride, however, the disparity is absent, and the decline in guinea pig cells is perhaps even slightly steeper than in ground squirrel cells (Fig. 3). Thus, a large rise in amiloride-sensitive influx in guinea pig cells offsets the usual inhibitory effect of cooling. In ground squirrel cells the amiloride-sensitive component is constant with temperature between 37°C and 20°C (Zhou & Willis, unpublished results). In guinea pig cells incubated at pH 8 cooling no longer causes activation of the amiloride-sensitive component.

When cytoplasmic pH is adjusted independently and pH in the medium held constant at 7·4, amiloride-sensitive influx into guinea pig red blood cells is activated by lower cytoplasmic pH at 37°C and becomes more sensitive to H ion at

J. S. WILLIS

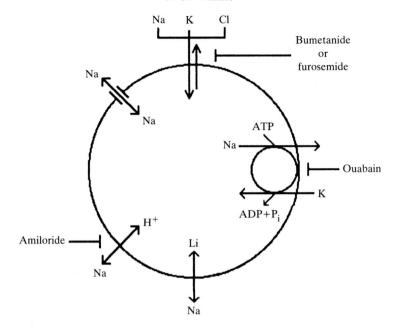

Fig. 2. Pathways of Na permeation in red blood cells, based on findings in human and rodent red blood cells. Symbolism same as in Fig. 1. In addition to the pathways shown, Na may also enter the cell by the anion transport pathway complexed as carbonate ion.

20°C, whereas in ground squirrel cells amiloride-sensitive influx is *inhibited* by decreased pH at 37°C and is insensitive to pH at 20°C. Thus, the low overall sensitivity of Na influx to cooling in guinea pig cells is attributable to a *reverse* temperature response in this one component, the Na-H exchange mechanism, owing (for some unknown reason) to increased sensitivity to H ion with cooling. In ground squirrel red cells the mechanism appears to be either lacking or suppressed.

Thus, the difference in passive permeability between red blood cells of ground squirrel and guinea pig which leads to the latter being leakier at low temperature can be assigned to specific mechanisms for Na and K, different for the two ions and neither of them involving the basic pore-like residual permeability.

Regulation of Ca ion

According to the paradigm outlined above, a first step is to establish whether there is loss of maintenance of an ionic gradient at low temperature before attempting to analyse the causes of failure. This has not been done for Ca ion, even though elegant methods now exist based on absorption or fluorescence of Ca binding agents for estimating the concentration of Ca in living cells (Tsien, 1983). In red blood cells, where so much of the information about membrane transport has been obtained, such fluorescence or absorption measures would be complicated by the intrinsic fluorescence or absorption of haemoglobin.

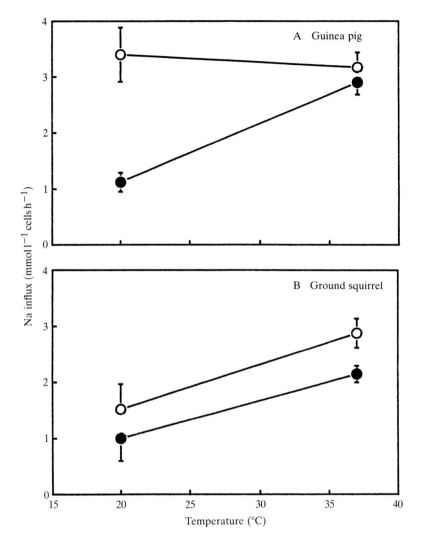

Fig. 3. Effects of temperature and amiloride on Na influx in red blood cells of (A) guinea pig and (B) ground squirrel. Cells were incubated in medium containing (mM): 150 NaCl, 7·5 KCl, 10 glucose and 10 MOPS buffer adjusted to pH 7·4 at each temperature. Open circles, control; closed circles, 1 mM amiloride present in the medium.

Nevertheless, the activation of the Gardos (Ca-activated) K channel at 5°C in red blood cells of guinea pigs and other species (Hall & Willis, 1984; see above) has been taken as *prima facie* evidence of increased concentration of Ca ion. In guinea pig red blood cells uptake of ^{45}Ca at 5°C is rapid and is not increased in ATP-depleted cells, whereas in red blood cells of hedgehog (a hibernator), there is very little uptake at 5°C unless cells are ATP depleted first (Ellory & Hall, 1983). Since the pool of Ca in the cell is so small, such uptake probably denotes a change in concentration of total Ca, and presumably an increase in ionic Ca as well.

Hall, Wolowyk, Wang, Ellory & Ma (1986) have recently found that the ATP-dependent efflux of Ca from guinea pig red blood cells was slow compared with that from red blood cells of Richardson's ground squirrel, implying a more rapid pump at the low temperature in the hibernator.

Summary

To summarize the findings regarding ion regulation in cold-tolerant cells, we have the most complete picture in red blood cells, particularly for the comparison of the cold-tolerant ground squirrel with the cold-sensitive guinea pig cells. In this case the maintenance of the K gradient in the cold-tolerant cells at low temperature is accounted for both by a faster pump and slower leaks. The explanation of the less reduced passive permeability in the cold-sensitive guinea pig cells is different for Na and K. K leak is elevated over what it would normally be because of loss of Ca extrusion and opening of the Ca-activated K channel. Na permeability remains high at reduced temperature because of stimulation of the Na/H exchange, partly because of increased sensitivity to H ion with cooling. Loss of Ca regulation in guinea pig cells and its retention in ground squirrel cells probably hinges on the difference in Ca pump activity.

To what extent is this picture based on red blood cells typical of cells in general? Na/K gradients are mainly preserved *in vivo* during hibernation and are commonly decreased during forced hypothermia in cold-sensitive forms. *In vitro*, Na/K gradients are usually better maintained at low temperature in cells from hibernators, but not always. Both Na-K pump and passive permeability at least to K have usually been found to account for this difference, but competent pumps can be found in cells with failing ion regulation. Little is known about Na permeability at low temperature for most cells. Studies of regulation of Ca ion have hardly begun.

This review has only dealt with the general issue of membrane handling of ions to the neglect of diverse and diverging topics concerned with specialized and cell-specific roles of membrane function such as nerve conduction and epithelial transport. Of these specific topics, probably that of cardiac excitability has been the most actively studied (Duker, Sjoquist, Svenson, Wohlfahrt & Johansson, 1986) and is perhaps the most relevant. Others, such as maintenance of transepithelial transport (which has never adequately been demonstrated at low temperature in a hibernator system), have been almost totally ignored.

Relationship of cold tolerance to other adaptations

Cold is not the only circumstance which may enforce minimal metabolism on the cell and place ion balance under constraint: two other likely possibilities are hypoxia and starvation. There are two reasons why consideration of adaptation to these other states is relevant to a discussion of cold tolerance: the first is that the cellular mechanisms may be similar among the three states and easier to analyse in the other two. The second reason is that amongst different species, cells may face a combination of these challenges. Thus, peripheral tissues of diving mammals may face both hypoxia and hypothermia and must be adapted to both, whereas core tissues of deep hibernators might not have to face hypoxia but frequently must cope with long term starvation. The possibilities of such diversity may account for some of the variation in response observed amongst different tissues and species.

Hypoxia

Hochachka (1986) has noted the involvement of reduced passive permeability in adaptation to hypothermia and has speculated upon its potential significance for conserving ATP utilized by ionic pumps for tissues exposed to hypoxia. On this basis, he proposes a unified hypothesis for explaining the two kinds of adaptation. Insofar as this hypothesis calls for a specific suppression of pump activity in the face of reduced ATP availability, this hypothesis would seem to be in conflict with both the facts and the uncertainties outlined above (i.e. pumping is well maintained in most cold-tolerant and some cold-sensitive cells; status of ATP availability in the cold is still unclear). There appears to be at the present only indirect evidence about the maintenance of K gradients in hypoxic resistant tissues (and even that only in whole, ectothermic organisms) and no evidence whatever about differences in membrane transport. Clearly, then, this is an area ripe for exploration. There is also a large literature on studies of ischaemic damage which represents an approach similar to that of hypothermic organ storage, but the possible parallels between the two have seldom been examined. Such findings, for example, as those showing loss of epithelial cell polarity of membrane transport as a cause of ischaemic renal damage (Molitoris, Wilson, Schrier & Simon, 1985) beg to be tested in hypothermia.

Starvation

The possible savings in energy of reduced Na/K pumping, with or without reduced ion leakage, during energy starvation has not been overlooked. In a sense it might be regarded as an extension of a briefly popular, now controversial, hypothesis that a cause of obesity is reduced rate of ion turnover (Deluise, Blackburn & Flier, 1980; Bray, Kral, Bjorntorp, 1981; Beutler, Kuhl & Sacks, 1983). In human starvation, cell Na concentrations are often elevated (Metcoff, 1966). In human infants recovering from 'marasmic' malnutrition (i.e. not complicated by oedema, a condition which may imply complex response of transport mechanisms), the rate of ouabain-sensitive K influx and the amount of ouabain binding rises in red blood cells as does ouabain-sensitive Na efflux in white blood cells (Patrick & Golden, 1977; Willis & Golden, 1985). Residual (i.e. ouabain-and-bumetanide-insensitive) K influx and Na influx are both increased in red blood cells of malnourished children compared with control, nutritionally normal children. In severely fasted rats there is a reduction in residual K leak, in

Na-stimulated ouabain-sensitive K influx (Willis, Layman, Glore, Bechtel & Sobolewski, 1982; Zhao & Willis, unpublished results) and in Na-K ATPase activity (Pimplikar & Kaplay, 1981) compared with controls. Recent studies in my laboratory indicate that in red blood cells of individual prairie dogs tested before and after a period of enforced starvation the Na-K pump rate falls during starvation as do both the Na influx and the residual K influx.

Thus, in human cells capacity for active transport falls, but leaks rise as does intracellular Na, leading to undiminished transport. So, where is the energy saving? In rodents, on the other hand (or at least in individual prairie dogs), leaks fall, cytoplasmic concentrations remain the same and pump rates decline. There are two ways one might view these (as yet skimpy) results. The human infants may not be adapted, at least with respect to membrane function, and the situation observed may represent one of the threats of malnutrition to cell function (e.g. increased membrane leakiness due to compromised antioxidant competence, Golden, 1985). Prairie dogs, which can endure six weeks without food or water without observable distress (Pfeiffer, Reinking & Hamilton, 1979), may be supposed to be able to adapt rapidly to periods of starvation and the membrane function of their cells to represent a stable state.

Alternatively, both cases may represent different sorts of adaptation. In human infants the first line of defence to energy deprivation is growth retardation. Diminished gradients may be a part of the inhibition of amino acid accumulation and protein synthesis. Then, too, maintenance of low passive permeability to ions may be a more energy demanding activity than has been realized, requiring reducing equivalents for maintenance of low reduced glutathione (GSH) and ATP for Ca ion regulation (Hynes & Willis, 1986), which as we have seen is connected with both passive K and Na permeability. ATP may even be a direct requirement for low ion permeability in some cases (Noma, 1983).

Cold tolerance and the organizational hierarchy

After the elements responsible for differences in temperature sensitivity have been identified (e.g. pumps, specific leak paths, specific metabolic components), the next task is to understand the mechanisms that account for the differences in the behaviour of those elements. This is not simply a matter of analysing molecular mechanisms, though that is an important step, because cold tolerance may well involve relationships at various levels of organization. Consequently, reliance solely upon isolated molecular (or even cellular) components may be misleading.

Twenty years ago a discussion of this subject of organizational levels focused upon the then newly described Na-K ATPase as a possible case-in-point (Willis, 1967). This has proven fortuitous, for in the interim it has not only been shown that rupture of cell membrane often greatly increases its temperature sensitivity (Willis et al. 1978), but also its kinetics of ATP activation (Soltoff & Mandel, 1984), its species- and temperature-related interaction with ouabain (Willis & Ellory, 1983), and its inhibition by Ca ion (Yingst, 1982). Such cautionary tales are the common lot of reductionist biologists but, as pointed out in that earlier review, the change in temperature behaviour with progressive isolation might also be used analytically to pinpoint the locus of adaptation.

Since the job of identification of responsible elements is far from complete, that of identifying mechanisms is in most cases even less well developed, except for the issue of lipid involvement, where the mechanistic analysis got perilously ahead of the descriptive one. Consequently, most of what follows is mere 'futurology' based on currently recognized possibilities.

Molecule/membrane

At the molecular level and in a membrane context there appears a clear dichotomy of choice for localizing temperature resistance or sensitivity: lipids and proteins. Sadly, most of the enthusiasm has been centred on the first. The reasons for this were seductive, and – up to a point – persuasive. For Na-K ATPase (a heavily worked molecule), the developments in the past 20 years have shown that this important integral enzyme depends upon a lipid matrix, that its temperature sensitivity can be governed by lipids, and attempts have been made to connect specific lipids with specific steps of its reaction sequence (Ellory & Willis, 1981; Goldman & Albers, 1975). More generally, the homeoviscous hypothesis held sway in the field of ectothermic and procaryote adaptation (Cossins, 1981; 1987).

Attempts to extend this concept to mammals and to the cold adaptation of hibernators have generally dealt with mitochondrial inner membranes and with seasonal comparisons within a single species. Mitochondrial membranes lack cholesterol and, therefore, may be expected to display simpler responses to temperature than, say, plasma membranes. The problem with seasonal comparisons is, as argued above, that differences in temperature sensitivity at the level of functioning, intact cells have usually not been demonstrated and may not exist (Willis *et al.* 1981).

The subject of the involvement of lipids and membrane fluidity in cold tolerance of hibernators has been reviewed exhaustively and nothing can be added here to the conclusion of the most recent and extensive of these reviews that 'the process of "homeoviscous adaptation", which occurs in procaryotes and some eucaryotes acclimating to low temperature, does not seem to be operative in the membrane of hibernating mammals' (Aloia & Raison, 1987). This is not to say, of course, that the lipids are not important factors in determining the thermal behaviour of the membrane proteins. Indeed, the change in temperature sensitivity of Na-K ATPase that occurs with membrane rupture may well be due to the loss of asymmetry of specific phospholipid distribution (Willis *et al.* 1978; Tanaka & Ohnishi, 1976). Nevertheless, the *differences* in cold sensitivity of plasma membrane function observed between species or with season do not seem to be due to different lipids, at least not on the scale of large, bulk phase effects.

If lipids are not the answer, then what is it about the proteins that is different? Again, there would appear to be two choices: quality and numbers. Here the problem is that to analyse the system you must 'first, catch your protein'. Among the various effector proteins associated with the permeation pathways discussed above, only two have been 'caught' to any significant degree in the sense of having been isolated, purified, numbers counted in membranes, etc., namely Na-K ATPase and (plasma membrane) Ca-ATPase, and of these we know very little about the involvement of the Ca-ATPase generally in cold adaptation.

Perhaps the simplest possibility is regulation of numbers of active sites within the membrane. This has been shown to be a possibility for cultured cells depleted of cytoplasmic K, primarily by means of decrease in turnover in (i.e. reduced rate of removal from) the membrane (Cook, Karin, Fishman, Tate, Pollack & Hayden, 1985) and is also thought to occur in tissue cells within organisms (Chan & Sanslone, 1969). No large increase in ouabain-binding (the method for estimating numbers of Na-K pump sites) with cooling was found in ground squirrel red blood cells (nor any decrease in guinea pig cells, Ellory & Willis, 1982). On the other hand, red blood cells are not known to regulate pump sites numbers over short time spans and probably lack the submembrane vesicular pool which would serve as a reservoir of additional pumps.

Changes in Na-K site numbers reported have been rather slow (hours, see also Pumplin & Fambrough, 1982, on acetylcholine receptors), but regulation of channel density can be quite rapid as found by Abramcheck, van Driesche & Helman (1985) in frog skin, using noise analysis of short circuit current to estimate numbers of amiloride-sensitive Na channels. Amiloride inhibition of the channels caused as much as a four fold increase in density within only a few minutes. (For other examples, see Hochachka, 1986.) The finding of such an apparent 'autoregulation' may be of particular significance in terms of trying to understand resistance to inhibition by cold.

Alteration of the quality of the molecules, as implied by different kinetics (Fig. 3), could be achieved in various ways, by different primary structure of the protein, by post-translational modification (e.g. phosphorylation) or by impinging modifiers within the membrane. Examples of such controls are sketchy for integral membrane proteins, but Ca-ATPase is known to be activated by phosphorylation mediated by calmodulin (Schatzman, 1983) and Na-K ATPase can be inhibited by phosphorylation (Schulz & Cantley, 1985).

Well documented examples of altered kinetics of Na-K pump are rare, but the case of low K sheep and goat red blood cells is well studied and the high Na cells of dog (puppy) and bear red blood cells may have some points of resemblance (Ellory, 1977; Miles & Lee, 1972). In the case of the LK ruminant cells the pump kinetics for intracellular K and Na are altered by an intrinsic, presumably adjacent, protein antigen (Sachs, Ellory, Kropp, Dunham & Hoffman, 1974). (Incidentally, this intrinsic suppression of activity also decreases temperature sensitivity, Joiner & Lauf, 1979.) Another example of an altered protein of Na-K ATPase is the ouabain-resistant form which appears in certain rodent cells and appears to be due to a change in structure of the alpha subunit (judging from altered electrophoretic properties) but might also involve a second intrinsic protein factor (Sweadner, 1983; Lelievre, Zachowski, Charlemagne, Laget & Paraf, 1979; Lelievre, Piascik, Potter, Wallick & Schwartz, 1983). It seems unlikely that differences in cold sensitivity would be based upon expression of different genes in view of the widespread occurrence of cold resistance of some tissues even within cold sensitive organisms, but this possibility cannot be ruled out.

Cell/membrane

Cells as a whole regulate their volume and ionic composition, somehow integrating the diverse activities of a large number of permeability pathways. This is evident in the growing number of studies of volume regulation in cultured cells and vertebrate red blood cells. But it is also apparent from the consideration of the extreme example that dog red blood cells progress from ordinary, high-K, low-Na cells with active Na-K pumps to high-Na, low-K cells lacking useful Na-K pumps and kept in osmotic balance by a different array of mechanisms. What are the interactions at the cellular level among all these factors that permit such integration? The answer to this question would seem to be of paramount importance in view of the above demonstration of a diversity of cases of resistance variably relying upon pumps and leaks at low temperature and in starvation.

At the simplest level, kinetic effects of the ions involved would allow some regulation, as mentioned above for Na-K transport (i.e. increased Na activates pumping because in many cases steady-state Na concentration is far below half-saturation of the system for Na). Another simple, though less obvious, factor is the role of fixed charge within the cells as a determinant of cell volume and, therefore, a possible means of modulating volume by means of masking or unmasking with H or Ca ion (Jakobsson, 1980). Such effects, however, would not explain numerous examples of 'cross-talk' between apical and basal membranes of transporting epithelia involving transcellular effects on passive permeability (Diamond, 1982; Kristensen, 1980). Such effects are comparatively easy to identify in epithelial cells, but could be more generally shared in other cell types. To explain these we presumably have to turn to the growing and complex catalogue of regulatory and second messenger networks involving Ca ion, H ion, calmodulin, and inositol triphosphate. Disruption of these factors has been implicated in damage due to ischaemia (Mathys, Patel, Kreisberg, Stewart & Venkatachalam, 1984), but it remains to be seen if they can be shown to be functional in adaptation to hypoxia and hypothermia.

Cell/organism

It is obvious that the survival of hypothermia depends upon integrated functioning of many tissues, so that cold resistance in only some of the tissues would not guarantee survival. It is not so clear that organismic integrity is necessary for cold adaptation in tolerant forms.

Exceptionally, tissues from hibernators show little or no cold tolerance upon removal from the organism, even though they clearly function at low temperature during natural hibernation or forced hypothermia. For example, brain slices from hamsters do not reaccumulate K well at low temperature no retain it better than those from rat (Goldman & Willis, 1973a), yet intact brain tissue of both hypothermic and hibernating hamsters retains K quite well, much better than that of hypothermic rats (Willis, Goldman & Fang, 1971; Musacchia, 1984). Similarly, hamster liver slices are totally unable to regulate K at low temperature (i.e. either to reaccumulate it or to retain it, Becker et al. 1982), yet in the intact hibernating hamster liver retains K well at 5°C (Willis et al. 1971). In neither case is the apparent loss of cold tolerance due simply to tissue disruption, because in both cases the slices perform adequately at 37°C. To complicate matters further, when hepatocytes of hamster are made and cultured for 18h they exhibit capacity to retain K at low temperature superior to that of guinea pig hepatocytes (Becker et al. 1982).

Such results suggest that organismic factors help to determine, or at least govern the expression of, cellular cold resistance. Hormonal effects, either positively or negatively, are the most obvious possibility. Many hormones and other humoral agents are implicated with control of membrane permeability. For Na-K active transport alone, a short list would include aldosterone, thyroid hormones, insulin and postulated endogenous inhibitor. The liver/hepatocyte results described above might only reflect an artefact of catecholamine release during the killing of the animal (thereby opening catecholamine receptor driven permeability pathways and making the leak too large for the pump to cope with at low temperature), the effect of longer term incubation of hepatocytes being simply to wash out the catecholamines.

The interaction between self-determination of an adaptive state and the control from organismic centres is an intriguing one. In the case of capacity adaptation in ectotherms it seems likely that both are involved. Cultured hepatocytes of green sunfish acclimated to different temperatures exhibit some aspects of cold acclimation seen in liver of intact fish, but not all (Koban, 1986). Mammalian cells in culture can respond to changes in availability of required solutes in their environment by regulating membrane carriers for those solutes (Na-K ATPase in low K, as mentioned above, sugar and amino acid carriers, Kalckar, 1977), yet regulation of these carriers is under hormonal control in the organism. Perhaps in the cell, dwelling within the organism, the regulation of hormone receptors related to the synthesis and deployment of carriers, is substituted for the direct control of the carriers themselves. That way both the cell and the organism have a 'vote' in the status of the cell. Whether such a mechanism plays a role in cold adaptation remains to be seen. Certainly, there are already several known cases of fasting influencing the receptor levels for metabolically relevant hormones (Bonen, Clune & Tan, 1986; van der Hayden, Docter, van Toor, Wilson, Hennemann & Krenning, 1986).

Unpublished research described in this paper done by or in collaboration with the author was supported by NIH grant GM 11494.

References

ABRAMCHECK, F. J., VAN DRIESCHE, W. & HELMAN, S. I. (1985). Autoregulation of apical membrane Na permeability of tight epithelia. *J. gen. Physiol.* **85**, 555–582.

ALOIA, R. C. & RAISON, J. K. (1987). Membrane function in mammalian hibernation. *Biochim. biophys. Acta* (in press).

BECKER, J. H. (1987). Relative cold resistance of potassium cotransport and pump systems in ground squirrel and guinea pig kidney cultures. *J. therm. Biol.* (in press).

BECKER, J. H., WILLIS, J. S. & SOBOLEWSKI, J. (1982). Regulation of liver K at low temperature. *Fed. Proc.* **41**, 948.

BELKE, D. D., PEHOWICH, D. J. & WANG, L. C. H. (1987). Seasonal variation in calcium uptake by cardiac sarcoplasmic reticulum in a hibernator, the Richardson's ground squirrel. *J. therm. Biol.* (in press).

BEUTLER, E., KUHL, W. & SACKS, P. (1983). Sodium-potassium-ATPase activity is influenced by ethnic origin and not by obesity. *New Eng. J. Med.* **309**, 756–760.

BONEN, A., CLUNE, P. A. & TAN, M. H. (1986). Chronic exercise increases insulin binding in muscles but not liver. *Am. J. Physiol.* **251**, E196–E203.

BOWLER, K. & DUNCAN, C. (1969). The temperature characteristics of brain microsomal ATPases of the hedgehog: changes associated with hibernation. *Physiol. Zool.* **42**, 211–219.

BRAY, G. A., KRAL, J. G. & BJORNTORP, P. (1981). Hepatic sodium-potassium-dependent ATPase in obesity. *New Eng. J. Med.* **304**, 1580–1582.

BRUGNARA, C., CANESSA, M., CUSI, C. & TOSTESON, D. C. (1986). Furosemide-sensitive Na and K fluxes in human red cells. Net uphill Na extrusion and equilibrium properties. *J. gen. Physiol.* **87**, 91–112.

BULLARD, R. W., DAVID, G. W. & NICHOLS, C. T. (1960). The mechanisms of hypoxic tolerance in hibernating and non-hibernating mammals. In *Mammalian Hibernation* (ed. C. P. Lyman & A. R. Dawe), pp. 321–336. Harvard, Cambridge, Massachusetts: Museum of Comparative Zoology.

BURLINGTON, R. F., MEINENGER, G. A. & THURSTON, J. T. (1976). Effect of low temperatures on high energy phosphate compounds in isolated hearts from a hibernator and a non-hibernator. *Comp. Biochem. Physiol.* **55**B, 403–407.

BURTON, A. C. & ENDHOLM, O. G. (1955). *Man in a Cold Environment.* London: Edward Arnold. 273pp.

CANESSA, M., BRUGNARA, C., CUSI, D. & TOSTESON, D. C. (1986). Modes of operation and variable stoichiometry of the furosemide-sensitive Na and K fluxes in human red cells. *J. gen. Physiol.* **87**, 113–142.

CHAN, P. C. & SANSLONE, W. R. (1969). The influence of a low-potassium diet on rat-erythrocyte-membrane adenosine triphosphatase. *Arch. Biochem. Biophys.* **134**, 48–52.

CHARNOCK, J. S. (1978). Membrane lipid phase transitions: A possible biological response to hibernation? In *Strategies in Cold: Natural Torpidity and Thermogenesis* (ed. L. C. H. Wang & J. W. Hudson), pp. 417–460. New York: Academic Press.

CHARNOCK, J. S. & SIMONSON, L. P. (1978). Variations in (Na+K)-ATPase and Mg-ATPase of the Richardson ground squirrel renal cortex during hibernation. *Comp. Biochem. Physiol.* **60**B, 433–439.

CHATFIELD, P. O., LYMAN, C. P. & IRVING, L. (1953). Physiological adaptation to cold of peripheral nerve of the rat. *Am. J. Physiol.* **172**, 639–644.

COOK, J. S., KARIN, N. J., FISHMAN, J. B., TATE, E. H., POLLACK, L. R. & HAYDEN, T. L. (1985). Regulation of turnover of Na,K-ATPase in cultured cells. In *Regulation and Development of Membrane Transport Processes* (ed. J. S. Graves), pp. 3–20. New York: Wiley and Sons.

COSSINS, A. R. (1981). The adaptation of membrane dynamic structure to temperature. In *Effects of Low Temperatures on Biological Membranes* (ed. G. J. Morris & A. Clarke), pp. 83–106. London: Academic Press.

COSSINS, A. R. & RAYNARD, R. S. (1987). Adaptive responses of animal cell membranes to temperature. This volume.

DAWSON, M. J., GADIAN, D. G. & WILKIE, D. R. (1980). Mechanical relaxation rate and metabolism studied in fatiguing muscle by phosphorus nuclear magnetic resonance. *J. Physiol., Lond.* **299**, 465–484.

DAWSON, M. J. & WILKIE, D. R. (1984). Muscle and brain metabolism ^{31}P nuclear magnetic resonance. In *Recent Advances in Physiology*. Number 10 (ed. P. F. Baker), pp. 247–276. Edinburgh: Churchill Livingstone.

DE LUISE, M., BLACKBURN, G. L. & FLIER, J. S. (1980). Reduced activity of the red-cell sodium-potassium pump in human obesity. *New Eng. J. Med.* **303**, 1017–1022.

DIAMOND, J. R. (1982). Transcellular cross-talk between epithelial cell membranes. *Nature, Lond.* **300**, 683–685.

DUKER, G., SJOQUIST, P. O., SVENSSON, O., WOHLFAHRT, B. & JOHANSSON, B. W. (1986). Hypothermic effects on cardiac action potentials: difference between a hibernator, hedgehog, and a non-hibernator, guinea pig. In *Living in the Cold* (ed. H. C. Heller, X. J. Musacchia & L. C. H. Wang), pp. 565–572. New York: Elsevier.

ELLORY, J. C. (1977). The sodium pump in ruminant red cells. In *Membrane Transport in Red Cells* (ed. J. C. Ellory & V. L. Lew), pp. 363–382. London: Academic Press.

ELLORY, J. C. & HALL, A. C. (1983). Ca transport in hibernator and non-hibernator species' red cells at low temperature. *J. Physiol., Lond.* **345**, 148.

ELLORY, J. C. & WILLIS, J. S. (1981). Phasing out the sodium pump. In *Effects of Low Temperatures on Biological Membranes* (ed. G. J. Morris & A. Clarke), pp. 107–120. London: Academic Press.

ELLORY, J. C. & WILLIS, J. S. (1982). Kinetics of the sodium pump in red cells of different temperature sensitivity. *J. gen. Physiol.* **79**, 1115–1130.

ESCOBALES, N. & CANESSA, M. (1985). Ca-activated Na fluxes in human red cells, amiloride sensitivity. *J. biol. Chem.* **260**, 11 914–11 923.

ESCOBALES, N. & CANESSA, M. (1985). Amiloride sensitive Na transport in human red cells: Evidence for a Na:H exchange system. *J. Membrane Biol.* **90**, 21–28.

FISHER, K. C. & MANERY, J. F. (1967). Water and electrolyte metabolism in heterotherms. In *Mammalian Hibernation III* (ed. K. C. Fisher, A. R. Dawe, C. P. Lyman, F. Schonbaum & F. E. South), pp. 235–279. Edinburgh: Oliver and Boyd.

FRENCH, A. R. (1986). Patterns of thermoregulation during hibernation. In *Living in the Cold* (ed. H. C. Heller, X. J. Musacchia & L. C. H. Wang), pp. 393–402. New York: Elsevier.

FULLER, B. J., MARLEY, S. P. E. & GREEN, C. J. (1985). Gluconeogenesis in stored kidneys from normal and cold-acclimated rats. *Cryo-letters* **6**, 91–98.

GOLDEN, M. H. N. (1985). The consequences of protein deficiency in man and its relationship to the features of kwashiorkor. In *Nutritional Adaptation in Man* (ed. K. Blaxter & J. C. Waterlow), pp. 169–188. London: John Libbey.

GOLDMAN, S. S. & ALBERS, R. W. (1975). Cold resistance of the brain during hibernation: Temperature sensitivity of the partial reactions of the Na,K-ATPase. *Archs Biochem. Biophys.* **169**, 540–544.

GOLDMAN, S. S. & WILLIS, J. S. (1973). Cold resistance of the brain during hibernation. I. K transport in cerebral cortex slices. *Cryobiology* **10**, 212–217.

GREEN, C. J., MARLEY, S. P. E. & FULLER, B. J. (1985). Gluconeogenesis in stored kidneys from hibernating and non-hibernating ground squirrels. *Cryo-letters* **6**, 353–360.

HALL, A. C. & WILLIS, J. S. (1984). Differential effects of temperature on three components of passive permeability to potassium in rodent red cells. *J. Physiol., Lond.* **348**, 629–643.

HALL, A. C. & WILLIS, J. S. (1986). The temperature dependence of passive potassium permeability in mammalian erythrocytes. *Cryobiology* **23**, 395–405.

HALL, A. C. & WILLIS, J. S. (1987). Effects of temperature on (Na+K) cotransport in erythrocytes from various mammalian species. *Trans. Biochem. Soc. Lond.* (in press).

HALL, A. C., WOLOWYK, M. W., WANG, L. C. H., ELLORY, J. C. & MA, B. (1986). The effects of temperature on Ca transport in red cells from a hibernator (*Spermophilus richardsonii*). *J. therm. Biol.* (in press).

HOCHACHKA, P. W. (1986). Defense strategies against hypoxia and hypothermia. *Science* **231**, 234–241.

HUME, D. (1986). Role of calcium in pathogenesis of acute renal failure. *Am. J. Physiol.* **250**, F579–F589.

HYNES, T. R. & WILLIS, J. S. (1986). Metabolic regulation of low K permeability in cold-stored erythrocytes: role of calcium ion and reduced glutathione. *J. therm. Biol.* (in press).

JAKOBSSON, E. (1980). Interactions of cell volume, membrane potential, and membrane transport parameters. *Am. J. Physiol.* **238**, C196–C206.

JOINER, C. H. & LAUF, P. K. (1979). Temperature dependence of active K transport in cation dimorphic sheep erythrocytes. *Biochim. biophys. Acta* **552**, 540–545.

KALCKAR, H. M. (1977). Cellular regulation of transport and uptake of nutrients: an overview. *J. cell. Physiol.* **89**, 503–516.

KAMM, K. E., ZATZMAN, M. L., JONES, A. W. & SOUTH, F. E. (1979a). Maintenance of ionic gradients in the cold in aorta from rat and ground squirrel. *Am. J. Physiol.* **237**, C17–C22.

KAMM, K. E., ZATZMAN, M. L., JONES, A. W. & SOUTH, F. E. (1979b). Effects of temperature on ionic transport in aortas from rat and ground squirrel. *Am. J. Physiol.* **237**, C23–C30.

KAPLAN, J. H. & KENNEY, L. J. (1985). Temperature effects on sodium pump phosphoenzyme distribution in human red blood cells. *J. gen. Physiol.* **85**, 123–136.

KIMZEY, S. L. & WILLIS, J. S. (1971). Temperature adaptation of active Na-K transport and of passive permeability in erythrocytes of ground squirrels. *J. gen. Physiol.* **58**, 634–649.

KOBAN, M. (1986). Can cultured teleost hepatocytes show temperature acclimation? *Am. J. Physiol.* **250**, R211–R220.

KONDO, N. & SHIBATA, S. (1984). Calcium source for excitation-contraction coupling in myocardium of nonhibernating and hibernating chipmunks. *Science* **225**, 641–643.

KREGENOW, F. M. (1981). Osmoregulatory salt transporting mechanisms: control of cell volume in anisotonic media. *A. Rev. Physiol.* **43**, 493–505.

KRISTENSEN, L. O. (1980). Energization of alanine transport in isolated rat hepatocytes. Electrogenic Na-alanine cotransport leading to increased K permeability. *J. biol. Chem.* **225**, 5236–5243.

KRUUV, J., GLOFCHESKI, D. J. & LEPOCK, J. R. (1985). Factors influencing survival of mammalian cells exposed to hypothermia. III. Effects on stationary phase cells. *Cryo-letters* **6**, 99–106.

LELIEVRE, L. G., PIASCIK, M. T., POTTER, J. D., WALLICK, E. T. & SCHWARTZ, A. (1983). Involvement of calmodulin in the inhibition of Na,K-ATPase by ouabain. *Curr. Topics in Membranes and Transp.* **19**, 1023–1027.

LELIEVRE, L., ZACHOWSKI, A., CHARLEMAGNE, D., LAGET, P. & PARAF, A. (1979). Inhibition of (Na+K)-ATPase by ouabain involvement of calcium and membrane proteins. *Biochim. biophys. Acta* **557**, 399–408.

MATHYS, E., PATEL, Y., KREISBERG, J., STEWART, J. H., VENKATACHALAM, M. (1984). Lipid alterations induced by renal ischemia: pathogenic factors in membrane damage. *Kidney Internat.* **26**, 153–161.

METCOFF, J. (1966). Cellular energy metabolism in protein-calorie malnutrition. In *Protein-Calorie Malnutrition* (ed. R. E. Olson), pp. 65–72. New York: Academic Press.

MILES, P. R. & LEE, P. (1972). Sodium and potassium content and membrane transport properties in red cells from newborn puppies. *J. cell. Physiol.* **79**, 367–376.

MILLER, L. K. & IRVING, L. (1963). Alteration of peripheral nerve function in the rat after prolonged outdoor cold exposure. *Am. J. Physiol.* **204**, 359–362.

MOLITORIS, B. A., WILSON, P. D., SCHRIER, R. W. & SIMON, F. R. (1985). Ischemia produces partial loss of surface membrane polarity and accumulation of putative calcium ionophores. *J. Clin. Invest.* **76**, 2097–2105.

MORRISON, P. R. (1960). Some interrelationships between weight and hibernation function. In *Mammalian Hibernation* (ed. C. P. Lyman & A. R. Dawe), pp. 75–92. Harvard, Cambridge, Massachusetts: Museum of Comparative Zoology.

MUSACCHIA, X. J. (1984). Comparative physiological and biochemical aspects of hypothermia as a model of hibernation. *Cryobiology* **21**, 583–592.

NELSON, R. J., KRUUV, J., KOCH, C. J. & FREY, H. E. (1971). Effect of sub-optimal temperatures on survival of mammalian cells. *Expl Cell Res.* **68**, 247–252.

NOMA, A. (1983). ATP-regulated K channels in cardiac muscle. *Nature, Lond.* **305**, 147–148.

PARKER, J. C. (1977). Solute and water transport in dog and cat red blood cells. In *Membrane Transport in Red Cells* (ed. J. C. Ellory & V. L. Lew), pp. 427–466. London: Academic Press.

PATRICK J. & GOLDEN, M. (1977). Leukocyte electrolyte and sodium transport in protein energy malnutrition. *Am. J. Clin. Nutr.* **30**, 1478–1481.

PFEIFFER, E. W., REINKING, L. N. & HAMILTON, J. D. (1979). Some effects of food and water deprivation on metabolism in black-tailed prairie dogs, *Cynomys ludovicianus*. *Comp. Biochem. Physiol.* **63**A, 19–22.

PIMPLICKAR, S. W. & KAPLAY, S. S. (1981). Kidney, liver and erythrocyte membrane Na,K-adenosine triphosphatase in protein-energy malnourished rats. *Biochem. Med.* **26**, 12–19.

PUMPLIN, D. W. & FAMBROUGH, D. M. (1982). Turnover of acetylcholine receptors in muscle. *A. Rev. Physiol.* **44**, 319–336.

SACHS, J. R., ELLORY, J. C., KROPP, D. L., DUNHAM, P. B. & HOFFMAN, J. F. (1974). Antibody-induced alterations in the kinetic characteristics of the Na:K pump in goat red blood cells. *J. gen. Physiol.* **63**, 389–414.

SCHATZMANN, H. J. (1983). The red cell calcium pump. *A. Rev. Physiol.* **45**, 303–312.

SCHULZ, J. T. III, CANTLEY, L. C. JR (1985). Characterization of an endogenous, membrane-bound kinase that phosphorylates the Na,K-ATPase in Friend erythroleukemia cells. In *The Sodium Pump* (ed. I. M. Glynn & J. C. Ellory). Cambridge: The Company of Biologists Limited.

SOLTOFF, S. P. & MANDEL, L. J. (1984). Active ion transport in the renal proximal tubule. III. The ATP dependence of the Na pump. *J. gen. Physiol.* **84**, 643–662.

SOUTHARD, J. H., RICE, M. J. & BELZER, F. O. (1985). Preservation of renal function by adenosine-stimulated ATP synthesis in hypothermically perfused dog kidneys. *Cryobiology* **22**, 237–242.

STEWART, G. W., ELLORY, J. C. & KLEIN, R. A. (1980). Increased human red cell cation permeability below 12°C. *Nature, Lond.* **286**, 218–231.

SWEADNER, K. J. (1983). Possible functional differences between the two Na,K-ATPases of the brain. *Curr. Topics in Membrane Transport* **19**, 765–780.

TANAKA, K. I. & OHNISHI, S. I. (1976). Heterogeneity in the fluidity of intact erythrocyte membrane and its homogenization upon hemolysis. *Biochim. biophys. Acta* **426**, 218–231.

TESSITORE, N., SAKHRANI, L. M. & MASSRY, S. G. (1986). Quantitative requirement for ATP for active transport in isolated renal cells. *Am. J. Physiol.* **251**, C32–C40.

TSIEN, R. Y. (1983). Intracellular measurements of ion activities. *A. Rev. Biophys. Bioeng.* **12**, 91–116.

VAN DER HAYDEN, J. T. M., DOCTER, R., VAN TOOR, H., WILSON, J. H. P., HENNEMANN, G. & KRENNING, E. P. (1986). Effects of caloric deprivation on thyroid hormone tissue uptake and generation of low-T3 syndrome. *Am. J. Physiol.* **251**, E156–E163.

VAN ROSSUM, G. D. V. (1972). The metabolic coupling of ion transport. In *Hibernation and Hypothermia, Perspectives and Challenges* (ed.-in-chief F. E. South), pp. 191–218. New York: Elsevier.

WIETH, J. O. (1970). Paradoxical temperature dependence of sodium and potassium fluxes in human red cells. *J. Physiol.* **207**, 563–580.

WILLIS, J. S. (1967). Cold adaptation of activities of tissues of hibernating mammals. In *Mammalian Hibernation III* (ed. K. C. Fisher, A. R. Dawe, C. P. Lyman, E. Schonbaum & F. E. South), pp. 356–381. Edinburgh: Oliver & Boyd.

WILLIS, J. S. (1978). Cold tolerance of mammalian cells: prevalence and properties. In *Strategies in the Cold: Natural Torpidity and Thermogenesis* (ed. L. Wang & J. W. Hudson), pp. 317–415. New York: Academic Press.

WILLIS, J. S. (1979). Hibernation: cellular aspects. *A. Rev. Physiol.* **41**, 275–286.

WILLIS, J. S. (1982). Is there cold adaptation of metabolism in hibernators? In *Hibernation and Torpor in Mammals and Birds* (ed. C. P. Lyman, J. S. Willis, A. Malan & L. C. H. Wang), pp. 140–171. New York: Academic Press.

WILLIS, J. S. (1986). Membrane transport at low temperature in hibernators and non-hibernators. In *Living in the Cold* (ed. H. C. Heller, X. J. Musacchia & L. C. H. Wang), pp. 27–34. New York: Elsevier.

WILLIS, J. S. & ELLORY, J. C. (1983). Ouabain sensitivity: diversity and disparities. In *Current Topics in Membrane Transport*. 19. *Structure, Mechanism, and Function of the Na-K Pump* (ed. J. F. Hoffman & Bliss Forbush III), pp. 277–280. New York: Academic Press.

WILLIS, J. S. & ELLORY, J. C. (1985). Partial fluxes at low temperature. In *The Sodium Pump* (ed. I. M. Glynn & J. C. Ellory), pp. 559–562. Cambridge, England: The Company of Biologists Limited.

WILLIS, J. S. & ELLORY, J. C. (1985). Partial fluxes at low temperature: ouabain-sensitive Na:Na exchange. *Fed. Proc.* **44**, 841.

WILLIS, J. S., ELLORY, J. C. & BECKER, J. H. (1978). Na-K pump and Na-K ATPase: on the disparity of their temperature sensitivity. *Am. J. Physiol.* **235**, C159–C167.

WILLIS, J. S., ELLORY, J. C. & COSSINS, A. R. (1981). Membranes of mammalian hibernators at low temperatures. In *Effects of Low Temperatures on Biological Membranes* (ed. G. J. Morris & A. Clarke), pp. 121–144. London: Academic Press.

WILLIS, J. S., ELLORY, J. C. & WOLOWYK, M. W. (1980). Temperature sensitivity of the sodium pump in red cells from various hibernator and non-hibernator species. *J. comp. Physiol.* **138**, 43–47.

WILLIS, J. S., FANG, L. S. T. & FOSTER, R. F. (1972). The significance and analysis of membrane function in hibernation. In *Hibernation and Hypothermia: Perspectives and Challenges* (ed. F. E. South), pp. 123–147. New York: Elsevier.

WILLIS, J. S., FOSTER, R. F. & BEHRENDS, C. L. (1975). Cold-stored kidney tissue of hibernators: Effects of brief warming on K regulation. *Cryobiology* **12**, 255–265.

WILLIS, J. S. & GOLDEN, M. G. (1985). Active and passive membrane transport in intact erythrocytes of PCM children. *Proc. 13 Int. Congr. Nutr.* **13**, 47.

WILLIS, J. S., GOLDMAN, S. S. & FANG, L. S. T. (1971). Tissue K concentration in relation to the role of the kidney in hibernation and the cause of periodic arousal. *Comp. Biochem. Physiol.* **39**A, 437–445.

WILLIS, J. S., LAYMAN, D. K., GLORE, S., BECHTEL, P. & SOBOLEWSKI, J. (1982). Effect of starvation on K influx in erythrocytes of lean and obese Zucker rats. *J. Nutr.* **112**, xxvii.

WILLIS, J. S. & LI, N. M. (1969). Cold resistance of Na-K ATPase of renal cortex of the hamster, a hibernating mammal. *Am. J. Physiol.* **217**, 321–326.

WOLOWYK, M. W., LI, K. K. & WANG, L. C. H. (1986). Regulation of smooth muscle voltage sensitive Ca channels during hibernation in the ground squirrel (*Spermophilus richardsonii*). *J. therm. Biol.* (in press).

WOOD, L. & BEUTLER, E. (1967). Temperature dependence of sodium-potassium activated erythrocyte adenosine triphosphatase. *J. Lab. Clin. Med.* **70**, 287–294.

YINGST, D. R. (1982). The effect of cytoplasm Ca inhibition of the Na-K pump. *Fed. Proc.* **41**, 974.

ZEIDLER, R. B. & WILLIS, J. S. (1976). Cultured cells from renal cortex of hibernators and non-hibernators. Regulation of cell K at low temperature. *Biochim. biophys. Acta* **436**, 628–651.

ZHOU, Z. Q., DRYDEN, W. F. & WANG, L. C. H. (1986). Seasonal variation in myocardiac contractility in Richardson's ground squirrel (*Spermophilus richardsonii*). *J. therm. Biol.* (in press).

Printed in Great Britain © *Society for Experimental Biology 1987*   311

COLD SHOCK INJURY IN ANIMAL CELLS

P. F. WATSON

Department of Physiology, Royal Veterinary College, Royal College Street,
London NW1 0TU

and G. J. MORRIS

Cell Systems Ltd, Cambridge Science Park, Milton Road, Cambridge CB4 4FY

Summary

Cold shock injury (damage to cell structure and function arising from a sudden reduction in temperature) was for many years considered a phenomenon peculiar to certain cell-types. Only in recent years has it become apparent that widely different cell-types manifest cold shock injury. Thus, cold shock appears to be a more general phenomenon, differences between cell-types being quantitative (in the rate of cooling and temperature range at which injury is sustained) rather than qualitative.

Loss of particular cell functions depends on cell-type, but reflects the underlying structural and biochemical damage which has been inflicted by rapid cooling. In particular, membranes lose their selective permeability with the result that many cellular components are released including lipids, proteins and ions. Additionally, sodium and calcium gain access to the interior of the cell. Consequent upon this initial disruption, metabolic activities are diminished and further secondary changes ensue.

The possible mechanisms of cold shock injury include membrane thermotropism and protein denaturation. Susceptibility to cold shock is influenced by membrane composition, and much experimental evidence points to particular involvement of membrane lipids. One hypothesis implicates lipid phase changes in a cooling rate dependent loss of membrane integrity. Other recent hypotheses invoke biophysical concepts and cytoarchitectural features as considerations in a better understanding of cold shock.

Introduction

Cellular injury following rapid cooling has long been recognized in mammalian spermatozoa (reviewed by Watson, 1981*a*) where it was first called temperature shock (Milovanov, 1934). It is now apparent that a diverse range of cell types and tissues are sensitive to rapid cooling injury including microorganisms, eggs and embryos, blood cells and spermatozoa. (For a comprehensive literature survey, see Morris & Watson, 1984.) The phenomenon may indeed be a general one, different cell-types varying only in the temperature range and cooling rate at which injury is sustained. Even within cell-type, species differences in sensitivity are evident. For example, ungulate spermatozoa are more prone to injury on rapid cooling than rabbit, human or fowl spermatozoa (Wales & White, 1959). Most of the studies of cold shock in spermatozoa have used ram, bull or boar semen. Watson (1981a) proposed that a continuum of susceptibility exists between boar spermatozoa, the most sensitive, and fowl spermatozoa, apparently the most resistant yet studied.

Under certain conditions, cellular viability may be retained following rapid cooling but physiological processes are altered, e.g. the cold induced contraction of mammalian striated muscle (reviewed by Sakai & Kurihara, 1974), induction of polyploidy in eggs of fish (Purdom, 1972; Valenti, 1975; Lemoine & Smith, 1980), insects (Kawamura, 1979) and *Xenopus* (Kawahara, 1978). While this review will concentrate on cold shock in animal cells, the phenomenon is not confined to the animal kingdom. The extensive literature on cold shock in prokaryotes and plants has recently been reviewed elsewhere (Morris, 1986).

Injury following rapid cooling (direct chilling injury) has been defined as that which is expressed quickly after the temperature reduction, is dependent upon rate of cooling, and is largely independent of subsequent warming rate (Levitt, 1980; Morris, 1986).

It has been demonstrated in animal cells following two treatments: (i) Cold shock, i.e. rapid cooling of cells in isotonic media. (ii) Thermal shock or cold shock haemolysis. This phenomenon appears to be restricted to erythrocytes, which are

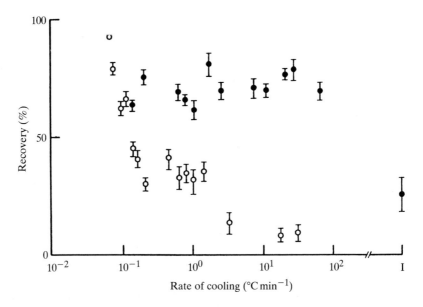

Fig. 1. Recovery (%) of *Amoeba* sp. strain Bor (○) and *A. proteus* (●) following cooling at different rates to −10°C. All cell suspensions were undercooled (i.e. in the absence of ice) and maintained at −10°C for 5 min before rewarming. I = Instantaneous cooling achieved by direct injection of 0·1 ml of a concentrated cell suspension into 10 ml of precooled medium. (Reproduced with permission from McLellan et al. 1984.)

sensitized to a reduction in temperature following exposure to hypertonic solutions.

Emphasis here will be placed on the phenomenon of cold shock and an analysis of the injury induced by this stress. Much of the experimental work on cold shock has been conducted on protozoa and mammalian spermatozoa, which accounts for the prominence of these two cell-types in this review. Thermal shock may be considered as a special case of cold shock injury, and a brief summary of the literature is included here; a more thorough treatment may be found in the recent review by Morris (1986).

Factors affecting cell viability after cold shock

The assessment of viability involves the measurement of particular cell functions and accordingly, appropriate measures vary with cell-type. Thus, spermatozoa have typically been assessed by the proportion able to exclude vital dyes as a measure of cell membrane function, and by their motility (Wales & White, 1959). The loss of selective permeability by the erythrocyte membrane, however, is demonstrated by haemolysis (see later). For cell cultures and free living protozoa the ability to divide and grow is a readily appreciable indicator of viability, and a further extreme illustration is the ability of mammalian embryos to implant and sustain embryogenesis and development to the birth of live young (Moore & Bilton, 1973).

Many variables determine the viability of cells following rapid cooling and rewarming:

Rate of cooling

Cell viability following a reduction in temperature is determined by the rate of cooling. Results such as those illustrated (Fig. 1) for *Amoeba* sp. demonstrate two important features of cold shock injury. Firstly cell injury is not due simply to exposure to a critical temperature, as all cells in the experiment described in Fig. 1 were exposed to the same final temperature; it is the rapidity of temperature change which is the critical factor. Secondly, the terms 'cold shock sensitive' and 'cold shock resistant' are not absolute and are valid only if the rate of cooling and the final temperature attained are defined. In this example at a rate of cooling of $0.1°C min^{-1}$ both strains of *Amoeba* demonstrate significant resistance, whilst following 'instantaneous cooling' both are very markedly sensitive. It is only within the range of intermediate cooling rates that a distinction between resistant and sensitive may be made.

Further, whilst cell injury following exposure to low temperatures is determined by the rate of cooling it is virtually independent of warming rate. Only when very slow rates of warming are employed is there a further reduction in viability

(McLellan *et al.* 1984). We can assume, therefore that the primary damage is induced during the cooling phase and may be compounded by significantly longer exposure to low temperature as occurs during very slow warming.

Similarly, the rate of cooling was very soon realized to be a relevant factor in spermatozoa; cooling bull and ram semen at rates no greater than about $10^{\circ}C\,h^{-1}$ permitted the recovery of high percentages of cells (Birillo & Pulhaljskii, 1936; Gladcinova, 1937). In fact, Quinn, Salamon & White (1968*b*) found that rapid cooling by dilution of ram semen directly into cold medium was more detrimental than dilution followed by rapid cooling of the diluted sample. It is likely that this result was a reflection of the greater cooling rates achieved by the former technique.

Length of incubation at reduced temperature

Following cooling, cell recovery is dependent on the period of incubation at the reduced temperature before rewarming. At any rate of cooling a greater loss of viability occurs with increasing time at the reduced temperature. With spermatozoa, functions are lost at different rates such that motility may be retained long after fertility has declined (see for example – Watson & Martin, 1976). This is probably attributable to the marked regional specialization of the spermatozoon but clearly demonstrates the differential effects of cold stress on different cellular functions.

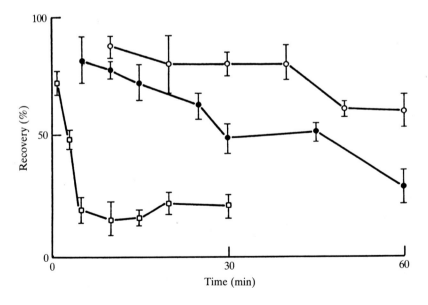

Fig. 2. Recovery (%) of *Amoeba* sp. strain Bor following exposure to -10° (□), -5° (●) or $0^{\circ}C$ (○) for different times before rewarming to $20^{\circ}C$. (Reproduced with permission from McLellan *et al.* 1984.)

Final temperature attained

Cell viability following cold shock is also dependent on the final temperature attained (Fig. 2). In these experiments with *Amoeba* the cellular reaction following rapid cooling to −10°C was typically that of cold shock, with maximum lethal injury being sustained by the population after only 5 min. In further experiments where cells were cooled as rapidly as possible to +10°C, loss of viability was only observed following an incubation period of days at this temperature. At intermediate temperatures (−5° and 0°C) there appears to be a continuum between these responses.

Again, spermatozoa reveal a similar phenomenon (Fig. 3). Clearly the final temperature attained is a significant determinant of survival. But in addition, the temperature interval of the cold shock is also a component; at lower temperatures the temperature interval to inflict a given degree of injury is less than that at higher temperatures (Quinn *et al.* 1968*b*).

It has been argued that all cell-types are sensitive to cold shock if they are cooled rapidly enough and to a sufficiently low temperature (Morris *et al.* 1983). This has important consequences as previously it has been considered that cells sensitive to cold shock were somehow atypical and that normal biological material was

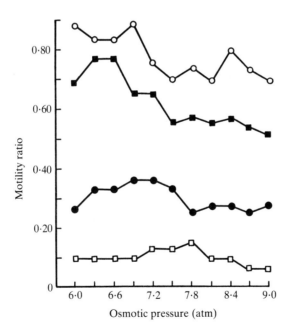

Fig. 3. Motility ratios of diluted ram spermatozoa subjected to a constant temperature interval cold shock of 22 °C in diluents of varying tonicity. Sperm motility was scored on a subjective scale of 0–5 using a microscope equipped with a warm stage at 37°C, and the motility ratio was calculated from the motility score after cold shock: motility score of control. The effect of tonicity was not significant. Temperature gradients: (○) 37°–15°C, (■) 32°–10°C, (●) 27°–5°C, (□) 22°–0°C. (Reproduced with permission from Quinn *et al.* 1968*b*.)

resistant. Indeed, much research into the basis of cold shock injury has been a study of the comparative biochemistry of cell-types considered to be resistant or sensitive. The widespread occurrence of cold shock injury also provokes the thought that the primary site of injury may be common to both prokaryotic and eukaryotic cells.

Temperature of growth

Maintenance of poikilothermic cells at reduced temperatures may modify the response to subsequent rapid cooling (Morris, 1986). However, this treatment does not induce resistance to cold shock, but alters the temperature at which such injury is induced. For example, when cells of *Tetrahymena pyriformis* grown at 20°C were cooled rapidly to +4°C there was a 50 % reduction in viability whereas for cultures maintained at 10°C the equivalent loss of viability occurred on cooling to −5·8°C (Fig. 4). Thus with both cultures a 50 % reduction in viability occurred at approximately 15°C below the temperature of growth and complete loss of viability (>99 %) was observed following rapid cooling to 20°C below the growth temperature.

No simple parallels can be drawn with vertebrate cells but it is perhaps relevant that cold shock injury has not been demonstrated in the spermatozoa of poikilotherms. Thus, for example, no reports exist of cold shock injury in fish spermatozoa and no particular precautions are necessary during cooling to 0°C for preservation, but ultra-rapid cooling rates have not been examined.

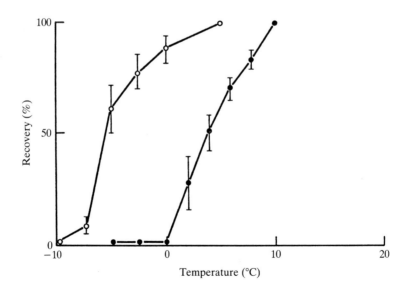

Fig. 4. Recovery (%) of *Tetrahymena pyriformis* following rapid cooling to different temperatures. Cells were from cultures either maintained at 20°C for 3 days (●), or following 3 days at 20°C transferred to 10°C for a further 3 days (○). (Reproduced with permission from Morris *et al.* 1984.)

Table 1. *Additives which reduce cold shock injury*

Cell type	Additive	References
Amoeba sp.	Ca^{2+}, glycerol, glucose	McLellan *et al.* 1984
Ram & bull spermatozoa	EDTA, Mg^{2+}	Quinn & White, 1968
Boar spermatozoa	Phosphatidyl serine	Butler & Roberts, 1975
Spermatozoa (general)	Egg yolk, lecithin, milk, casein, egg albumen, bovine serum albumin, testicular & sperm lipoproteins, butylated hydroxytoluene	Reviewed in Watson, 1981
Erythrocytes (Cold shock Haemolysis)	Lecithin	Lovelock, 1955
,,	Glycerol, dimethylsulphoxide	Morris, 1975
,,	Amphotericin B, valinomycin	Jung & Green, 1978
,,	Hexanol, ethanol, chlorpromazine	Dubbleman *et al.* 1979

Additives

A variety of additives has been demonstrated to reduce cold shock injury (Table 1). The efficacy of an additive is dependent on the rate of cooling. With *Amoeba* in the presence of glycerol, an optimal rate of cooling with respect to survival was observed (Fig. 5). The effect of glycerol in this case is to shift the response to cooling rates observed with control cells (Fig. 1) to faster values.

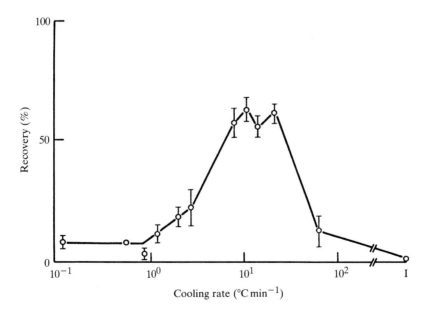

Fig. 5. Effect of pregrowth in glycerol ($0·07\,mol\,l^{-1}$ for 17 h) on the recovery (%) of *Amoeba* sp. strain Bor following cooling at different rates to $-10°C$ for 5 min before rewarming. I = 'Instantaneous' cooling (see Fig. 1). (Reproduced with permission from McLellan *et al.* 1984.)

The majority of cold shock protective agents, unlike glycerol, are non-permeating and thus probably exert their action at the plasma membrane. Egg yolk was demonstrated to have a protective effect for bull spermatozoa during cooling (Phillips & Lardy, 1940) and its role in reducing cold shock injury in bull and ram spermatozoa was soon investigated (Lasley, Easley & Bogart, 1942; Blackshaw, 1954). The active component has been traced to the low density lipoprotein fraction (Kampschmidt, Mayer & Herman, 1953; Pace & Graham, 1974; Watson, 1976; Foulkes, 1977) of which phosphatidylcholine, a phospholipid component of the lipoprotein micelles (Evans et al. 1973) is of major importance in cold shock protection (Quinn, Chow & White, 1980; Watson, 1981b). Its action appears to be via a membrane stabilizing role although it does not permanently modify the membrane composition (Kampschmidt et al. 1953; Quinn et al. 1980). With boar spermatozoa, even though they are highly susceptible to cold shock, the protection afforded by egg yolk is negligible (Benson, Pickett, Komarek & Lucas, 1967; Pursel, Johnson & Schulman, 1972), although phosphatidylserine (but not phosphatidylcholine or phosphatidylethanolamine) provided some protection (Butler & Roberts, 1975).

Other agents with a protective ability do not fall into a single class (see Table 1). It is possible that a number of them chelate divalent metal ions which may play a role in cold shock injury or may act to stabilize the plasma membrane proteins during thermal phase transition of the lipids (see Watson, 1981a).

Cell maturity

Immature spermatozoa taken from the upper or middle regions of the epididymis display far less susceptibility to cold shock injury (Lasley & Bogart, 1944; Lasley & Mayer, 1944; White & Wales, 1961; Quinn & White, 1967) and this has been linked to the changing lipid composition of the membranes as the spermatozoa mature (see Watson, 1981a). This response is probably limited to spermatozoa since they are considerably modified in their biochemistry and morphology during maturation.

Cell age

A further modification of boar spermatozoa has been observed to occur during incubation at 30°C or 37°C. They are profoundly sensitive to cold shock immediately after ejaculation; indeed, cooling below 15°C is greatly injurious irrespective of the rate of cooling. After several hours' incubation, however, a high proportion will survive cold shock and appear unaffected by the stress (Fig. 6). A similar but less pronounced change has been observed with ram spermatozoa (Quinn, Salamon & White, 1968a) and has been confirmed recently by calcium uptake studies (Robertson, 1986).

This acquired resistance of boar spermatozoa is influenced by pH, dilution and extender composition (Pursel, Johnson & Rampacek, 1972; Pursel, Johnson &

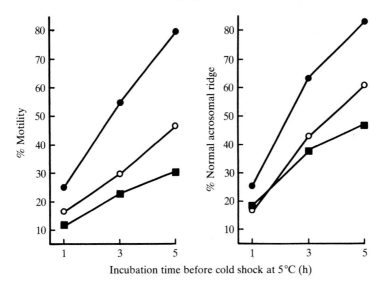

Fig. 6. The effect of incubation at 30°C on the susceptibility of boar spermatozoa to cold shock after dilution (semen:diluent) 1:2 (●), 1:6 (○), or 1:10 (■). (Drawn from data of Pursel *et al.* 1973.)

Schulman, 1972, 1973) and although there is no satisfactory explanation, it would appear likely that the plasma membrane is modified (see Watson, 1981*a*). There are no data to indicate whether this phenomenon is limited to the spermatozoa of certain species or is a more general one.

A further illustration of the effect of age or maturity is seen with embryos, and here again, species differences are apparent. Mouse embryos are relatively resistant to cooling even at early cleavage stages (Wilmut, 1972). Similarly, sheep embryos at 1–4 cell stage or at the morula stage tolerated cooling at $1°C min^{-1}$, but faster rates were not investigated (Moore & Bilton, 1973). In contrast, cow embryos before the late morula stage would not tolerate cooling at $10°C min^{-1}$ (Wilmut *et al.* 1975) but later stages did so (Trounson, Willadsen & Rowson, 1976*a*; Trounson, Willadsen, Rowson & Newcomb, 1976*b*). At the late morula stage a slight cooling rate dependency was evident (Trounson *et al.* 1976*a*), but blastocysts tolerated either fast ($>10°C min^{-1}$) or slow ($0·2°C min^{-1}$) cooling with equal success. No success has attended attempts to cool pig embryos below 15°C regardless of stage or cooling rate (Wilmut, 1972; Polge & Willadsen, 1978).

Cold shock haemolysis

Human erythrocytes suspended in plasma or isotonic salt solutions are not lysed by the stress of cold shock. However, pretreatments such as exposure to phospholipases (reviewed by Mollby, 1978), calmodulin antagonists (Takahashi, 1983) or hypertonic solutions (reviewed by Morris, 1986) sensitize the cells to a reduction in temperature. The following discussion will concentrate on the sensitization of erythrocytes to cold shock by exposure to hypertonic solutions, a phenomenon originally termed thermal shock (Lovelock, 1953). However, it is evidently a special case of cold shock injury and this relationship may be emphasized by redefining this stress as cold shock haemolysis.

Many variables determine the extent of haemolysis following hypertonic induced cold shock. Many of these factors are interrelated and all contribute to lysis, but they will be considered separately for ease of explanation:

Hypertonic solutions

(i) Concentration. There is a critical concentration necessary to sensitize erythrocytes to a reduction in temperature. No lysis is observed on cooling of cells suspended in solutions of sodium chloride $<1400\,\mathrm{mosmol\,kg^{-1}}$ but at osmolalities above this haemolysis occurs and is maximal following exposure to sodium chloride solutions of $1800\,\mathrm{mosmol\,kg^{-1}}$ (Fig. 7).

(ii) Composition. Cold shock haemolysis is induced by a variety of hypertonic solutions of non-permeating ionic and non-ionic compounds. Additives to which erythrocytes are freely permeable do not sensitize the cells to cold shock lysis. The extent of cold shock haemolysis induced following exposure to solutions of equal osmolality varies with the composition of the solution. The effects of different

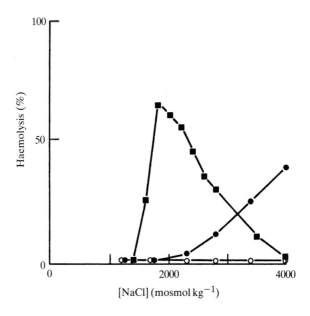

Fig. 7. Haemolysis (%) of human erythrocytes following exposure to hypertonic sodium chloride for 5 min at 37°C (○), upon resuspension into isotonic NaCl (●) and following rapid cooling to 0°C (■). (Redrawn from Lovelock, 1953; Takahashi & Williams, 1983.)

anions and cations in hypertonic solutions of approximately equal osmolality have been compared with that induced by sodium chloride (Morris, 1975).

(iii) Time and temperature of hypertonic exposure. The effects of length of exposure to hypertonic solutions before cooling are complex and depend both on the temperature of exposure and the composition of the hypertonic solution. For example, following exposure of erythrocytes at 25°C to solutions of sucrose the sensitivity to subsequent reduction in temperature increases continuously with increasing time of exposure. By contrast, cells exposed to hypertonic solutions of sodium chloride are initially at their most sensitive and increasing time of exposure reduces cold shock haemolysis.

The combined effects of temperature and period of exposure on the sensitization of erythrocytes to a reduction in temperature has been examined at a single concentration of sodium chloride (Takahashi & Williams, 1983). During exposure of erythrocytes to hypertonic solutions at 37° and 30°C, cells are maximally sensitive following short periods of exposure. By contrast, at 15°C cells are initially resistant to cold shock and, during increasing periods of exposure, sensitization to cold shock develops. No haemolysis upon cooling to 0°C was observed following exposure to hypertonic NaCl at 13°C for periods up to 90 min.

Rate of cooling

Following exposure to hypertonic solutions of sodium chloride a greater degree of haemolysis is observed following rapid cooling to 0°C than following slow cooling. The opposite is observed following short periods of exposure to hypertonic sucrose, i.e. maximum injury is induced following slow compared with rapid cooling (Fig. 8).

Final temperature attained

With erythrocytes exposed to hypertonic NaCl at 37°C no haemolysis is induced upon rapid cooling to temperatures >12·5°C, but cooling the cells from 37° to 10°C induces lysis, which is increased at lower temperatures. In addition, at any final temperature haemolysis is further increased as the period of isothermal incubation is extended (Takahashi & Williams, 1983).

Mechanism of injury of cold shock haemolysis

Unlike most other cell-types, erythrocytes in isotonic solutions are not sensitive to a reduction in temperature. For cold shock haemolysis to occur, cellular shrinkage is essential at all stages before and during cooling. Erythrocytes exposed to a hypertonic solution at 37°C and then resuspended into an isotonic solution at 37°C are not susceptible to a subsequent reduction in temperature. Any changes occurring in response to hypertonic solutions which sensitize the cells to the stress of rapid cooling are fully reversible. In addition, there is no hysteresis in the

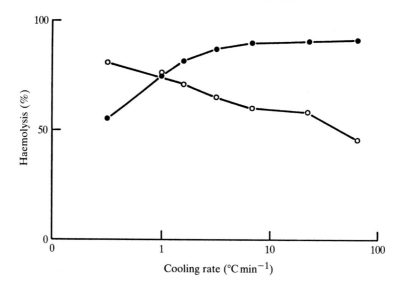

Fig. 8. Haemolysis (%) of human erythrocytes cooled from +25°C to 0°C at different cooling rates in $1 \cdot 2 \, mol \, l^{-1}$ NaCl (●) or 40% (w/v) sucrose (○). The erythrocytes were exposed to hypertonic conditions for 5 min at 25°C before cooling. (Redrawn from Morris & Farrant, 1973.)

system. If cells are exposed to a hypertonic solution at 0°C and then rewarmed to 37°C before rapid cooling to 0°C, haemolysis is reduced in comparison to addition of cells to hypertonic solutions at 37°C.

The extent of cold shock haemolysis is directly proportional to the reduction in cell volume before cooling rather than the osmolality of the solution (Takahashi & Williams, 1979). Additives which are freely permeable to erythrocytes, such as glycerol and dimethylsulphoxide, do not sensitize erythrocytes to the stress of rapid cooling. The storage of cells in hypertonic solutions can modify the sensitivity to cold shock; this is dependent on the composition of the hypertonic solution. For example, prolonged exposure to hypertonic sodium chloride would be expected to result in a loss of membrane selective permeability, after which cell volume would increase and the sensitivity to cold shock would decrease. In contrast, in hypertonic sucrose solutions, potassium would leak from the erythrocyte, the cell volume decrease and the sensitivity to rapid cooling would increase.

The processes by which a cell resistant to the stress of rapid cooling is converted, by cellular shrinkage, to a cell sensitive to cold shock are the key steps for an understanding of cold shock haemolysis. The membrane of the human erythrocyte has been extensively studied; the extrinsic protein spectrin in association with actin comprises a cytoskeletal network. This cytoskeleton is closely associated with the lipid bilayer and reduces the lateral mobility of lipids and proteins within the membrane. As a working hypothesis it has been suggested that cellular shrinkage either directly disrupts the cytoskeleton:membrane interactions or renders them susceptible to cooling (Green, Jung, Cuppoletti & Owens, 1981). With the removal of this physical restriction the membrane may be free to undergo thermotropic events (see below).

Morphology of cold shock injury

Light microscopy

With a specially modified light microscope, on which it is possible accurately to control the temperature and rate of change of temperature, the morphology of cold shock in *Amoeba* (McLellan *et al.* 1984) and ram spermatozoa (W. V. Holt, G. J. Morris & G. E. Coulson, in preparation) has been studied.

In *Amoeba* as the temperature is lowered at $10^{\circ}C\,min^{-1}$ cytoplasmic streaming is reduced and close to $10^{\circ}C$ ceases. There were no further observable morphological changes during undercooling to $-10^{\circ}C$. Upon rewarming, a contraction of the cytoplasm occurred together with gross deformation of the plasmalemma. Following cold shock, cellular shrinkage will still occur in response to hypertonic solutions, clearly demonstrating that the plasmalemma is osmotically intact. Any alterations which may occur in the physical state of the plasmalemma during cooling are apparently reversed upon rewarming.

In ram spermatozoa cooled rapidly ($10^{\circ}C\,min^{-1}$) motility decreased with decreasing temperature. At approximately $16^{\circ}C$ many spermatozoa displayed rigid bowing of the mid-piece which impaired flagellar wave generation and forward progression. Further cooling increased the proportion of such spermatozoa. A number (5–30 %) of these underwent a sudden displacement of the head through a complete arc, causing the flagellum to become deflected through an angle of 180° in the distal region of the mid-piece (Fig. 9); this was completed within $2\cdot5\,s$. Spermatozoa were observed to undergo this 'folding' process within the temperature range 16° to $8^{\circ}C$. On rewarming, these spermatozoa did not unfold and often regained motility in a reverse direction. A number of spermatozoa acquired localized swellings of the flagellum, especially in the vicinity of the end-piece and occasionally in the distal region of the mid-piece. Upon rewarming, these spermatozoa with swollen regions of the flagellum often recovered forward motility. Staining spermatozoa with Giemsa to demonstrate the acrosome (Watson, 1975) reveals that following cold shock a high proportion of spermatozoa have damaged acrosomal membranes and this can also be seen by phase contrast illumination.

In the presence of rhodamine 123, a probe considered to be specific for active mitochondria (Johnson, Walsh, Bockus & Chen, 1981), a bright fluorescence of the mid-piece was observed at all temperatures during cooling and rewarming. Although many spermatozoa failed to regain motility upon rewarming they all exhibited the same bright fluorescence.

In the presence of protective compounds no rigid bowing of the mid-piece occurred at $16^{\circ}C$ and forward progression was maintained into the temperature range 8° to $10^{\circ}C$. No flagellar folding occurred, although flagellar swellings were

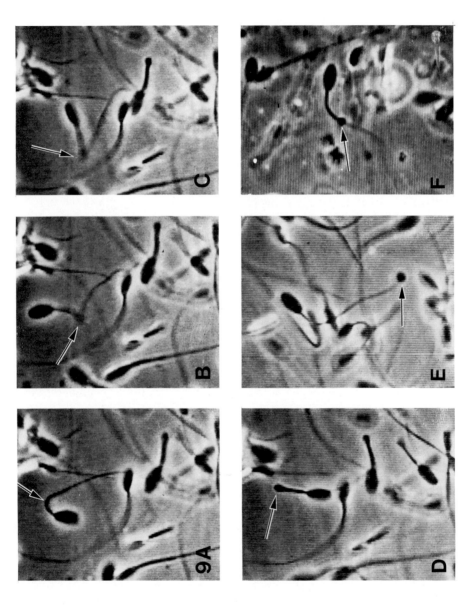

Fig. 9. (A)–(D). A series of photographs, taken directly from a video monitor, which show successive freeze-frame images of a single spermatozoon (arrows) undergoing the folding process described in the text. The time separation between the first and last video frame is less than 2 s. These events occurred at approximately 10°C. Upon rewarming to 30°C some spermatozoa developed a membranous swelling at either the end piece (E) or in the mid-piece (F). (By courtesy of Drs W. V. Holt, G. J. Morris & G. E. Coulson, reproduced with permission.)

still observed. On rewarming a large proportion of cells regained motility (W. V. Holt *et al.* in preparation).

Electron microscopy

Thin section electron microscopy of *Amoeba* following cold shock demonstrated an aggregation of actin, a compound known to be a significant component of the cytoskeleton in this cell-type (McLellan *et al.* 1984). This cellular response to cooling may be analogous to the cold-induced contraction of mammalian striated muscle (reviewed by Sakai & Kurihara, 1974). While elements of the cytoskeleton may be cold labile and thus responsible for the observed cell shrinkage, the trigger for cellular contraction following cold shock might also be due to alterations in intracellular ion concentrations, especially that of calcium, as a result of a loss of membrane selective permeability (see below).

With the electron microscope, the extent of the structural derangement of spermatozoa can be seen. The plasma membrane overlying the acrosome is particularly vulnerable especially in susceptible species and is readily lost. The initial change in the underlying acrosome is a wrinkling of the outer acrosomal membrane anterior to the equatorial segment which develops into vesiculation and/or a lifting away from underlying structures resulting in a 'ballooned' membrane (Fig. 10). Loss of acrosomal matrix occurs during this process; visible interruption of the continuity of the membrane is not necessary for this to occur. In all species studied the total loss of the anterior outer acrosomal membrane was uncommon and apparently unrelated to the cold stress. Throughout the changes in the anterior acrosome, the equatorial segment appears unaltered.

Membranes are not significantly elastic, i.e. they cannot increase or decrease in thickness (Kwok & Evans, 1981), and so large changes in surface area can occur only if material is added or removed from the plasmalemma. The observed expansion of the plasma and outer acrosomal membranes on cold shock raises the possibility of either loss of bilayer structure associated with distension or addition of membranous material from another source. No obvious source is available, except that in stallion semen there was evidence of numerous membrane fragments in the vicinity of the sperm heads (Watson, Plummer & Allen, 1987).

A quantitative assessment of sperm head membrane damage in three species subjected to a similar cold stress illustrated the species differences in susceptibility to cold shock (Table 2). With increasing cold stress, however, the sequence of membrane damage was similar; the acrosome was generally only damaged after the plasma membrane was broken or lost and progressed to the total loss of acrosomal matrix in severe stress. Both the proportion of damaged cells and the extent of membrane destruction was related to the degree of cold shock.

Of midpiece and tail structures, only the plasma membrane and mitochondria appeared to be affected. The plasma membrane frequently showed discontinuities and the mitochondria demonstrated occasional loss of electron-density and internal structure (Fig. 10). However, the latter were often seen adjacent to

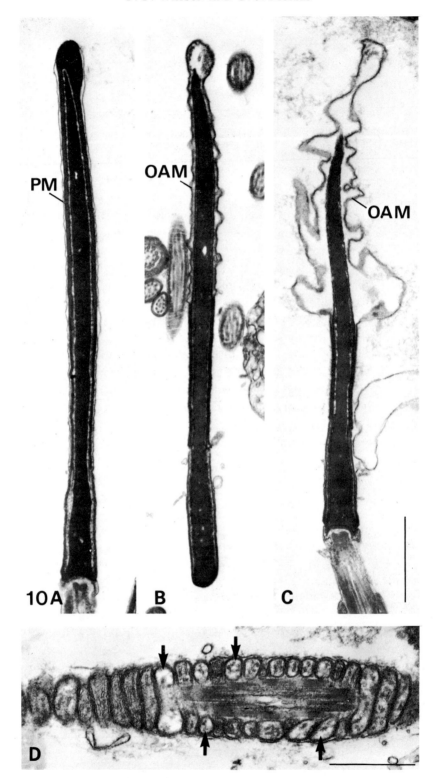

Table 2. *Percentages of stallion, ram and boar spermatozoa showing damage to head membranes after cold shock to 8 °C*

Membrane damage	Stallion		Ram		Boar	
	Control	Cold-shock	Control	Cold-shock	Control	Cold-shock
1. Intact	79	62	66	49	86	11
2. Broken PM	2	1	1	1	11	12
3. PM lost, Acr. Intact	9	12	20	41	2	49
4. PM lost, Acr. contents lost	7	20	12	9	0	27
5. PM lost, Acr. lost	4	5	1	0	1	0

PM = plasma membrane, Acr. = acrosome.

mitochondrial gyres of quite normal appearance, and the damage was independent of plasma membrane damage. The centrioles, outer dense fibres and axoneme appeared unaltered by cold shock, although the swellings and looping of the tails observed in the light microscope were occasionally seen.

The localization of calcium in cold-shocked ram spermatozoa by an electron microscopical histochemical technique (Plummer & Watson, 1985) revealed that access to binding sites on the outer acrosomal membrane and post-acrosomal dense lamina depended on a breach of the plasma membrane. There was a concentration of calcium binding just anterior to the equatorial segment, but when the outer acrosomal membrane swelled and the matrix was dispersing, the distribution of calcium binding sites was more uniform on the anterior acrosomal membrane. In contrast, boar spermatozoa revealed more calcium within the intact plasma membrane and no alteration in the apparently random distribution of calcium on the anterior outer acrosomal membrane when the acrosome was damaged (Watson, Plummer, Glossop & Robertson, 1985). In both species, the equatorial segment was devoid of calcium.

In the midpiece, large calcium deposits were seen outside and finer deposits within mitochondria; in addition, deposits were also seen in association with the outer dense fibres. In all cases, these midpiece and tail deposits were independent of plasma membrane damage and did not increase with cold shock. This was surprising in view of the known uptake of calcium by cold-shocked spermatozoa, and its expected localization in mitochondria (Babcock, First & Lardy, 1976). However, this histochemical technique may be incapable of detecting tightly bound calcium in this location.

Fig. 10. Transmission electron micrographs from cold-shocked boar spermatozoa, showing sagittal sections illustrating damage to head membranes, and a longitudinal section of the midpiece. (A) Undamaged membranes. (B) Plasma membrane (pm) lost, outer acrosomal membrane (oam) wrinkled and acrosomal contents partially dispersed. (C) Outer acrosomal membrane displaced (oam), and contents dispersed. (D) Midpiece showing breaks in the plasma membrane and loss of electron density in occasional mitochondria (arrows). Bars = 1 μm.

Biochemical alterations induced by cold shock

Many changes in the biochemistry of cells have been reported following cold shock but it is often difficult to distinguish between the primary effects which are responsible for the loss of cellular viability and secondary, pathological events.

Plasmalemma selective permeability

It is generally presumed that changes in the selective permeability properties of the plasmalemma occur as an early cold shock injury. In spermatozoa Na^+ and Ca^{2+} ions are gained by cells following cold shock whilst K^+ and Mg^{2+} ions are lost (reviewed by Watson, 1981a). In our recent studies with ram spermatozoa, calcium uptake was shown to be a function of cold stress (Robertson & Watson, 1986). As the temperature was reduced below 16°C so the uptake of calcium increased and this change in membrane function was reversible providing the cooling was slow ($0.125°C min^{-1}$). With rapid cooling a permanent change in membrane permeability occurred. These studies suggested that two independent effects were being induced:

(i) a temperature inhibition of membrane pumps independent of cooling rate, and

(ii) an increase in permeability to calcium ions caused by rapid cooling.

The latter may well be attributable to lipid phase transitions (see below) leading to a massive increase in intracellular calcium levels up to 10-fold in ram spermatozoa and 60-fold in boar spermatozoa. These situations were associated with gross membrane disruption provoking the thought that high calcium uptake is causally linked with the membrane damage. However, while calcium uptake could be prevented by the presence of EGTA (ethylene glycol-bis(β-amino ethyl ether)N,N,N',N'-tetra-acetic acid), considerable membrane damage still occurred (Robertson & Watson, unpublished) implying that calcium is not the only, or even the major, factor in severe cold shock injury. In both ram and boar spermatozoa, incubation prior to cold shock significantly reduced the permeability changes with cold shock (Robertson, Plummer & Watson, in preparation). Whether or not calcium is involved in the gross disruption of sperm membranes, there is no doubt that a high intracellular calcium content is metabolically undesirable. The cold-induced contraction of certain mammalian striated muscle cells is initiated by alterations in the free calcium level rather than changes in the calcium affinity of the contractile apparatus (Jeacocke, 1982).

Following cold shock a wide range of cytoplasmic compounds may be released (Table 3). These include pigments from *Blepharisma* (Giese, 1973) and cytoplasmic enzymes from spermatozoa (see Watson, 1981a). Some, at least, of these enzymes may originate from cytoplasmic droplets (Harrison & White, 1972) which are membrane-bound cytoplasmic remnants shed by spermatozoa at ejaculation but present in the semen. Undoubtedly, however, enzyme loss from spermatozoa is a consequence of cold shock.

Table 3. *Reported changes in biochemical parameters of spermatozoa following cold shock*

Biochemical parameter	Change with cold shock	Reference
Respiratory activity	decreased	Blackshaw & Salisbury, 1957
Glycolysis	decreased	
Fructolysis	decreased	Mann & Lutwak-Mann, 1955
Intracellular ATP level	decline (no resynthesis)	
Intracellular protein	released	
Membrane lipid & protein	released	
Lactate dehydrogenase, glutamic oxaloacetic transaminase, acid and alkaline phosphatase, glucose-6-phosphate dehydrogenase, glucose phosphate isomerase	released	Murdoch & White, 1968; Pursel, Johnson & Gerrits, 1970; Harrison & White, 1972; Moore, Hall & Hibbitt, 1976
Acrosin	released	Church & Graves, 1976
Plasmalogen	released	Hartree & Mann, 1959
Lipid phosphorus/phospholipid	released	Blackshaw & Salisbury, 1957; Darin-Bennett, Poulos & White, 1973
DNA	released	Quinn, White & Cleland, 1969
Polysaccharide		
Intracellular K^+ & Mg^{2+} ions	decreased	Blackshaw & Salisbury, 1957; Quinn & White, 1966
Intracellular Na^+ & Ca^{2+} ions	increased	Robertson & Watson, 1986

Respiration

Anaerobic glycolysis and respiratory activity are diminished in ram and bull spermatozoa following rapid cooling and rewarming (Table 3). Cytochrome *c* is released from cold-shocked spermatozoa indicating a breakdown of the electron-transport chain, although this is considered to be a secondary effect (Mann & Lutwak-Mann, 1955). Levels of ATP diminish immediately and are not resynthesized (Mann & Lutwak-Mann, 1955). The profound decline in respiratory activity correlates with the poor motility after cold shock and suggests that the occasional very obviously damaged mitochondria seen in the electron microscope may be only the visible aspect of a more general mitochondrial defect.

Other biochemical changes

A wide variety of other biochemical changes has been demonstrated in cold shocked cells (Table 3) including alterations in carbohydrate metabolism, an increase in ammonia formation indicating a breakdown of intracellular protein and progressive degeneration of DNA (reviewed in Watson, 1981*a*). In addition, studies with spermatozoa have revealed release of membrane constituents; Darin-Bennett, Poulos & White (1973) demonstrated that the particulate nature of the phospholipid material released from ram spermatozoa correlated with the morphological evidence of membrane fragmentation. Many of these reactions, together with some of the metabolic changes outlined above, are probably secondary reactions occurring as the consequence of the loss of selective permeability of the plasma membrane. Distinguishing between primary and secondary events in cold shock is quite difficult (see Watson, 1981a).

Mechanisms of cold shock injury

Cellular proteins and membranes are potentially vulnerable to rapid cooling injury. Although the intracellular sites of injury may vary, the stresses inducing damage are probably interrelated. For convenience, however, they will be considered separately.

Denaturation of proteins

During reduction in temperature, proteins may spontaneously unfold or, in the case of a multisubunit structure, dissociate into biologically inactive species which may or may not reassemble upon rewarming. Considering the complex energy balance responsible for the conformational stability of proteins such effects are not surprising. All the separate interactions including hydrophobic forces, hydrogen bonding and electrostatic contributions, vary in different ways with temperature. It has been suggested that all proteins should be stable only within a limited temperature range and cold denaturation should be as universal as heat denaturation (Franks, 1982).

Two general cases will be discussed here, namely cytoplasmic enzymes and the cytoskeleton, the special case of membrane proteins will be examined in a later section.

Metabolic injury

Loss of viability following rapid cooling cannot be due solely to metabolic imbalances resulting from different temperature coefficients of enzyme reactions, enzyme denaturation or the accumulation of toxic byproducts. Maximum damage is observed following rapid cooling during which cells are exposed to low temperatures for the shortest period. Longer times of exposure, as occur at slower rates of cooling are protective. However, impaired metabolism or the inability to repair damage at low temperatures may be a contributory factor to injury in cells damaged by other stresses during cooling.

Cytoskeleton

Chromosome movements, cell motility and the maintenance of cell shape in eukaryotic cells are functions generally considered to involve the cytoskeleton which is made up of three dynamic, independently controlled but interacting arrays of protein filaments: microtubules, intermediate filaments and microfilaments. Each system is composed of distinct proteins and each has an array of regulatory proteins. The *in vitro* activity of these regulatory proteins is variously modified by physical and chemical factors. For example, the properties of some proteins from amoebae which cross-link filamentous actin have been demonstrated to alter with temperature; specifically, a 90 000 Da protein is Ca^{2+} sensitive at 28 °C but not 0 °C, whilst two polypeptides exhibit a Ca^{2+} sensitive crosslinking activity at 0 °C but not at 28 °C (Hellewell & Taylor, 1979).

(i) *Microtubules.* That certain elements of the cytoskeleton, specifically microtubules, may depolymerize at low temperatures has been appreciated for many years. Microtubules have a wide range of functions in cells and there is evidence that individual classes may respond differently to low temperatures (Behnke & Forer, 1967).

(a) Cytoplasmic microtubules. The low temperature depolymerization of cytoplasmic microtubules has been demonstrated in many cell-types both by electron microscopy (e.g. Behnke & Forer, 1967) and immunofluorescence (e.g. Weber, Pollack & Bibring, 1975) and has been employed as a diagnostic feature to identify microtubules. However, this response is not universal, and a number of mammalian cell-types have been reported to have cold-stable cytoplasmic microtubules (Bershadsky, Gelfand, Suitking & Tint, 1979).

In spite of such data surprisingly few attempts have been made to link the effects of low temperatures on cells directly with the depolymerization of cytoplasmic microtubules. An exception to this is a study on the cell wall-less zoospores of the alga *Chlorosarcinopsis gelatinosa*, which are ellipsoidal. Upon cooling to 2 °C they become spherical but revert to their original shape following rewarming. These changes in shape correlate with the disappearance and reappearance of peripheral cytoplasmic microtubules (Melkonian, Kroger & Marquart, 1980).

(b) Microtubules associated with the mitotic apparatus. The mitotic spindle is composed predominantly of microtubules and in many cell types this structure has been reported to disappear at low temperatures (Inoue, 1952; Roth, 1967) resulting in a complete inhibition of chromosome movements (Lambert & Bayer, 1977). However, in some cell types exposure to low temperatures does not induce complete breakdown of the mitotic apparatus, only a reduction in the number of microtubules (Brinkley & Cartwright, 1975; Lambert & Bayer, 1977). There is evidence that individual microtubules within the mitotic apparatus may have a differential sensitivity to low temperatures. For example, following exposure of mammalian tissue culture cells to 0 °C for 1 h, the pole to pole microtubules were disrupted whilst the pole to chromosome tubules remained intact (Welsh, Pedman, Brinkley & Means, 1979).

(c) Microtubules of cilia and flagella. These are generally considered to be stable and do not depolymerize at low temperatures. Our own studies of mammalian spermatozoa confirm that the axoneme appears to be unaltered by cold shock.

It is evident that a variety of responses are observed in microtubule-containing structures upon exposure to low temperatures. These have been attributed either to differences in polypeptide pattern between cold-labile and cold-resistant microtubules (Hesketh, 1984), or to reflect different physical environments e.g. anchorage to non-microtubule structures (Lambert & Bayer, 1977).

(ii) *Intermediate filaments*. The intermediate filaments of mammalian tissue culture cells persist after cold treatments which depolymerize the microtubules (Virtanen, Lehto, Lehtonen & Brudleg, 1980; Maro, Sauron, Paulin & Bornens, 1983). By contrast, the intermediate filaments of epidermal cells of teleost fish (tonofilaments) dissociate at low temperatures and this is reversed within minutes of rewarming the cells. This is unrelated to the cold-induced depolymerization of microtubules observed in the same cells, since neither colchicine or vinblastin, which depolymerize microtubules have any effect on tonofilaments (Schliwa & Euteneuer, 1979).

(iii) *Microfilaments*. Cytoplasmic streaming is mediated by actin microfilaments and upon a reduction in temperature the rate of streaming may slow down or cease (Patterson & Graham, 1977). These changes occur within minutes of cooling and it has been suggested that they are a direct effect of a disassembly of actin filaments. However, the rate of cytoplasmic streaming is dependent on the synthesis of ATP and is also very sensitive to Ca^{2+} ion concentration. It is thus probable that the changes in the rate of cytoplasmic streaming observed at low temperatures may also reflect alterations in cellular metabolism or the intracellular concentration or localization of calcium (reviewed by Minorsky, 1985).

Membrane thermotropic behaviour

It is a fundamental property of phospholipid bilayers that they exhibit thermotropism, i.e. a pure phospholipid in aqueous suspension undergoes an abrupt change from a disordered fluid or liquid crystalline state to a highly ordered hexagonal lattice of fatty acyl chains (the gel state) over a specific temperature range. The temperature at the midpoint of this phase change is the transition temperature (Tc) and the change in state has been variously called the lipid phase transition, the gel-liquid crystalline transition or the order-disorder transition. From extensive studies using model systems the relationship between composition and thermotropic behaviour has been examined (reviewed by Morris & Clarke, 1986). Biological membranes contain a mixture of phospholipid species, other lipids and proteins, and much of our understanding of the effects of reduced temperatures on membranes derives from studies on bilayers of mixed phospholipids where variations in chain length, degree of unsaturation, head group and insertion of proteins have been experimentally controlled. Under equilibrium conditions (i.e. slow cooling) protein molecules are excluded from the growing phospholipid lattice and are concentrated into regions of fluid lipid. Low temperatures induce a lateral phase separation into regions of crystalline phospho-lipid and domains of high protein concentration containing trapped phospholipid.

Thermotropic behaviour has been studied extensively in the ciliated protozoan *Tetrahymena pyriformis*. Freeze-fracture studies of the intramembranous particle distribution of cells cooled 'slowly' to different temperatures showed that a lateral phase separation was initiated within the plane of the membrane at approximately 10°C below the growth temperature (Martin *et al*. 1976). This was complete, as judged by electron microscopy, 20°C below the growth temperature. The occurrence of a phase change has been confirmed by other techniques. Using wide angle X-ray diffraction, the transition temperature for pellicle lipids of cells grown at 39°C was 26°C (Nakajama *et al*. 1983). In addition, Arrhenius plots of the ATPase activity of the pellicle membrane from cells grown at 39°C showed a discontinuity at 28°C (Nozawa, 1980). The injuries sustained by *Tetrahymena* subjected to cold shock show a parallel to this membrane phase behaviour. The initial loss of viability becomes evident 12 to 15°C below the growth temperature and increases as the temperature is further reduced suggesting a direct correlation with altered membrane structure.

Similarly, Holt & North (1984) have demonstrated by freeze-fracture studies particle-free areas of plasma membrane of ram spermatozoa cooled slowly to 5°C and interpreted these as evidence of lipid phase separation. Moreover, the Arrhenius plot of a calcium activated Mg-ATPase isolated from ram sperm membranes showed a discontinuity at 22–24°C (Holt & North, 1985). Further studies have revealed several other discontinuities, one occurring at 17°C (Holt & North, 1986) which was the temperature below which cooled ram spermatozoa became unable to export calcium and progressively accumulated calcium (Robert-son & Watson, 1986). These observations are consistent with membrane lipid thermotropism resulting in changes in membrane function.

To a first approximation, at any temperature below the nucleation point of the membrane lipids the same amount of gel phase lipid will be formed following either rapid or slow cooling. As cellular viability following cold shock is evidently a function of cooling rate it may not simply be the presence of gel phase lipid within the membrane which determines cold shock injury but rather the more specific aspects of the nature and pattern of crystal growth and the redistribution of membrane proteins. These are considered in turn:

(i) Membrane proteins. Following slow cooling to temperatures below a lipid phase separation, membrane enzymes are concentrated into regions of lipid which are still fluid. This frequently results in a reduction in enzyme activity, which may be apparent in an Arrhenius plot as a discontinuity. Such breakpoints have been variously ascribed to the onset of a phase separation, the temperature at which a phase separation occurs or alternatively some intrinsic property of the enzyme. Whilst the aggregation of proteins within the plane of the membrane has many complex effects on their activity it is not damaging in the short term (reviewed by Morris & Clarke, 1986). By definition, cold shock injury occurs following rapid cooling with insufficient time for lateral diffusion to occur and proteins will be 'set' into solid phase lipid, where the activity of proteins may be inhibited.

(ii) Membrane lipid. During rapid cooling, events equivalent to undercooling are observed both within biological membranes and in model systems of lipid bilayers. This undercooling is clearly demonstrated by differential scanning calorimetry and is shown to be reduced at lower rates of cooling (Black & Dixon, 1981). Many nucleation sites will occur within membranes following rapid cooling. Crystal growth would be rapid, and there would be insufficient time for a significant redistribution of components within the plane of the membrane. Studies using differential scanning calorimetry combined with X-ray diffraction indicate that following rapid cooling, bilayers are crystalline but domains of hexagonally packed crystalline phospholipid are small (Melchior, Bruggeman & Stein, 1982).

As rapid cooling will result in multiple nucleation sites and consequently small gel phase domains, extensive grain boundaries and numerous packing faults would be expected between adjacent regions of hexagonally packed phospholipids and such structures have been demonstrated in bilayers by electron diffraction (Hui, Parsons & Cowden, 1974). The occurrence of lipids with different packing characteristics or the presence of proteins (Chapman, Gomez-Fernandez & Goni, 1979) would increase the incidence of imperfections in gel lipid structure. The two halves of the bilayer nucleate independently (Sillerud & Barnett, 1982) and if packing faults within the two monolayers are coincident, then at such sites leakage of small molecules across the membrane could occur (see Morris, Coulson & Clarke, 1984, for a fuller discussion).

As the rate of cooling is reduced the probability of undercooling and consequently the number of nucleation sites within the membrane will decrease. At slower rates of cooling there is greater time for gel phase growth; both lipid lateral phase separations and diffusion of proteins can be clearly demonstrated by freeze-fracture electron microscopy. The areas of gel phase lipid formed will be larger than those nucleated following rapid cooling and consequently the incidence of grain boundaries, packing faults and dislocations will be reduced.

The processes of nucleation and crystal growth will depend on the composition of the membranes and the occurrence (if any) of structure within the membrane which promote or depress nucleation. In addition, membrane thermotropic behaviour will be influenced by compounds which interact with membrane components. Membrane composition has been compared in spermatozoa of different species, and correlations with cold shock sensitivity observed. Thus, highly sensitive species have a low cholesterol:phospholipid ratio and a high proportion of long chain fatty acids in the phospholipid fraction (see Watson, 1981a). The presence of high cholesterol proportion would have the effect of broadening phase transitions and thus limiting phase separation phenomena (see Watson, 1981a, for full discussion).

Thermoelastic stress

Another hypothesis of cold shock injury has been developed recently which describes the dynamic tension which results within a fluid membrane vesicle as the temperature is reduced rapidly (McGrath, 1984). During cooling, a condensation of membrane lipid surface area will occur; for example, with a lipid monolayer on a Langmuir trough there is approximately a 0·8% decrease in area per °C reduction in temperature. The surface area of a membrane bound vesicle will therefore be decreasing in area at a faster rate than the aqueous contents decrease in volume and consequently an increase in membrane tension will result. Tensions known to cause damage to liposomes will develop for temperature decreases of 10–20°C, larger temperature decreases will create greater tensions, as will faster rates of cooling. Folded membrane surfaces, possessing 'excess' surface area, may experience membrane area condensation without developing a critical membrane tension. This interesting proposal merits further investigation as a key stress during chilling. It should be noted that this model does not require a phase transition to occur for injury to be expressed. However, during a lipid phase separation a major reduction in surface area occurs which would further increase membrane tension.

In spermatozoa, differences in head shape appear to be correlated with cold shock sensitivity (Watson & Plummer, 1985). The sperm head profiles from transmission micrographs suggest that resistance to cold shock is associated with smaller, more convex shapes whereas the larger paddle-shaped, bilaterally flattened profiles are associated with vulnerability. This striking correlation provokes the thought that some feature of this latter cytoarchitectural arrangement predisposes to cold shock. Cytoskeletal proteins, actin and spectrin, have been demonstrated in spermatozoa (Virtanen, Bradley, Paasuivo & Lehto, 1984) but clear cytoskeletal structures have not been identified except in the equatorial segment of the acrosome (Russell, Peterson & Freund, 1980). Thus, the sequence of membrane deterioration especially evident in cold-shocked ungulate spermatozoa may be a consequence of the absence of cytoskeletal attachments to membrane expanses over the broad faces of the head. Spermatozoa of other species may be less vulnerable owing to smaller size or the greater curvature of the surface membranes.

Conclusion

It is only in the last few years that possible mechanisms of cold shock injury have been considered. The focus on membrane phenomena is appropriate since the plasma membrane forms the essential barrier between the cell interior and the external environment. It should be borne in mind, however, that internal membranes may also be affected by cooling injury. The task is always to determine the primary effects of rapid cooling on cell structure and function, and to distinguish them from secondary effects which develop as a consequence. The growing evidence that rapid cooling injury is a more generalized phenomenon should encourage further efforts to understand the mechanisms. These studies will be of value not only to develop our understanding of thermal responses of animal cells, but also may shed light on cooling phenomena in nature. Their particular application, however, will be to the prevention of cooling injury during cryopreservation, an area of increasing importance in biotechnology.

We are indebted to Dr J. M. Plummer, Royal Veterinary College for the electron micrographs in Fig. 10.

References

BABCOCK, D. F., FIRST, N. L. & LARDY, H. A. (1976). Action of ionophore A23187 at the cellular level. Separation of effects at the plasma and mitochondrial membranes. *J. biol. Chem.* **251**, 3881–3886.

BEHNKE, O. & FORER, A. (1967). Evidence for four classes of microtubules in individual cells. *J. Cell Sci.* **2**, 169–192.

BERSHADSKY, A. D., GELFAND, V. I., SUITKING, T. M. & TINT, I. S. (1979). Cold stable microtubules in the cytoplasm of mouse embryo fibroblasts. *Cell Biology International Reports* **3**, 45–50.

BENSON, R. W., PICKETT, B. W., KOMAREK, R. J. & LUCAS, J. J. (1967). Effect of incubation and cold shock on motility of boar spermatozoa and their relationship to lipid content. *J. Anim. Sci.* **26**, 1078–1081.

BIRILLO, I. M. & PULHALJSKII, L. H. (1936). Problems of prolonged storage of bull and ram sperm. *Problemy Zhivotnovodstva* No. **10**, 24–30 (in Russian).

BLACK, J. G. & DIXON, G. S. (1981). A.C. calorimetry of dimyristoyl phosphatidylcholine multilayers: Hysteresis and annealing near the gel to liquid-crystal transition. *Biochemistry* **20**, 6740–6744.

BLACKSHAW, A. W. (1954). The prevention of temperature shock of bull and ram semen. *Aust. J. biol. Sci.* **7**, 573–582.

BLACKSHAW, A. W. & SALISBURY, G. W. (1957). Factors influencing metabolic activity of bull spermatozoa. II. Cold-shock and its prevention. *J. Dairy Sci.* **40**, 1099–1106.

BRINKLEY, B. R. & CARTWRIGHT, J. (1975). Cold-labile and cold-stable microtubules in the mitotic spindle of mammalian cells. *Ann. N.Y. Acad. Sci.* **253**, 428–439.

BUTLER, W. J. & ROBERTS, T. K. (1975). Effects of some phosphatidyl compounds on boar spermatozoa following cold shock or slow cooling. *J. Reprod. Fert.* **43**, 183–187.

CHAPMAN, D., GOMEZ-FERNANDEZ, J. P. & GONI, F. M. (1979). Intrinsic protein-lipid interactions: Physical and biochemical evidence. *FEBS Lett.* **98**, 211–228.

CHURCH, V. E. & GRAVES, C. N. (1976). Loss of acrosin from bovine spermatozoa following cold shock: protective effect of seminal plasma. *Cryobiology* **13**, 341–346.

DARIN-BENNETT, A., POULOS, A. & WHITE, I. G. (1973). The effect of cold shock and freeze-thawing on release of phospholipids by ram, bull and boar spermatozoa. *Aust. J. biol. Sci.* **26**, 1409–1420.

DUBBLEMAN, T. M., DE BRUIJNE, A. W., CHRISTIANSE, K. & VAN STEVENICK, J. (1979). Hypertonic cryohaemolysis of human red blood cells. *J. Membrane Biol.* **50**, 225–240.

EVANS, R. J., BAUER, D. H., BANDEMER, S. L., VAGHEFI, S. B. & FLEGAL, C. J. (1973). Structure of egg yolk very low density lipoprotein. Polydispersity of the very low density lipoprotein and the role of lipovitellenin in the structure. *Archs Biochem. Biophys.* **154**, 493–500.

FRANKS, F. (1982). Physiological water stress. In *Biophysics of Water* (ed. F. Franks & S. F. Matthias), pp. 279–294. Chichester: John Wiley and Sons.

FOULKES, J. A. (1977). Separation of lipoproteins from egg yolk and their effect on the motility and integrity of bovine spermatozoa. *J. Reprod. Fert.* **49**, 277–284.

GIESE, A. C. (1973). Blepharisma: *The Biology of a Light-Sensitive Protozoan*, pp. 163–164. Stanford: Stanford University Press.

GLADCINOVA, E. F. (1937). The influence of marked drops in temperature on survival of spermatozoa. *Uspekhi Zootekhnicheskikh Nauk* **4**(3), 56–64 (in Russian).

GREEN, F. A., JUNG, C. J., CUPPOLETTI, J. & OWENS, N. (1981). Hypertonic cryo-haemolysis and the cytoskeleton system. *Biochem. biophys. Acta* **648**, 225–230.

HARRISON, R. A. P. & WHITE, I. G. (1972). Glycolytic enzymes in the spermatozoa and cytoplasmic droplets of bull, boar and ram and their leakage after shock. *J. Reprod. Fert.* **30**, 105–115.

HARTREE, E. F. & MANN, T. (1959). Plasmalogen in ram semen and its role in sperm metabolism. *Biochem. J.* **71**, 423–434.

HELLEWELL, S. B. & TAYLOR, D. L. (1979). The contractile basis of amoeboid movement. 6. The solation: construction coupling hypothesis. *J. Cell Biol.* **83**, 633–648.

HESKETH, G. J. M. (1984). Differences in polypeptide composition and enzyme activity between cold-stable and cold-labile microtubules and the study of microtubule alkaline phosphatase activity. *FEBS Lett.* **169**, 313–318.

HOLT, W. V. & NORTH, R. D. (1984). Partially irreversible cold-induced lipid phase transitions in mammalian sperm plasma membrane domains: Freeze-fracture study. *J. exp. Zool.* **230**, 473–483.

HOLT, W. V. & NORTH, R. D. (1985). Determination of lipid composition and thermal phase transition temperature in an enriched plasma membrane fraction from ram spermatozoa. *J. Reprod. Fert.* **73**, 285–295.

HOLT, W. V. & NORTH, R. D. (1986). Thermotropic phase transitions in the plasma membrane of ram spermatozoa. *J. Reprod. Fert.* **78**, 447–457.

HUI, S. W., PARSONS, D. F. & COWDEN, M. (1974). Electron diffraction of wet phospholipid bilayers. *Proc. natn. Acad. Sci. U.S.A.* **71**, 5068–5072.

INOUE, S. (1952). Effect of low temperature on the birefringence of the mitotic spindle. *Biological Bulletin* **103**, 316–324.

JEACOCKE, R. E. (1982). Calcium efflux during cold-induced contraction of mammalian striated muscle fibres. *Biochem. biophys. Acta* **682**, 238–244.

JOHNSON, L. V., WALSH, L. M., BOCKUS, B. J. & CHEN, L. B. (1981). Localization of mitochondria in living cells with rhodamine 123. *Proc. natn. Acad. Sci. U.S.A.* **77**, 990–994.

JUNG, C. J. & GREEN, F. A. (1978). Hypertonic haemolysis: Ionophore and pH effects. *J. Membrane Biol.* **39**, 273–284.

KAMPSCHMIDT, R. F., MAYER, D. T. & HERMAN, H. A. (1953). Lipid and lipo-protein constituents of egg yolk in the resistance and storage of bull spermatozoa. *J. Dairy Sci.* **36**, 733–742.

KAWAHARA, H. (1978). Production of triploid and gynogenetic diploid *Xenopus* by cold treatment. *Devel. Growth Different.* **20**, 227–236.

KAWAMURA, N. (1979). Cytological studies on mosaic silkworms induced by low temperature treatment. *Chromosoma* **74**, 179–188.

KWOK, R. & EVANS, E. (1981). Thermoelasticity of large lecithin bilayer vesicles. *Biophys. J.* **35**, 637–652.

LASLEY, J. F. & BOGART, R. (1944). A comparative study of the epididymal and ejaculated spermatozoa of the boar. *J. Anim. Sci.* **3**, 360–370.

LASLEY, J. F. & MAYER, D. T. (1944). A variable physiological factor necessary for the survival of bull spermatozoa. *J. Anim. Sci.* **3**, 129–135.

LASLEY, J. F., EASLEY, G. T. & BOGART, R. (1942). Some factors influencing the resistance of bull sperm to unfavourable environmental conditions. *J. Anim. Sci.* **1**, 79.

LAMBERT, A. M. & BAYER, A. S. (1977). Microtubule distribution and reversible arrest of chromosome movements induced by low temperatures. *Eur. J. Cell Biol.* **15**, 1–23.

LEMOINE, M. L. & SMITH, L. T. (1980). Polyploidy induced in brook trout by cold shock. *Trans. Amer. Fish Soc.* **109**, 626–631.

LEVITT, J. (1980). Responses of plants to environmental stress. In *Chilling, Freezing and High Temperature Stress*, 2nd edn, vol. 2, pp. 23–64. New York: Academic Press.

LOVELOCK, J. E. (1953). The haemolysis of human red blood cells by freezing and thawing. *Biochem. biophys. Acta* **10**, 414–426.

LOVELOCK, J. E. (1955). Haemolysis by thermal shock. *Brit. J. Haematol.* **1**, 117–129.

McGRATH, J. J. (1984). Effect of thermoelastic stress on thermal shock and the freezing response of cell-size, unilamellar liposomes. *Cryobiology* **21**, 696–697.

McLELLAN, M. R., MORRIS, G. J., COULSON, G. E., JAMES, E. R. & KALININA, L. V. (1984). Role of cytoplasmic proteins in cold shock injury in *Amoeba*. *Cryobiology* **21**, 44–59.

MANN, T. & LUTWAK-MANN, C. (1955). Biochemical changes underlying the phenomenon of cold shock in spermatozoa. *Archs Sci. biol.* **39**, 578–588.

MARO, B., SAURON, M. E., PAULIN, D. & BORNENS, M. (1983). Further evidence for interaction between microtubules and vimetin filaments: Taxol and cold effects. *Biol. Cell* **47**, 243–246.

MARTIN, C. E., HIRAMITSUI, K., NOZAWA, Y., SKRIVER, L. & THOMPSON, G. A. (1976). Molecular control of membrane properties during temperature acclimation. Fatty acid desaturase regulation of membrane fluidity in acclimating *Tetrahymena* cells. *Biochemistry* **15**, 5218–5227.

MELCHIOR, P. L., BRUGGEMAN, E. P. & STEIN, J. M. (1982). The physical state of quick frozen membranes and lipids. *Biochem. biophys. Acta* **690**, 81–88.

MELKONIAN, M., KROGER, K.-H. & MARQUART, K.-G. (1980). Cell shape and microtubules in zoospores of the green alga *Chlorosarcinopsis gelatinosa*: Effects of low temperatures. *Protoplasma* **104**, 283–293.

MILAVANOV, V. K. (1934). [Artificial insemination of livestock.] *Iskustvennoe osemenenie s.-l.* Zivotnyh. Moscow: Seljhozgiz (in Russian).

MINORSKY, P. V. (1985). An heuristic hypothesis of chilling injury in plants: A role for calcium as the primary physiological transducer of injury. *Plant Cell and Environment* **8**, 75–94.

MOLLBY, R. (1976). Effect of staphylococcal Beta-hemolysin (Sphingomyelinase C) on cell membranes. In *Staphylococci and Staphylococcal diseases* (ed. J. Jeljaszewick), pp. 665–667. Stuttgart: Gustav Fischer.

MOORE, N. W. & BILTON, R. J. (1973). The storage of fertilized sheep ova at 5°C. *Aust. J. biol. Sci.* **26**, 1421–1427.

MOORE, H. D. M., HALL, G. A. & HIBBITT, K. G. (1976). Seminal plasma proteins and the reaction of spermatozoa from intact boars and from boars without seminal vesicles to cooling. *J. Reprod. Fert.* **47**, 39–45.

MORRIS, G. J. (1975). Lipid loss and haemolysis by thermal shock: Lack of correlation. *Cryobiology* **12**, 192–201.

MORRIS, G. J. (1986). Direct chilling injury. In *The Effects of Low Temperatures on Biological Systems* (ed. B. W. W. Grout & G. J. Morris), pp. 120–146. London: Edward Arnold.

MORRIS, G. J. & CLARKE, A. (1986). Cells at low temperatures. In *The Effects of Low Temperatures on Biological Systems* (ed. B. W. W. Grout & G. J. Morris), pp. 72–119. London: Edward Arnold.

MORRIS, G. J. & FARRANT, J. (1973). Effect of cooling rate on thermal shock haemolysis. *Cryobiology* **10**, 119–125.

MORRIS, G. J. & WATSON, P. F. (1984). Cold shock injury – a comprehensive bibliography. *Cryo-Lett.* **5**, 352–372.

MORRIS, G. J., COULSON, G. E. & CLARKE, A. (1984). Cold shock injury in *Tetrahymena pyriformis*. *Cryobiology* **21**, 664–671.

MORRIS, G. J., COULSON, G. E., MEYER, M. A., McLELLAN, M. R., FULLER, B. J., GROUT, B. W. W., PRITCHARD, H. W. & KNIGHT, S. C. (1983). Cold shock: A widespread cellular reaction. *Cryo-Lett.* **4**, 179–192.

MURDOCH, R. N. & WHITE, I. G. (1968). Studies of the distribution and source of enzymes in mammalian semen. *Aust. J. biol. Sci.* **21**, 483–490.

NAKAJAMA, H., GOTO, M., OHKI, K., HITSUI, T. & NOZAWA, Y. (1983). An X-ray diffraction study of phase transition temperatures of various membranes isolated from *Tetrahymena pyriformis* cells grown at different temperatures. *Biochim. biophys. Acta* **730**, 17–24.

NOZAWA, Y. (1980). Modification of membrane lipid composition and membrane fluidity in *Tetrahymena*. In *Membrane Fluidity: Biophysical Techniques and Cellular Regulation* (ed. M. Kates & A. Kuksis), pp. 399–418. New York: The Humana Press.

PACE, M. M. & GRAHAM, E. F. (1974). Components in egg yolk which protect bovine spermatozoa during freezing. *J. Anim. Sci.* **39**, 1144–1149.

PATTERSON, B. D. & GRAHAM, D. (1977). Effect of chilling temperatures on the protoplasmic streaming of plants from different climates. *J. exp. Bot.* **28**, 736–743.

PHILLIPS, P. H. & LARDY, H. A. (1940). A yolk-buffer pabulum for the preservation of bull sperm. *J. Dairy Sci.* **23**, 399–404.

PLUMMER, J. M. & WATSON, P. F. (1985). Ultrastructural localization of calcium ions in ram spermatozoa before and after cold shock as demonstrated by a pyroantimonate technique. *J. Reprod. Fert.* **75**, 255–263.

POLGE, C. & WILLADSEN, S. M. (1978). Freezing eggs and embryos of farm animals. *Cryobiology* **15**, 370–373.

PURDOM, C. E. (1972). Induced polyploidy in plaice (*Pleuronectes platessa*) and its hybrid with the flounder (*Platichthys flesus*). *Heredity* **29**, 11–24.

PURSEL, V. G., JOHNSON, L. A. & GERRITS, R. J. (1970). Distribution of glutamic oxaloacetic transaminase and lactic dehydrogenase activities in boar semen after cold shock and freezing. *Cryobiology* **7**, 141–144.

PURSEL, V. G., JOHNSON, L. A. & RAMPACEK, G. B. (1972). Acrosome morphology of boar spermatozoa incubated before cold shock. *J. Anim. Sci.* **34**, 278–283.

PURSEL, V. G., JOHNSON, L. A. & SCHULMAN, L. L. (1972). Interactions of extender composition and incubation period on cold shock susceptibility of boar spermatozoa. *J. Anim. Sci.* **35**, 580–584.

PURSEL, V. G., JOHNSON, L. A. & SCHULMAN, L. L. (1973). Effect of dilution seminal plasma and incubation period on cold shock susceptibility of boar spermatozoa. *J. Anim. Sci.* **37**, 528–531.

QUINN, P. J. & WHITE, I. G. (1966). The effect of cold shock and deep-freezing on the concentration of major cations in spermatozoa. *J. Reprod. Fert.* **12**, 263–270.

QUINN, P. J. & WHITE, I. G. (1967). Phospholipid and cholesterol content of epididymal and ejaculated ram spermatozoa and seminal plasma in relation to cold shock. *Aust. J. biol. Sci.* **20**, 1205–1215.

QUINN, P. J. & WHITE, I. G. (1968). The effect of pH, cations and protective agents on the susceptibility of ram spermatozoa to cold shock. *Expl Cell Res.* **49**, 31–39.

QUINN, P. J., CHOW, P. Y. W. & WHITE, I. G. (1980). Evidence that phospholipid protects ram spermatozoa from cold shock at a plasma membrane site. *J. Reprod. Fert.* **60**, 403–407.

QUINN, P. J., SALAMON, S. & WHITE, I. G. (1968a). The effect of cold shock and deep-freezing on ram spermatozoa collected by electrical ejaculation and by artificial vagina. *Aust. J. agric. Res.* **19**, 119–128.

QUINN, P. J., SALAMON, S. & WHITE, I. G. (1968b). Effect of osmotic pressure and temperature gradients on cold shock in ram and bull spermatozoa. *Aust. J. biol. Sci.* **21**, 133–140.

QUINN, P. J., WHITE, I. G. & CLELAND, K. W. (1969). Chemical and ultrastructural changes in ram spermatozoa after washing, cold shock and freezing. *J. Reprod. Fert.* **18**, 209–220.

ROBERTSON, L. (1986). Studies of intracellular calcium and membrane changes during dilution and cooling of ram spermatozoa. Ph.D. Thesis, University of London, 1986.

ROBERTSON, L. & WATSON, P. F. (1986). Calcium transport in diluted or cooled ram semen. *J. Reprod. Fert.* **77**, 177–185.

ROTH, L. E. (1967). Electron microscopy of mitosis in *Amoeba*. 3. Cold and urea treatment: A basis for test of direct action of mitotic inhibitors on microtubule formation. *J. Cell Biol.* **34**, 47–59.

RUSSELL, L., PETERSON, R. N. & FREUND, M. (1980). On the presence of bridges linking the inner and outer acrosomal membranes of boar spermatozoa. *Anat. Rec.* **198**, 449–459.

SAKAI, T. & KURIHARA, S. (1974). A study of rapid cooling contracture from the viewpoint of excitation-contraction coupling. *Jikei Medical J.* **21**, 47–88.

SCHLIWA, M. & EUTENEUER, U. (1979). Structural transformations of epidermal tonofilaments upon cold treatment. *Expl Cell Res.* **122**, 93–101.

SILLERUD, L. O. & BARNETT, R. E. (1982). Lack of transbilayer coupling in phase transitions of phosphatidylcholine vesicles. *Biochemistry* **21**, 1756–1760.

TAKAHASHI, T. (1983). Calmodulin antagonists induce isotonic thermal shock haemolysis in human erythrocytes. *Cryobiology* **20**, 726.

TAKAHASHI, T. & WILLIAMS, R. J. (1979). Thermal shock in red cells. *Cryobiology* **16**, 588–589.

TAKAHASHI, T. & WILLIAMS, R. J. (1983). Thermal shock haemolysis in human red cells. I. The effects of temperature, time and osmotic stress. *Cryobiology* **20**, 507–520.

340 P. F. WATSON AND G. J. MORRIS

TROUNSON, A. O., WILLADSEN, S. M. & ROWSON, L. E. A. (1976a). The influence of *in vitro* culture and cooling on the survival and development of cow embryos. *J. Reprod. Fert.* **47**, 367–370.

TROUNSON, A. O., WILLADSEN, S. M., ROWSON, L. E. A. & NEWCOMB, R. (1976b). The storage of cow eggs at room temperature and at low temperatures. *J. Reprod. Fert.* **46**, 173–178.

VALENTI, P. J. (1975). Induced polyploidy in *Tilapia anrea* (Steindachner) by means of temperature shock treatment. *J. Fish Biol.* **7**, 519–528.

VIRTANEN, I., LEHTO, V.-P., LEHTONEN, E. & BRUDLEG, R. A. (1980). Organisation of intermediate filaments in cultured fibroblasts upon disruption of microtubules by cold treatment. *Eur. J. Cell Biol.* **23**, 80–84.

VIRTANEN, I., BRADLEY, R. A., PAASIVUO, R. & LEHTO, V.-P. (1984). Distinct cytoskeleton domains revealed in sperm cells. *J. Cell Biol.* **99**, 1083–1091.

WALES, R. G. & WHITE, I. G. (1959). The susceptibility of spermatozoa to temperature shock. *J. Endocr.* **19**, 211–220.

WATSON, P. F. (1975). Use of a giemsa stain to detect changes in acrosomes of frozen ram spermatozoa. *Vet. Rec.* **97**, 12–15.

WATSON, P. F. (1976). The protection of ram and bull spermatozoa by the low density lipoprotein fractions of egg yolk during storage at 5°C and deep-freezing. *J. thermal Biol.* **1**, 137–141.

WATSON, P. F. (1981a). The effects of cold shock on sperm cell membranes. In *The Effects of Low Temperatures on Biological Membranes* (ed. G. J. Morris & A. Clarke), pp. 189–218. London: Academic Press.

WATSON, P. F. (1981b). The roles of lipid and protein in the protection of ram spermatozoa at 5°C by egg-yolk lipoprotein. *J. Reprod. Fert.* **62**, 483–492.

WATSON, P. F. & MARTIN, I. C. A. (1976). Artificial insemination of sheep: the fertility of semen extended in diluents containing egg yolk and inseminated soon after dilution or stored at 5°C for 24 or 48 hours. *Theriogenology* **6**, 553–558.

WATSON, P. F. & PLUMMER, J. M. (1985). Responses of boar sperm membranes to cold shock and cooling. In *Deep-freezing of Boar Semen* (ed. L. A. Johnson & K. Larsson), pp. 113–127. Uppsala: Swedish Univ. Agric. Sci.

WATSON, P. F., PLUMMER, J. M. & ALLEN, W. E. (1987). Quantitative assessment of membrane damage in cold-shocked equine spermatozoa. *J. Reprod. Fert. Suppl.* **35** (in press).

WATSON, P. F., PLUMMER, J. M., GLOSSOP, C. E. & ROBERTSON, L. (1985). A comparison of calcium accumulation in boar and ram spermatozoa following cold shock. In *Deep Freezing of Boar Semen* (ed. L. A. Johnson & K. Larsson), p. 266. Uppsala: Swedish Univ. Agric. Sci.

WEBER, C., POLLACK, R. & BIBRING, T. (1975). Antibody against tubulin: the specific visualization of cytoplasmic microtubules in tissue culture cells. *Proc. natn. Acad. Sci. U.S.A.* **72**, 459–463.

WELSH, M. J., PEDMAN, J. R., BRINKLEY, B. R. & MEANS, A. R. (1979). Tubulin and calmodulin. Effects of microtubule and microfilament inhibitors on localization of the mitotic apparatus. *J. Cell Biol.* **81**, 624–634.

WHITE, I. G. & WALES, R. G. (1961). Comparison of epididymal and ejaculated semen of the ram. *J. Reprod. Fert.* **2**, 225–237.

WILMUT, I. (1972). The low temperature preservation of mammalian embryos. *J. Reprod. Fert.* **31**, 513–514.

WILMUT, I., POLGE, C. & ROWSON, L. E. A. (1975). The effect on cow embryos of cooling to 20°, 0°, and −196°C. *J. Reprod. Fert.* **45**, 409–411.

Printed in Great Britain © *Society for Experimental Biology 1987*   341

STORAGE OF CELLS AND TISSUES AT HYPOTHERMIA FOR CLINICAL USE

BARRY J. FULLER

Academic Department of Surgery, Royal Free Hospital Medical School,
London NW3 2QG, UK

Summary

The ability of cells and tissues to withstand periods of removal from, or severe changes in, their normal environment is a necessary consequence of many surgical and medical therapies, particularly in the growing area of transplantation. The sequence of changes in mammalian cells during ischaemia is discussed. Following from this, the ability of hypothermia to slow all cell metabolic processes is described, and the concept is introduced that hypothermia itself eventually causes cell damage. Current knowledge on hypothermic damage is outlined, and the biological basis for therapeutic measures to minimize this damage is described. Finally, brief descriptions are given for hypothermic storage regimes in use clinically at present.

Introduction

The concept of treating disease by introducing to the sick patient normal living tissues from another source, which can assume the function of the atrophied tissue, has been a physician's dream for centuries. One of the earliest recorded attempts to transfer living tissues is that of Hunter in 1778 who reported on ovarian and testicular grafts. By the beginning of this century (Carrel & Guthrie, 1905) surgical and anaesthetic techniques had progressed sufficiently to put solid organ grafting (in this case of kidneys) tantalizingly close to becoming the valuable clinical procedure which we recognize it to be today. However, several major obstacles remained to be overcome before this progression could be made. One of these was the necessity to combat the tendency of the recipient patient's own immune system to recognize the graft as 'foreign', and to so activate immune destruction of the graft; this is beyond the scope of the current chapter. The other problem was that the grafted cells or tissues of necessity experienced a period of ischaemia (cessation of blood supply) whilst they were removed from the donor, taken to the recipient patient and transfused or transplanted. Obviously the grafts must be maintained in a viable condition during the *ex-vivo* period so that they can resume normal function quickly. With the pattern of modern clinical practice, and the establishment of computerized lists of waiting recipients of known immunological type, it is frequently the case that the *ex-vivo* storage period for donated tissues and organs may be many hours or, in some cases, days. This requires that exacting attention be paid to preventing ischaemic degeneration of the graft. One fundamental way to slow the rate of degeneration of mammalian cells and tissues is to cool their environment. This observation has been known to mankind for many thousands of years in relation to the putrefaction of meats and animal food products, and was the subject of a treatise by Boyle (1683). However, the scientific basis for the cause of cell death during ischaemia, and its amelioration by hypothermia, has only begun to be understood at a biochemical level during the past thirty years, and still many areas remain either controversial or to be verified. In the first part of this chapter we shall consider the mechanisms of ischaemic injury; subsequent sections will deal with the molecular basis for the protection by cooling, and the specific problems caused themselves by hypothermia; finally we shall discuss the present 'state of the art' for hypothermic storage of the various cells, tissues and organs commonly used in transfusion/transplantation.

A. The harmful effects of ischaemia

The term 'homeostasis' (originally attributed to Cannon) – the regulation of stable internal body environment – is a very useful concept when applied to mammalian systems because in a single word is conveyed a complexity of interrelated metabolic processes essential for normal vitality; as Claude Bernard so eloquently hypothesized: 'La fixité du milieu intérieur est la condition de la vie libre, indépendante.' The circulatory system bathes the component cells of all tissues and organs with a nutrient-rich plasma, in which ions, dissolved gases and metabolic substrates are maintained within closely-defined limits. Hydrogen ion concentration is regulated such that pH remains within the range 7·31 – 7·43, and the osmolality of the plasma remains close to $300\,mosmoles\,kg^{-1}$ water. The vascular system also acts as a heat exchanger, maintaining a constant body temperature close to 37°C. Within this controlled environment cells of tissues and organs are able to function and replicate (where necessary) with a high degree of efficiency. To carry out these vital processes, a continuous supply of energy is required, which is derived from the metabolic breakdown of substrates (e.g. glucose, fatty acids, ketone bodies, amino acids) supplied from the plasma, along with dissolved oxygen. Substrates are oxidized by a series of well-documented enzymic transformations (for review see Stryer, 1981) to carbon dioxide, with a concomitant reduction of the cellular pool of nicotinamide adenine dinucleotides (NAD, NADP). Energy is liberated from these enzymatic transformations, with the greatest energy yield being produced by reconversion of the reduced nucleotides to the oxidized forms *via* the mitochondrial electron transfer system. As is well known, this process requires molecular oxygen to yield the final end product (water), and the liberated energy is harnessed and stored by the cell in the energy-rich phosphate bonds of adenosine triphosphate (ATP) for subsequent use e.g. in synthetic reactions, contractility etc. Energy is also expended in the maintenance of the intracellular environment, or more correctly, environments (since most cell types exhibit a degree of intracellular compartmentalization which can be recognized both structurally and functionally). For example, the intra-

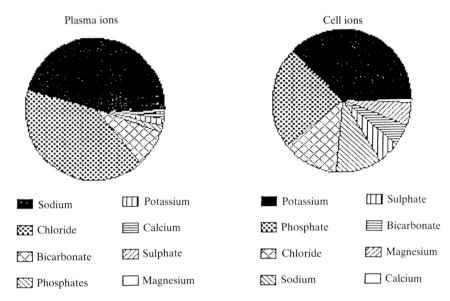

Plasma ions   Cell ions

■ Sodium   ⊞ Potassium   ■ Potassium   ⊞ Sulphate

▨ Chloride   ≣ Calcium   ▨ Phosphate   ≣ Bicarbonate

▨ Bicarbonate   ▨ Sulphate   ▨ Chloride   ▨ Magnesium

▨ Phosphates   ☐ Magnesium   ▨ Sodium   ☐ Calcium

Fig. 1. Pye chart representation of the differences between plasma and intracellular ion contents. In plasma sodium and chloride ions predominate, bicarbonate is the next most plentiful, and there are low levels of potassium, calcium, magnesium, sulphate and phosphates. In the intracellular environment potassium is now seen to be the most plentiful ion; phosphates and sulphate are present in proportionately higher concentrations than found in plasma whilst sodium is much reduced. The values represented are total concentrations and make no distinction between bound and 'free' species.

lysozomal environment is very different from that in the general body of the cell. Indeed, Clegg (1984) has argued that cells may be divided into a much larger number of compartments than considered before, by the presence of a very fine microtrabechular lattice which can only be visualized by high voltage electron microscopy. Even considering the cell as a single unit, it is known (Mudge, 1951) that the ionic environment is very different from that of plasma (see Fig. 1). Plasma contains relatively high concentrations of sodium and chloride, and low concentrations of potassium and magnesium. Conversely, the intracellular environment is rich in potassium and magnesium, with low sodium and chloride concentrations. The divalent cations calcium and magnesium are very important to cell function (see Campbell, 1983) and the concentration of the free ions, and their intracellular localization, are exquisitely controlled. Other multivalent metals (transition metals such as iron and copper) are also important biologically and are subject to similar strict controls of distribution (see Halliwell & Gutteridge, 1985). Mammalian cells expend large quantities of metabolically-derived energy on the control of the intracellular environments, by the activity of transmembrane pumps such as the $Na^+ K^+$ ATPase, and the $Ca^{2+} Mg^{2+}$ ATPase, and the control of internal stores, such as the mitochondrial accumulation of calcium. An excellent review of the importance of the membrane Na^+-K^+ ATPase in controlling intracellular ionic contents was given in, 1964 by Judah & Ahmed. The same authors also demonstrated in liver that a transmembrane pump was at least partly responsible for control of cell calcium, although we now realize that other factors also play a role. The membrane pumps contribute towards the net negative charge of the membrane potential, and to cell volume control. Sodium ions from the plasma, and potassium from the intracellular compartment, tend to diffuse across the cell membrane down their respective concentration gradients. By the activity of the membrane ATPase, sodium is translocated out of the cell, and potassium into the cell (on a stoichimetric basis of three sodium ions per two potassium ions). The major plasma anion (chloride) distributes freely across the membrane, and the negative charge of the membrane potential dictates that less chloride will partion into the intracellular compartment, given that some intracellular anion results from charges on protein. Under conditions of normal cellular metabolism these diffusional movements are self-limiting so that if the negative charge increases (too great a cation loss), the further diffusion of positively charged potassium ions out of the cell is opposed. The net result of the $Na^+–K^+$ ATPase activity is the movement of sodium and chloride out of the cell. Water distributes freely across the cell membrane, and its movement is dictated by the relative osmotic forces across the membrane. The $Na^+–K^+$ ATPase by controlling sodium chloride expulsion from the cell, also controls internal osmolality, and following from this, cell volume.

From these brief descriptions of cell functionality, it will be obvious that anything which upsets the balance of homeostasis will have severe and far reaching effects. In particular ischaemia, induced by interruption of normal blood circulation and removal of a tissue from its physiological site, immediately jeopardizes the ability of individual cells to carry out vital life processes. The essential supply of energy substrates and dissolved oxygen is at once interrupted. Many cells have metabolic reserves which can be mobilized to provide energy substrates (e.g. glucose from glycogen), but it is the absence of supplies of oxygen which is most immediately noticeable. At normal body temperatures the available oxygen, present as dissolved gas or bound to haemoglobin in the red corpuscles in the vascular bed, is used up very quickly once blood flow is interrupted. At a gross level this is manifest as a transition in colour of an ischaemic organ e.g. kidney, from 'pink' (normal content of corpuscular oxy-haemoglobin) to 'purple' (deoxy-haemoglobin predominant) once ischaemia exceeds about two minutes. Since oxygen is essential to mitochondrial electron transport, it would be supposed that this process would show an early derangement in function, and this is exactly what has been shown in a variety of studies. There is an inability to maintain the cellular stores of high-energy phosphates (e.g. ATP, creatine phosphate) necessary for normal metabolism. For example, in studies on the isolated rat heart by phosphorus-nuclear-magnetic-resonance Garlick et al. (1979) demonstrated that once the organ was rendered ischaemic, in 10 minutes phosphocreatine was completely exhausted, and ATP concentrations had fallen by 50 %. In non-muscle tissues, which lack phosphocreatine reserves, the fall in ATP is more rapid. The failure of mitochondrial electron transport also means that reduced nucleotides

(NADH, NADPH) cannot be reoxidized and these rapidly accumulate. This, coupled with the loss of ATP generation, inhibits many of the enzymatic processes of normal oxidative metabolism. Many cells have the capability to perform some anaerobic metabolism e.g. glycolysis, which can partially restore ATP production by non-mitochondrial substrate transformation, but this is a very inefficient process (liberating only 3 % of the available chemical energy from a molecule of glucose) compared to normal oxidative metabolism (which can yield about 50 % of available energy). Additionally glycogenolysis to yield lactate *via* glycolysis causes a net accumulation of hydrogen ions in tissues. In studies on rat kidney ischaemia, it was demonstrated that intracellular pH dropped from around 7·3 to 6·7 in 20 min (Sehr *et al.* 1979). Lactate formation may not be the sole source of hydrogen ion accumulation during ischaemia (Rouslin *et al.* 1986). Whatever the cause, it is known that the resultant pH shift inhibits many metabolic processes, including eventually glycolysis.

The abolition of all available methods for cell energy trapping, and the subsequent depletion of ATP, ensure that many of the homeostatic mechanisms considered above are threatened. Transmembrane pumping mechanisms (dependent on ATP supply) are unable to operate, causing large changes in ionic content of intracellular compartments as ion gradients are dissipated by diffusion. Sodium ions diffuse into, and potassium ions out of, the cells. Further, the presence of intracellular impermeant anion results in a larger sodium influx than potassium efflux, effectively reducing the transmembrane negative charge, which facilitates entry of chloride with the sodium. This net movement of sodium chloride into the cell results in water uptake (and cell swelling) by osmotic forces (Leaf, 1955). Once mitochondrial electron transport (and maintenance of the proton gradient) begin to fail, the ability to maintain normal very low intracellular calcium is lost. It was reported by Wallach *et al.* (1966) that liver slices and perfused liver tended to accumulate calcium, particularly if incubated in metabolically-unfavourable conditions. There appear to be two interrelated problems in control of cell calcium during ischaemia. Firstly as with the Na^+–K^+ ATPase, the transmembrane pumping of calcium from the cytosol requires consumption of ATP and so is impaired when ATP concentrations fall. However, another consequence of falling mitochondrial ATP concentrations appears to be loss of sequestered calcium to the cytosol (Bygrave, 1978). Mechanisms of cell damage resulting from failure to control cytosolic free calcium remain contentious (see Campbell, 1983 for review). It has been postulated that in cells which contain intracellular networks of actin-containing microfilaments, abnormal free calcium concentrations may act to cause disruption (Uyeda & Furuya, 1986). Phospholipase A_2, an enzyme which can act to destroy the phospholipid components of biological membranes, has been shown to be activated by abnormal free calcium ion concentrations, both for the plasma and mitochondrial membrane species (Bygrave, 1978). Whatever the true sequence of events are, calcium disbalance has been shown to occur hand in hand with functional impairment in many types of cell damage (Jewell *et al.* 1982).

The falling intracellular pH may have severe consequences for cell metabolism *via* impaired control of other metal ions, in this case of the transition metal iron. Under normal conditions of aerobic metabolism, cells undertake tetravalent reduction of oxygen *via* the mitochondrial electron transfer processes to liberate water. However, it is considered that this sequence of transfers is not totally efficient, and that small quantities of highly reactive oxygen species (which have undergone only partial or univalent reduction) are liberated (see Fig. 2). These reactive oxygen species include superoxide and hydroxyl radical, which can react with a wide variety of biological molecules in a non-enzymic fashion to bring about changes in structure and function (see Halliwell & Gutteridge, 1985). Not only does the alteration of cell macromolecules itself cause damage, e.g. by changing protein conformation or phospholipid fluidity characteristics, but also many of the breakdown products are toxic. In the case of lipid peroxidation, malonyl dialdehyde (one of the detectable fragmentation products) and lipid hydroper-oxides are known to cause the mitochondrial release of calcium (Richter & Frei, 1985). Cells normally have a variety of defence mechanisms for neutralizing these damaging radical species, including enzymatic (e.g. superoxide dismutase, cata-lase) and non-enzymatic (e.g. reduced glutathione, *a*-tocopherol) scavengers. It has been shown that during ischaemia and also during the immediate post-ischaemic period when blood flow is re-established, the ability of these natural defence mechanisms is impaired, and harmful free-radical effects may occur (Parks *et al.* 1982; Green *et al.* 1985). Again the sequence of events remains to be fully elucidated, but it is considered that the transition metal iron plays a central role in propagating molecular damage from radical interaction (Fujimoto *et al.* 1984). Free iron can catalyse the breakdown of many of the intermediates of biolmolecules and reactive oxygen species by redox cycling (Aust, 1986). This can

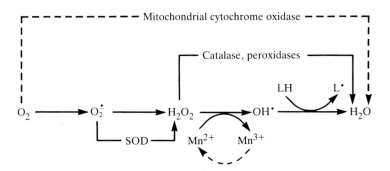

Fig. 2. Oxygen normally undergoes tetravalent reduction *via* the mitochondrial cytochrome oxidase complex. Under certain circumstances univalent reduction can proceed to liberate superoxide ($O^{-\cdot}$), which dismutates to hydrogen peroxide spontaneously or by interaction with super-oxide dismutase (SOD). In the presence of transition metals hydroxyl radical ($OH^{\cdot}$) can be formed, and this process can be ongoing if redox cycling of the transition metal takes place. $OH^{\cdot}$ will react with almost any biomolecule e.g. lipid (LH) to produce a lipid radical ($L^{\cdot}$) and water.

convert a relatively minor insult into a chain reaction of degradation, as has been shown for destruction of polyunsaturated fatty acid components of membranes (Poli *et al.* 1985). Cells normally strictly control iron metabolism by binding the metal to protein stores and carriers (transferrin, ferritin, haemosiderin etc.), but it has been shown that in acid pH, free iron can be liberated from some of those sources (O'Connell *et al.* 1985). This, coupled with a propensity to degrade some nucleotide precursors *via* a radical-generating sequence during ischaemia may have disastrous consequences for cell and tissue viability (Adkison *et al.* 1986). The end result of this cascade of events is a tissue which cannot resume normal function once blood supply is restored. The time course of this lethal progression is such that most organs are severely damaged by one hour of ischaemia at 37 °C, and for some specialized tissues such as cardiac muscle, sensitivity is even higher.

It must also be pointed out that whilst we have considered the cellular effects of ischaemia, for organized tissues and organs there is another damage component which concerns the vascular bed. If the vasculature has been damaged in such a way that restoration of normal blood supply is compromised after the ischaemic insult, then the tissue will be subjected to a further period of anoxia. This has been termed the 'no-reflow' phenomenon and is thought to result from a complex interaction of events (see Fig. 3). There is some evidence that red blood cells trapped in the vascular bed during ischaemia are themselves metabolically damaged and become less deformable (Weed *et al.* 1969), so blocking microcirculatory perfusion. It is also recognized that some of the altered lipid adducts liberated by phospholipase action and free-radical interaction are potent vasocative agents, capable of causing platelet activation and again blocking reflow (Demopoulos *et al.* 1980).

B. Protection by cooling – a balance between benefit and harm.

The ability of reduced temperatures to inhibit metabolic processes has been recognized since the earliest attempts to use *in vitro* incubations of cells and tissues for study. Molecular interactions which form the basis of metabolism require energies of activation, and in a simplistic fashion it can be said that removal of energy in the form of heat will slow these reactions. The relationship is further complicated when metabolism in any case requires an input of energy (most often in the form of ATP), the supply of which itself is highly temperature dependent. It is most common to express the relationship between reaction rate and temperature in the form of an Arrhenius plot (see Fig. 4). An alternate method of depicting the temperature dependence is to compute the Q_{10} of a particular reaction i.e. the decrease in a given metabolic process for a reduction of 10 °C. For example, Zimmerman *et al.* (1982) have shown that the Q_{10} for ketogenesis by the perfused rat liver was 5·2 over the range 20–10 °C (i.e. the reaction proceeded 5 times slower at 10 °C). Thus it can be seen that reducing cell temperature close to 0 °C is a powerful method for slowing metabolic changes which result from ischaemia. However, it must be stressed that cell metabolism is so varied and complex,

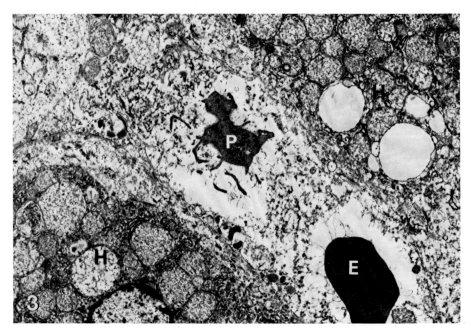

Fig. 3. A transmission electron micrograph of pig liver which has been subjected to ischaemia at 37°C for 2h, and then blood supply to the organ has been restored for 5 min. In the sinusoidal space between two hepatocytes (H) can be seen a trapped and distorted erythrocyte (E) and a platelet (P) which has been activated and is also contributing to vascular blockage (×3400).

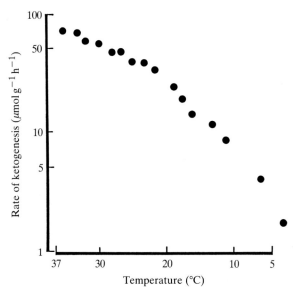

Fig. 4. An Arrhenius plot depicting the change in rate of ketogenesis in perfused rat livers as temperature is reduced. (Data redrawn from Zimmerman *et al.* 1982.)

requiring for different processes different substrates, enzymes and organelle function, that it would be naive to expect all processes to be slowed by the same degree. Using data again from Zimmerman *et al.* (1982) the Q_{10} for glucose production from gluconeogenesis by perfused rat livers was 8·5 over the same temperature range as for the 5·2-fold reduction in ketogenesis. It is obvious that during cooling that there could be a rate-dependent dislocation of metabolic processes which occur normally in a coupled fashion at 37°C, leading to depletion of particular substrates, or atypical accumulation of intermediates and end products. This type of metabolic 'disjointedness' is largely beyond the influence of the investigator. Another problem is that passive diffusional processes are much less affected by temperature reduction to 0°C than is the active metabolism. As has been discussed in the earlier part of this chapter, maintenance of intracellular ionic composition and volume regulation depend on 'pump-leak' interactions at the membrane level where membrane-bound enzymes transport various ions and solutes into and out of cells, opposing passive diffusion driven by differences in chemical potential gradients. Applied hypothermia inhibits pump activity both by direct effects on enzyme activity and on supply of ATP energy (as mitochondrial energy transduction fails), whilst appreciable trans-membrane diffusion takes place at temperatures close to 0°C. For example, Ellory & Willis (1981) demonstrated that in a variety of species erythrocyte Na^+-K^+ATPase activity at 5°C was only 1 % of that measured at 37°C. Berthon and co-workers (1980) demonstrated that in isolated hepatocytes cooled from 38°C to 1°C for 90 min, there was an approximate 25 % reduction in cell potassium, a doubling of cell sodium and 50 % increase in chloride ion. Concomitant with these changes was a 20 % increase in cell water content. The overall effect was to reduce membrane net negative charge. The ability of reduced temperatures to allow the diffusional redistribution of ions between intra- and extracellular fluid has been recognized from early studies (see Leaf, 1955). In the short term these solute movements are readily reversible – in the hepatocyte studies of Berthon *et al.* (1980) restoration to 37°C caused a rapid return to normal cell ion homeostasis. However, as with warm ischaemia there is a gradual accumulation of damaging changes which eventually become irreversible. Now, however, the time scale has been extended and the changes take place in some cases for different reasons than seen in warm ischaemia. Changes in sodium and potassium themselves may not cause many irreversible alterations, but the lack of maintenance of normal trans-membrane sodium gradients will adversely affect many secondary transport systems such as for glucose and amino acids and also electrical events in excitable tissues. Most cells have a variety of inter-related cation transport systems depending on energy supply, all of which will be affected by reduced temperature (see Ellory & Willis, 1981; Ellory, this volume). It has been recognized that changes in divalent cations (calcium and magnesium) also follow as a consequence of cooling, and particularly for calcium these have been assigned more damaging roles. Van Rossum (1970) showed that a net increase in calcium and decrease in magnesium occurred in liver slices maintained at hypothermia over the same time scale (120 min) as for sodium

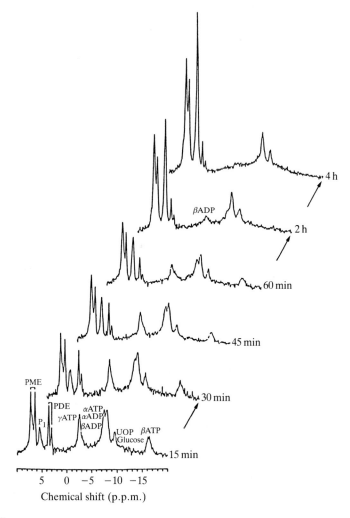

Fig. 5. ^{31}P nuclear magnetic resonance spectra for rat liver stored at hypothermia after vascular flushing, taken at intervals of up to 8 h. As cold storage time increases, signals for ATP and ADP disappear, and in a parallel fashion the inorganic phosphate content (Pi) increases. (See Busza et al. 1986.)

and potassium changes. There still remains some controversy as to whether the increase in total tissue calcium reflects an increased cytosolic free calcium or represents binding to other intracellular sites. Such binding in any case may represent damage. As was mentioned previously, mitochondria can sequester calcium in a way that causes uncoupling of energy transduction (Bygrave, 1978). This process is facilitated by loss of phosphorylation and increase in anions such as inorganic phosphate (Reed & Bygrave, 1975). We have shown from ^{31}P NMR studies on rat liver during hypothermia (see Fig. 5) that there is a loss of ATP and ADP from the tissues, which coincides with a rising inorganic phosphate signal (Busza et al. 1986). Again we can see the complexity of interdependence between the various processes taking place in the cold tissues. Another example of this is the importance placed by Hall & Willis (1984) on increased cell calcium in facilitating further net potassium flux across cell membranes during hypothermia by a calcium activated 'Gardos' channel. This phenomenon has also been highlighted as a possible mechanism for the anomalous potassium flux seen in studies on some cells e.g. dog erythrocytes (Elford, 1975), which actually increased as cells were cooled from 12°C to 5°C. However, a second school of thought exists which places importance on the temperature-induced separation of membrane protein and lipid constituents as the gel transition temperatures of the various lipid classes are passed during cooling. This can result in lipid-rich domains with exclusion of integral proteins into protein-rich areas (see Quinn, 1985) with production of packing faults at boundary regions. It is suggested that these packing faults could represent areas of membrane 'leakiness' although most evidence for this results from studies of defined model membranes, not cellular systems. Nevertheless, phase separation of mammalian membranes have been demonstrated at temperatures of ~10°C and below (Hochli & Hackenbrook, 1976), which would fit the trend of increased permeabilities at hypothermia, not only to ions but to molecules as large as cytosolic proteins (Fuller & Attenburrow, 1976). Whether proteins could diffuse through such faults with the imposed constraints of size and charge is still to be demonstrated. Another possibility may be that hypothermic cell swelling (and changes in divalent cation contents which normally act to stabilize bilayer structure) could cause loss of very small vesicles of membrane material entrapping cytosolic constituents. In other forms of cold-induced membrane damage such losses of membrane phospholipids have been well documented (see Watson & Morris, this volume).

The continuing cell metabolism has marked consequences for another of the important homeostatic mechanisms of cell survival, namely pH regulation. There is a consistently reported increase in hydrogen ion concentration in the majority of mammalian tissues studied during hypothermic storage. For example Chan *et al.* (1983) reported studies using ^{31}P NMR on human kidneys which demonstrated an approximate 0·6–0·8 pH unit fall during 24 h storage in ice. We have demonstrated a similar degree of increasing hydrogen concentration in rat livers over an 8 h study time. It has been suggested that the falling pH, in combination with the altered ionic environment during hypothermia, may serve to destabilize lysozomes with release of harmful proteases into the tissue (Calman *et al.* 1973). Certainly, lysozomal enzyme release has been documented in some experimental organ storage experiments (Johnson *et al.* 1979). The source of increased hydrogen ion concentrations has been attributed to a continuing production of lactate from glycolysis but as in warm ischaemia hydrolysis of ATP may play a significant role (Rouslin *et al.* 1986). Another damaging interaction of falling pH may involve release of free iron and catalysis of oxidative stress, as highlighted in the section on warm ischaemia. We have shown that lipid peroxidation can proceed during hypothermic storage of kidney (Gower *et al.* 1986) and liver tissue. In isolated hepatocytes malonyl dialdehyde concentration was shown to increase during 24 h at 2°C (see Fig. 6). This increase in products of lipid peroxidation could be

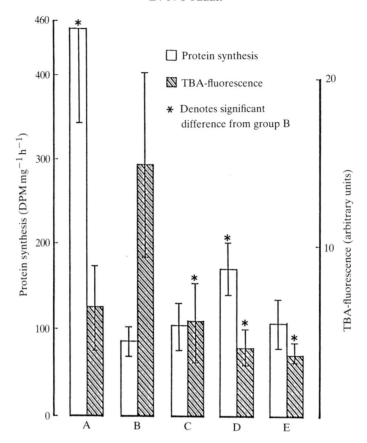

Fig. 6. The content of lipid peroxidation products (TBA-fluorescence) and protein synthesis in isolated rat hepatocytes either immediately upon isolation (A) or after 24 h storage in ice (B–E). In group C, the iron chelator desferrioxamine was added at a concentration of $1 \, \text{mmol} \, l^{-1}$, in group D it was at $5 \, \text{mmol} \, l^{-1}$ concentration, and in group E at $10 \, \text{mmol} \, l^{-1}$ concentration. It can be seen that the chelation of free iron reduced lipid peridation during cold storage (groups C–E compared to group B) and desferrioxamine at $5 \, \text{mmol} \, l^{-1}$ concentration allowed a small improvement if post-storage protein synthesis.

prevented by chelating free iron, from which followed a small but significant protection of post-storage cell metabolism. The mechanisms of induction of oxidative stress during hypothermia again may differ from those prevailing in warm ischaemia, and a review of these can be found in Fuller & Green (1986). There may also be a relationship between free radical interaction and vascular damage seen when blood supply is restored to organs after hypothermia. In particular, we have shown a more significant increase of malonyl dialdehyde in the medulla of rabbit kidneys after cold storage (Gower et al. 1987), which may have some bearing on the reported sensitivity of the blood supply in the cortico-medullary junction to cold ischaemia (Norlen et al. 1978). This is especially so because inhibitors of cyclo-oxygenase can prevent this increased medullary malonyl dialdehyde content – (cyclo-oxygenases are involved in prostaglandin synthesis which produces a variety of vasoactive substances). The production proceeds *via* a series of radical-intermediates, and there appears to be an intimate relationship between lipid peroxidation and prostaglandin synthesis (see Halliwell & Gutteridge, 1984). The demonstrated reduction in some cellular anti-oxidants during hypothermic storage (Green *et al.* 1986) may exacerbate the tendency towards oxidative stress.

It will be obvious from the two preceding sections (A, B) that many of the indicators of damage are common to both warm and cold ischaemia. Individual stress markers may occur in a different sequence in either condition, and of course the overall time course for progression of the syndrome is slower at hypothermia. However the similarities merely serve to reinforce the concept that loss of intracellular homeostasis is the hall mark of dying cells in either environment.

C. Strategies to maximize the benefit of cooling

Two methods have been developed to extend hypothermic storage of tissues and organs to provide clinically-useful times. One involves bathing the tissues or flushing the vascular bed of an organ with cold solutions of specialized composition (flush storage), followed by storage in ice. The second depends upon continuously perfusing a synthetic solution containing substrates and oxygen through the organ at intermediate hypothermia (10–15 °C), with the aim of stimulating low levels of metabolic function and preventing cell deterioration.

(i) *Immersion or flush storage*

In flush storage one of the first strategies devised to minimize cold damage was based on prevention of the progressive ionic redistributions described above. If ionic changes are driven by passive diffusion down chemical potential gradients, then abolishing the gradients should remove the driving force. This stimulated Keeler *et al.* (1966) to use solutions of high potassium and magnesium content to perfuse rat kidneys at hypothermia, and this modification enhanced successful storage. The idea was further developed by Collins *et al.* (1969) who made extensive studies on kidney storage after flushing the renal artery with solutions containing elevated concentrations of potassium, magnesium and phosphate (see Table 1) and also made hyperosmolar with glucose to minimize cell water uptake. Sodium and chloride contents were much reduced compared with those in normal plasma. This type of preservation solution became labelled as an 'intracellular'-type flush because the content to a large extent reflected that of intracellular fluid, although this is not strictly true. In any case this solution, and derivatives of it, allowed significant prolongation of preservation to periods of 48 h or longer in kidneys harvested under optimal conditions. Since then other solutions have been developed (Sacks *et al.* 1973; Ross *et al.* 1976) and found to be similarly successful, each based on the original premise of minimizing diffusion processes, but with differences in relative concentrations of ions or species of impermeant molecules.

Table 1. *The constituents of three flush solutions used for organ storage; A – Collins C2 (Euro-Collins); B – Sack's; C – Marshall's Citrate*

Contents per litre		A	B	C
Na^+	(mEq)	10	15	80
K^+	,,	115	143	80
Mg^{2+}	,,	—	16	70
Cl^-	,,	15	16	—
SO_4^{2-}	,,	—	—	70
$HPO_4^{(-)}$	,,	100	120	—
HCO_3^-	,,	10	38	—
Citrate	,,	—	—	165
Mannitol	,,	—	206	150
Glucose	,,	190	—	—
Osmolality	(mosm kg^{-1})	375	430	400
pH		7·25	7·0	7·1

Studies have been undertaken to try and establish which of the constituents of these storage solutions provides the significant enhancing effect, but without clearcut answers. Green & Pegg (1979) undertook extensive studies in the isolated rabbit kidney and arrived at the following conclusions: the most important factor was prevention of cellular oedema by inclusion of slowly permeating species of non-ionic (glucose, sucrose, mannitol) or ionic (phosphate, sulphate) derivation. Phosphate was also important in providing maintenance of pH by its buffering capacity, although organic buffers such as Hepes also can provide this control. The efficacy of other additives and drugs remains to be proved unequivocally although there are several on-going lines of research which in future may provide improved flush storage. One such line has centred on the problems of calcium redistribution alluded to above in Section B. It has been recognized that some flush constituents such as phosphate and citrate are good calcium chelating agents. More recently specific calcium channel blocking agents (Shapiro *et al.* 1986) and inhibitors of the transport protein calmodulin (Asari *et al.* 1984) have been shown to enhance cold storage of kidneys. The exact mechanisms of action here remain to be elucidated.

The use of organic buffers has been shown to be effective in controlling pH drop in studies on rat kidney storage (Bore *et al.* 1981). Improved buffering capacity correlated with higher post-storage function. Such agents have yet to be tried clinically.

The use of antioxidants and free radical scavengers in tissue storage is an area of increasing study. As described in Section B above, one way to use such therapy is to add the agents to the storage solutions and this has proved effective in minimizing the oxidative stress in kidney storage experiments (Fuller *et al.* 1985). Another mode is to infuse scavenging agents into the recipient just before re-establishment of the blood flow to the stored organ. This method has proved effective in studies to test the theory that most free radical stress occurs at the time of rewarming and reoxygenating the cells. As argued previously (Fuller & Green, 1986) it is likely that therapy will be required both during storage and after reflow for maximum protection. Radical scavenging agents have also proved effective during continuous hypothermic perfusion (Atalla *et al.* 1985). Again, the transition from experimental to clinical application has yet to take place.

The search for agents to minimize storage-induced diffusion of ions and water has also continued. Gluconate has been proposed as an anion for replacement of chloride in preservation solutions to prevent cell swelling (Southard & Belzer, 1980). There is some suggestion that isosmolar storage solutions are more protective than hypertonic solutions, provided agents such as citrate or gluconate are used to prevent cell oedema. A method which involves gaseous persufflation *via* the renal vein during hypothermic storage of kidneys has been developed (Rolles *et al.* 1984) which overcomes the additive damaging effect of a period of warm ischaemia prior to cold flushing. The mechanism of this protection is at present unclear; it appears to require oxygen although no change in high energy adenine nucleotides was detected as a consequence of persufflation.

(ii) *Continuous hypothermic perfusion*

This method is applicable to tissues and organs, and as the method suggests, involves continuous perfusion of the vascular bed, almost always with an acellular solution. In general the composition of the solutions tends to be closer to that of plasma rather than of intracellular fluid. Temperatures are maintained at about 10 °C; oxygen and metabolic substrates are added to the perfusate to support the reduced metabolic activities taking place at these temperatures. It has been shown that lipid substrates, in particular short-chain fatty acids, can be utilized during hypothermic kidney perfusion (Huang *et al.* 1971) to provide energy. Metabolism is such under these conditions that kidneys are able to resynthesize tissue reserves of high-energy adenine nucleotides depleted by a prior warm ischaemic insult (Pegg *et al.* 1984). In studies on continuous hypothermic perfusion of liver we have similarly shown that a better maintenance of ATP reserves was achieved (Attenburrow *et al.* 1981) and that this could stimulate partial activity of trans-membrane Na^+-K^+ ATPase which was absent in livers stored by flushing alone. As for the kidney, livers subjected to ischaemic damage during harvesting could be 'recharged' by the perfusion period, such that post-storage protein synthesis was maintained better than if flushing alone was used for storage (Fuller & Attenburrow, 1979a) in studies on rat liver. Similar improvement in prolongation of storage and in post-storage function have been reported for biochemical studies of dog liver (D'Alessandro *et al.* 1986) and transplantation of rat liver (Kamada *et al.* 1980).

Perfusion pressures utilized (usually 40–60 mmHg) are much lower than normal arterial pressure and are designed to be the minimum required for provision of adequate tissue perfusion. To balance this applied hydrostatic pressure and to prevent interstitial oedema, oncotic agents such as albumin (Belzer, 1977) or synthetic macromolecular agents (Fuller & Attenburrow, 1979b) have been used. Perfusion is advantageous in that accumulation within the tissue of harmful

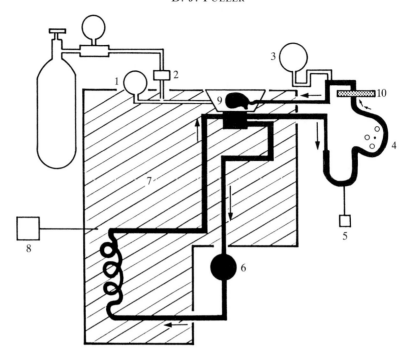

Fig. 7. A schematic view of a perfusion circuit suitable for hypothermic perfusion of kidneys in clinical practice. 1 – gas pressure gauge; 2 – gas filter; 3 – perfusate pressure gauge; 4 – roller pump; 5 – flow meter; 6 – secondary pump for cooling system; 7 – cold chamber; 8 – temperature monitor; 9 – kidney in sterile chamber; 10 – membrane filter. Perfusate is pumped into the renal artery in a controlled fashion by pump 4. Effluent perfusate passes to the reservoir, from where it is passed through a heat-exchanging circuit by pump 6.

metabolites (e.g. hydrogen ions, ammonia) can be prevented, and it allows for continual removal of any residual blood cells, so clearing the microcirculation over a period of hours. It has been shown that even prolonged flushing (up to 30 min) of rabbit kidneys at the time of harvesting did not remove all blood cells (Foreman et al. 1982).

Continuous hypothermic perfusion of kidneys is preferred in the clinical situation if the organs have been damaged by prior warm ischaemia during harvesting. Experimentally, the most consistent prolonged storage of kidneys (3 days) has been achieved by continuous perfusion (see Johnson, 1983, for review). However, the method is more expensive than simple flush storage because of the requirements for large volumes of perfusates and reliable perfusion machines. A typical perfusion circuit is depicted in Fig. 7. The use of sterile, non-toxic and largely disposable tubing with oxygenators and filters, in addition to a reliable pump, are some of the reasons for the high cost of the circuitry. Experienced personnel are required to supervise the perfusion procedure, since it has been shown that bad perfusion technique will cause organ damage (Cerra et al. 1977). An interesting experimental approach to prolongation of successful perfusion has been to reconnect the organ to an intermediate host blood supply or normothemic perfusion circuit (Van der Wijk *et al.* 1980) in the middle of a period of hypothermic perfusion. This has been found to allow consistent 6 day survival of canine kidneys. At present this method has not found clinical application, but the possibility that immuno-reactiveness of an organ could be modulated by various agents during prolonged perfusion storage (Pegg & Taylor, 1984) may give further impetus to this type of approach.

In general it will be obvious from the above descriptions that hypothermic perfusion is the most likely candidate as a method to increase storage time and yield superior post-storage organ function. It must be pointed out however, that constraints of finance and personnel, especially in the United Kingdom and Europe, have acted to prevent adoption of the method. Increased activity in the field of transplantation may eventually change this outlook.

D. Some examples of current methods of hypothermic storage of cells, tissues and organs in clinical application

(i) *Blood cells*

In the United Kingdom whole blood is routinely stored at 4 °C after dilution with a solution based on acid-citrate-dextrose (Beutler, 1972), with a maximum permitted storage of 21 days. Platelet-rich plasma, which is often required when bleeding complications occur in patients, is much more storage sensitive and is generally used within 24 h after harvesting and maintenance at 4 °C, although some centres may prolong storage to about 72 h by keeping the platelets in autologous plasma and gently agitated at 22 °C.

(ii) *Corneas*

Corneal grafting to treat various types of blindness depends upon the current practice of harvesting excised cornea from deceased donor patients. The cornea are stored in a special solution at 4 °C for up to 4 days (Taylor, 1986).

(iii) *Skin*

Split thickness skin for grafting may be placed dermal side down on a sterile nylon mesh. The whole can then be rolled and stored in tubes filled with chilled sterile tissue culture medium for up to 7 days (Ninneman & Fisher, 1981).

(iv) *Kidney*

In Europe and the United Kingdom the majority of kidneys for transplantation are harvested after flushing with pre-chilled Collins solution (Euro-Collins) or Marshall's Citrate solution. The organs are often flushed *in situ* first, because excessive handling during removal may cause a vasoconstrictor response and subsequent poor perfusion. The organs are then packed in ice for transport, along with small samples of donor tissue for immunological matching. Storage periods often range from 24–48 h.

(v) *Heart*

Only shorter preservation periods are acceptable in clinical transplantation, where 6 h has not been exceeded. The organ is harvested after flushing with a high potassium content cardioplegic solution, and stored in this, packed in ice. Similarly brief total ischaemic periods are essential in heart and lung transplants. Such circumstances make logistics of transplantation and effective use of donor organs difficult and expensive to organize.

(vi) *Liver*

Again the predominant method of storage is by packing carefully in a fluid-filled bag in ice, after vascular flush with one of the 'intracellular'-type flush solutions (Euro-Collins or Citrate) or a high potassium high magnesium plasma-based flush. A point of note is that these solutions must be flushed out of the organ just before blood supply is restored in the recipient as they are also excellent cardioplegics and will cause cardiac arrest in the paitent. 12-h storage has not been exceeded in clinical practice.

(vii) *Pancreas*

Once again, flush preservation with one of the common storage solutions, followed by ice packaging for periods not in excess of about 8 h is the current clinical procedure.

General comments

It must be stressed that the practices outlined in this section are the ones most commonly encountered in present clinical transplantation. There are numerous examples of experimental preservation protocols which have reported considerably longer successful preservation periods e.g. for 24 h in livers stored by continuous hypothermic. It is outside the scope of the present discussion to examine these variations in technique and details can be found in several recent publications (Karow & Pegg, 1981; Marberger & Dreikorn, 1983).

From the data and concepts described in this chapter several points emerge. On the one hand, given the structural and metabolic complexities of mammalian cells it is perhaps surprising that by simple storage protocols we have been able to achieve the success we already have in storing organs and tissues *ex vivo*. On the other hand the insult produced by removing the cells from their normal environment is so severe that the quality of early post-transplant function is poor (increasing overall risk to the patient) as storage time lengthens, and for some organs (heart, lung, liver) the 'safe' time is very short. Therefore there is considerable impetus for further research into the various mechanisms of cell damage at hypothermia, and methods which could be employed to circumvent these.

References

ADKISON, D., HOLLWORTH, M., BENOIT, J., PARKS, D. A., McCORD, J. & GRANGER, D. (1986). The role of free radicals in ischaemia-reperfusion injury to the liver. *Acta Physiol. Scand. Suppl.* **548**, 101–107.

ASARI, H., ANAISE, D., BACHAROFF, R., SATO, T. & RAPPAPORT, F. (1984). Preservation techniques for organ transplantation. 1. Protective effects of calmodulin inhibitors in cold-preserved kidneys. *Transplant.* **37**, 113–114.

ATALLA, S., TOLEDO-PEREYRA, L., MACKENZIE, G. & CEDERNA, J. (1985). Influence of oxygen-derived free radical scavengers on ischaemic livers. *Transplant.* **40**, 584–590.

ATTENBURROW, V., FULLER, B. J. & HOBBS, K. E. F. (1981). Effects of temperature and method of hypothermic preservation on hepatic energy metabolism. *Cryo-Lett.* **2**, 15–20.

AUST, S. (1986). The role of iron in lipid peroxidation. In *Free Radicals, Cell Damage & Disease* (ed. C. Rice-Evans), pp. 15–27. London: Richelieu Press.

BELZER, F. O. (1977). Renal preservation by continuous hypothermic perfusion; past, present and future. *Transplant. Proc.* **9**, 1543–1546.

BERTHON, B., CLARET, M., MAZET, J. L. & POGGIOLI, J. (1980). Volume- and temperature-dependent permeabilities in isolated rat liver cells. *J. Physiol.* **305**, 267–277.

BEUTLER, E. (1972). Preservation of erythrocytes. In *Haematology* (ed. W. Williams, E. Beutler, A. Ersler & R. Rundles), pp. 1299–1300. New York: McGraw-Hill.

BORE, P. J., SEHR, P. A., CHAN, L., THULBORN, K., ROSS, B. D. & RADDA, G. K. (1981). The importance of pH in renal preservation. *Transplant. Proc.* **13**, 707–708.

BOYLE, R. (1683). *New experiments and observations touching cold.* London: R. Davis.

BUSZA, A. L., FULLER, B. J., PROCTOR, E. & GADIAN, D. G. (1986). 31 P nmr of *ex vivo* rat liver flush-stored at hypothermia for transplantation studies. *Biochem. Soc. Trans.* (in press).

BYGRAVE, F. (1978). Mitochondria and the control of intracellular calcium. *Biol. Rev.* **53**, 43–79.

CALMAN, K. C., QUINN, R. O. & BELL, P. R. F. (1973). Metabolic aspects of organ storage and the prediction of viability. In *Organ Preservation* (ed. D. E. Pegg), pp. 225–240. Edinburgh: Churchill Livingstone.

CAMPBELL, A. K. (1983). *Intracellular Calcium.* New York: John Wiley & Sons.

CARREL, A. & GUTHRIE, C. (1905). The transplantation of veins and organs. *Amer. Med.* **10**, 1011–1035.

CERRA, F. B., RAZA, S., ANDRES, G. A. & SIEGEL, J. (1977). The endothelial damage of pulsatile renal preservation and its relationship to perfusion pressure and colloid osmotic pressure. *Surgery* **81**, 534–541.

CHAN, L., BORE, P. & ROSS, B. D. (1983). Possible new approaches to organ preservation. In *Renal Preservation* (ed. M. Margberger & K. Dreikorn), pp. 323–337. Baltimore: Williams & Wilkins.

CLEGG, J. (1984). Intracellular water and the cytomatrix: some methods of study and current views. *J. Cell Biol.* **99**, 167s–171s.

COLLINS, G. M., BRAVO-SHUGARMAN, M. & TEASAKI, P. I. (1969). Kidney preservation for transplantation. *Lancet* **2**, 1219–1221.

D'ALESSANDRO, A., SOUTHARD, J., KALAYOGLU, M. & BELZER, F. (1986). Effect of drug treatment on liver slice function following 72 hour hypothermic perfusion. *Cryobiology* **23**, 415–421.

DEMOPLOLOUS, H., FLAMM, E., PIETRONIGRO, D. & SELIIGMAN, M. (1980). The free radical pathology and the microcirculation in the major central nervous system disorders. *Acta Physiol. Scand. Suppl.* **492**, 91–119.

ELFORD, B. C. (1975). Interactions between temperature and tonicity on cation transport in erythrocytes. *J. Physiol., Lond.* **246**, 371–395.

ELLORY, J. C. (1987). This volume.

ELLORY, J. C. & WILLIS, J. S. (1981). Phasing out the sodium pump. In *Effects of Low Temperatures on Biological Membranes* (ed. G. J. Morris & A. Clarke), pp. 107–120. London: Academic Press.

FOREMAN, J. C., WUSTEMAN, M. C. & PEGG, D. E. (1982). Washout of red blood cells from kidneys damaged by warm ishaemia. In *Organ Preservation; Basic and Applied Aspects* (ed. D. Pegg, I. Jacobsen & N. Halasz), pp. 183–190. Lancaster: MTP Press.

FUJIMOTO, Y., MARUTA, S., YOSHIDA, A. & FUJITA, T. (1984). The effect of transition metals on lipid peroxidation of rabbit renal cortical mitochondria. *Res. Comm. Chem. Pathol. Pharmacol.* **44**, 495–498.

FULLER, B. J. & ATTENBURROW, V. D. (1976). Experimental studies on hypothermic storage of liver in rat. *Les Colloques de l'INSERM* **62**, 393–398.

FULLER, B. J. & ATENBURROW, V. D. (1979a). The effects of hypothermic storage by continuous perfusion and simple portal flushing on hepatic protein synthesis and urea production in the rat. In *Organ Preservation 2* (ed. D. Pegg & I. Jacobsen), pp. 278–292. Edinburgh: Churchill Livingstone.

FULLER, B. J. & ATTENBURROW, V. D. (1979b). The effects of increasing the osmotic and oncotic pressure of the perfusate on bloodless hypothermic liver perfusion in the rat. *Cryobiol.* **15**, 279–289.

FULLER, B. J. & GREEN, C. J. (1986). Oxidative stress in organs stored at low temperatures for transplantation. In *Free Radicals, Cell Damage & Disease* (ed. C. Rice-Evans), pp. 223–240. London: Richelieu Press.

FULLER, B. J., GREEN, C. J., HEALING, G., MARLEY, S., SIMPKIN, S. & LUNEC, J. (1985). The role of iron in oxygen free radical damage during kidney storage at hypothermia. *Cryobiol.* **22**, 614.

GARLICK, P. B., RADDA, G. K. & SEELEY, P. J. (1979). Studies of acidosis in the ischaemic heart by phosphorus nuclear magnetic resonance. *Biochem. J.* **184**, 547–554.

GOWER, J., FULLER, B. J. & GREEN, C. J. (1987). Lipid peroxidation in the cortex and medulla of rabbit kidneys subjected to cold ischaemia and the value of protective agents. *Free Radical Res. Comm.* (in press).

GREEN, C. J., HEALING, G., LUNEC, J., FULLER, B. J. & SIMPKIN, S. (1986). Evidence of free radical-induced damage in rabbit kidneys after simple hypothermic preservation and autotransplant. *Transplant.* **41**, 161–165.

GREEN, C. J., HEALING, G., SIMPKIN, S., LUNEC, J. & FULLER, B. J. (1986). Increased susceptibility to lipid peroxidation in rabbit kidneys: a consequence of warm ischaemia and reperfusion. *Comp. Biochem. Physiol.* **83B**, 603–606.

GREEN, C. J. & PEGG, D. E. (1979). The effect of variation in electrolyte composition and osmolality of solutions for infusion and hypothermic storage of kidneys. In *Organ Preservation 2* (ed. D. E. Pegg & I. A. Jacobsen), pp. 86–101. Edinburgh: Churchill Livingstone Press.

HALLIWELL, B. & GUTTERIDGE, J. (1985). *Free Radicals in Biology and Medicine*. Oxford: Clarendon Press.

HALL, A. & WILLIS, J. S. (1984). Differential effects of temperature on three components of passive permeability to potassium in rodent red cells. *J. Physiol.* **348**, 629–643.

HOCHLI, M. & HACKENBROCK, C. (1976). Fluidity in mitochondrial membranes: thermotropic lateral translation motion of intramembrane particles. *Proc. natn. Acad. Sci. U.S.A.* **73**, 1636–1640.

HUANG, J. S., DOWNES, G. L. & BELZER, F. O. (1971). Utilisation of fatty acids in perfused hypothermic dog kidney. *J. Lipid Res.* **12**, 622–627.

HUNTER, J. (1778). *A practical treatise on diseases of the teeth*. London: Johnson.

JEWELL, S., BELLOMO, G., THOR, H., ORRENIUS, S. & SMITH, M. (1982). Bleb formation in hepatocytes during metabolism is caused by disturbances in thiol and calcium homeostasis. *Science* **217**, 1257–1259.

JOHNSON, R. W. (1983). The current status of continuous perfusion for renal preservation. In *Renal Preservation* (ed. M. Marberger & K. Dreikorn), pp. 244–260. Baltimore: Williams & Wilkins.

JOHNSON, R. W., COHEN, G. L. & BALLARDIE, F. D. (1979). The limitations of continuous perfusion with plasma protein fraction. In *Organ Preservation* (ed. D. Pegg & I. Jacobsen), pp. 18–32. Edinburgh: Churchill Livingstone Press.

JUDAH, J. & AHMED, K. (1964). The biochemistry of sodium transport. *Biol. Rev.* **39**, 160–193.

KAMADA, N., CALNE, R. Y., WIGHT, D. G. & LINES, J. G. (1980). Orthotopic rat liver transplantation after long-term preservation by continuous perfusion with fluorcarbon emulsions. *Transplant.* **30**, 43–46.

KAROW, A. & PEGG, D. E. (1981). *Organ Preservation for Transplantation.* New York: Marcel Dekker.

KEELER, R., SWINNEY, J., TAYLOR, R. M. R. & ULDALL, M. B. (1966). The problem of renal preservation. *Br. J. Urol.* **38**, 653–655.

KOYAMA, I., BULKLEY, G., WILLIAMS, G. & IM, J. (1985). The role of oxygen free radicals in mediating the reperfusion injury of cold-preserved ischaemic kidneys. *Transplant.* **40**, 590–595.

LEAF, A. (1955). On the mechanism of fluid exchange of tissues in vitro. *Biochem. J.* **62**, 241–248.

LI, M., INNES, G., FULLER, B., HOBBS, K., GRIFFITH, J. & DORMANDY, T. L. (1986). The effect of a free radical scavenger on the decline of tissue linoleic acid in stored and transplanted rat hearts. *Eur. Surg. Res.* **18** (Suppl 1), 33.

MARBERGER, M. & DREIKORN, K. (1983). *Renal Preservation.* Baltimore: Williams and Wilkins.

MUDGE, G. (1951). Electrolyte and water metabolism of rabbit kidney slices: effect of metabolic inhibitors. *Am. J. Physiol.* **167**, 206–223.

NINNEMAN, J. & FISHER, J. (1981). Skin and Chorioamnion. In *Organ Preservation for Transplantation* (ed. A. Karrow & D. Pegg), pp. 411–425. New York: Marcel Dekker.

NORLEN, B. J., ENGBERG, A., KALLSKROG, O. & WOLGAST, M. (1978). Intrarenal haemodynamics in the transplanted rat kidney. *Kidney Int.* **14**, 1–9.

O'CONNELL, M. J., WARD, R. J., BAUM, H. & PETERS, T. J. (1985). The role of iron in ferritin- and haemosiderin-mediated lipid peroxidation in liposomes. *Biochem. J.* **229**, 135–139.

PARKS, D. A., BULKLEY, G. B., GRANGER, D., HAMILTON, S. & McCORD, J. J. (1982). Ischaemic injury to the cat small intestine: role of superoxide radicals. *Gastroenterol.* **82**, 9–15.

PEGG, D. E. & TAYLOR, M. J. (1984). Immunological modification of grafts during preservation. In *Transplantation Immunology* (ed. R. Y. Calne), pp. 391–409. Oxford: Oxford University Press.

PEGG, D. E., FOREMAN, J. & ROLLES, K. (1984). Metabolism during preservation and viability of ischaemically-injured canine kidneys. *Transplant.* **38**, 78–81.

POLI, G., CHEESEMAN, K. H., DIANZANI, M. & SLATER, T. F. (1985). *Free Radicals in Liver Injury.* Oxford: IRL Press.

QUINN, P. J. (1985). A lipid-phase separation model of low temperature damage to biological membranes. *Cryobiol.* **22**, 128–146.

REED, K. C. & BYGRAVE, F. L. (1975). A kinetic study of mitochondrial calcium transport. *Eur. J. Biochem.* **55**, 497–504.

RICHTER, C. & FREI, B. (1985). Calcium movements induced by hydroperoxides in mitochondria. In *Oxidative Stress* (ed. H. Sies), pp. 221–241. London: Academic Press.

ROLLES, K., FOREMAN, J. & PEGG, D. E. (1984). Preservation of ischaemically-injured canine kidneys by retrograde oxygen persufflation. *Transplant.* **38**, 102–106.

ROSS, H., MARSHALL, V. C. & ESCOTT, M. L. (1976). 72 hour canine kidney preservation without continuous perfusion. *Transplant.* **21**, 498–501.

ROUSLIN, W., ERICKSON, J. & SOLARO, J. (1986). Effects of oligomycin and acidosis on the rates of ATP depletion in the ischaemic heart muscle. *Am. J. Physiol.* **250**, H503–508.

SACKS, S. A., PETRITSCH, P. & KAUFMAN, J. J. (1973). Canine kidney preservation using a new perfusate. *Lancet* **1**, 1024–1028.

SEHR, P., BORE, P. J., PAPATHEOFANIS, J. & RADDA, G. K. (1979). Non-destructive measurement of metabolites and tissue pH in the kidney by 31P nuclear magnetic resonance. *Br. J. expl Pathol.* **60**, 632–641.

SHAPIRO, J., CHEUNG, C., ITABISHI, A., CHAN, L. & SCHRIER, R. W. (1985). The effect of verapamil on renal function after warm and cold ischaemia in the isolated perfused rat kidney. *Transplant.* **40**, 596–600.

Southard, J. & Belzer, F. O. (1980). Control of canine kidney cortex slice volume and ion distribution at hypothermia by impermeable anions. *Cryobiol.* **17**, 540–548.

Stryer, L. (1981). *Biochemistry.* San Francisco: W. Freeman & Co.

Taylor, M. J. (1986). Clinical cryobiology of tissues: preservation of cornea. *Cryobiol.* **23**, 332–353.

Uyeda, T. & Furuya, M. (1986). Effects of low temperature and calcium on microfilament structure in flagellates of Physarum polycephalum. *Expl Cell Res.* **165**, 461–472.

Van der Wijk, J., Sloof, M., Rijkmans, B. & Koostra, G. (1980). Successful 96- and 144-hour experimental kidney preservation: a combination of standard machine preservation and newly-developed normothermic ex vivo perfusion. *Cryobiol.* **17**, 473–477.

Van Rossum, G. D. (1970). Net movements of calcium and magnesium in slices of rat liver. *J. gen. Physiol.* **55**, 18–32.

Wallach, S., Reizenstein, D. & Bellavia, J. (1966). The cellular transport of calcium in rat liver. *J. gen. Physiol.* **49**, 743–751.

Watson, P. F. & Morris, G. J. (1987). This volume.

Weed, R. I., La Celle, P. I. & Merrill, E. W. (1969). Metabolic dependence of red cell deformability. *J. Clin. Invest.* **48**, 795–809.

Zimmerman, F. A., Dietz, H. G., Kohler, Ch., Kilian, N., Kosterhon, J. & Scholz, R. (1982). Effect of hypothermia on anabolic and catabolic processes and on oxygen consumption in perfused rat livers. In *Organ Preservation: Basic & Applied Aspects* (ed. D. E. Pegg, I. A. Jacobsen & N. Halasz), pp. 121–126. Lancaster: MTP Press.

Printed in Great Britain © *Society for Experimental Biology 1987*

MECHANISMS OF FREEZING DAMAGE

D. E. PEGG

MRC Medical Cryobiology Group, University Department of Surgery, Douglas House, Trumpington Road, Cambridge CB2 2AH

Summary

Freezing of aqueous systems involves numerous simultaneous changes but this review concentrates on direct effects of the formation of ice and the consequent concentration of solutes in the remaining liquid phase. It is generally believed that cell injury at low cooling rates is principally due to the concentration of both intracellular and extracellular electrolytes and that cryoprotectants act by reducing this build-up. New experimental data are presented to support this explanation; we find that the extent of damage to human red blood cells during freezing in solutions of sodium chloride/glycerol/water can be quantitatively accounted for by the increase in solute concentration. However, we also show that a given degree of damage occurs at lower concentrations of solute in the presence of higher concentrations of glycerol; it appears that glycerol contributes an element of damage itself.

Recently published studies from Mazur's laboratory have suggested that the dominant damaging factor at low cooling rates is actually the reduction of the quantity of unfrozen water rather than the corresponding increase in salt concentration that accompanies freezing. These data are re-evaluated, and it is argued that the experimental results could equally well be explained by a susceptibility of cells to shrinkage and re-expansion as the concentration of external impermeant solutes first increases during freezing and then decreases during thawing. It is concluded that external ice probably has no directly damaging effect upon dilute suspensions of cells. However, it is also argued that ice *is* directly damaging whenever it forms intracellularly, and also when it forms extracellularly in densely packed cell suspensions. In the latter case the damage is probably due to recrystallization of the ice masses during thawing. Extracellular ice also has a directly damaging effect when tissues and organs are frozen.

The difficulties of designing experimental methods that will yield unequivocal results is emphasized, and consequently the above conclusions must be regarded as tentative at the present time.

Introduction

Freezing of any aqueous system involves numerous simultaneous changes: the temperature is reduced, viscosity increases, ice crystallizes, and the remaining solution is reduced in volume; as the remaining solution becomes more saturated, some solutes may precipitate or, if gases, may form bubbles. Physicochemical parameters that are dependent on more than one of these variables (for example, pH and osmolality are affected by changes in temperature, concentration and composition) may change in complex and sometimes surprising ways. When cells are included in the system, the complexity increases: we now have a multi-compartment system, in which the contents of the compartments differ and the properties of the membranes dividing the compartments may also be dependent upon those variables that change during freezing; moreover, events that have a random statistical basis, like the nucleation of ice, will not affect all compartments equally. This matrix of inter-related phenomena should be kept in mind when evaluating experimental evidence purporting to prove this or that theory of freezing injury. It is all too easy to overlook concurrent phenomena, and, equally, it must be recognized that it is usually not possible to design clean experiments to examine just one factor under the conditions that obtain during freezing.

The earliest theories of freezing injury concentrated on the action of ice itself. It was supposed that ice crystals could pierce or tease apart cells and intracellular structures, destroying them by direct mechanical action, and as a result vitrifi-cation, which is amorphous solidification, became the goal of workers like Luyet (1952). Nevertheless it was realized very early on that one consequence of ice formation was an increase in the concentration of the remaining solution and that *this* might be the fundamental cause of freezing injury (Moran, 1929). The work of Lovelock in the 1950s (Lovelock, 1953a,b, 1954a,b, 1955) firmly established this explanation, and since it is still generally accepted, those experiments will be reviewed in some detail. But first we must consider the elements of the physical chemistry that is involved.

Physicochemical considerations

When an aqueous solution is progressively cooled below its equilibrium freezing point, the probability that ice will nucleate increases. If sufficient time is allowed, the probability of nucleation reaches unity at the homogeneous nucleation temperature, which is $-39\,°C$ for pure water but is lowered by the presence of dissolved solutes. However, if cooling is rapid nucleation may not occur until a lower temperature is reached, but since the rate of formation of nucleation centres increases as the temperature is lowered it is extremely difficult to cool sufficiently rapidly to avoid nucleation altogether. In the example illustrated in Fig. 1, cooling would have to be no slower than $300\,°C\,s^{-1}$ in order to avoid detectable nucleation. With pure water, it is estimated that the necessary cooling rate is of the order of $10^{6}\,°C\,s^{-1}$ (Uhlmann, 1972).

Let us then assume that ice nuclei do form in our system. The rate of subsequent growth of the ice crystals is directly dependent on temperature and it is important to notice that the optimum temperature for crystal growth is much higher than the temperature at which the likelihood of nucleation is maximal. In the example shown in Fig. 1, a sample cooled at $80\,°C\,s^{-1}$ would be nucleated but ice would grow exceedingly slowly unless the sample was rewarmed to say −55°C or higher whereas samples cooled more slowly than say 20°C s^{-1} would nucleate *and* exhibit ice crystal growth. The fastest cooling that is practicable for typical biological samples is usually much slower than that, which means that under ordinary conditions the ice crystals do grow once they have nucleated. These phenomena are discussed in detail by Franks, Mathias & Trafford (1984) and by MacFarlane (1986).

The next important point to note is that, as soon as ice starts to separate, the remaining liquid becomes more concentrated and therefore its freezing point is reduced: thus the system is stabilized by negative feed-back and at each temperature only sufficient ice will form to concentrate the remaining solution until its freezing point is equal to that temperature. Hence, cooling a given solution progressively increases the amount of water that is converted to ice, reduces the volume of liquid that remains, and increases the concentration of the solute dissolved in it. In fact, the familiar freezing point depression curve (Fig. 2A), which describes the relationship between concentration and temperature when that system is frozen, can be used to calculate the proportional increase in concentration of any given solution as it is frozen to progressively lower temperatures (e.g. Fig. 2B).

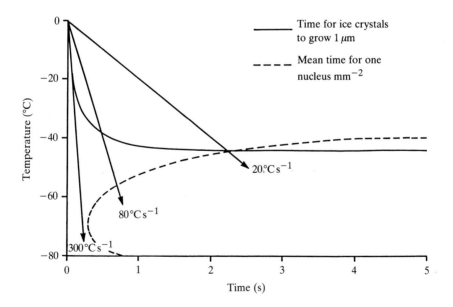

Fig. 1. Diagram constructed from data for a 50 % solution of polyvinylpyrrolidone (Luyet, 1967) showing the time- and temperature-dependence of ice crystal growth and ice nucleation in a thin film of solution. These are *experimental* data, and it should be noted that nucleation was detectable only when some crystal-growth also occurred. The arrows indicate cooling trajectories that appear to avoid nucleation (300°C s^{-1}), produce nucleation without significant growth of ice crystals (80°C s^{-1}) and produce nucleation with subsequent crystal growth (20°C s^{-1} or slower).

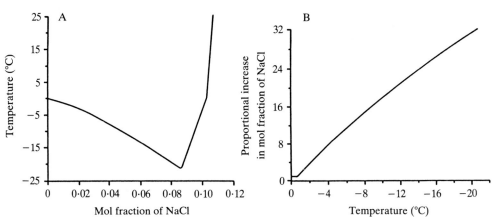

Fig. 2. (A) Freezing point depression curve for the binary system sodium chloride/water. (B) Graph showing the rise in concentration of sodium chloride in the remaining liquid phase when an initially isotonic solution (0·0027 mole fraction NaCl) is progressively frozen.

Damage due to solute concentration

Lovelock took the data of Fig. 2, which is the freezing point depression curve for the system sodium chloride/water, and compared the extent of haemolysis of red blood cells that were frozen at and then thawed from temperatures in the range $-2\,^{\circ}C$ to $-15\,^{\circ}C$, with the haemolysis that occurred when similar cells were exposed to sodium chloride solutions with freezing points equal to the selected temperatures and were then returned to isotonic saline. The agreement was excellent (Lovelock, 1953a). We have recently repeated Lovelock's experiments with the results shown in Fig. 3. The correspondence is impressive but it should be noted that the *temperature* of exposure was different in the two experiments and that the formation of ice in one case caused the cells to be sequestered into ever-narrowing channels during the freezing process, no doubt subjecting them to mechanical stresses that the cells in the salt-exposure experiment did not experience. Conversely, thawing involved resuspension in a larger volume of liquid than that in which the cells were ultimately frozen, whereas the volume in which the salt-exposed cells were suspended was never reduced. Thus, the correspondence of the curves must certainly be accepted as strong evidence but not as proof of a causal relationship between salt concentration and haemolysis.

Lovelock continued his argument by showing that the addition of non-toxic neutral solutes to the solution in which the cells were suspended reduced the haemolysis produced by freezing and also had the effect of controlling the rise in salt concentration that occurs during progressive freezing (Lovelock, 1953b, 1954a). This happens simply because the additional solute further depresses the freezing point of the solution or, in other words, reduces the amount of ice in equilibrium with the liquid phase at any given sub-zero temperature. Since the quantity of ice is inversely related to the volume of liquid remaining, and since the concentration of solute is inversely related to the volume of liquid, more added solute (which means less ice) also means lower salt concentrations: the point is illustrated in Fig. 4. Lovelock also carried out experiments in which red cells were frozen in saline containing various amounts of glycerol and showed that the temperature at which haemolysis started was reduced as the glycerol concentration was increased but always corresponded to the same sodium chloride concentration (mole fraction of sodium chloride = 0·014) (Lovelock, 1953*b*). We have recently carried out an extensive comparison of the effect of exposure to high salt concentrations and of freezing on the haemolysis of red cells in the presence of various glycerol concentrations. In effect this is the original Lovelock experiment, but now *in the presence of glycerol*, and again the correspondence between freezing and solute-exposure is excellent (see Fig. 5). This suggests, in agreement with Lovelock, that the effect of glycerol is to lower the temperature at which a given salt concentration is achieved without altering the relationship between hae-molysis and salt concentration in that glycerol concentration. However, the same degree of haemolysis occurs at a lower salt concentration as the concentration of

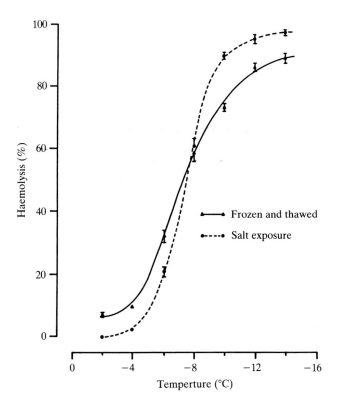

Fig. 3. Experimental results when human red blood cells were frozen to the indicated temperatures and then thawed (▲——▲) compared with an experiment in which similar cells were exposed to nominally equivalent salt concentrations and returned to isotonic conditions (●–––●). The equivalent salt concentrations were calculated from the phase diagram data shown in Fig. 2 but are nominal in that no allowance was made for dilution by the small amount of trapped fluid in the packed red cells or the water that was withdrawn from the cells. The correspondence of the two curves is striking.

glycerol is increased; for example, the 50% haemolysis point occurred at 4·6 g sodium chloride $100 \, g^{-1}$ solution when the solution contained a glycerol/salt ratio of 10 (i.e. R = 10) but at 11·0 g sodium chloride $100 \, g^{-1}$ solution when R = 0 (see Fig. 6). These concentrations correspond to sodium chloride mole fractions of 0·024 and 0·037, respectively. These findings suggest that glycerol *also* has a damaging effect and may explain an observation originally made by Fahy & Karow (1977), who pointed out that, although Lovelock had reported that haemolysis started when the mole fraction of sodium chloride was 0·014 whatever the glycerol concentration in which the cells were frozen, Lovelock's own data actually showed that a given degree of haemolysis occurred at progressively lower salt concentrations as the concentration of glycerol was increased (see Fig. 7). We have previously pointed out that this effect could be due to susceptibility of the cells either to a reduction in temperature or to an increase in glycerol concentration (Pegg *et al.* 1979), but it now seems to be due to the latter.

Thus we have strong circumstantial evidence that red cells are damaged by exposure to high salt concentrations and, to a lesser extent, by high concentrations of the cryoprotective solute glycerol. An additional point that can be made using the experimental data reviewed thus far concerns the temperature dependence of this type of damage. Examination of Fig. 5 shows that when the bath in which the cell samples were cooled in the freezing experiment was colder than about $-30\,°C$

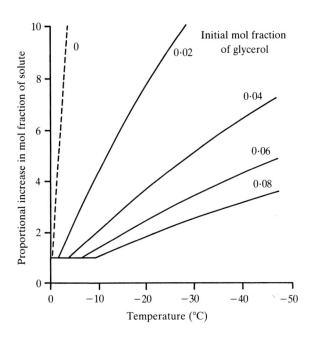

Fig. 4. Diagram to illustrate the effect of increasing initial concentrations of glycerol on the proportional increase in solute concentration in the liquid phase when ternary solutions of sodium chloride/glycerol/water, all initially containing 0·0027 mole fraction sodium chloride, are progressively frozen.

the extent of damage was less than expected. It should be noted that the cells in the R = 10 experiment that were cooled to −47°C had to cool through −27°C to reach that temperature, yet haemolysis was reduced to less than one-third. Thus, whatever the mechanism of damage it must be kinetic in nature, which means that maximum damage from a given treatment takes a finite time to accumulate, a time which is dependent upon temperature. Hence, this form of damage has a cooling-rate dependence: we shall return to this point later.

Damage due to reduction in the unfrozen fraction

As we considered in the introduction to this chapter, many changes occur simultaneously during the freezing of biological systems and this makes it very difficult to design experiments that separate the dependent variables. Mazur has recently emphasized the problem of separating the effect of reduced liquid volume from that of increased salt concentration: these are in fact rigidly coupled if the

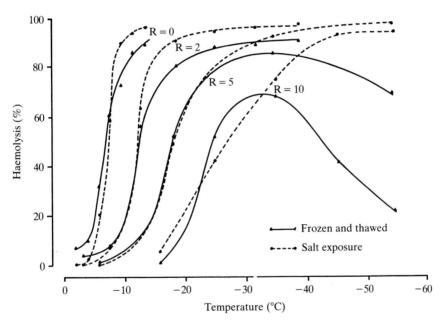

Fig. 5. Experimental results when human red blood cells were frozen to the indicated temperatures and then thawed (▲——▲), compared with experiments in which similar cells were exposed to nominally equivalent solutions and returned to isotonic conditions (●---●). The solution compositions were calculated as described by Pegg (1983). Four sets of solutions were used, designated R = 0, 2, 5 or 10 where R is the weight ratio of glycerol to sodium chloride: the R = 0 solution contained no glycerol and the data are the same as that shown in Fig. 3. Note that as the glycerol concentration was increased so the equivalent degree of haemolysis occurred at a lower temperature, and above −30°C the haemolysis closely corresponded to that produced by the equivalent salt-exposure. Below −30°C, the extent of damage in the freezing experiment was less than that in the salt-exposure experiment.

initial salt concentration is held constant, at say 0·15 molal. However, since the physical chemistry requires that the salt concentration be determined by temperature during the progressive freezing of a solution of given R-value, it is possible to uncouple the unfrozen fraction of liquid (or water) from the rise in salt concentration if the experimenter starts with different total solute concentrations. Mazur has performed such experiments and has concluded that it is the unfrozen fraction, not the salt concentration, that is the main damaging influence first with red blood cells (Mazur, Rall & Rigopoulos, 1981), and now with early mouse embryos also (Mazur, personal communication). This was a surprising conclusion, to Mazur himself as much as to others: if it is accepted, then one has to accommodate the Lovelock experiments, and our subsequent observations, as simply coincidence. That is certainly possible, but it seems improbable. Let us then look at some of Mazur's experiments. Mazur expressed his salt concentrations in terms of the mass of water present, and the unfrozen fraction, for consistency, as the fraction of water that remained unfrozen. His data for R = 11·3 (glycerol/ sodium chloride) are plotted in the form of a contour diagram in Fig. 8A. Mazur has also published data for R = 5·4 and we have been able to obtain consonant results for R = 5 and R = 10 solutions (Pegg & Diaper, 1984).

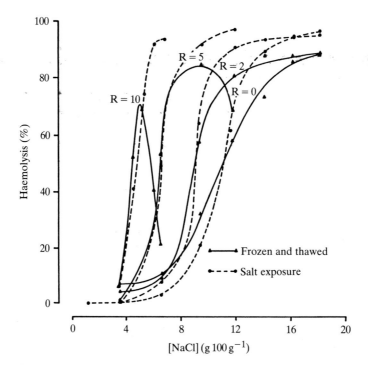

Fig. 6. The same data as those shown in Fig. 5 are plotted with sodium chloride concentration, rather than temperature, on the abscissa. It is clear that equivalent haemolysis occurred at progressively lower salt concentrations as the glycerol concentration was increased.

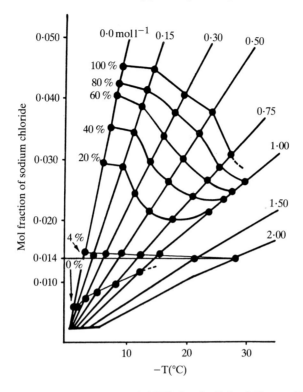

Fig. 7. The recalculation of Lovelock's (1953b) data by Fahy & Karow (1977), showing that when isohaemolysis lines are superimposed on a plot similar to Fig. 4, the expected set of lines parallel to the abscissa were not obtained: rather, it was found that either glycerol concentration or reduced temperature augmented the damage produced by high concentrations of sodium chloride. (Reproduced with the permission of the authors, the Editor of *Cryobiology* and Academic Press.)

Inspection of Fig. 8(B), which shows isohaemolysis contours on a surface grid of unfrozen water (U) *versus* salt concentration for solutions of glycerol, sodium chloride and water where R = 11·3, certainly appears to confirm Mazur's conclusion: the dominant trend of the contour lines is horizontal, consistent with an effect of unfrozen fraction, although there is also a tendency for greater damage to be observed at high rather than low sodium chloride concentrations; that is, it is possible to tolerate lower U-values for a given degree of haemolysis at low sodium chloride concentrations than at high sodium chloride concentrations. However, it is important to note the conditions under which these experiments were carried out: in order to separate the effects of volume of ice from increase in salt concentration it was necessary to cool samples to similar temperatures in different initial salt concentrations, which is to say with different initial cell volumes. Fig. 8(B) in fact shows the calculated initial volumes of the red cells (Pegg, 1984) on the same grid surface. Comparison of Fig. 8(A) with Fig. 8(B) suggests that the greatest apparent susceptibility to low U-fraction occurred with cells that started the experiment swollen, while cells of normal initial volume (87×10^{-15} litre) or less than normal tended to survive better. However, consideration of any possible effect of starting volume is complicated by the fact that the final volume also differed according to the final temperature to which that sample was cooled. Mazur has pointed out (Mazur & Rigopoulos, 1983; Mazur & Cole, 1985) that

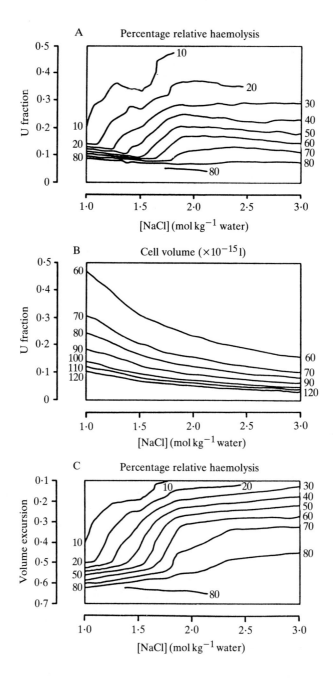

each sample joins the same volume/temperature curve but joins it at a different point depending upon the starting solution composition and passes along it for different distances according to the final temperature: thus the extent to which cells shrink and then re-expand varies. A measure of this change in volume can be called the proportional volume excursion (V_e) which is defined by the expression

$$V_e = \frac{V_i - V_f}{V_i},$$

where V_i is the initial volume and V_f is the final volume. When haemolysis is plotted as contours on the surface where V_e replaces U (see Fig. 8(C)) the resemblance to Fig. 8(A) is striking. Thus, the point has to be made that, although Mazur's experiment was *designed* to investigate the effect of U, it is simply impossible to vary U without varying other parameters of the system in a way in which they do not vary during the actual freezing of cell suspensions in practice. Mazur was forced to vary initial cell volume, with the result that the choice of U to attach to the vertical axis in the plot of Fig. 8(A) is arbitrary: volume excursion is no less valid. The question then arises, if the volume excursion of cells is varied in the same way as in the above experiment but in the absence of ice (although to do this, one cannot vary temperature in the same way as it varies in the freezing experiment) whether similar haemolysis data are generated; preliminary indications are affirmative. If further experiments show this to be the case, a fairly consistent picture will begin to emerge: the concentration of *salts* is the factor most likely to damage cells during freezing but this damage is mediated not simply by osmotic shrinkage of the cells as proposed by Meryman (1970) but by shrinkage and re-expansion, susceptibility to this volume excursion being dependent upon the salt concentration reached. Without being too specific, it seems reasonable to propose that changes in cell membrane structure might be produced by increased salt concentrations such that susceptibility to volume excursions have the observed dependence. There are parallels between such a mechanism and that originally proposed by Lovelock (1954b, 1955) and the more recent 'critical volume increment' theory proposed by Steponkus (1984) to explain freezing injury to isolated protoplasts.

Fig. 8. Data for red cells in R = 11·3 solutions of sodium chloride/glycerol/water, published by Mazur, Rall & Rigopoulos (1981). (A) Isohaemolysis contours on a grid of unfrozen water fraction (U) *versus* sodium chloride concentration. Note that the contours are predominantly horizontal, suggesting a major influence of U, but low U-values were better tolerated at low salt concentrations. (B) As explained in the text, the experimental results shown in A were obtained by varying the initial concentration of the R = 11·3 solution in which the red cells were suspended. This meant that their initial *volumes* differed, and this diagram shows the initial volumes of the cells used to obtain the data shown in A. (C) Here, isohaemolysis contours are shown on a grid of proportional volume excursion and salt concentration: it is clear that the response surface is quite similar to that shown in A, and could be interpreted as indicating that cell damage during freezing is mainly due to the cycle of contraction and re-expansion of the cells, with the rider that cells are less susceptible to such volume changes in the presence of low salt concentrations.

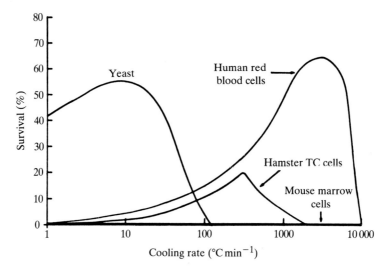

Fig. 9. Graph showing the effect of cooling rate on the survival of yeast cells, some tissue culture cells (hamster TC cells), human red blood cells and mouse bone marrow cells. Note that the first three cells have different optimum cooling rates, while less than 2 % of the bone marrow cells survive at any cooling rate. (Reproduced with the permission of IPC Press Ltd.)

Direct injury due to ice

It must not be supposed from the above that ice has no direct relevance to freezing injury under all circumstances. Ice quite probably is relevant in three specific situations that we shall now consider.

(1) *Intracellular freezing*

It was pointed out above that injury caused by the mechanisms that we have been discussing thus far has a kinetic nature and essentially ceases below some ill-defined temperature in the region of $-30°$ to $-40°C$ for red blood cells. It follows that this form of injury must have a cooling-rate dependence; when this is investigated it is discovered that many cells exhibit responses, some of which are illustrated in Fig. 9, that are similar in form but show optimal survival at different cooling rates. As cooling rate is accelerated the rising part of the curve is easy to explain on the basis we have already discussed. The subsequent fall has been very convincingly explained by the occurrence of *intracellular* freezing. Mazur (1963) provided a mathematical treatment of water-loss from red blood cells during freezing with a range of cooling rates and showed that the cooling rate at which water loss ceased to keep pace with the increase in external osmolality (that is, the cooling rate at which there was significant super-cooling) coincided with the cooling rate at which survival started to decrease. It is at this point that intracellular freezing would be expected. Calculations showed a similar correspondence in the case of sea urchin eggs which have far different water permeability and size and also have a different optimum cooling rate for survival. Still more compelling evidence was provided by the cryomicroscope which showed that the fall-off in recovery occurred at cooling rates at which intracellular freezing could actually be observed in three very different cells; with ova it was at 2°C per minute, with HeLa cells at 50°C per minute and with red blood cells at 800°C per minute (Mazur, 1977). It will be recalled that nucleation is maximal at a lower temperature than ice crystal growth: hence it is possible, with very rapid cooling, to obtain highly nucleated but essentially unfrozen cells. When such cells are rewarmed they pass through the zone of rapid crystal growth and if warming is insufficiently rapid they may then freeze internally, and such freezing is lethal (Mazur, 1977). Thus, there is very convincing evidence that under most conditions the occurrence of intracellular freezing is lethal.

(2) *Extracellular freezing in densely packed systems*

A second situation where ice is probably responsible for directly inflicting damage occurs when cells are frozen in a densely packed suspension, specifically when the packed cell volume or 'cytocrit' exceeds 60 %. We have studied this situation in the case of human red blood cells equilibrated with 2 molar glycerol, by comparing a 2 % haematocrit with a 75 % haematocrit (Pegg *et al.* 1984). It turns out that the effect of packing depends upon the cooling and warming rates used. In Fig. 10 we show our results in the form of a contour diagram: haemolysis in the 75 % haematocrit sample is normalized to the results of the 2 % haematocrit sample, that is the graph shows the additional effect of packing over and above the effect of freezing and thawing at the low haematocrit. When the cooling rate was less than 100°C per minute, the magnitude of the packing effect was entirely controlled by the warming rate, being reduced by rapid warming. However, we were able to show that these cells lack any intracellular ice, and were not even nucleated, so that intracellular recrystallization during slow warming was not responsible for this effect. However, cryomicroscopy showed extensive *extracellular* recrystallization during slow rewarming but not during rapid warming: the rates used were those that had been found to give low and high survival, respectively, in the haemolysis experiments. Thus, it seems likely that the cells at low temperatures, restricted within narrow liquid-filled channels, are subjected to mechanical stresses by the recrystallization process and that such stresses are exacerbated by cell-crowding.

(3) *Extracellular freezing in tissues and organs*

A third situation in which ice appears to have a directly damaging influence is in the case of tissues and organs, three-dimensional functional assemblies of cells in which the interconnections and extracellular structures may be just as vital as the cells themselves. Work in our laboratory has shown that both the occurrence and the precise location of extracellular ice in smooth muscle tissue can have a dramatic effect on its post-thaw function (Hunt, Taylor & Pegg, 1982; Taylor &

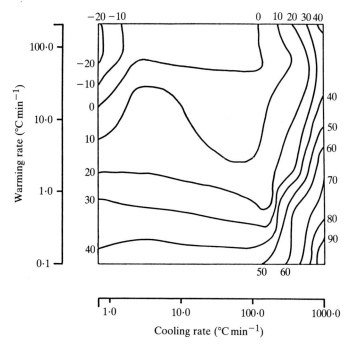

Fig. 10. Relative haemolysis of red cells frozen and thawed in $2\,mol\,l^{-1}$ glycerol at a haematocrit of 75%, relative to cell survival in samples frozen and thawed at a haematocrit of 2%. The contours show haemolysis ascribable to packing, separated from the overall level of cryoprotection. (Reproduced with the permission of the Editor of *Cryobiology* and Academic Press.)

Pegg, 1983). Cooling to $-21\,^{\circ}\mathrm{C}$ with sufficient dimethyl sulphoxide to prevent freezing gave 75% contractile functional recovery. Cooling to the same temperature with less dimethyl sulphoxide, so that ice was formed, gave a lesser degree of function that was dependent on cooling rate and was correlated with the morphology of the ice that was formed. Cooling at $2\,^{\circ}\mathrm{C}$ per minute produced a random location of ice and 21% functional recovery, whereas cooling at $0\cdot3\,^{\circ}\mathrm{C}$ per minute produced ice that was confined to the connective tissues surrounding the muscle bundles and the contractile response rose to 53% of control. The possibility that intracapillary ice might damage the microcirculation of an organ also seems real, and again we have evidence, in this case from rabbit kidneys permeated with 2 molar glycerol, that this is indeed so (Jacobsen *et al.* 1984).

Conclusions

The mechanisms of injury during the freezing of isolated mammalian cells depend upon the cooling rate used and the rate at which the cell in question can lose water in response to an increase in external osmolality. If cooling is too rapid, intracellular freezing will occur and the cell will be damaged, probably destroyed. At slow cooling rates, injury is probably due to the cycle of cell shrinkage during cooling and re-expansion during warming. Penetrating cryoprotectants such as glycerol reduce the build-up of salt concentration during freezing and therefore reduce the degree of cell shrinkage, but they also contribute directly to cell damage, though to a much lesser extent than they protect against the build-up of salt concentration. Extracellular freezing is probably innocuous to dilute suspensions of single cells but it may damage densely packed cell suspensions through recrystallization processes during the warming phase. Extracellular ice is frankly harmful to organized cells and tissues. These conclusions are tentative. The difficulty of conducting experiments in such a way that the effects of variables can be unequivocally separated must be noted.

The new experimental data reported in this paper were obtained by Mr M. P. Diaper who also prepared the illustrations.

References

Fahy, G. M., & Karow, A. M., Jr (1977). Ultrastructure-function correlative studies for cardiac cryopreservation. V. Absence of a correlation between electrolyte toxicity and cryoinjury in the slowly frozen, cryoprotected rat heart. *Cryobiology* **14**, 418–427.

Franks, F., Mathias, S. F. & Trafford, K. (1984). The nucleation of ice in undercooled water and aqueous polymer solutions. *Colloids and Surfaces* **11**, 275–285.

Hunt, C. J., Taylor, M. J. & Pegg, D. E. (1982). Freeze-substitution and isothermal freeze-fixation studies to elucidate the pattern of ice-formation in smooth muscle at 252K (−21°C). *J. Microscopy* **125**, 177–186.

Jacobsen, I. A., Pegg, D. E., Starklint, H. & Diaper, M. P. (1984). Effect of cooling rate and warming rate on glycerolised rabbit kidneys. *Cryobiology* **21**, 637–653.

Lovelock, J. E. (1953a). The haemolysis of human red blood cells by freezing and thawing. *Biochim. biophys. Acta* **10**, 414–426.

Lovelock, J. E. (1953b). The mechanism of the protective action of glycerol against haemolysis by freezing and thawing. *Biochim. biophys. Acta* **11**, 28–36.

Lovelock, J. E. (1954a). The protective action by neutral solutes against haemolysis by freezing and thawing. *Biochem. J.* **56**, 265–270.

Lovelock, J. E. (1954b). Physical instability and thermal shock in red cells. *Nature, Lond.* **173**, 659–660.

Lovelock, J. E. (1955). Haemolysis by thermal shock. *Br. J. Haemat.* **1**, 117–129.

Luyet, B. J. (1952). Survival of cells, tissues and organs after ultra-rapid freezing. In *Freezing and Drying* (ed. R. J. C. Harris), pp. 77–98. London: Institute of Biology.

Luyet, B. J. (1967). On the possible biological significance of some physical changes encountered in the cooling and the rewarming of aqueous solutions. In *Cellular Injury and Resistance in Freezing Organisms* (ed. E. Asahina). Japan: Institute of Low Temperature Science, Hokkaido.

MacFarlane, D. R. (1986). Devitrification in glass-forming aqueous solutions. *Cryobiology* **23**, 230–244.

Mazur, P. (1963). Kinetics of water loss from cells at subzero temperatures and the likelihood of intracellular freezing. *J. gen. Physiol.* **47**, 347–369.

Mazur, P. (1977). The role of intracellular freezing in the death of cells cooled at supraoptimal rates. *Cryobiology* **14**, 251–272.

Mazur, P. & Cole, K. W. (1985). Infuence of cell concentration on the contribution of unfrozen fraction and salt concentration to the survival of slowly frozen human erythrocytes. *Cryobiology* **22**, 505–536.

Mazur, P., Rall, W. F. & Rigopoulos, N. (1981). The relative contributions of the fraction of unfrozen water and of salt concentration to the survival of slowly frozen human erythrocytes. *Biophys. J.* **36**, 653–675.

MAZUR, P. & RIGOPOULOS, N. (1983). Contributions of unfrozen fraction and of salt concentration to the survival of slowly frozen human erythrocytes: influence of warming rate. *Cryobiology* **20**, 274–289.

MERYMAN, H. T. (1970). The exceeding of a minimum tolerable cell volume in hypertonic suspension as a cause of freezing injury. In *The Frozen Cell* (ed. G. E. W. Wolstenholme & M. O'Connor), pp. 51–64. London: Churchill.

MORAN, T. (1929). Critical temperature of freezing – living muscle. *Proc. R. Soc. Ser.* B. **105**, 177–197.

PEGG, D. E. (1983). Simple equations for obtaining melting points and eutectic temperatures for the ternary system glycerol/sodium chloride/water. *Cryo-Letters* **4**, 259–268.

PEGG, D. E. (1984). Red cell volume in glycerol/sodium chloride/water mixtures. *Cryobiology* **21**, 234–239.

PEGG, D. E. & DIAPER, M. P. (1984). The possible roles of unfrozen liquid volume, solute concentration and low temperature in the causation of freezing injury. *Cryobiology* **21**, 692.

PEGG, D. E., DIAPER, M. P., SKAER, H. LE B. & HUNT, C. J. (1984). The effect of cooling rate and warming rate on the packing effect in human erythrocytes frozen and thawed in the presence of 2M glycerol. *Cryobiology* **21**, 491–502.

PEGG, D. E., JACOBSEN, I. A., ARMITAGE, W. J. & TAYLOR, M. J. (1979). Mechanisms of cryoinjury in organs. In *Organ Preservation* II, chapter 11 (ed. D. E. Pegg & I. A. Jacobsen), pp. 132–144. Edinburgh: Churchill Livingstone.

STEPONKUS, P. L. (1984). Role of the plasma membrane in freezing injury and cold acclimation. *A. Rev. Plant Physiol.* **35**, 543–584.

TAYLOR, M. J. & PEGG, D. E. (1983). The effect of ice formation on the function of smooth muscle tissue stored at −21 °C or −60 °C. *Cryobiology* **20**, 36–40.

UHLMANN, D. R. (1972). The kinetic theory of glass formation. *J. non-cryst. Solids* **7**, 337.

Printed in Great Britain © Society for Experimental Biology 1987

CRYOPRESERVATION OF ANIMAL CELLS

W. J. ARMITAGE*

Department of Ophthalmology, University of Bristol, UK

Summary

A wide range of cells of biological, medical and agricultural importance can be cryopreserved at $-196°C$ for many years in a stable state. The survival after freezing and thawing of different types of cell varies markedly, and depends on the ability of the cells to withstand a variety of stresses imposed by the physical and physico-chemical changes occurring in the bathing medium during cooling to and warming from the storage temperature. In most instances, cells will survive freezing only when a cryoprotectant (an additive such as glycerol or dimethyl sulphoxide that can protect cells against freezing injury) is included in the bathing medium: moreover, survival is dependent on the rates at which cells are cooled and warmed. Each cell type has an optimum cooling rate at which survival is maximal. This optimum cooling rate is determined to a large extent by the water permeability of the cell; but it is also influenced by the presence of cyroprotectants, and the warming rate. Therefore, when devising a method of cryopreservation for a particular cell type, the most important variables are the rates of cooling and warming, and the type and concentration of cryoprotectant. In general, as the concentration of cryoprotectant is increased, cell survival improves and the optimum cooling rate is reduced. However, high concentrations of cryoprotectants can have detrimental toxic and osmotic effects on cells. Therefore, the way in which cryoprotectants are added to the bathing medium before freezing and, in particular, the procedure used to return the cells to a medium without cryoprotectants after thawing, can be crucial to the survival of the cells.

Vitrification could be a way to avoid the mechanisms of injury associated with the formation of ice. When an aqueous solution contains a sufficiently high concentration of solute (e.g. >40 % w/w of the cyroprotectant propane-1,2-diol), the large increase in viscosity that occurs during cooling inhibits ice formation, and the solution eventually becomes an amorphous, glassy solid. There is no formation of a crystalline solid, nor is there a progressive rise in solute concentration, both of which occur as a result of ice formation during cooling. Vitrification could, therefore, prove to be a useful alternative method of cryopreservation, particularly for organized tissues and organs. There are, however, several problems that have to be overcome, which include: (a) the deleterious effects on cells of very high concentrations of cryoprotectants; (b) the occurrence of cracks in the vitreous solid; and (c) extensive ice formation during warming (devitrification).

*Address for correspondence: UK Transplant Service, Southmead, Bristol BS10 5ND.

Cryopreservation

Cryopreservation in liquid nitrogen at $-196\,°C$ enables living cells to be stored for many years in a biologically stable state. Since no chemical reactions can take place below about $-130\,°C$, the only limitation to the length of storage in liquid nitrogen is thought to be the accumulation of damage caused by background ionizing radiation (Ashwood-Smith & Friedmann, 1979; Mazur, 1984). Cells stored at higher temperatures (e.g. $-70\,°C$ to $-80\,°C$) tend to be less stable, although useful long-term storage can be attained depending on the type of cell and the conditions under which the cells are frozen.

The bias towards the development of techniques for the cryopreservation of mammalian cells is a reflection of the impact that the ability to store cells for long periods has had in agriculture and medicine, as well as in biology. For example, the cryopreservation of bovine sperm (Polge & Lovelock, 1952) and embryos (Wilmut & Rowson, 1973; Leibo, 1983) has revolutionized the cattle breeding industry. Similarly, the cryopreservation of erythrocytes (Pyle, 1964; Rowe & Lenny, 1981), bone marrow (Pegg, 1964; Buckner, Appelbaum & Thomas, 1981), and leucocytes (Pegg, 1965; Weiner, 1976) has been of great value to haematologists and transplantation immunologists. In biology, the cryopreservation of embryos is important in a number of areas, including mammalian genetics and the preservation of rare mammalian breeds (Elliott & Whelan, 1977). Also, the cryopreservation of tissue culture cells has simplified the problems of maintaining large numbers of cell lines.

The survival of cryopreserved cells depends on their ability to cope with a range of physical, physico-chemical, physiological and biochemical stresses encountered during freezing and thawing. Uncontrolled freezing is usually lethal to cells, but many cell types will survive freezing and thawing when a cryoprotectant (a chemical that can protect cells against freezing injury) is present in the bathing medium, and when the rate of cooling to the storage temperature and the subsequent rate of warming are controlled. Different types of cell vary widely in their response to freezing, and some, for example human granulocytes, are highly sensitive to freezing and have not yet been successfully cryopreserved. In contrast to isolated cells, however, there has been a marked lack of success in the cryopreservation of organized tissues and, in particular, large vascularized organs. Pegg, Jacobsen, Armitage & Taylor (1979) and Mazur (1984) have suggested that the high degree of structural organization of tissues and organs, the fixed size and shape, the variety of cell types and the high cell density are factors that increase the susceptibility of organs to freezing and thawing.

Avoidance of freezing injury

The variables of greatest importance to cryopreservation are the rates of cooling and warming, and the type and concentration of cryoprotectant. These variables are interdependent and, moreover, a particular combination of cryoprotectant, cooling rate and warming rate that permits high survival of a given type of cell can be lethal to a different type of cell. The treatment of cells before freezing and the post-thaw handling conditions can also affect the survival of cryopreserved cells. For example, the method used to separate human T lymphocytes from a mixed suspension of mononuclear cells appears to increase the susceptibility of the T cells to the effects of freezing and thawing. Thus, T cells survive freezing and thawing better when they are frozen in a mixed suspension of mononuclear cells and separated from the other cells after thawing, than when they are separated before freezing and cryopreserved as a purified suspension (Venkataraman & Westerman, 1986). Moreover, freezing may sensitize cells to stresses encountered after thawing, such as dilution of the bathing medium to remove the cryoprotectants, and centrifugation (Lovelock, 1953*a*; Farrant & Morris, 1973).

Cooling rate

Cells generally have an optimum rate of cooling at which survival is maximal (Mazur, 1970). Cooling at rates higher than the optimum increases the likelihood of intracellular ice formation, and the consequences of the presence of intracellular ice during subsequent warming are usually lethal (Farrant, 1977; Mazur, 1977). When a cell suspension is cooled, ice initially forms almost exclusively in the extracellular solution, and the cells are exposed to a progressive increase in external solute concentration as the temperature falls. In 1963, Mazur demonstrated quantitatively that it is the rate at which water can leave a cell relative to the rate of increase in external solute concentration (which is directly related to cooling rate) that determines whether osmotic equilibrium is maintained across the plasma membrane by an efflux of water from the cell or by intracellular freezing. Thus, at low cooling rates, cells equilibrate by losing water (i.e. they behave as osmometers); but at high cooling rates, they cannot lose water quickly enough to maintain osmotic equilibrium, and they become increasingly supercooled until they eventually freeze. Thus, cells cooled sufficiently slowly are shrunken and do not contain intracellular ice, whereas cells cooled too rapidly do not shrink appreciably, but they do contain ice. The rate of water loss from a cell is determined by the water permeability of the plasma membrane, the membrane surface area, and the cell volume. The water permeability of the membrane is dependent on temperature, and can also be affected by changes in solute concentration (Armitage, 1986*a*). Large cells with a low ratio of surface area to volume, or cells with a low water permeability, freeze intracellularly at lower cooling rates than small cells with a high surface area to volume ratio, or cells with a high water permeability.

Some types of cell survive rapid cooling when they are cooled by the so-called two-step method (Luyet & Keene, 1955). Cells are initially cooled to a relatively high subzero temperature (e.g. $-25\,^{\circ}\mathrm{C}$), held at that temperature for a few minutes, and then cooled rapidly by immersion in liquid nitrogen. The period spent at the intermediate temperature allows the cells time to dehydrate sufficiently to avoid intracellular ice formation during subsequent rapid cooling in liquid nitrogen. The presence of a cryoprotectant is thought to permit survival of the cells at the intermediate temperature. Survival is determined by the temperature of the intermediate step, the time spent at that temperature, and the rates of cooling and warming. The intermediate temperature is also a function of cryoprotectant concentration (McGann & Farrant, 1976a,b).

Cooling at rates lower than the optimum rate results in damage to cells caused by prolonged exposure at relatively high subzero temperatures ($>-50°C$) to the adverse changes occurring in the extracellular solution. Several mechanisms have been proposed for slow-cooling injury: namely, direct damage to the membrane caused by high electrolyte concentrations (Lovelock, 1953a); excessive cellular shrinkage resulting in damaging stresses in the plasma membrane (Meryman, 1968, 1971); thermal shock and dilution shock (Farrant & Morris, 1973); and changes in the volume of the liquid phase in the ice lattice (Mazur, Rall & Rigopoulos, 1981). Although the plasma membrane is highly likely to be a site of injury (Farrant & Woolgar, 1972; Morris & Clarke, 1981), the removal of water could also disrupt the internal organization of a cell and may adversely affect metabolism (Clegg, Seitz, Seitz & Hazlewood, 1982). This proposal is perhaps supported by the observation that granulocytes and platelets are damaged by exposure to osmolalities greater than about $0·6\,osmol\,kg^{-1}$, yet these cells continue to behave as osmometers at osmolalities in excess of $2\,osmol\,kg^{-1}$, indicating the retention of normal semipermeable membrane properties (Armitage & Mazur, 1984a; Armitage, Parmar & Hunt, 1985; Armitage, 1986a).

Warming rate

The recovery of cells after freezing and thawing is also dependent on the rate of warming (Miller & Mazur, 1976; Akhtar, Pegg & Foreman, 1979; Pegg, Diaper, Skaer & Hunt, 1984). Cells that are cooled at rates likely to cause some intracellular ice formation survive better when the warming rate is high than when it is low. Conversely, the survival of cells cooled more slowly than the optimum rate is higher when they are also warmed slowly. It was the recognition of this interaction between cooling and warming rates that led to the first successful recovery of viable cryopreserved embryos, which survived only when they were both cooled and warmed slowly (Whittingham, Leibo & Mazur, 1972; Wilmut, 1972).

Control of cooling and warming rates

Linear rates of cooling and warming can be achieved with sophisticated controlled-rate cooling equipment (e.g. Hayes, Pegg & Kingston, 1974). However, a wide range of cooling rates can be obtained by cooling samples in insulated containers immersed in liquid nitrogen, or in alcohol cooled to $-79°C$ with solid CO_2 (e.g. Miller & Mazur, 1976; Pegg et al. 1984). These methods cool samples at non-linear rates; therefore, cooling rates are calculated over the range of temperature where controlled cooling is believed to be essential for survival (i.e. about $-10\,°C$ to $-60\,°C$): indeed, the cooling rate is often reasonably linear within this range. Samples are often warmed by transferring them from liquid nitrogen to a stirred water bath; thus, the warming rate can be varied by altering the temperature of the water and the amount of insulation around the samples. The rates of cooling and warming under a given set of conditions are also affected by the sample volume and the nature of the container (i.e. glass *vs* plastic).

Cryoprotectants

Many types of cell do not survive freezing and thawing unless a cryoprotectant is present in the bathing medium. In 1949, Polge, Smith & Parkes reported that fowl sperm survived freezing when the suspending medium contained glycerol, and ten years later the cryoprotective properties of dimethyl sulphoxide (Me_2SO) were discovered by Lovelock & Bishop (1959). These two compounds have since remained the most commonly used additives for the cryopreservation of cells, although there is a wide variety of chemically very diverse compounds that allow cells to survive freezing (Nash, 1962; Shlafer, 1981). Cryoprotectants are broadly divided on the basis of whether they permeate cells; for example, cells are usually permeable to glycerol and Me_2SO, but are impermeable to dextran and sucrose. The only common features that these compounds seem to possess are high solubility in water and low toxicity to cells. Although direct interactions between cryoprotectants and cells cannot be excluded, the main effect of additives such as glycerol is to reduce colligatively both the amount of ice formed and the rise in electrolyte concentration at any given temperature (Lovelock, 1953*b*). The high molecular weight of polymers such as polyvinylpyrrolidone precludes a significant colligative effect, but their extremely non-ideal behaviour at high concentrations can reduce the amount of water that crystallizes (Farrant, 1969; Franks, 1982).

Cryoprotectants do not mitigate the effects of intracellular freezing. Indeed the presence of a cryoprotectant can reduce the cooling rate at which intracellular nucleation first occurs (Diller, 1979). Cryoprotectants are, therefore, only effective at protecting cells against slow-cooling injury, and as the initial concentration of cryoprotectant is increased, survival improves and the optimum cooling rate reduces (Mazur, Leibo, Farrant, Chu, Hanna & Smith, 1970).

Although cooling rate becomes less important in determining survival the higher the concentration of cryoprotectant, there are problems associated with incorporating high concentrations of cryoprotectants into cell suspensions, and particularly with the subsequent removal of the cryoprotectants in order to return the cells to a normal physiological medium. Damage caused by permeating cryoprotectants can be due to chemical toxicity, or can be the result of detrimental osmotic effects, which are a consequence of cellular membranes being more permeable to water than to the cryoprotectant. When a cell is exposed to a step increase in permeating cryoprotectant in the bathing medium, the cell initially shrinks as water moves out

of the cell to restore osmotic equilibrium. As the cryoprotectant begins to enter
the cell, it is accompanied by water, and the cell returns asymptotically towards its
normal volume. Conversely, if the external concentration of cryoprotectant is then
reduced abruptly, the cell swells as water enters to restore osmotic equilibrium.
The cell then returns to its normal volume as cryoprotectant and water leave the
cell simultaneously (Davson & Danielli, 1952). The magnitude and duration of
these volume excursions are dependent on the size of the concentration gradient
across the plasma membrane, the water permeability coefficient, the solute
permeability coefficient and the reflexion coefficient of the membrane (House,
1974). Therefore, the osmotic effects of permeating cryoprotectants can be
lessened by reducing both the rate of addition of the cryoprotectant to the bathing
medium, and the rate of subsequent dilution of the medium to remove the
cryoprotectant. Because membrane permeability increases with increasing temperature, the osmotic effects of addition and dilution are less at, for example, 25 °C
than at 0 °C. However, any benefit gained by increasing temperature to reduce
osmotic stress may be lost because of increased toxicity of the additive at the
higher temperature. Cells are generally less permeable to glycerol, hence the
osmotic effects are greater, than to Me_2SO, but glycerol is less toxic (Schlafer,
1981). Methanol has been shown to afford protection to tissue culture cells during
freezing (Ashwood-Smith & Lough, 1975), and the very high permeability
coefficient of this additive should present minimal osmotic problems, though again
this has to be balanced against potential chemical toxicity.

   The development of procedures for the addition and dilution of cryoprotectants
and the separation of osmotic and toxic effects can be a protracted task if done
solely by trial and error, particularly for cells that are very susceptible to osmotic
stress. Alternatively, the extent to which a given cell type can tolerate changes in
cell volume can be determined by exposing the cells to a range of concentrations of
non-permeating solute (Armitage & Mazur, 1984a; Armitage et al. 1985). Stepwise
addition and dilution protocols that maintain cell volume changes within those
limits can then be calculated providing that the permeability characteristics of the
cell have been determined (Frim & Mazur, 1983; Armitage & Mazur, 1984b). For
example, human platelets do not tolerate more than about 0·7 mol l^{-1} glycerol
when the addition and dilution of glycerol are abrupt (Rowe & Lenny, 1981;
Armitage, 1982; Armitage & Hunt, 1982), and these cells are also very sensitive to
osmotic stress (Armitage et al. 1985). Estimations of the coefficients of water
permeability and glycerol permeability showed platelets to be about 4000-fold
more permeable to water than to glycerol; consequently, the osmotic effects of the
addition and dilution of high concentrations of glycerol would be considerable
(Armitage, 1986a,b). However, platelet function was found to be substantially
improved after exposure to 1 mol l^{-1} glycerol when the rates of glycerol addition
and subsequent dilution were such that cell volume remained within 60 % to 130 %
of normal volume (Armitage, 1986c).

   An alternative method for removing high concentrations of permeating cryoprotectants from the cell suspending medium is to dilute the medium with a

solution containing a hyperosmotic concentration of non-permeating solute. The reduction in external cryoprotectant concentration is osmotically balanced by the increase in non-permeating solute (e.g. sucrose) concentration; thus, cell swelling is limited (Leibo & Mazur, 1978). Ashwood-Smith (1975) overcame the problems associated with the use of high concentrations of permeating additives by combining low concentrations (0·5%–1% v/v) of glycerol or Me_2SO with 9%–9·5% w/v dextran, which does not enter cells. The survival of Chinese hamster fibroblasts after freezing and thawing was dependent on the molecular weight of the dextran, but the separate cryoprotective effects of the permeating and non-permeating compounds were found to be additive. However, the combination of glycerol and non-permeating cryoprotectants, including dextran, did not result in a better survival of platelets after freezing and thawing than when glycerol was used as the sole cryoprotectant (Armitage, Brodthagen & Parmar, 1983; Brodthagen, Armitage & Parmar, 1985).

Interactions between cryobiological variables

The survival of cells after freezing and thawing depends on complex interactions between cooling rate, warming rate, and the type and concentration of cryoprotectant. Therefore, an optimum cooling rate for a given cell type has no practical meaning in terms of cell survival unless the other cryobiological variables have also been defined. A change in any one of the variables can profoundly reduce survival, although high survival can sometimes be regained by adjustments to the other variables. Consequently, manipulation of the major cryobiological variables can lead to the development of one or more methods of cryopreservation for a given type of cell. The scope for manipulation, however, is limited by the individual biological characteristics of each cell type, such as cell size, water permeability, tolerance of osmotic stress and tolerance of cryoprotectants, and also by the practical reasons for cryopreserving the cells (e.g. the need to store large volumes of cell suspension). These general points will be illustrated briefly with reference to erythrocytes and lymphocytes.

Erythrocytes can survive freezing and thawing in the absence of any cryoprotectant and, because of their very high permeability to water, survival of about 60% of the cells is achieved at an optimum cooling rate of about $3000°C\,min^{-1}$ when they are warmed rapidly (Rapatz, Sullivan & Luyet, 1968). This cooling rate is clearly not feasible with the large volumes of blood that are required for clinical use, nor is 40% haemolysis clinically acceptable; consequently, two methods of red cell cryopreservation have been developed for use in blood banks (see Pegg, 1976; Rowe, Lenny & Mannoni, 1980). In the first method to be developed, erythrocytes were equilibrated with 40%–50% v/v glycerol and cooled slowly at an uncontrolled rate to −80°C. The main problems associated with this method are those of adding and, particularly, removing such high concentrations of glycerol without damaging the cells by osmotic stress. Therefore, a second method was developed that used a lower concentration of glycerol (15%–18% v/v) and a higher cooling rate (*ca* 90°C min^{-1}). However, when the cells do not have to be preserved in large volumes, high survival can be attained by using non-permeating cryoprotectants and a high cooling rate. For example, Pegg, Diaper, Scholey & Coombs (1982) have described a convenient technique for the preservation of trypsin-treated antibody-coupled red cells for use in haemagglutination assays. The cells are suspended in a solution containing 0·3 g dl^{-1} NaCl, 7 g dl^{-1} sucrose and 10 g dl^{-1} dextran, and the suspension allowed to drop from a pipette onto the surface of liquid nitrogen. The frozen droplets (*ca* 20 μl) are thawed a few at a time in warm phosphate-buffered saline, and survival is greater than 80 % for ovine, bovine and human cells.

High recoveries of human lymphocytes are attainable when the cells are equilibrated with 10 % v/v Me$_2$SO and cooled at 1°C min^{-1} (Pegg, 1965; Weiner, 1976). However, lymphocytes also survive rapid two-step cooling when they are equilibrated with 5 % v/v Me$_2$SO, cooled to −25°C for 10 min and then immersed in liquid nitrogen (Farrant, Knight, McGann & O'Brien, 1974). Moreover, lymphocytes suspended in serum can be cryopreserved without Me$_2$SO or glycerol by cooling to −10°C for 2 min followed by rapid cooling in liquid nitrogen (Knight, Farrant & McGann, 1977).

Avoidance of ice formation

Liquid storage at low temperatures

In 1965, Farrant proposed that freezing, and hence the increase in electrolyte concentration, could be prevented by gradually increasing the amount of Me$_2$SO during cooling, such that the concentration of Me$_2$SO was always just sufficient to prevent freezing on a colligative basis. Strips of smooth muscle were equilibrated with 10 % v/v Me$_2$SO in Kreb's solution at 37°C. During slow cooling, the concentration of Me$_2$SO was progressively increased to 55 % v/v, which was sufficient to prevent freezing at −79°C: during slow warming, the Me$_2$SO concentration was gradually reduced, until the tissue was returned to a medium without cryoprotectant at 37°C. Strips of smooth muscle contracted in response to histamine after cooling by this method to −79°C, whereas smooth muscle *frozen* to −79°C in 10 % v/v Me$_2$SO and warmed rapidly showed no contractile response to histamine. Recovery was subsequently found to be improved when the Kreb's solution was modified to contain a high potassium concentration; moreover, the extent of recovery was dependent on the nature of the major anion in the solution and the pH (Elford & Walter, 1972; Taylor, Elford & Walter, 1978; Taylor, 1981).

Vitrification

Water and aqueous solutions do not necessarily solidify by crystallization when cooled, instead they can vitrify to form amorphous, glassy solids (Franks, 1982; MacFarlane, 1986). The vitrification of pure water requires a cooling rate of 10^6 to 10^7 K s^{-1}, which is clearly unattainable in cell suspensions. However, that critical rate is reduced in the presence of solutes, and some cryoprotectants are also good glass formers, for example propane-1,2-diol (Boutron & Kaufman, 1979). Thus, when an aqueous solution containing a sufficiently high concentration of propane-1,2-diol (*ca* 40 % w/w) is cooled, it will initially supercool well below its freezing point. As the temperature falls, there is a large increase in viscosity of the solution that inhibits ice formation and, eventually, the solution becomes a glass, which has the random molecular organization of a liquid rather than the regular molecular arrangement of a crystalline solid. The advantage of vitrification for the preservation of cells and, in particular, organized tissues is that the mechanisms of injury currently associated with the formation of ice are avoided (Pegg *et al.* 1979; Taylor & Pegg, 1983). There are, however, several problems associated with the application of vitrification to biological materials. First, cells have to tolerate exposure to the necessarily very high concentrations of solute. Second, cracks can form in vitrified solutions that are cooled or warmed too rapidly below the glass transition temperature. Third, extensive and potentially harmful ice crystallization can occur during warming. This phenomenon, which is called devitrification, can be reduced or avoided by increasing the initial solute concentration, or by increasing the rates of cooling and warming. Finally, a reduction in temperature *per se* can be detrimental to some cells. Fahy, MacFarlane, Angell & Meryman (1984) have suggested vitrification as a means of attaining the cryopreservation of organs, and they have investigated extensively the glass-forming properties of various combinations of cryoprotectants. In 1985, Rall & Fahy achieved the successful vitrification of mouse embryos using a mixture of 20·5 % w/v Me_2SO, 15·5 % w/v acetamide, 10 % w/v propane-1,2-diol and 6 % w/v polyethylene glycol in tissue culture medium. Vitrification of monocytes has subsequently been reported using the same combination of cryoprotectants (Takahashi, Hirsh, Erbe, Bross, Steere & Williams, 1986), and the vitrification of embryos has also been reported using mixtures of glycerol and propane-1,2-diol (Scheffen, Van Der Zwalmen & Massip, 1986).

Viability

Evaluation of the survival of cells after freezing and thawing is an area fraught with difficulties. Merely counting the number of cells recovered after freezing and thawing provides no information about the viability of the cells. There are numerous tests of membrane integrity, metabolic activity, and physiological function that can be used, but the extent to which any one of those tests is an indicator of overall viability is open to question. The results of some tests are very sensitive to the length of time that elapses after thawing before the test is carried out. Sometimes, latent freezing injury is not expressed until some time after thawing; conversely, some changes that are apparent immediately after thawing may be reversible through the operation of cellular repair mechanisms. In tissue culture cells, for example, there can be considerable delays before division starts, and some cells, which never divide, remain intact for many hours before they lyse (Frim, Snyder, McGann & Kruuv, 1978).

Nevertheless, when a large number of variations in cooling and warming conditions are being examined, it is useful to have a screening test that is simple, quick to perform, and convenient; for example, the haemolysis of erythrocytes. Trypan blue or eosin have been used in dye exclusion tests for many types of cell. Cells that take up the dye are considered to be non-viable, which is probably true; but the assumption that cells that exclude dye are viable is somewhat tenuous and can lead to erroneously high values of cell survival (e.g. Knight & Farrant, 1978). The number of cells in a suspension that exclude the dye can be affected by the presence of serum, the concentration of the dye, and the cell concentration. The number of dye-excluding cells also becomes time dependent in suspensions that have been exposed to damaging stresses such as freezing and thawing (Black & Berenbaum, 1964; Tennant, 1964).

Fluorescein diacetate (FDA) has been used as an indicator of membrane integrity for a number of cell types, including fertilized ova and granulocytes. The principle of this assay is that cells are permeable to the non-fluorescent FDA, which is broken down by intracelluar esterase activity to form fluorescein. If the plasma membrane is intact, the fluorescein is retained and the cell fluoresces green when excited with light at a suitable wavelength (450–490 nm) (Rotman & Papermaster, 1966). Ethidium bromide can be used as a counter stain for FDA: cells with intact membranes are impermeable to ethidium bromide and, therefore, fluoresce green, but damaged cells take up ethidium bromide and their nuclei fluoresce red (e.g. Dankberg & Persidsky, 1976). Acridine orange is another fluorescent stain that can be used for assessing cellular integrity (e.g. Smith, Ashwood-Smith & Young, 1963; Hofmann, Till, Hofmann, Michel & Quiess, 1979). The FDA assay has been used to assess the survival of frozen and thawed fertilized mouse ova, and has shown a high degree of correlation with the ability of the ova to develop in culture (Jakowski, 1977, quoted by Leibo & Mazur, 1978). However, apparently high survivals of granulocytes, which are very sensitive to freezing injury, can be observed using FDA when the assay is performed immediately after thawing, or when the cells are held at 0°C after removal of the cryoprotectant. Such a result bears no relation whatsoever to the results of more rigorous assays of functional integrity, such as chemotaxis, which show a marked lack of cellular function (Frim & Mazur, 1983). The same is true for exposure to glycerol and to osmotic stress. However, when granulocytes are incubated at 37°C for an hour or more before the FDA assay is performed, the results are more in accord with chemotactic activity (Frim & Mazur, 1983; Armitage & Mazur, 1984a,b; Takahashi, Hammett & Cho, 1985). This suggests that the cellular lesion is expressed as a loss of membrane integrity only after a period of metabolic activity at physiological temperature.

Histochemical stains for enzyme activity, such as nitroblue tetrazolium (NBT), can also be misleading, and exact details of the conditions under which the assay is performed are necessary for correct interpretation of the results. Taylor (1986) has given a clear example of the difficulties in interpretation of the NBT assay when it is used to assess the viability of corneal endothelium after freezing and thawing.

Depending on the conditions under which the assay is performed, the reduction of NBT to a dark diformazan precipitate in the endothelial cells can indicate that the cells are viable or that the cells are dead.

Some assays of cellular function, such as the incorporation of [³H]-thymidine into DNA by stimulated lymphocytes, are dependent both on the level of activity of individual cells as well as on the proportion of active cells within the sample. Knight & Farrant (1978) were able to distinguish between these two factors by comparing the uptake of [³H]-thymidine as a function of cell concentration in samples of frozen and non-frozen lymphocytes. They showed that the reduction in uptake by the frozen and thawed cells was the result of a loss of active cells rather than a reduction in the efficiency of uptake by individual cells.

The choice of viability assay is governed to a large extent by cell type; but the reliability of any assay is dependent on many factors, some of which are intrinsic properties of the assay itself, and others that are characteristic of the cell and the conditions under which the assay is performed. Simple assays, such as FDA, are useful for screening, but whenever possible they should be used in conjunction with one or more assays of cellular function. Cell division in culture is a clear indicator of viability. Although functional tests in non-dividing cells, such as chemotaxis and phagocytosis in granulocytes and platelet aggregation, assess properties that are essential for normal *in vivo* function, they do not necessarily indicate that the cells will function normally *in vivo*.

Conclusions

Despite the range of stresses that cells encounter during freezing and thawing, many types of cell of medical, agricultural and biological importance can be successfully cryopreserved (Ashwood-Smith & Farrant, 1980). However, different cells vary markedly in their ability to tolerate those stresses. Some types of cell, for example human granulocytes, are highly sensitive to freezing injury and have not yet been successfully cryopreserved, whereas other cell types, such as erythrocytes and lymphocytes, can be cryopreserved by a variety of methods. Cell parameters, such as water and cryoprotectant permeability coefficients, are not available for many cell types; thus, the scope for devising protocols to minimize osmotic stress during the addition and dilution of cryoprotectants, and for predicting optimum rates of cooling, is limited. Consequently, most cryopreservation methods have to be developed empirically, and an awareness of the interrelationship between rates of cooling and warming, and the type and concentration of cryoprotectant, is necessary particularly for cells that prove to be sensitive to freezing.

Vitrification is a possible alternative method of cryopreservation at $-196°C$; but to attain vitrification at cooling rates that are feasible with biological materials, the presence of very high concentrations of solute is necessary. When insufficient solute is present, or when the rates of cooling and warming are not high enough, devitrification during warming can result in the growth of significant amounts of ice.

References

Akhtar, T., Pegg, D. E. & Foreman, J. (1979). The effect of cooling and warming rates on the survival of cryopreserved L-cells. *Cryobiology* **16**, 424–429.

Armitage, W. J. (1982). Transport of 5-hydroxytryptamine by human platelets incubated in glycerol. *Transfusion* **22**, 203–205.

Armitage, W. J. (1986a). Effects of solute concentration on intracellular water volume and hydraulic conductivity of human blood platelets. *J. Physiol. (Lond.)* **374**, 375–385.

Armitage, W. J. (1986b). Permeability of human blood platelets to glycerol. *J. cell. Physiol.* **128**, 121–126.

Armitage, W. J. (1986c). Osmotic stress as a factor in the detrimental effect of glycerol on human platelets. *Cryobiology* **23**, 116–125.

Armitage, W. J., Brodthagen, U. A. & Parmar, N. (1983). Human platelet aggregation after freezing in glycerol and dextran. *Cryo-Letters* **4**, 349–354.

Armitage, W. J. & Hunt, C. J. (1982). The effect of glycerol on aggregation and ultrastructure of human platelets. *Cryobiology* **19**, 110–117.

Armitage, W. J. & Mazur, P. (1984a). Osmotic tolerance of human granulocytes. *Am. J. Physiol.* **247** (*Cell Physiol.* **16**), C373–C381.

Armitage, W. J. & Mazur, P. (1984b). Toxic and osmotic effects of glycerol on human granulocytes. *Am. J. Physiol.* **247** (*Cell Physiol.* **16**), C382–389.

Armitage, W. J., Parmar, N. & Hunt, C. J. (1985). The effects of osmotic stress on human platelets. *J. cell. Physiol.* **123**, 241–248.

Ashwood-Smith, M. J. (1975). Current concepts concerning radioprotective and cyroprotective properties of dimethyl sulfoxide in cellular systems. *Ann. N.Y. Acad. Sci.* **243**, 246–256.

Ashwood-Smith, M. J. & Farrant, J. (1980) (eds). *Low Temperature Preservation in Medicine and Biology*. Tunbridge Wells: Pitman.

Ashwood-Smith, M. J. & Friedmann, G. B. (1979). Lethal and chromosomal effects of freezing, thawing, storage time, and X-irradiation on mammalian cells preserved at $-196°C$ in dimethyl sulfoxide. *Cryobiology* **16**, 132–140.

Ashwood-Smith, M. J. & Lough, P. (1975). Cryoprotection of mammalian cells in tissue culture with methanol. *Cryobiology* **12**, 517–518.

Black, L. & Berenbaum, M. C. (1964). Factors affecting the dye exclusion test for cell viability. *Expl Cell Res.* **35**, 9–13.

Boutron, P. & Kaufmann, A. (1979). Stability of the amorphous state in the system water-1,2-propanediol. *Cryobiology* **16**, 557–568.

Brodthagen, U. A., Armitage, W. J. & Parmar, N. (1985). Platelet cryopreservation with glycerol, dextran and mannitol: recovery of 5-hydroxytryptamine transport and hypotonic stress response. *Cryobiology* **22**, 1–9.

Buckner, C. D., Appelbaum, F. R. & Thomas, E. D. (1981). Bone marrow and fetal liver. In *Organ Preservation for Transplantation* (ed. A. M. Karow & D. E. Pegg), pp. 355–375. New York: Dekker.

Clegg, J. S., Seitz, P., Seitz, W. & Hazlewood, C. F. (1982). Cellular responses to extreme water loss: the water replacement hypothesis. *Cryobiology* **19**, 306–316.

Dankberg, F. & Persidsky, M. D. A. (1976). A test of granulocyte membrane integrity and phagocytic function. *Cryobiology* **13**, 430–432.

Davson, H. & Danielli, J. F. (1952). *The Permeability of Natural Membranes*. Cambridge: Cambridge University Press.

Diller, K. R. (1979). Intracellular freezing of glycerolized red cells. *Cryobiology* **16**, 125–131.

Elford, B. C. & Walter, C. A. (1972). Effects of electrolyte composition and pH on the structure and function of smooth muscle cooled to $-79°C$ in unfrozen media. *Cryobiology* **9**, 82–100.

Elliott, K. & Whelan, J. (1977) (eds). *The Freezing of Mammalian Embryos. Ciba Foundation Symposium No. 52*. Amsterdam: Elsevier.

Fahy, G. M., MacFarlane, D. R., Angell, C. A. & Meryman, H. T. (1984). Vitrification as an approach to cryopreservation. *Cryobiology* **21**, 407–426.

Farrant, J. (1965). Mechanism of cell damage during freezing and thawing and its prevention. *Nature, Lond.* **205**, 1284–1287.

FARRANT, J. (1969). Is there a common mechanism of protection of living cells by polyvinylpyrrolidone and glycerol during freezing? *Nature, Lond.* **222**, 1175–1176.

FARRANT, J. (1977). Water transport and cell survival in cryobiological procedures. *Phil. Trans. R. Soc. Ser.* B **278**, 191–205.

FARRANT, J., KNIGHT, S. C., McGANN, L. E. & O'BRIEN, J. (1974). Optimal recovery of lymphocytes and tissue culture cells following rapid cooling. *Nature, Lond.* **249**, 452–453.

FARRANT, J. & MORRIS, G. J. (1973). Thermal shock and dilution shock as the causes of freezing injury. *Cryobiology* **10**, 134–140.

FARRANT, J. & WOOLGAR, A. E. (1972). Human red cells under hyperosmotic conditions: a model system for investigating freezing damage. 1. Sodium chloride. *Cryobiology* **9**, 9–15.

FRANKS, F. (1982). The properties of aqueous solutions at subzero temperatures. In *Water: A Comprehensive Treatise*, vol. 7 (ed. F. Franks), pp. 215–338. New York: Plenum.

FRIM, J. & MAZUR, P. (1983). Interactions of cooling rate, warming rate, glycerol concentration, and dilution procedure on the viability of frozen-thawed human granulocytes. *Cryobiology* **20**, 657–676.

FRIM, J., SNYDER, R. A., McGANN, L. E. & KRUUV, J. (1978). Growth kinetics of cells following freezing in liquid nitrogen. *Cryobiology* **15**, 502–516.

HAYES, A. R., PEGG, D. E. & KINGSTON, R. E. (1974). A multirate small-volume cooling machine. *Cryobiology* **11**, 371–377.

HOFMANN, J., TILL, U., HOFMANN, B., MICHEL, E. & QUIESS, M. (1979). The use of acridine orange for testing blood platelet integrity. *Acta biol. med. germ.* **38**, 1149–1157.

HOUSE, C. R. (1974). *Water Transport in Cells and Tissues*. London: Arnold.

KNIGHT, S. C. & FARRANT, J. (1978). Comparing stimulation of lymphocytes in different samples: separate effects of numbers of responding cells and their capacity to respond. *J. immunol. Methods* **22**, 63–71.

JACKOWSKI, S. C. (1977). Ph.D. Dissertation, University of Tennessee, Knoxville.

KNIGHT, S. C., FARRANT, J. & McGANN, L. E. (1977). Storage of human lymphocytes by freezing in serum alone. *Cryobiology* **14**, 112–115.

LEIBO, S. P. (1983). Field trials of one-step diluted frozen-thawed bovine embryos. *Cryobiology* **20**, 742.

LEIBO, S. P. & MAZUR, P. (1978). Methods for the preservation of mammalian embryos by freezing. In *Methods of Mammalian Reproduction* (ed. J. C. Daniel), pp. 179–201. New York: Academic Press.

LOVELOCK, J. E. (1953a). The haemolysis of human red cells by freezing and thawing. *Biochim. biophys. Acta* **10**, 414–426.

LOVELOCK, J. E. (1953b). The mechanism of the protective action of glycerol against haemolysis by freezing and thawing. *Biochim. biophys. Acta* **11**, 28–36.

LOVELOCK, J. E. & BISHOP, M. W. H. (1959). Prevention of freezing damage to living cells by dimethyl sulphoxide. *Nature, Lond.* **183**, 1394–1395.

LUYET, B. J. & KEANE, J. (1955). A critical temperature range apparently characterized by sensitivity of bull semen to high freezing velocity. *Biodynamica* **7**, 281–292.

MACFARLANE, D. R. (1986). Devitrification in glass-forming aqueous solutions. *Cryobiology* **23**, 230–244.

MAZUR, P. (1963). Kinetics of water loss from cells at subzero temperatures and the likelihood of intracellular freezing. *J. gen. Physiol.* **47**, 347–369.

MAZUR, P. (1970). Cryobiology: the freezing of biological systems. *Science* **168**, 939–949.

MAZUR, P. (1977). The role of intracellular freezing in the death of cells cooled at supraoptimal rates. *Cryobiology* **14**, 251–272.

MAZUR, P. (1984). Freezing of living cells: mechanisms and implications. *Am. J. Physiol.* **247** (*Cell Physiol.* **16**), C125–C142.

MAZUR, P., LEIBO, S. P., FARRANT, J., CHU, E. H. Y., HANNA, M. G. & SMITH, L. H. (1970). Interactions of cooling rate, warming rate and protective additive on the survival of frozen mammalian cells. In *Ciba Foundation Symposium on the Frozen Cell* (ed. G. E. W. Wolstenholme & M. O'Connor), pp. 69–88. London: Churchill.

MAZUR, P., RALL, W. F. & RIGOPOULOS, N. (1981). Relative contributions of the fraction of unfrozen water and of salt concentration to the survival of slowly frozen human erythrocytes. *Biophys. J.* **36**, 653–675.

McGANN, L. E. & FARRANT, J. (1976a). Survival of tissue culture cells frozen by a two-step procedure to −196 °C. I. Holding temperature and time. *Cryobiology* **13**, 261–268.

McGANN, L. E. & FARRANT, J. (1976b). Survival of tissue culture cells frozen by a two-step procedure to −196 °C. II. Warming rate and concentration of dimethyl sulfoxide. *Cryobiology* **13**, 269–273.

MERYMAN, H. T. (1968). Modified model for the mechanism of freezing injury in erythrocytes. *Nature, Lond.* **218**, 333–336.

MERYMAN, H. T. (1971). Osmotic stress as a mechanism of freezing injury. *Cryobiology* **8**, 489–500.

MILLER, R. H. & MAZUR, P. (1976). Survival of frozen-thawed human red cells as a function of cooling and warming velocities. *Cryobiology* **13**, 404–414.

MORRIS, G. J. & CLARKE, A. (1981) (eds). *Effects of Low Temperatures on Biological Membranes.* London: Academic Press.

NASH, T. (1962). The chemical contribution of compounds which protect erythrocytes against freezing damage. *J. gen. Physiol.* **46**, 167–175.

PEGG, D. E. (1964). Freezing of bone marrow for clinical use. *Cryobiology* **1**, 64–71.

PEGG, D. E. (1976). Long-term preservation of cells and tissues: a review. *J. clin. Path.* **29**, 271–285.

PEGG, D. E., DIAPER, M. P., SCHOLEY, S. E. & COOMBS, R. R. A. (1982). Droplet freezing of antibody-linked indicator red cells of sheep, ox and human origin. *Cryobiology* **19**, 573–584.

PEGG, D. E., DIAPER, M. P., SKAER, H. LeB. & HUNT, C. J. (1984). The effect of cooling rate on the packing effect in human erythrocytes frozen and thawed in the presence of 2 M glycerol. *Cryobiology* **21**, 491–502.

PEGG, D. E., JACOBSEN, I. A., ARMITAGE, W. J. & TAYLOR, M. J. (1979). Mechanisms of freezing injury in organs. In *Organ Preservation* II (ed. D. E. Pegg & I. A. Jacobsen), pp. 132–146. Edinburgh: Churchill Livingstone.

PEGG, P. J. (1965). The preservation of leucocytes for cytogenetic and cytochemical studies. *Br. J. Haemat.* **11**, 586–591.

POLGE, C. & LOVELOCK, J. E. (1952). Preservation of bull semen at −79 °C. *Vet. Rec.* **64**, 396–397.

POLGE, C., SMITH, A. U. & PARKES, A. S. (1949). Revival of spermatozoa after vitrification and dehydration at low temperatures. *Nature, Lond.* **164**, 666.

PYLE, H. M. (1964). Glycerol preservation of red cells. *Cryobiology* **1**, 57–60.

RALL, W. F. & FAHY, G. M. (1985). Ice-free cryopreservation of mouse embryos at −196 °C by vitrification. *Nature, Lond.* **313**, 573–575.

RAPATZ, G., SULLIVAN, J. J. & LUYET, B. J. (1968). Preservation of erythrocytes in blood containing various cryoprotective agents, frozen at various rates and brought to a given final temperature. *Cryobiology* **15**, 18–25.

ROTMAN, B. & PAPERMASTER, B. W. (1966). Membrane properties of living mammalian cells as studied by enzymatic hydrolysis of fluorogenic esters. *Proc. natn. Acad. Sci. U.S.A.* **55**, 134–141.

ROWE, A. W. & LENNY, L. L. (1981). *Platelets.* In *Organ Preservation for Transplantation* (ed. A. M. Karow & D. E. Pegg), pp. 335–354. New York: Dekker.

ROWE, A. W., LENNY, L. L. & MANNONI, P. (1980). Cryopreservation of red cells and platelets. In *Low Temperature Preservation in Medicine and Biology* (ed. M. J. Ashwood-Smith & J. Farrant), pp. 85–120. Tunbridge Wells: Pitman.

SCHEFFEN, B., VAN DER ZWALMEN, P. & MASSIP, A. (1986). A simple and efficient procedure for the preservation of mouse embryos by vitrification. *Cryo-Letters* **7**, 260–269.

SHLAFER, M. (1981). Phamacological considerations in cryopreservation. In *Organ Preservation for Transplantation* (ed. A. M. Karow & D. E. Pegg), pp. 177–212. New York: Dekker.

SMITH, A. U., ASHWOOD-SMITH, M. J. & YOUNG, M. R. (1963). Some *in vitro* studies on rabbit corneal tissue. *Expl Eye Res.* **2**, 71–87.

TAKAHASHI, T., HAMMETT, M. F. & CHO, M. S. (1985). Multifaceted freezing injury in human polymorphonuclear cells at high subfreezing temperatures. *Cryobiology* **22**, 215–236.

TAKAHASHI, T., HIRSH, A., ERBE, E. F., BROSS, J. B., STEERE, R. L. & WILLIAMS, R. J. (1986). Vitrification of human monocytes. *Cryobiology* **23**, 103–115.

TAYLOR, M. J. (1981). The meaning of pH at low temperatures. *Cryo-Letters* **2**, 231–239.

TAYLOR, M. J. (1986). Clinical cryobiology of tissues: preservation of corneas. *Cryobiology* **23**, 323–353.

TAYLOR, M. J., ELFORD, B. C. & WALTER, C. A. (1978). The pH-dependent recovery of smooth muscle from storage at $-13\,°C$ in unfrozen media. *Cryobiology* **15**, 452–460.

TAYLOR, M. J. & PEGG, D. E. (1983). The effect of ice formation on the function of smooth muscle tissue stored at -21 or $-60\,°C$. *Cryobiology* **20**, 36–40.

TENNANT, J. R. (1964). Evaluation of the trypan blue technique for determination of cell viability. *Transplantation* **2**, 685–694.

VENKATARAMAN, M. & WESTERMAN, M. P. (1986). Susceptibility of human T cells, T-cell subsets, and B cells to cryopreservation. *Cryobiology* **23**, 199–208.

WEINER, R. S. (1976). Cryopreservation of lymphocytes for use in *in vitro* assays of cellular immunity. *J. immunol. Methods* **10**, 49–60.

WHITTINGHAM, D. G., LEIBO, S. P. & MAZUR, P. (1972). Survival of mouse embryos frozen to $-196\,°C$ and $-269\,°C$. *Science* **178**, 411–414.

WILMUT, I. (1972). Effect of cooling rate, warming rate, cryoprotective agent and stage of development on survival of mouse embryos during cooling and thawing. *Life Sci.* **11**, 1071–1079.

WILMUT, I. & ROWSON, L. E. (1973). Experiments on the low temperature preservation of cow embryos. *Vet. Rec.* **92**, 686–690.

MECHANISMS OF CRYOPROTECTANT ACTION

M. J. ASHWOOD-SMITH

Physiological Laboratory, Cambridge and Department of Biology, University of Victoria, Victoria, British Columbia, Canada

Introduction

The majority of cells are destroyed when subjected to the many stresses associated with cooling to and thawing from low temperatures (Ashwood-Smith & Farrant, 1980). Only a very small fraction, perhaps less than 1 % or so of nucleated animal cells survive freezing and thawing. Microorganisms and viruses are, in general, more resistant than eukaryotic cells.

Changes in the rate of cooling and thawing modify the survival of all cells but the most important factor influencing survival is the presence or absence of molecules that possess cryoprotective properties. When present in substantial amounts, usually in the range of $0 \cdot 5$ to $2 \, \text{mol} \, l^{-1}$, cryoprotectants can increase survival values dramatically thus permitting the successful cryopreservation of many cells. Several of the chemicals used for the laboratory preservation of red cells, bone marrow cells, lymphocytes, embryos, spermatozoa etc., have been found to have a similar function in nature in that they confer frost resistance to some invertebrates (Ring, 1980) and in one instance an amphibian (Storey & Storey, 1984). The simple polyhydric alcohol, glycerol, is one good example of this phenomenon and the non-reducing sugar, trehalose, another.

The mechanisms of frost resistance in plants (Li & Sakai, 1978) are less well understood than in animals and are complex. It is now becoming increasingly clear that supercooling (undercooling), perhaps with the aid of large amounts of glycerol or a simple sugar, is not the only way to avoid the destructive effects of intracellular ice formation. Some insects (Ring, 1980) and certain plants utilize the deliberate initiation of controlled extracellular ice formation with specific protein ice nucleators to prevent supercooling. Considerable attention has recently been focussed on this problem with the cloning of the ice nucleating protein from so called 'ice + bacteria' (Orser *et al.* 1985).

Freezing damage is minimized when controlled slow rates of freezing occur as severe supercooling is unlikely and thus the possibility of intracellular ice formation is lessened. The slow formation of extracellular ice initiated by nucleating agents thus permits a gentle desiccation to take place. This situation parallels that which occurs in the standard laboratory procedures for the cryobiological preservation of many cells with glycerol or dimethyl sulphoxide (DMSO).

Another mechanism by which frost damage in nature may be minimized is by changing the manner in which ice, once formed, crystallizes. In this instance, specific proteins known as thermal hysteresis proteins (THPs), induce a difference between the observed freezing point and melting point of an aqueous solution (Franks, 1985).

There are, therefore, a number of ways in which freezing damage may be avoided. For the standard preservation of cells and tissues large amounts of intracellular cryoprotective agents such as either glycerol or DMSO in conjunction with controlled rates of freezing and thawing are employed (Ashwood-Smith, 1980). Attempts to utilize procedures based on processes occurring in nature, other than those involving glycerol, have not been extensively investigated thus far.

The total avoidance of ice formation during freezing and thawing has been a dream of biologists for many years. Luyet and his colleagues (1938, 1939, 1940) experimented with vitrification (literally glass formation) and recently Rall & Fahy (1985) have reintroduced this concept for cell preservation, specially in relation to the cryopreservation of embryos.

I shall now discuss in detail some of the considerations raised in this brief introduction.

Freezing damage in cellular systems

Other contributors to this symposium have covered this topic (Armitage; Pegg: this volume) and except when a discussion of freezing damage is apposite to explain the action of cryoprotectants it will not be alluded to further.

Chemical characteristics of cryoprotectants

A list of chemicals which have been shown to protect a variety of cells from damage associated with freezing and thawing is given in Table 1. In the vast majority of cases the molecules are simple, low molecular weight chemicals of high water solubility and low cellular toxicity. All the low molecular weight compounds are capable of considerable hydrogen bonding with water molecules and thus tend to have high heats of solution. Even some polymers such as polyvinyl pyrrolidone (PVP) produce easily measured increases in temperature when dissolved in water. In this latter instance, however, PVP has first to be freeze-dried so that it dissolves almost instantly otherwise the slow rate of solution makes it difficult to observe easily the heat of solution. The toxicity of cryoprotectants may vary with the cell type although in general most of these molecules are, by pharmacological standards, remarkably non-toxic. When given to mice by I.P. or I.V. injection the observed toxicity is often related to osmotic effects associated with the molar quantities injected. Both DMSO and glycerol, for example, cause considerable alterations in peripheral blood pressure, effects clearly caused by osmotic imbalances (Ashwood-Smith, 1962).

The observations that one cell type is better able to tolerate one cryoprotectant rather than another is often the result of a difference in permeability and not an inherent difference in toxicity. Methanol will penetrate most cells so rapidly that it is often difficult to measure the two step osmotic change produced when cells are exposed to $1–3\,\text{mol}\,l^{-1}$ solutions (cell contraction followed by expansion). If these parameters are not appreciated then it is easy to ascribe 'toxicity' to a cryoprotective molecule which would not be seen under careful conditions of its addition or removal. For example, propanediol penetrates most cells less fast than methanol but faster than DMSO and much faster than glycerol.

Penetration rates are also profoundly affected by temperature much more so for some cryoprotectants than others. The most striking example is probably that of glycerol and the rabbit oocyte. Diffusion is much faster at body temperature than at lower temperatures (Smith, 1952). Unawareness of this fact may well have contributed to less than optimum results in the cryopreservation of human blastocysts with glycerol (Ashwood-Smith, 1986).

Sometimes toxicity is not apparent except after freezing and thawing. A good instance of this is the case with the simplest alcohol, methanol. At first sight the possibility of methanol as a cryoprotective agent might give rise to consideration apprehension to biologists brought up with the knowledge that methanol is extremely toxic to most vertebrates. However at the cell level this is not so and most cells will divide quite happily in the presence of 1% methanol! It is an exceptionally good cryoprotective agent for a number of cells (Ashwood-Smith & Lough, 1975) and it is very effective for the cryopreservation of mouse embryos (Rall *et al.* 1984). The problem of its cellular toxicity arises as a result of its concentration during the freezing process as water separates as ice. This, of course, is not unique to methanol, it happens with all cryoprotectants and with

Table 1. *Cryoprotective agents*

Alcohols	*Sugars*
Ethanol	Adnitol
Ethylene glycol	Erythritol
Glycerol***	Fructose
Methanol**	Galactose
Polyethylene glycol*	Glucose
Propylene glycol	Inositol
(propane diol)**	Lactose
	Maltose
Polymers	Raffinose
Polyethylene glycol (PEG)*	Sorbitol
Polyvinyl pyrrolidone (PVP)*	Sucrose**
Dextran (Mol wt from $5\times10^4\rightarrow$	Xylitol
5×10^5 daltons)*	
Hydroxyethyl starch (HES)*	*Miscellaneous compounds*
Serum proteins*	Acetamide
	Dimethyl acetamide
	Dimethyl sulphoxide (DMSO)***

* Effective to a limited extent in nucleated cells.
** Moderately effective or not extensively investigated.
*** Very effective and the most commonly used agents.

polymers concentration during freezing may play an important role in their mechanism of protection.

The toxicity of cryoprotectants to cells and tissues has been discussed in detail by Fahy (1986). He makes the interesting, but perhaps debatable point that 100 % protection should be theoretically obtainable when standard cryoprotective agents such as glycerol or DMSO are used. Fahy believes that the reason for the disparity rests with the chemical toxicity of the cryoprotectants themselves. Acetamide, for example is claimed to modify the toxic effects of exposure of cells to high levels of DMSO (Rall & Fahy, 1985).

A central problem which arises when discussing the adverse effects of cryoprotectants is related to the time and temperature of exposure. Bone marrow cells, for instance, suffer a time and temperature dependent inhibition of many biochemical processes such as lipid, protein and DNA synthesis when exposed to 10 % DMSO but these effects are almost completely reversible if the exposure period is relatively short (Ashwood-Smith, 1967).

The well documented effects of DMSO as a derepressor of genes and an inducer of single strand breaks in mammalian DNA is a good example of the detrimental effects of DMSO only manifesting themselves, in general, over a 24 h exposure period and at concentrations of 1–2 %. The short exposure periods involved in cell cryopreservation are without genetic effects (Ashwood-Smith, 1985) and are discussed in more detail later.

The orthodox view of cryoprotectant action

Lovelock in a series of papers in the mid-fifties (1953a,b, 1954, 1957) concluded that the haemolysis of red blood cells after freezing and thawing could be accounted for by the greatly increased concentration of sodium chloride prior to the eutectic point (−21·5°C) causing denaturation of lipoproteins in the cell membrane thus leading to haemolysis. This theory of cell destruction following freezing and thawing has been widely accepted although Meryman has argued that salts, per se, are not responsible. According to Meryman (1968) and Meryman et al. (1977) every cell has a minimum volume beyond which it cannot shrink during the freezing process as the increased solutes in the presence of extracellular ice slowly desiccate the cell causing it to collapse in on itself. This theory has become known as the 'minimum cell volume theory' as opposed to the 'salt denaturing' theory of Lovelock. In essence the first concept is concerned with passive events and the second with active events, the denaturating of the lipoproteins of the cell membranes as a direct result of high salt concentrations (Lovelock, 1957).

In either case the cellular destruction can be avoided by three strategies. (1) Vitrification achieved by cooling so rapidly that water changes directly into the glass or vitreous state. (2) Rapid freezing so that the exposure of cells to high extracellular concentrations of solutes is brief. This works well with red cells but with most cells rapid rates of freezing and thawing are nearly always accompanied by the formation of intracellular ice which is invariably lethal. (3) The addition of non-toxic molecules in molar amounts such that the freezing point is depressed and the amount of salt in solution is decreased. As the concentration of glycerol or DMSO is increased so the amount of salt dissolved in the unfrozen water/ice mix is lowered. This means that cells are exposed to lower amounts of salts at any specific temperature in the presence of cryoprotectants. This strategy when used in conjunction with cooling and thawing at optimum rates has been utilized in the present day techniques for the successful cryopreservation of most animal and plant cells (Ashwood-Smith, 1980; Morris, 1980).

There is nothing remarkable about strategy three. In essence, any non-toxic molecule at high concentration and soluble in water is capable of cryoprotection. Lovelock originally maintained that the chosen molecule should be able to penetrate cells in order to be active. This belief was, in part, responsible for the introduction of DMSO as a cryoprotectant, initially with red blood cells (Lovelock & Bishop, 1959) and later as an alternative to glycerol (Ashwood-Smith, 1967). However doubts still remain as to whether or not penetration of a cryoprotectant is necessary for activity. Mazur & Miller (1976) working with human red cells and glycerol, came to the conclusion that permeation was not a prerequisite for survival after freezing and thawing. In these experiments they also made observations that were not consistent with the 'minimum cell volume hypothesis' of Meryman. Notwithstanding these experiments, however, the consensus still is that small molecular weight molecules that penetrate cells are better than similar molecules that do not. Sucrose and other sugars are usually far less effective as cryoprotective agents than the penetrating polyhyric alcohols.

Polymers as cryoprotectants

How can we explain the cryoprotective properties of polymeric molecules such as PVP, hydroxyethyl starch (HES), proteins and dextran? They clearly cannot penetrate cells, although small quantities may enter by endocytotic processes, and yet they possess limited cryoprotective properties with nucleated cells (Connor & Ashwood-Smith, 1973) and are very effective agents for the cryopreservation of red blood cells (Rowe *et al.* 1980). The addition of serum will increase the survival of most cell types and Knight *et al.* (1977) have published protocols for the successful cryopreservation of human lymphocytes with serum as the only cryoprotective agent.

It has always been considered that the freezing point depression produced by a 5% or 10% solution of most polymers used in cryopreservation procedures is totally inadequate to account, on a molar basis (the average molecular weight being in the region of 70000 daltons), for the observed protection. However as solutions containing polymers progressively freeze so the concentration of polymer in the unfrozen water increases with concomitant and appreciable effects on the freezing point depression. Measurements of these deviations from expected linear relationship in a non-frozen PVP solution have been published (Connor & Ashwood-Smith, 1973).

Thus it is possible, at least in part, to demonstrate that polymers have a similar effect on the freezing point depression as small molecules such as DMSO and glycerol. If, in fact, a concentration of DMSO is chosen that has about the same cryoprotective effect as a 10 % solution of PVP towards chinese hamster tissue culture cells, namely 2·5 %, then it can be demonstrated that the concentration of sodium chloride in the unfrozen water during the freezing process is approximately the same in both cases. In other words, the polymer has acted colligatively and has reduced the ion concentration at any particular temperature above the eutectic in a quantitatively similar way to DMSO (Ashwood-Smith, 1986).

This is not the whole story as polymers have other effects during the freezing of aqueous ionic solutions which may be related to their cryoprotective action. PVP, dextran and HES for example 'bind' a certain amount of water during freezing as determined by NMR studies and this bound water is presumably not available for the solution of solutes. HES has been shown by Korber et al. (1982) to influence ice formation and to cause a proportion of water to remain unfrozen. Calculations indicated that perhaps as much as 0·52 g of water remained unfrozen per gram of HES; PVP and polyethylene glycol behaved in a similar manner. A recent study by Baust et al. (1986) using pulsed NMR indicates that simple and complex mixtures of cryoprotectants permit greater proton mobility between $-20°C$ and $-30°C$.

The study by Korber et al. (1982) confirmed and extended an earlier study by Korber & Scheiwe (1980) with HES in which they concluded that, 'a certain proportion of water is absorbed by HES and kept from freezing, i.e. appears to be thermally inert within the range of temperatures studied. The protective action of HES against solution effects is attributed to its water absorption capacity and kinetics instead of a postponement of lethal salt enrichment to lower temperatures as caused by DMSO and glycerol'. The latter part of their conclusions are not, of course, necessarily at variance with the observations already mentioned that ascribe colligative properties to polymer solutions. More than one protective mechanism may operate with cryoprotectant molecules.

Protection in noncellular systems

The enzyme catalase is destroyed by repeated cycles of freezing and thawing but can be protected from inactivation by glycerol, DMSO and PVP. The concentrations necessary to observe these protective effects are considerably less than with cellular systems (Ashwood-Smith & Warby, 1972) and denaturation effects on lipoproteins are excluded. Certain photochemical events are far more efficient in the frozen state than in aqueous solutions. The formation of thymine dimers, for example, occurs with a much higher yield when frozen solutions of thymine are irradiated with U.V. (2540 nm radiation). The photochemical reaction which produces dimers is more efficient because the thymine molecules are closer together in the frozen solution but the effect can be abolished by adding 1–2 % DMSO, glycerol or ethanol to the solution (Ashwood-Smith, unpublished observations). This is highly suggestive that the positioning of an inert molecule between the target molecules is sufficiently effective to 'distance' the target molecules from each other. In much the same way colloids in suspension (latex particles, Indian ink for example) are protected from freezing damage, manifested as aggregation, when cryoprotective molecules are added prior to freezing and thawing.

In attempting to explain the action of cryoprotective molecules the concept of 'bound water' (water which is thermally inactive and non-freezable) is constantly encountered. Franks (1983) has discussed this question with special reference to the polyhydric alcohols. The hydroxyl groups in a number of polymers such as HES and proteins share this property with the usual cryoprotectants such as glycerol, methanol and propane-diol. Sugars, many of which have some cryoprotective action also possess hydroxyl groups. All these compounds participate in active hydrogen bonding as mentioned in the introduction. Franks (1983) states that aqueous solutions of polydydroxy compounds 'have the tendency to supersaturation and incomplete freezing, manifestations of the phenomena known as bound water which is, however, a misnomer'. Franks believes that these effects result in gel formation, liquid crystals, and protection from dehydration.

Other biological actions of cryoprotective molecules

Cryoprotectant molecules possess interesting properties which may or may not be related to their abilities to modify damage associated with freezing and thawing although from time to time speculations have been made linking some of these properties.

Radioprotective actions of DMSO and glycerol

DMSO and glycerol are radical scavengers *in vitro* when aqueous solutions are X-irradiated (Ashwood-Smith, 1967, 1975) and both these molecules protect cellular systems from X-ray damage by a factor of between 2 and 3 by mechanisms which appear to be independent of the oxygen effect. DMSO is also an excellent radioprotective agent *in vivo* conferring a dose reduction factor of about 1·6 to X-irradiated mice. In this instance it is possible that the mechanism of radioprotection is pharmacological rather than physico-chemical as the amount of DMSO in the tissues is considerably lower than that necessary to demonstrate radioprotection *in vitro*.

In the past, suggestions have been made that freezing damage especially during storage could be related to oxygen mediated radical effects (Heckly & Quay, 1983) leading to DNA breakage etc (Ashwood-Smith, 1985). However this idea has not been substantiated and we are left with clear evidence that both DMSO and glycerol are radioprotective and cryoprotective but without a persuasive mechanistic link. It is, of course, entirely possible that none exists.

Gene activation of DNA effects of DMSO

A substantial number of references on the genetic or epigenetic effects of exposing cells to relatively large concentrations of DMSO for lengthy periods of time are on record. Cells that contain repressed genetic information may be derepressed by treatment with DMSO. DNA strand breaks, hypomethylation and subsequent gene expression result (Ashwood-Smith, 1985). So characteristic and repeatable is this effect that DMSO is routinely used for transformation of erythroid leukaemic cell lines (Pulito *et al.* 1983). The mechanism by which DMSO affects these changes is not understood although one may postulate that it acts by interferring with hydrogen bonding between repressor molecules on DNA strands. No genetic effects of DMSO *per se* have been observed during short exposure periods and with the concentrations used in standard cryopreservation procedures (Ashwood-Smith, 1985). DMSO is routinely used as a solvent for chemicals being tested in either tests *in vitro* for mutagens as putative carcinogens (Ames tests. etc.) or in carcinogenicity tests (*in vivo*). Apart from instances when DMSO may chemically react with the test chemical, as it may do with some chlorinated chemicals or metals, the overwhelming evidence points to its lack of activity as a mutagen, carcinogen or teratogen.

Cell fusion action of DMSO

Both DMSO and polyethylene glycol (PEG) under specific conditions can cause the membranes of mammalian cells to fuse and the combination of both molecules is used for the preparation of hybridomas. Thus the inclusion of both PEG and DMSO in vitrification solutions (Rall & Fahy, 1985) appears to be, *a priori*, particularly worrying (Ashwood-Smith, 1985). It should be stated, however, that combinations of polymers with low molecular weight cryoprotective agents often give excellent additive cryoprotection without additive toxicity problems (Ash-wood-Smith, 1975) and it may be that the cell fusion properties of DMSO and PEG are much modified by the presence of other components of the vitrification solutions such as acetamide (Fahy *et al.* 1984).

In nature this mixing of high and low molecular weight cryoprotectants is the rule as proteins and glycoproteins are often associated with the high molar concentrations of sugars and polyhydric alcohols which may be considered as the primary cryoprotectants. Also, in laboratory situations DMSO or glycerol is added to standard tissue culture solutions that invariably contain between 10 and 15 % serum.

Mechanical protection and cryoprotectant polymers

Although animal red blood cells are almost as well protected from freezing damage by polymers such as PVP, HES and dextrans as they are by glycerol and DMSO this is not the case with most nucleated animal cells. DMSO or glycerol are considerably more effective than polymers. There are other differences.

Strangely, both PVP and HES afford good protection to human red cell from mechanical injury (Ben-David & Gavendo, 1972). Mechanical injury was produced by vigorously shaking red cell samples with steel balls. Both PVP (mol. wt. average 30 000 daltons) and HES protected red cells from mechanical damage but only PVP reduced osmotic induced fragility. The suggestion is made that PVP is effective in preventing damaged cells from disruption. Although such a proposition is attractive one might expect some direct evidence for PVP binding, at least to some extent, with cell membranes. No evidence for the direct binding of PVP has been reported (Connor & Ashwood-Smith, 1973). No effect of PVP on the uptake of amino acids by cells has been demonstrated which would also indicate no interference with membrane function.

However, the interaction of polymers with surfaces has been studied by Williams (1983) who demonstrated that molecules such as PVP, dextran and HES lower the surface energy of haemoglobin droplets thus forming a stable interface. Williams postulated that the leakage of haemoglobin from red cells injured during fast freezing and thawing is hindered by the polymer/red cell membrane interface.

Suppression and induction of ice formation in nature

(a) *Suppression*

In general the most widely encountered method in nature for the avoidance of freezing damage is associated with the production of large quantities of glycerol or similar molecules. Recently, however, it has become apparent that two other strategies may be widespread although less than fully appreciated (Devries & Lin, 1977).

Fish that live in arctic waters often swim around in water that is several degrees below the freezing point of their blood. The melting point of the blood is, however, not lowered. In other words there is a difference of several degrees between the observed freezing point and melting point and thus blood always supercools. Lethal freezing is thereby avoided. This phenomenon which may be more common than realized is related to the presence of specific antifreeze glycoproteins (AFGP) (Dunan, 1982). As they act by depressing the freezing point without changing the melting point their action cannot be colligative (Franks, 1985).

A recent article by Franks *et al.* (1987) has investigated the properties of these molecules isolated from the serum of polar fish and overwintering insects (also see Ring, 1980). Homogeneous nucleation (spontaneous) rates of ice at temperatures of 233 K were measured in PVP solutions and solutions containing AFGP. It was noted that both molecules at low concentrations affected the rates of homogeneous nucleation and crystallization. It was quite remarkable that AFGP inhibited ice crystal growth with an efficiency of 10^7 more than PVP suggesting a differing mode of action of this molecule. Suggestions have been made that AFGP functions by sorption on ice as it grows thus causing its growth to be suppressed.

(b) *Nucleation*

The discovery that ice + bacteria (Lindow *et al.* 1978) produce a specific protein molecule, since cloned (Orser *et al.* 1985), that prevents supercooling has focussed attention on another mechanism operating in nature which, although functioning on a 'knife edge' allows some insects and plants to deliberately use the physical characteristics of ice as a protective shield.

In essence specific protein molecules act as ice nucleators and *prevent* supercooling. The controlled formation of extracellular ice thus ensures that excessive supercooling with the probability of lethal intracellular ice is minimized (Ashwood-Smith, 1986). Also, the vapour pressure difference between the new formed extracellular ice and the non-frozen intracellular contents permits a gentle and controlled desiccation of the non-frozen cells to occur just as it does in controlled freezing experiments in the laboratory. Whether or not heat released during the phase change of water to ice is important in this context is debatable.

This particular strategy adopted by some insects and plants assumes a predictable and changing temperature profile. A too drastic temperature change would probably precipitate a damaging sequence of events that could have been avoided by organisms containing large amounts of glycerol.

The alpine plant, *Lobelia teleki* (Krog *et al.* 1979) produces a carbohydrate molecule which can initiate ice crystal formation. Recent studies of a series of naturally occurring molecules have indicated very efficient ice nucleation at temperatures no more than 1–3 degrees below the theoretical freezing point of a number of complex biological solutions. This work, which is in progress (Ashwood-Smith & Morris), suggests that the suppression of supercooling may be widespread in nature.

General conclusions

The formation of intracellular ice is invariably fatal to the vast majority of animal and plant cells (Ashwood-Smith, 1986). Prior to this event and to a large extent depending on the cooling rate osmotic stresses are imposed on cells because of the formation of extracellular ice. The kinetics and temperature profiles of these events are complicated but are modified by the present of cryoprotective agents, functioning in the main by simple colligative action. The prevention of freezing by antifreeze proteins or the initiation of ice formation by nucleating agents are unusual ways of modifying freezing damage. The total avoidance of ice formation by vitrification requires extremely rapid rates of cooling and warming in conjunction with very high molar amounts of cryoprotective chemicals. Devitrification during thawing of vitrified solutions is a particular problem (MacFarlane, 1986). With the exception of vitrification the standard methods of circumventing ice formation in the laboratory have been shown to occur in nature.

Vitrification processes could perhaps occur in aerosols created in outer space, however!

References

ASHWOOD-SMITH, M. J. (1962). *Biological Properties of Dimethyl Sulphoxide*. Ph.D. Thesis, University of London, UK.

ASHWOOD-SMITH, M. J. (1967). Radioprotective and cryoprotective properties of dimethyl sulfoxide in cellular systems. *Ann. N.Y. Acad. Sci.* **141**, 45–62.

ASHWOOD-SMITH, M. J. (1975). Current concepts concerning radioprotective and cryoprotective properties of dimethyl sulfoxide in cellular systems. *Ann. N.Y. Acad. Sci.* **243**, 246–256.

ASHWOOD-SMITH, M. J. (1980). *Low Temperature Preservation of Cells, Tissues and Organs*. In Low Temperature Preservation in Medicine and Biology (ed. M. J. Ashwood-Smith & J. Farrant), pp. 19–44. Tunbridge Wells: Pitmans Medical.

ASHWOOD-SMITH, M. J. (1985). Genetic damage is not produced by normal cryopreservation procedures involving either glycerol or dimethyl sulfoxide: a cautionary note, however, on possible effects of dimethyl sulfoxide. *Cryobiology* **22**, 427–433.

ASHWOOD-SMITH, M. J. (1986). The cryopreservation of human embryos. *Human Reprod.* **1**, no. 5, 319–332.

ASHWOOD-SMITH, M. J. & FARRANT, J. (1980). (eds) *Low Temperature Preservation in Medicine and Biology*. Tunbridge Wells: Pitmans Medical.

ASHWOOD-SMITH, M. J. & LOUGH, P. (1975). Cryoprotection of mammalian cells in tissue culture with methanol. *Cryogiology* **12**, 517–518.

ASHWOOD-SMITH, M. J. & WARBY, C. (1972). Protective effect of low and high molecular weight compounds on the stability of catalase subjected to freezing and thawing. *Cryobiology* **9**, 137–140.

ASHWOOD-SMITH, M. J., WARBY, C., CONNOR, K. W. & BECKER, G. (1972). Low temperature preservation of mammalian cells in tissue culture with polyvinyl pyrrolidone (PVP), dextrans and hydroxyethyl starch (HES). *Cryobiology* **9**, 441–449.

BAUST, J. G., WASYLYK, J. M., SZYMANSKI, R. L., HANSEN, T. N. & TICE, A. R. (1986). Nonidealistic behaviour of cryoprotective mixtures as revealed by pNMR analysis. *Cryobiology* **23**, 562–573.

BEN-DAVID, A. & GAVENDO, S. (1972). The protective effect of polyvinyl pyrrolidone and hydroxethyl starch on noncryogenic injury to red blood cells. *Cryobiology* **9**, 192–197.

CONNOR, W. & ASHWOOD-SMITH, M. J. (1973). Cryoprotection of mammalian cells in tissue culture with polymers: possible mechanisms. *Cryobiology* **10**, 488–496.

DEVRIES, A. L. & LIN, Y. (1977). Structure of a peptide antifreeze and mechanism of absorption to ice. *Biochim. biophys. Act.* **495**, 388–392.

DUMAN, J. G. (1982). Insect antifreezes and ice-nucleating agents. *Cryobiology* **19**, 613–627.

FAHY, G. H. (1986). The relevance of cryoprotectant 'toxicity' to cryobiology. *Cryobiology* **23**, 1–13.

FAHY, G. M., MACFARLANE, D. R., ANGELL, C. A. & MERYMAN, H. T. (1984). Vitrification as an approach to cryopreservation. *Cryobiology* **21**, 407–426.

FRANKS, F. (1983). Solute-water interactions: Do polyhydroxy compounds alter the properties of water? *Cryobiology* **20**, 335–345.

FRANKS, F. (1985). Complex aqueous systems at subzero temperatures. In *Properties of Water in Foods* (ed. D. Simatos & J. Multon), pp. 479–509. Nato ASI Series, Lancaster: Martinus Myhoff.

FRANKS, F., DARLINGTON, J., SCHENZ, T., MATHIAS, S. F., SLADE, L. & LEVINE, H. (1987). Antifreeze activity of antarctic fish glycoprotein and a synthetic polymer. *Nature, Lond.* **325**, 146–147.

HECKLEY, R. J. & QUAY, J. (1983). Adventitious chemistry at reduced water activities: free radicals and polyhydroxy agents. *Cryobiology* **20**, 613–624.

KNIGHT, S. C., FARRANT, J. & MCGANN, L. E. (1977). Storage of human lymphocytes by freezing in serum alone. Cryobiology **14**, 112–115.

KORBER, C. & SCHEIWE, M. W. (1980). The cryoprotective properties of hydroxyethyl starch investigated by means of differential thermal analysis. *Cryobiology* **17**, 54–65.

KORBER, C. H., SCHEIWE, M. W., BOUTRON, P. & RAU, G. (1982). The influence of hydroxyethyl starch on ice formation in aqueous solutions. *Cryobiology* **19**, 478–492.

Krog, J. O., Zachariassen, K. E., Larsen, B. & Smidsrod, O. (1979). Thermal buffering in Afro-alpine plants due to nucleating agents induced water freezing. *Nature, Lond.* **282**, 300–301.

Li, P. H. & Sakai, A. (1978) (eds). *Plant Cold Hardiness and Freezing Stress*. New York: Academic Press.

Lindow, S. E., Arny, D. C., Upper, C. D. & Barchet, W. R. (1978). The role of bacterial ice nuclei in first injury to sensitive plants. In *Plant Cold Hardiness and Freezing Stress* (ed. P. H. Li & A. Sakai), pp. 249–263. New York: Academic Press.

Lovelock, J. E. (1953a). The haemolysis of human red blood cells by freezing and thawing. *Biochim. biophys. Acta.* **10**, 414–426.

Lovelock, J. E. (1953b). The mechanism of the protective action of glycerol against haemolysis by freezing and thawing. *Biochim. biophys. Acta.* **11**, 28–36.

Lovelock, J. E. (1954). The protective action of neutral solutes against haemolysis by freezing and thawing. *Biochem. J.* **56**, 265–270.

Lovelock, J. E. (1957). Denaturation of lipid-protein complexes as a cause of damage by freezing. *Proc. R. Soc. Ser. B* **147**, 427–433.

Lovelock, J. E. & Bishop, M. W. H. (1959). Prevention of freezing damage to cells by dimethyl sulphoxide. *Nature, Lond.* **183**, 1394–1395.

Luyet, B. J. & Gehenio, P. M. (1938). The survival of moss vitrified in liquid air and its relation to water content. *Biodynamica* **42**, 1–17.

Luyet, B. J. & Gehenio, P. M. (1940). *Life and Death at Low Temperatures*, pp. 203–228. Normandy, Missouri: Biodynamica.

Luyet, B. J. & Thoennes, G. (1939). The survival of plant cells immersed in liquid air. *Science* **88**, 284–285.

MacFarlane, D. R. (1986). Devitrification in glass-forming aqueous solutions. *Cryobiology* **23**, 231–244.

Mazur, P. & Miller, R. H. (1976). Survival of frozen-thawed human red cells as a function of the permeation of glycerol and sucrose. *Cryobiology* **13**, 523–536.

Meryman, H. T. (1968). Modified model for the mechanisms of freezing injury of erythrocytes. *Nature, Lond.* **218**, 333–336.

Meryman, H. T., Williams, R. T. & Douglas, M. St. J. (1977). Freezing injury from 'solution effects' and its prevention by natural or artificial cryoprotectives. *Cryobiology* **14**, 287–302.

Morris, G. J. (1980). Plant cells. In *Low Temperature Preservation in Medicine and Biology* (ed. M. J. Ashwood-Smith & J. Farrant), pp. 253–283. Tunbridge Wells: Pitmans Medical.

Orser, C., Staskawicz, B. J., Panopoulos, N. J., Dahlbeck, D. & Lindow, S. E. (1985). Cloning and expressions of bacterial ice nucleation genes in *Escherichia coli. J. Bact.* **164**(1), 359–366.

Pulito, V. L., Miller, D. L., Sassa, S. & Yamane, T. (1983). DNA fragments in friend erythroleukemia cells induced by DMSO. *Proc. natn. Acad. Sci. U.S.A.* **80**, 5912–5915.

Rall, W. F., Czlonkowska, M., Barton, S. C. & Polge, C. (1984). Cryoprotection of day-4 mouse embryos by methanol. *J. Reprod. Fertil.* **70**, 293–300.

Rall, W. F. & Fahy, G. M. (1985). Ice free cryopreservation of mouse embryos at −196°C by vitrification. *Nature, Lond.* **313**, 573–575.

Ring, P. A. (1980). *Insects and their cells*. In *Low Temperature Preservation in Medicine and Biology* (ed. M. J. Ashwood-Smith & J. Farrant), pp. 187–217. Tunbridge Wells: Pitmans Medical.

Rowe, A. W., Lenny, L. L. & Manoni, P. (1986). Cryopreservation of red cells and platelets. In *Low Temperature Preservation in Medicine and Biology* (ed. M. J. Ashwood-Smith & J. Farrant), pp. 85–120. Tunbridge Wells: Pitmans Medical.

Smith, A. U. (1952). Behaviour of fertilized rabbit eggs exposed to glycerol and to low temperatures. *Nature, Lond.* **170**, 374.

Storey, K. B. & Storey, J. M. (1984). Biochemical adaptation for freezing tolerance in the Winter Frog, *Rana sylvatica. J. comp. Physiol. & Biochem. Syst. Envrn. Physiol.* **155**(1), 29–36.

Williams, R. J. (1983). The surface activity of PVP and other polymers and their antihemolytic activity. *Cryobiology* **20**, 521–526.

BIOCHEMICAL AND FUNCTIONAL ASPECTS OF RECOVERY OF MAMMALIAN SYSTEMS FROM DEEP SUB-ZERO TEMPERATURES

ROBERT DE LOECKER and FREDDY PENNINCKX

Department of Gastroenterological Surgery, Faculty of Medicine, University of Leuven, B-3000 Leuven, Belgium

Summary

The viability of isolated mammalian systems is, apart from possible morphological changes, essentially conditioned by the biochemical modifications from normal physiological conditions to an artificial environment where blood supply is interrupted leading to ischaemia and where the temperature is lowered.

In order to survive freezing and thawing, mammalian systems have to be protected by cryoprotectants, which apart from some inherent toxicity, may also interact with vital metabolic mechanisms (Conover, 1969, 1975; Fahy, 1986; Fahy *et al.* 1984; Jacobs & Herschler, 1986; Karow, 1982; Penninckx *et al.* 1983; Polge *et al.* 1949; Rowe *et al.* 1980; Schlafer, 1981; Taylor & Pignat, 1982). Cellular volume changes as a result of modifications in extra- and intracellular osmolality occurring during freezing and thawing prove particularly detrimental to the normal functioning of the cellular membranes (Crowe *et al.* 1983; Farrant, 1980; Farrant *et al.* 1977*b*; Karow, 1982; Mazur & Rigopoulos, 1983; Meryman, 1970; Meryman *et al.* 1977; Nei, 1976; Santarius & Giersch, 1983). Furthermore intracellular ice formation enhances structural and metabolic injury to subcellular particles (Farrant *et al.* 1977*a*; Fink, 1986; Fishbein & Griffin, 1976; Fujikawa, 1981; Fuller & De Loecker, 1985; Lazarus *et al.* 1982; Malinin, 1972; Mazur, 1984; Pavlock *et al.* 1984; Penninckx *et al.* 1984; Persidsky & Ellet, 1971; Rubinacci *et al.* 1986; Shikama, 1965; Steponkus & Wiest, 1979; Strauss & Ingenito, 1980; Takehara & Rowe, 1971; Tamiya *et al.* 1985).

Even with the protection of structural integrity, the preservation of energy production and the maintenance of the specific intracellular medium are essential to secure viability (Pegg, 1981).

Preservation of energy production

Energy production processes

With an efficiency of around 40 % for metabolic use, the rest appearing as heat, the complete oxidation of glucose to CO_2 and water produces $2867\,kJ\,mol^{-1}$, while the high-energy phosphate bond of ATP represents $30.5\,kJ\,mol^{-1}$ (Newsholme & Leech, 1983). Four major pathways are to be considered (Fig. 1).

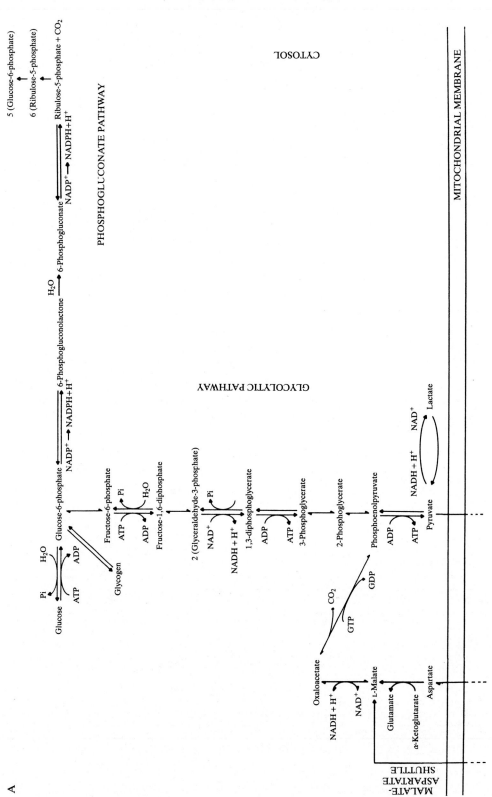

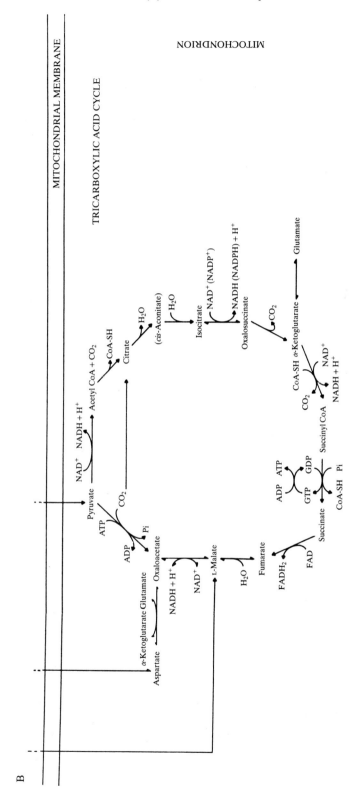

Fig. 1. (A,B) Metabolic pathways: glycolysis, phosphogluconate pathway, tricarboxylic acid cycle, malate-aspartate shuttle and respective location in cytosol and mitochondria.

Glycolytic pathway

Phosphorylation of glucose with the production of glucose-6-phosphate requires 1 mole of ATP, while in the cytoplasm a further mole of ATP is necessary for the generation of fructose-1,6-diphosphate, thus leading to the consumption of 2 moles of ATP with a loss of $71\,kJ\,mol^{-1}$ of glucose. Phosphofructokinase is the principal regulating enzyme of glycolysis and is inhibited by moderate levels of ATP, citrate and by the acidity of the medium. The cleavage of fructose-1,6-diphosphate yields 2 moles of glyceraldehyde-3-phosphate and with the required removal of 2 protons and 2 electrons generates 1 mole of reduced coenzyme NADH, liberating 1 mole of H^+ into the cytosol. The following two steps in the degradation towards pyruvate generate ATP. Since 2 moles of glyceraldehyde-3-phosphate are derived from each mole of glucose, 4 moles of ATP are generated resulting in a net production of 2 moles of ATP representing only a fraction of the total energy capacity of glucose.

Subsequent oxidation of NADH (nicotinamide adenine dinucleotide) in the electron transport chain produces an additional 6 moles of ATP, requiring apart from the transportation of NADH into the mitochondria, ADP, inorganic phosphate and molecular oxygen. Without the regeneration of NAD^+ from NADH, glycolysis remains blocked at the glyceraldehyde-3-phosphate level. In the absence of oxygen, NAD^+ is regenerated by the reduction of pyruvate to lactate, the end product of anaerobic glycolysis. Glycogen represents a substrate for glycolysis and produces 3 ATP molecules per glucose residue.

Tricarboxylic acid cycle (citric acid cycle, Krebs cycle)

Through the oxidation of pyruvate inside the mitochondria, each mole of glucose generates another 6 moles of NADH, 2 moles of NADPH, 2 moles of reduced flavoprotein ($FADH_2$) and 2 moles of ATP obtained by donation of the terminal phosphate group of formed GTP to ADP. The regeneration of oxidized coenzymes is also necessary here for the continuation of the reaction sequences. CO_2, perhaps the most important intermediate compound of the tricarboxylic acid cycle to be used in other biosynthetic pathways and needed for the formation of oxaloacetate from pyruvate, is regenerated in the three oxidative decarboxylation steps which also generate the reduced coenzymes.

Phosphogluconate pathway (pentose phosphate pathway, hexose monophosphate shunt)

Glucose-6-phosphate becomes oxidatively decarboxylated to ribulose-5-phosphate generating 2 moles of NADPH per mole of glucose. Since 6 moles of ribulose-5-phosphate can be converted into 5 moles of glucose-6-phosphate by a series of carbon exchange reactions, this pathway generates 6 moles of CO_2 and 12 moles of NADPH, upon complete oxidation of 1 mole of glucose which are mainly used for biosynthetic reactions but could also be converted to NADH and thus serve for ATP production. This pathway accounts for 30% of liver- and for less than 10% of red blood cell glucose metabolism and is even much less important in other tissues.

The electron-transport chain

Inside the mitochondrial electron-transport chain, NADH is reoxidized by the respiratory carriers, cytochromes and Fe-S compounds producing ATP. The malate-aspartate shuttle allows the functional penetration of NADH located in the cytoplasm through the mitochondrial membranes. Cytoplasmic NADH is oxidized to NAD^+ by the coupled reduction of oxaloacetate to malate, followed by the malate diffusion into the mitochondria where it becomes oxidized to oxaloacetate with the generation of NADH. Oxaloacetate is transaminated with glutamate to aspartate which is exchanged across the membrane for glutamate. Thus a high NAD^+ concentration is maintained in the cytoplasm, necessary to promote the glycolysis of glyceraldehyde-3-phosphate, while inside the mitochondria high levels of NADH stimulate the electron transfer reactions.

Protons and electrons from NADH enter the electron transport chain situated at the internal mitochondrial membrane by combining with flavoprotein and further transfer to ubiquinone (Q) from which protons are released into the medium and the electrons are transferred down the cytochrome chain ($b \rightarrow c_1 \rightarrow c \rightarrow a \rightarrow a_3$) until they finally combine with the free protons and molecular oxygen to form water. As the electron transport chain consists of a series of increasingly oxidized states, each electron transfer releases energy by which high-energy phosphate bonds are formed. The coupling between energy production and utilization appears intimately linked to the proton translocation and the ATPase sites resulting in a proton gradient and the transduction of this gradient energy into ATP synthesis (Mitchell, 1976). In this series of events NADH yields 3 moles of ATP, and reduced flavoproteins only 2 moles of ATP (Fig. 2).

Conclusion. Complete oxidation of 1 mole of glucose leads to the direct formation of 4 moles of ATP, the net formation of 10 moles of NADH generating 30 moles of ATP, 2 moles of $FADH_2$ generating an additional 4 moles of ATP, making up a total of 38 moles of ATP. Each mole of glucose oxidized along the

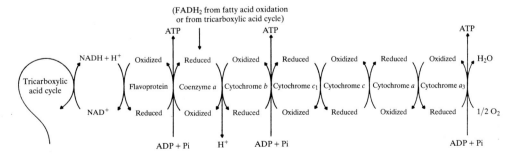

Fig. 2. Schematic representation of the electron transport chain.

phosphogluconate pathway produces 12 moles of NADH capable of yielding 36 moles of ATP. The overall efficiency of energy conservation during oxidative metabolism amounts to around 40 % while anaerobic glycolysis only presents an efficiency of 3 % (Newsholme & Leech, 1983; Pegg, 1981).

Fatty acids as an alternative source of energy for some tissues, can be metabolized by progressive β-oxidation, releasing 2-carbon fragments as acetyl coenzyme A (acetyl-CoA) which can be metabolized in the tricarboxylic acid cycle with the formation of 1 mole of reduced flavoprotein and 1 mole of NADH. The recovery of energy from fatty acids is aerobic and the glycolytic pathway of glucose is the only mechanism by which energy is obtained anaerobically (Feinberg, 1982).

Effects of low-temperature preservation on energy production

A reduced blood circulation leads to a restriction in supply of essential metabolites, to an inefficient removal of catabolic compounds and is essentially responsible for anoxia. Deprivation of molecular oxygen results in an inhibition of the electron transport chain, leading to an accumulation of $NADH + H^+$ and a depletion of NAD^+ and ATP. As NAD^+ availability is restricted, the tricarboxylic acid cycle loses efficiency while the increased $NADH + H^+$ concentrations inhibit the conversion of L-malate to oxaloacetate. As a direct consequence, the activity of the malate-aspartate shuttle is reduced, resulting in increasing cytosolic $NADH + H^+$ levels. As $NADP^+$ levels will also be reduced the phosphogluconate pathway becomes unimportant. However, the regeneration of NAD^+ remains possible by the reduction of pyruvate to lactate. Without this the glycolytic pathway would stop at the glyceraldehyde-3-phosphate level thus preventing ATP from being generated. Glycolysis requires the availability of a hexose monophosphate and of ATP in the regulatory phosphofructokinase reaction to produce fructose-1,6-diphosphate.

As in many tissues, glycogen represents an adequate supply of glucose units already phosphorylated; ATP may thus be conserved. Although the enzymatic phosphofructokinase reaction is stimulated by the increased ADP and inorganic phosphate levels occurring during ischaemia, the accumulation of H^+ will finally lead to the metabolic block of this reaction.

In ischaemic conditions, the ATP/ADP ratios drop dramatically and the hydrolysis of ATP may further lead to a generation of protons.

$$Mg - ATP^{2-} \rightarrow Mg - ADP^- + P_i^{2-} + H^+$$

The accumulations of protons and lactate are major toxic factors leading to structural damage to subcellular particles, lysosomes and mitochondria (Lochner et al. 1976). When this structural injury remains restricted, the oxidative phosphorylation may be resumed, with restoration of the normal ATP/ADP/AMP ratios. However as dephosphorylated nucleosides may be lost through diffusion, compensation of the depleted nucleotides may be difficult to achieve.

Cooling is essentially carried out to reduce the biochemical reaction rates, resulting in a considerable reduction in the demands for metabolites and in a conservation of energy levels. However the slowing down of metabolic processes by cooling may not be harmless as some reaction pathways are differently affected by cooling and may well be irreversibly uncoupled.

Maintenance of the intracellular medium

Ionic pump mechanisms

The maintenance of the intracellular medium is one of the main functions of the lipid-rich cellular membranes, being hydrophobic on the exterior – and hydrophylic at the interior sides (Dewey & Barr, 1970). Although the permeability of K^+ and Cl^- are up to one hundred times greater than of Na^+, the cells maintain a stable internal ionic composition which differs considerably from the surrounding fluids. Through the $(Na^+ - K^+)$ ATPase pump mechanism coupled to the hydrolysis of ATP, Na^+ is actively transported out of and K^+ into the cells. By this mechanism, intracellular Na^+ is exchanged for extracellular K^+ at a ratio of 3 to 2. Each ion diffuses back according to their concentration gradients, thus Na^+ enters the cell while K^+ leaves it. Although, as K^+ permeates more rapidly than Na^+, the passive loss of K^+ ions does not remain balanced by an equal gain of Na^+. This leads to a net loss of cations from the cells causing the content to become negatively charged with respect to the extracellular fluid. This transport effect however is self-limiting. Indeed as the internal cation deficit increases, the membrane charge increases and counteracts the continuous loss of K^+. The internal relative negative charge will then again attract cations leading to an equilibrium which is obtained with a membrane potential difference of about 10–100 mV between the internal and external surfaces of the cell, which is already achieved by the minute ion concentration of 10^{-11} mol cm^{-2}.

The membrane potential gradient or voltage gradient also acts on intracellular anions, but mostly on impermeable organic molecules. Cl^- diffuses freely and is distributed across the cell membranes according to the membrane potential. The internal Cl^- concentration will be low due to the internal negative charge, while the external Cl^- concentration will be high. Thus the $(Na^+ - K^+)$ ATPase pump extrudes NaCl because the membrane potential generated by the back diffusion of K^+ causes Cl^- to leave the cell. The adequate functioning of this pump system is an absolute requirement for a number of amino acid transport systems. The actively transported sugars equally depend on the pump mechanism for their transport (Schwarz *et al.* 1972). Since the cell membrane is highly permeable to water, the osmolalities of intra- and extracellular fluids should equal around 300 mosm kg^{-1}. This implies that each Na^+ or Cl^- leaving the cell is accompanied by 180 molecules of water as the ratio of water molecules to solute molecules amounts to 180 to 1. Thus the $(Na^+ - K^+)$ ATPase pump action apart from being responsible for the ionic distribution, controls cell volume.

Ca^{2+} ions are essentially extruded across the plasma membranes by a specific $(Ca^{2+} - Mg^{2+})$-ATPase system requiring the presence of Mg^{2+} intracellularly, or can be exchanged for external Na^+. In the mitochondria Ca^{2+} can be accumulated against a gradient by the energy provided either by ATP hydrolysis or by respiratory substrate oxidation or by membrane potential electrochemical gradients which can be established among other factors mainly by the extrusion of protons (Akerman & Nicholls, 1983). Mn^{2+}, Fe^{2+} and Sr^{2+} but not Mg^{2+} may be similarly accumulated. The normally very low concentrations of free Ca^{2+} $(0.1–1.0\,\mu\mathrm{mol\,l}^{-1})$ in the cytosol, important in regulating various cell activities, are obtained by the efflux of Ca^{2+} across the cell membranes, by the equilibrium between energy-dependent uptake of Ca^{2+} by the mitochondria and its energy-independent release, by the Ca^{2+} binding to cell components particularly proteins, by the buffering capacity of the cytosol and by the energy-dependent accumulation of Ca^{2+} by the endoplasmic reticulum (Fig. 3).

The intracellular homeostasis is inevitably affected by fluctuations in the extracellular conditions. The semipermeable membranes allow water to restore osmotic pressures and it is observed that cell swelling proves invariably more damaging than cell shrinkage. The osmotic stresses become particularly pronounced with the addition of cryoprotectants and during freezing and thawing

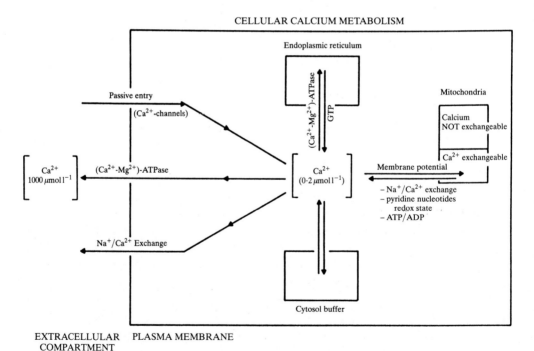

Fig. 3. Schematic representation of the control of cytosolic free calcium ion concentrations. The intracellular calcium controllers, plasma membrane, endoplasmic reticulum and mitochondria maintain the intracellular free calcium levels against concentration gradients.

modifying the soluble water content and thus the extra- and intracellular electrolyte concentrations. Although pH is precisely regulated and maintained at pH 7·3–7·4, tissues can more easily survive variations between pH 6·6 and 7·8 than more alkaline conditions.

Effects of low-temperature preservation on the intracellular medium

In ischaemic conditions, ATP levels will be reduced and the $(Na^+ - K^+)$ ATPase pump mechanism will be inhibited resulting in a passive ionic exchange by which Na^+ will enter and K^+ will leave the cells. As intracellular Na^+ levels will increase faster than the intracellular K^+ levels will be reduced due to the impermeant intracellular anions, the membrane potential will equally be reduced, allowing Cl^- to enter the cells. This net increase of intracellular NaCl will lead to water accumulations to maintain the osmotic equilibrium resulting in cell swelling. Thus during anoxia, the colloid osmotic pressure of the intracellular impermeant anions is responsible for cell swelling (Flores *et al.* 1972; Pegg, 1978). As ATP levels are reduced in ischaemic cells, calcium previously chelated to ATP will become free to be bound to the cellular membranes increasing their rigidity. However the most damaging increase in free Ca^{2+} is most likely due to the inhibition of the $(Ca^{2+} - Mg^{2+})$-ATPase pump mechanism which in general is the most important factor to control intracellular Ca^{2+} concentrations. The diminished blood reflow observed in organs after a period of ischaemia seems to be caused not only by the swelling of endothelial cells but also by structural changes to the cellular membranes. Cooling equally contributes to a blocking of the different ATPase dependent ionic pump activities although through a different mechanism than during ischaemia. Upon cooling the enzyme-dependent pump mechanisms are unable to utilize ATP and to keep the intracellular ionic composition constant leading to destruction.

Assessment of viability

After preservation at deep sub-zero temperatures, the biological material to be used for transplantation should be viable. Although the different manipulations, ischaemia, cooling and thawing inevitably lead to varying degrees of injury, viability should at least be preserved to such a degree that essential physiological and biochemical functions characteristic for each individual cell type, tissue or organ examined, can be effectively resumed.

Even if the initial functional efficiency will rarely be optimal, recuperation of essential mechanisms may well resume upon oxygenation when the structural damage is limited. The prediction of the degree of injury incurred or the viability preserved remains problematic. The only adequate and satisfying evaluation of viability will indeed be the assurance of a normal functioning unit when replaced in physiological conditions.

Different methods and approaches allow to some extent to obtain an indication of the degree of preserved viability by analysing specific aspects of morphology and metabolism.

Morphological integrity

Histological analysis by light- and electron microscopical techniques, although in most respects unable to determine any degree of tissue viability or functional integrity, will provide a reasonable evaluation of the structural injury. However even considerable morphological changes may still be reversible when replaced in an adequate oxygenated environment (Schlafer & Karow, 1972).

Most useful to determine modifications in membranous components, histo-chemical and autoradiographic techniques may indicate some shifts in enzyme distribution (Abouna, 1974; Jacob et al. 1985; Knight, 1980).

Functional integrity

Cells and tissues – general approach

In principle the most reliable index of viability and survival in proliferating cells is the demonstration of reproductive integrity. However, as often a loss of specific and essential cell functions rather than the loss of reproductive capacities are predominant, viability assessment of preserved cells and tissues can be based on the evaluation of the remaining metabolic activities. The degree of preservation of membrane permeability, membrane ionic pump activities and possibly changes in the cell membrane potentials, reflect to a great extent the integrity of the cell. Inhibitions of essential synthetic functions reveal injury at an intracellular level.

Some tests of viability such as the uptake of dye by dead cells or the release of ^{51}Cr or fluorescein from pre-labelled cells are intended to predict survival rather than measure viability. Most of these assays imply breakdown in membrane integrity and allow dead cells to be easily identified rather than long-term survival to be estimated.

Abnormal volume changes, shrinkage and particularly cell swelling appear detrimental to survival. Volumetric measurements may thus indicate or exclude the possibilities of a viable preparation. The evaluation of membrane-attached enzymatic systems which preserve their activity even when the cellular entity has disappeared, is only contributing to the viability analysis in morphologically intact cells (Fahy, 1981; Wilson, 1986).

Reproductive capacity. Although increases in total protein levels, in cell number as well as in protein synthetic capacities can all be assumed to imply proliferative ability, survival assays on cultured cells are essentially directed towards their reproductive capacity by measuring cloning efficiency in monolayer or in soft agar. Various approaches are also available to evaluate cellular injury by analysing the growth of multicellular spheroids, the tissue-like structures induced *in vitro* by reaggregating different cell types. Relative changes in volume of treated and control spheroids, cloning efficiency in soft agar of disaggregated spheroids and cell proliferation from spheroids adherent to culture surfaces illustrate the degree of survival (Freyer & Sutherland, 1980; Wilson, 1986).

An increase in the number of cells in a proliferating cell line is a promising indication allowing growth curves as well as the doubling time during exponential growth to be assessed. Cell growth rate, however, can only be interpreted as a rough indication while accurate measurements of cell survival and cell prolifer- ation should be made on the basis of the efficiency to form colonies indicating survival and the colony size expressing the proliferative capacity. Indeed an inhibition of the doubling time expresses injury in a whole population without distinguishing between an early cell loss and a prolonged adaptation period (Allen *et al.* 1984; Bickford *et al.* 1980; Fahy, 1981: Maack *et al.* 1982; Roper & Drewinko, 1976; Wilson, 1986). In this way conclusions derived from the growth of preserved cells placed on tissue culture medium appear rather unreliable.

Membrane integrity. Evaluation of the structural integrity of the membranes allows the estimation of instantaneous injury or progressive damage over a period of several hours (Quinn, 1985).

^{51}Cr release is estimated in cells previously labelled with ^{51}Cr in covalent binding to basic amino acids of the intracellular proteins. Upon membrane damage the labelled proteins leak out of the cells at a rate proportional to the damage incurred. However as slower natural leakage of ^{51}Cr from undamaged cells also takes place, this assay can only be used over short periods of time (Praus *et al.* 1980).

Dye exclusion tests after staining with viability dyes, trypan blue, eosin Y, naphthalene black, nigrosin (green), erythrosin B and fast green are more suited to suspensions than to monolayers as dead cells detach from the monolayer and are not accounted for in this assay. Cells however having lost their capacity to divide or to reproduce or presenting an impaired clonogenicity may well show a normal viability according to the dye exclusion tests (van der Meulen *et al.* 1981).

Viability parameters include the evaluation of the $(Na^+ - K^+)$ ATPase and $(Ca^{2+} - Mg^{2+})$ ATPase pump mechanisms and the membrane transport activities of amino acids, α-methylglucoside, p-aminohippurate (PAH), ^{131}I-hippuran, and the non-metabolisable amino acid analogue α-aminoisobutyric acid (De Loecker *et al.* 1980; Fahy, 1981; Rubinacci *et al.* 1986; Takehara & Rowe, 1971).

Metabolic integrity – respiration. Injury to respiration and glycolysis can be followed by quantitative Warburg manometry but involves extensive and compli- cated technical manipulations. Modifications in the dehydrogenase activity may be estimated by the reduction of methylene blue, added to the medium. The reduction or the absence of reduction of the dye, allows a qualitative evaluation of cell injury or cell death (Wilson, 1986).

Although the tetrazoliumbromide dye reduction related to the cell respiration, appears of limited value, the evaluation of the fluorescent dye, fluorescein appears more interesting (Rotman & Papermaster, 1966). The non-fluorescent derivative of fluorescein (fluorescein diacetate) unlike fluorescein is freely permeable. Upon reaction with intracellular esterases, fluorescein is released and retained in the cells as long as the plasma membranes remain intact. Oxygen consumption as an expression of the electron transfer chain reaction, as well as the residual ATP, ADP and AMP levels but particularly the evaluation of the ATP generating capacity after freezing and thawing contribute to the viability assessment (Buhl et al. 1976; Calman, 1974a,b; De Loecker & De Wever, 1979; Fahy, 1981; Southard et al. 1985).

Metabolic integrity – synthetic activity. Measurement of the incorporation of radioactive labelled precursors of metabolism provides an adequate approach to estimate specific metabolic activities. By evaluating the [^{3}H] thymidine incorporation into DNA and the [^{3}H] uridine incorporation into RNA, the nucleotide synthetic capacities of the cell are reflected. However with these kinds of experiments, care should be taken to distinguish between changes related to modifications in the intracellular nucleotide pool and the true modifications in nucleotide synthesis. Particularly the DNA assay is conditioned by the active reproductivity of the cells. To measure the rate of DNA synthesis, the gamma emittor isotope ^{125}I labelled iododeoxyuridine ([^{125}I]UdR) can be used as a specific stable label for newly synthesized DNA. Reutilization of the isotope is limited which makes it possible to be used in long-term experiments for up to 24 h. However this compound presents a variable degree of toxicity to different cell populations. The rate of release of [^{32}P] phosphate into the medium from pre-labelled cells is a function of the cell type analysed and increases when damage occurs. The incorporation of ^{32}P into nucleotides has also been used as a valuable parameter of intact nucleotide metabolism.

The preservation of energy metabolism can be functionally evaluated by following the metabolic fate of glucose breakdown products during *in vitro* incubation of radioactive ^{14}C and ^{3}H labelled hexoses and pentoses (Abouna, 1974; Fahy, 1981; Lundstam et al. 1977).

As protein synthesis represents an essential metabolic parameter, the incorporation of labelled amino acids into the proteins provide an important approach to the evaluation of the metabolic integrity (Lundstam et al. 1977; Penninckx et al. 1984; Penninckx & De Loecker, 1980). [^{14}C] and [^{3}H] labelled amino acids incorporated into the proteins are measured by liquid scintillation spectrometry while [^{35}S] methionine incorporation can be followed by autofluorography.

Analysis of functional changes in subcellular structures, mitochondria and microsomes may illustrate more localized cryoinjury (Frim & Mazur, 1980; Tsvetkov et al. 1985).

Cells and tissues – specific approach to individual cellular characteristics

For various individually functioning units, specific characteristic parameters can be analysed to approach the problems of viability.

With *erythrocytes* the appearance and the amount of haemoglobin released into the medium expresses the degree of haemolysis (Fujikawa, 1981; Mazur &

Rigopoulos, 1983; Meryman, 1970; Nei, 1976; Rotman & Papermaster, 1966; Rubalcava *et al.* 1969).

For *polymorphonuclear cells*, the recovery of phagocytosis, microbicidal ability and chemotaxis appear useful parameters to examine (Takahashi *et al.* 1985; van der Meulen *et al.* 1981). However when stored *leucocytes*, *lymphocytes* and *granulocytes* show a normal dye exclusion test and an active thymidine incorporation into DNA, this still does not guarantee that phagocytosis will be possible or *vice versa* (Frim & Mazur, 1980, 1983; Knight, 1980).

Due to the complexity of their functions (*platelets* participate in the formation of a platelet plug during haemostasis, perform a role in initiating the intrinsic system of coagulation and furthermore participate in phagocytosis) the evaluation of the *in vitro versus in vivo* viability may prove extremely problematic (Rowe & Lenny, 1981; Rowe *et al.* 1980).

Endothelial cell functions can be evaluated *in vitro* after the formation of a confluent monolayer (Pearson *et al.* 1982). Viability is assessed on the basis of their specific function in blood coagulation, producing prostacyclin, thrombin, plasminogen activator and its inhibitor, glycosaminoglycans, heparin and dermatan sulphates and releasing the procoagulant Factors VII and VIII (MacIntyre *et al.* 1978; Moncada *et al.* 1976; Rowe & Lenny, 1981).

After cryopreservation *haematopoietic stem cells* should retain their capacity to replicate and to repopulate the haematopoietic systems after bone marrow transplantation. This repopulation potential of stored bone marrow cells into lethally irradiated patients is the most efficient basis to estimate the viability of the preparation. Viability may be quantitatively evaluated after intravenous injection of stored stem cells into irradiated rodents, by counting the colonies which develop in the spleen (Rybka *et al.* 1980; Schaefer, 1980; Schlunk *et al.* 1981).

The metabolic integrity of isolated *pancreatic islets of Langerhans cells* can be evaluated after cryopreservation by assessing their ability to produce insulin determined by the radioimmune insulin assay. Transplantation of isolated cells into diabetic animals may provide further indication of their metabolic potentials (Ashwood-Smith, 1980; Kemp *et al.* 1977; McKay & Karow, 1983; Sandler & Andersson, 1984).

The metabolic activity of *thyroid cells* after cryopreservation has been assessed by the capacity of the long-acting thyroid stimulator (LATS) and the human thyroid stimulator (HTS) to activate cyclic-AMP production (Knox *et al.* 1977).

The *parathyroid activity* preserved after freezing and thawing has been measured after transplantation into hypocalcaemic rats (Wells & Christiansen, 1974).

Apart from the trypan blue exclusion method, the ability to bind ^{125}I-insulin, to form bilirubin conjugates as an expression of the microsomal detoxification and to synthesize glycogen, viability of isolated *liver cells* has also been examined by analysing selective enzyme systems, aryl hydrocarbon hydrolase (AHH) associated with the activation of some carcinogens, and tyrosine aminotransferase (TAT) activity. However it should be pointed out that some enzyme systems such as the microsomal enzymes P450 and P448 remain stable to freezing and thawing even in the absence of cryoprotectants (Bombeck *et al.* 1968; Fishbein & Griffin, 1976; Fuller & De Loecker, 1985; Fuller *et al.* 1982*a*). The ability of isolated hepatocytes to accumulate [^{99}Tc]HIDA(*N*-[2,6-dimethyl phenyl carbamoyl-methyl] iminodiacetic acid) before excretion can be estimated by gamma radio-spectrometry and relates to the retained viability (Fuller *et al.* 1982*b*).

For isolated and cultured *heart cells* the most demonstrative indication for survival and viability is obtained by assessing the pattern and strength of the specific heartbeat (Lochner *et al.* 1976; Offerijns *et al.* 1969; Offerijns & Ter Welle, 1974; Schlafer & Karow, 1972).

The viability assessment of *smooth muscle* preparations is usually centred on the muscular contractility while for *vein grafts* apart from the smooth muscle contractility, the ability to preserve an intact and functional endothelium has to be considered (Pearson *et al.* 1982; Penninckx *et al.* 1986; Taylor & Pegg, 1983; Weber *et al.* 1975, 1976).

The retention of *bone cell* viability after experimental cryopreservation has been followed by the *in vitro* incorporation of radioactive labelled nucleotides into DNA of calvarian explants of rodents and has been visualized by autoradiography techniques (Wezeman & Guzzino, 1986).

After being isolated from experimental animals, *tooth germs* have been frozen and after thawing reimplanted while histology and mitotic activity are assayed in the transplants where they should develop to normal dental tissue structures (Bartlett & Reade, 1972).

The viability of *skin tissue* after cryopreservation has been evaluated by analysing different metabolic activities involving synthetic functions and membrane transport efficiency (De Loecker *et al.* 1980; Maack *et al.* 1982; Praus *et al.* 1980).

An essential prerequisite for successful cryopreservation of *cornea tissue*, is the preservation of the intact functioning endothelium. Histological staining with nitro-blue tetrazolium to evaluate dehydrogenase and diaphorase activities and trypan blue staining are useful to assess the endothelial function. As with all these methods the cells examined are destroyed and subsequent transplantation becomes impossible. As it is suggested that the corneal endothelium pumps bicarbonate anions and presumably hydroxyl ions into the aqueous humour, relatively low bicarbonate concentrations in MK medium (TC 199 medium with 5 % dextran and antibiotics) may indicate an appreciable corneal survival (Hodson & Miller, 1976; McCarey *et al.* 1976; Taylor & Hunt, 1981).

For *spermatozoa*, the degree of post-thaw motility and acrosomal maintenance prove specific parameters (Fiser *et al.* 1986; Polge, 1980). Although reports on survival of freeze-thaw *oocytes* are still sporadic, survival assays on cryopreserved *mammalian embryos* are well established (Chen, 1986; Whittingham, 1980). The most reliable *in vitro* viability assay appears to be the development to the blastocyst stage and to a lesser degree the passage through one or two cleavage divisions. Where development in culture cannot be achieved, the embryos may be treated with fluorescein diacetate to evaluate the plasma membrane integrity

(Whittingham, 1980). The ultimate and final proof of ideally preserved spermatozoa, oocytes and embryos will always be the production of a normal offspring.

Organs

The assessment of life-sustaining functioning becomes more complicated when instead of individual cells and tissues, complex organs are considered, where circulation of blood constitutes an essential condition for survival. Even without freezing, hypothermia and ischaemia will easily result in vascular injury accompanied by progressive vascular resistance as a consequence of endothelial damage and leading to out-flow block when the organ is reimplanted. The initial events, triggering out-flow block may well be volume changes of the endothelial cells upon dilution of the cryoprotectant. These volume changes although not necessarily irreversible will lead to vascular occlusions before the original volume can be resumed (Pegg, 1982; Pegg *et al.* 1979).

If ever a whole organ could be frozen and thawed in such a way that the final out-flow block could be avoided, even allowing various periods of increased vascular resistance, the analysis of the perfusate composition might contribute to the evaluation of the activities of the cell types involved (Pegg *et al.* 1978). Apart from different enzyme systems released in the perfusate, lactate dehydrogenase (LDH), glutamic oxaloacetic transaminase (GOT), multiple other enzyme systems, as well as glucose, lactate, triglycerides, free fatty acids, potassium, ammonia and other components may provide some indication of the cellular integrity. As the acidification by increased lactate levels are detrimental, changes in pH and pCO_2 as well as the ratio NAD/NADH (nicotinamide adenine dinucleotide) as an expression of the oxidation-reduction (redox) potential of the perfusate and the presence of hypoxanthine or other degradation products of ATP may equally contribute to the evaluation of viability (Fahy, 1981; Lundstam *et al.* 1977). As a non-invasive method analysing several aspects of energy metabolism, the examination by phosphorus nuclear magnetic resonance may prove useful, as preservation results in a loss of ATP causing membrane instability and activation of membrane-bound phospholipases (Gadian & Radda, 1981; Wilkinson & Robinson, 1974).

On the other hand, localized injury may produce considerable extracellular concentrations of biological constituents, even when normal functioning cells remain present and possibly capable of eventually taking over the life-sustaining functions. The combined evaluation of the physiological and biochemical functions of normothermic perfused organs after preservation in response to different pharmacological stimulation may well, in principle, be the most efficient *in vitro* estimation of viability. Although most rewarding from an experimental point of view, the exposure of the organ for longer periods of time to artificial perfusion conditions, reduces its usefulness for eventual transplantation. Thus the glomerular filtration rate of kidneys, the ventricular pressure in heart preparations in function of the pharmacological responsiveness as well as the electrical activity of brain, the insulin release from pancreatic preparations after glucose tolerance and multiple biochemical parameters to evaluate liver functions, are all used to approach the notion of viability (Ashwood-Smith, 1980; Bombeck et al. 1968; Fahy, 1981; Kemp et al. 1977; Lundstam et al. 1977; Mazur et al. 1976; Pegg, 1978, 1982; Penninckx & De Loecker, 1980; Southard et al. 1985; Suda et al. 1974).

It is obvious, viability being an elusive concept, that evaluation or assessment of this essential characteristic of a surviving cell remains ambiguous. Indeed unmeasurable changes may lead to fatal injury while some gross abnormalities incurred during the course of the freeze-thaw cycle appear reversible. The many different viability assays represent interesting and subtle approaches to our concern which is to predict the degree by which a cell, a tissue or an organ survives.

Conclusions

In order to achieve survival of cryopreserved mammalian systems, the cellular morphology should be protected and kept intact, while the development of intracellular ice crystals should be avoided. Even if the cellular integrity is unaffected, the freeze-thaw cycle, through ischaemia and cooling seriously affects the energy production processes and furthermore disturbs the intracellular environment. Both factors prove detrimental to the normal biochemical functions. As energy production steps are affected, essential biochemical processes, necessary to maintain metabolism and repair functions, rapidly lose their efficiency leading to irreversible structural injury. The maintenance of the intracellular medium to a large extent conditions the metabolic efficiency for which the different membrane pump mechanisms are responsible.

In order to evaluate the preserved or retained metabolic activity and to establish the viability or survival after cryopreservation, several morphological, and functional assays have been devised. Apart from the general synthetic reactions inherent to all living mammalian structures, some specialized units presenting specific functions may be used to approach the problems of viability. As biological systems become more complex and whole organs are considered for preservation, the problems involving survival and viability increase dramatically. Individual assays by which the degree of viability can be deduced may be useful in the analysis of individual cell structure, although a single approach is usually insufficient to arrive at a reliable evaluation. The ultimate and surest proof of preserved viability can only be provided when the tissues or the organs examined are capable of restarting their characteristic physiological and biochemical functions upon transplantation into a normal environment.

The authors are indebted to the Belgian National Foundation for Medical Research (FGWO) for a grant to the laboratories.

References

ABOUNA, G. J. M. (1974). Viability assays in organ preservation. In *Organ Preservation for Transplantation* (ed. A. M. Karow, Jr, G. J. M. Abouna & A. L. Humphries, Jr), pp. 108–126. Boston: Little, Brown.

AKERMAN, K. E. O. & NICHOLLS, D. G. (1983). Physiological and bioenergetic aspects of mitochondrial calcium transport. *Rev. physiol. biochem. Pharmac.* **95**, 149–201.

ALLEN, E. D., GAU, T. C. & NATALE, R. B. (1984). *In vitro* colony formation by cryopreserved primary human tumor cells. *Cryobiology* **21**, 240–245.

ASHWOOD-SMITH, M. J. (1980). Low temperature preservation of cells, tissues and organs. In *Low Temperature Preservation in Medicine and Biology* (ed. M. J. Ashwood-Smith & J. Farrant), pp. 18–44. Bath: Pitman Medical.

BARTLETT, P. F. & READE, P. C. (1972). Cryopreservation of developing teeth. *Cryobiology* **9**, 205–211.

BICKFORD, H. R., SHEPARD, M. L. & COCKS, F. H. (1980). Postfreeze viability of human renal epithelial carcinoma cells in quaternary solutions. *Cryobiology* **17**, 363–370.

BOMBECK, C. T., BIAVA, C., CONDON, R. E. & NYPHYS, L. M. (1968). Parameters of normal liver function in isolated, perfused bovine liver. In *Organ Perfusion and Preservation* (ed. J. C. Norman), pp. 579–608. New York: Appleton-Century-Crofts.

BUHL, M. R., KEMP, G. & KEMP, E. (1976). Adenosine stimulated postimplantation regeneration of 5′-adenine nucleotides in rabbit kidney grafts. *Life Sci.* **19**, 1889–1896.

CALMAN, K. C. (1974a). The prediction of organ viability. I. An hypothesis. *Cryobiology* **11**, 1–6.

CALMAN, K. C. (1974b). The prediction of organ viability. II. Testing an hypothesis. *Cryobiology* **11**, 7–12.

CHEN, C. (1986). Pregnancy after human oocyte cryopreservation. *Lancet* **1**, 884–886.

CONOVER, T. E. (1969). Influence of organic solutes on the reactions of oxidative phosphorylation. *J. biol. Chem.* **244**, 254–259.

CONOVER, T. E. (1975). Influence of nonionic organic solutes on various reactions of energy conservation and utilization. *Ann. N.Y. Acad. Sci.* **243**, 24–37.

CROWE, J. H., CROWE, L. M. & MOURADIAN, R. (1983). Stabilization of biological membranes at low water activities. *Cryobiology* **20**, 346–356.

DE LOECKER, W. & DE WEVER, F. (1979). Oxidative phosphorylation in rat skin during preservation. *Cryobiology* **16**, 517–525.

DE LOECKER, W., DE WEVER, F. & PENNINCKX, F. (1980). Metabolic changes in human skin preserved at −3°C and at −196°C. *Cryobiology* **17**, 46–53.

DEWEY, M. M. & BARR, L. (1970). Some considerations about the structure of cellular membranes. In *Current Topics in Membranes and Transport* (ed. F. Bronner & A. Kleinzeller), vol. 1, pp. 1–33. New York and London: Academic Press.

FARRANT, J. (1980). General observations on cell preservation. In *Low Temperature Preservation in Medicine and Biology* (ed. M. J. Ashwood-Smith & J. Farrant), pp. 1–18. Bath: The Pitman Press.

FARRANT, J., WALTER, C. A., LEE, H. & MCGANN, L. E. (1977a). The use of two-step cooling procedure to examine factors influencing cell survival following freezing and thawing. *Cryobiology* **14**, 273–286.

FARRANT, J., WALTER, C. A., LEE, H., MORRIS, G. J. & CLARKE, K. J. (1977b). Structural and functional aspects of biological freezing techniques. *J. Microsc.* **111**, 17–34.

FAHY, G. M. (1981). Viability assessment. In *Organ Preservation for Transplantation* (ed. A. M. Karow, Jr & D. E. Pegg), pp. 53–73. New York and Basel: Marcel Dekker Inc.

FAHY, G. M. (1986). The relevance of cryoprotectant 'toxicity' to cryobiology. *Cryobiology* **23**, 1–13.

FAHY, G. M., MACFARLANE, D. R., ANGELL, C. A. & MERYMAN, H. T. (1984). Vitrification as an approach to cryopreservation. *Cryobiology* **21**, 407–426.

FEINBERG, H. (1982). Energetics and mitochondria. In *Organ Preservation, Basic and Applied Aspects* (ed. D. E. Pegg, I. A. Jacobsen & N. A. Halasz), pp. 3–18. Lancaster, Boston, The Hague: MTP Press Ltd.

424   R. DE LOECKER AND F. PENNINCKX

FINK, A. L. (1986). Effects of cryoprotectants on enzyme structure. *Cryobiology* **23**, 28–37.

FISER, P. S., FAIRFULL, R. W. & MARCUS, G. J. (1986). The effect of thawing velocity on survival and acrosomal integrity of ram spermatozoa, frozen at optimal and suboptimal rates in straws. *Cryobiology* **23**, 141–149.

FISHBEIN, W. N. & GRIFFIN, J. L. (1976). Studies on the mechanism of freezing damage to mouse liver. IV. Effects of ultrarapid freezing on structure and function of isolated mitochondria. *Cryobiology* **13**, 542–556.

FLORES, J., DI BONA, D. R., FREGA, N. & LEAF, A. (1972). Cell volume regulation and ischaemic tissue damage. *J. Membr. Biol.* **10**, 331–343.

FREYER, J. P. & SUTHERLAND, R. M. (1980). Selective dissociation and characterization of cells from different regions of multicell tumor spheroids. *Cancer Res.* **40**, 3956–3965.

FRIM, J. & MAZUR, P. (1980). Approaches to the preservation of human granulocytes by freezing. *Cryobiology* **17**, 282–286.

FRIM, J. & MAZUR, P. (1983). Interaction of cooling rate, warming rate, glycerol concentration and dilution procedure on the viability of frozen-thawed human granulocytes. *Cryobiology* **20**, 657–676.

FUJIKAWA, S. (1981). The effect of different cooling rates on the membrane of frozen human erythrocytes. In *Effects of Low Temperatures on Biological Membranes* (ed. G. J. Morris & A. Clarke), pp. 323–334. London, New York: Academic Press.

FULLER, B. J. & DE LOECKER, W. (1985). Changes in the permeability characteristics of isolated hepatocytes during slow freezing. *Cryo-Lett.* **6**, 361–370.

FULLER, B. J., GROUT, B. W. & WOODS, R. J. (1982a). Biochemical and ultrastructural examination of cryopreserved hepatocytes in rats. *Cryobiology* **19**, 493–502.

FULLER, B. J., WOODS, R. J., NUTT, L. H. & ATTENBURROW, V. D. (1982b). Survival of hepatocytes upon thawing from −196°C: Functional assessment after transplantation. In *Organ Preservation; Basic and Applied Aspects* (ed. D. E. Pegg, I. A. Jacobsen & N. A. Halasz), pp. 381–383. Lancaster, Boston, The Hague: MTP Press Ltd.

HODSON, S. & MILLER, F. (1976). The bicarbonate ion pump in the endothelium which regulates the hydration of rabbit cornea. *J. Physiol., Lond.* **263**, 563–577.

JACOB, S. W. & HERSCHLER, R. (1986). Pharmacology of DMSO. *Cryobiology* **23**, 14–27.

JACOB, G., KURZER, M. N. & FULLER, B. J. (1985). An assessment of tumor cell viability after *in vitro* freezing. *Cryobiology* **22**, 417–426.

GADIAN, D. G. & RADDA, G. K. (1981). NMR studies of tissue metabolism. *Ann. Rev. Biochem.* **50**, 69–83.

KAROW, A. M. (1982). Biophysical and chemical considerations in cryopreservation. In *Organ Preservation, Basic and Applied Aspects* (ed. D. E. Pegg, I. A. Jacobsen & N. A. Halasz), pp. 113–141. Lancaster, Boston, The Hague: MTP Press Ltd.

KEMP, J. A., MAZUR, P., MULLEN, Y., MILLER, R. H., CLARK, W. & BROWN, J. (1977). Reversal of experimental diabetes by fetal rat pancreas. I. Survival and function of fetal rat pancreas frozen to −196°C. *Transplant. Proc.* **9**, 325–328.

KNIGHT, S. C. (1980). Preservation of leucocytes. In *Low Temperature Preservation in Medicine and Biology* (ed. M. J. Ashwood-Smith & J. Farrant), pp. 121–138. Bath: The Pitmann Press.

KNOX, A. J. S., VON WESTARP, C., ROW, V. V. & VOLPÉ, R. (1977). The use of cryopreserved human thyroid tissue for the *in vitro* assay of thyroid stimulators. *Cryobiology* **14**, 543–548.

LAZARUS, H. M., WARNICK, C. T. & HOPFENBECK, A. (1982). DNA strand breakage after kidney storage. *Cryobiology* **19**, 129–135.

LOCHNER, A., KOTZÉ, J. C. N. & GEVERS, W. (1976). Mitochondrial oxidative phosphorylation in myocardial ischaemia. Effects of glycerol, glucose and insulin on anoxic hearts perfused at low pressure. *J. mol. cell. Cardiol.* **8**, 575–584.

LUNDSTAM, S., JONSSON, O., PETTERSSON, S. & SCHERSTEN, T. (1977). Glucose, fatty acid, and amino acid utilization during hypothermic perfusion of dog kidneys. *Transplant. Proc.* **9**, 1561–1563.

MAACK, P., RIEGER, P. & RÜDIGER, H. W. (1982). Recovery after freezing and thawing of cultured human diploid fibroblasts. *Cryobiology* **19**, 10–15.

MACINTYRE, D. E., PEARSON, J. D. & GORDON, J. L. (1978). Localization and stimulation of prostacyclin production in vascular cells. *Nature, Lond.* **271**, 549–551.

MALININ, T. I. (1972). Injury of human polymorphonuclear granulocytes frozen in the presence of cryoprotective agents. *Cryobiology* **9**, 123–130.

MAZUR, P. (1984). Freezing of living cells: mechanisms and implications. *Am. J. Physiol.* **247**, C125–C142.

MAZUR, P., KEMP, J. A. & MILLER, R. H. (1976). Survival of fetal rat pancreases frozen to $-78°$ and $-196°C$. *Proc. natn. Acad. Sci. U.S.A.* **73**, 4105–4109.

MAZUR, P. & RIGOPOULOS, N. (1983). Contributions of unfrozen fractions and of salt concentration to the survival of slowly frozen human erythrocytes: Influence of warming rate. *Cryobiology* **20**, 274–289.

McCAREY, B. E., MEYER, R. F. & KAUFMAN, H. E. (1976). Improved corneal storage for penetrating keratoplasties in humans. *Ann. Ophthalmol.* **8**, 1488–1495.

McKAY, D. B. & KAROW, A. M. (1983). Factors to consider in the assessment of viability of cryopreserved islets of Langerhans. *Cryobiology* **20**, 151–160.

MERYMAN, H. T. (1970). The exceeding of a minimum tolerable cell volume in hypertonic suspension as a cause of freezing injury. In *The Frozen Cell* (ed. G. E. W. Wolstenholme & M. O'Connor), pp. 51–64. London: J. & A. Churchill.

MERYMAN, H. T., WILLIAMS, R. J. & DOUGLAS, M. J. (1977). Freezing injury from 'solution effects' and its prevention by natural or artificial cryoprotection. *Cryobiology* **14**, 287–302.

MITCHELL, P. (1976). Vectorial chemistry and the molecular mechanics of chemiosmotic coupling: Power transmission by proticity. *Biochem. Soc. Trans.* **4**, 399–430.

MONCADA, S., GRYGLEWSKI, R., BUNTING, S. & VANE, J. R. (1976). An enzyme isolated from arteries transforms prostaglandin endoperoxides to an unstable substance that inhibits platelet aggregation. *Nature, Lond.* **263**, 663–665.

NEI, T. (1976). Freezing injury to erythrocytes: I. Freezing patterns and post-thaw hemolysis. *Cryobiology* **13**, 278–286.

NEWSHOLME, E. A. & LEECH, A. R. (1983). Thermodynamics in metabolism. In *Biochemistry for the Medical Sciences*, pp. 10–45. Chichester: John Wiley & Sons.

OFFERIJNS, F. G. J., FREUD, G. E. & KRIJNEN, H. W. (1969). Reanimation of myocardial cells preserved in the frozen state. *Nature, Lond.* **222**, 1174.

OFFERIJNS, F. G. J. & TER WELLE, H. F. (1974). The effect of freezing, of supercooling and of DMSO on the function of mitochondria and on the contractility of the rat heart. *Cryobiology* **11**, 152–159.

PAVLOCK, G. S., SOUTHARD, J. H., STARLING, J. R. & BELZER, F. O. (1984). Lysosomal enzyme release in hypothermically perfused dog kidneys. *Cryobiology* **21**, 521–528.

PEARSON, J. D., HUTCHINGS, A. & GORDON, J. L. (1982). Endothelial cell function and organ preservation *ex vivo*. In *Organ Preservation: Basic and Applied Aspects* (ed. D. E. Pegg, I. A. Jacobsen & N. A. Halasz), pp. 43–54. Lancaster, Boston, The Hague: MTP Press Ltd.

PEGG, D. E. (1978). An approach to hypothermic renal preservation. *Cryobiology* **15**, 1–17.

PEGG, D. E. (1981). The biology of cell survival in vitro. In *Organ Preservation for Transplantation* (ed. A. M. Karow, Jr & D. E. Pegg), pp. 31–52. New York: Marcel Dekker Inc.

PEGG, D. E. (1982). The principles of organ storage procedures. In *Organ Preservation. Basic and Applied Aspects* (ed. D. E. Pegg, I. A. Jacobsen & N. A. Halasz), pp. 55–66. Lancaster, Boston, The Hague: MTP Press Ltd.

PEGG, D. E., GREEN, C. J. & WALTER, C. A. (1978). Attempted canine renal cryopreservation using dimethyl sulphoxide helium perfusion and microwave thawing. *Cryobiology* **15**, 618–626.

PEGG, D. E., JACOBSEN, I. A., ARMITAGE, W. J. & TAYLOR, M. J. (1979). Mechanism of cryoinjury in organs. In *Organ Preservation* II (ed. D. E. Pegg & I. A. Jacobsen), pp. 132–146. Edinburgh, London, New York: Churchill Livingstone.

PENNINCKX, F., CHENG, N., KERREMANS, R., VAN DAMME, B. & DE LOECKER, W. (1983). The effects of different concentrations of glycerol and dimethylsulfoxide on the metabolic activities of kidney slices. *Cryobiology* **20**, 51–60.

PENNINCKX, F. & DE LOECKER, W. (1980). Metabolic changes in dog kidney during cryopreservation. *Cryobiology* **17**, 549–556.

PENNINCKX, F., POELMANS, S., KERREMANS, R. & DE LOECKER, W. (1984). Erythrocyte swelling after rapid dilution of cryoprotectants and its prevention. *Cryobiology* **21**, 25–32.

Penninckx, F., Vandekerckhove, Ph., De Loecker, W. & Kerremans, R. (1986). Preservation of *Taenia coli* by freezing and storage at −196°C. *Cryobiology* **23**, 222–229.

Persidsky, M. D. & Ellet, M. H. (1971). Lysosomes and cell cryoinjury. *Cryobiology* **8**, 345–349.

Polge, C. (1980). Freezing of spermatozoa. In *Low Temperature Preservation in Medicine and Biology* (ed. M. J. Ashwood-Smith & J. Farrant), pp. 45–64. Bath: Pitman Medical.

Polge, C., Smith, A. U. & Parkes, A. S. (1949). Revival of spermatozoa after vitrification and dehydration at low temperatures. *Nature, Lond.* **164**, 666.

Praus, R., Böhm, F. & Dvorak, R. (1980). Skin cryopreservation. I. Incorporation of radioactive sulfate as a criterion of pigskin graft viability after freezing to −196°C in the presence of cryoprotectants. *Cryobiology* **17**, 130–134.

Quinn, P. J. (1985). A lipid-phase separation model of low-temperature damage to biological membranes. *Cryobiology* **22**, 128–146.

Roper, P. R. & Drewinko, B. (1976). Comparison of *in vitro* methods to determine drug-induced cell lethality. *Cancer Res.* **36**, 2182–2188.

Rotman, B. & Papermaster, B. W. (1966). Membrane properties of living mammalian cells as studied by enzymatic hydrolysis of fluorogenic esters. *Proc. natn. Acad. Sci. U.S.A.* **55**, 134–141.

Rowe, A. W. & Lenny, L. L. (1981). Platelets. In *Organ Preservation for Transplantation* (ed. A. M. Karow, Jr & D. E. Pegg), pp. 335–354. New York and Basel: Marcel Dekker Inc.

Rowe, A. W., Lenny, L. L. & Mannoni, P. (1980). Cryopreservation of red cells and platelets. In *Low Temperature Preservation in Medicine and Biology* (ed. M. J. Ashwood-Smith & J. Farrant), pp. 85–120. Bath: The Pitman Press.

Rubalcava, B., Marinez de Munoz, D. & Gitler, C. (1969). Interaction of fluorescent probes with membranes. I. Effect of ions on erythrocyte membranes. *Biochemistry* **8**, 2742–2747.

Rubinacci, A., Fuller, B., Wuytack, F. & De Loecker, W. (1986). Ca^{2+} transport and permeability in inside-out red cell membrane vesicles after freezing. *Cryobiology* **23**, 134–140.

Rybka, W. B., Mittermeyer, K., Singer, J. W., Buckner, C. D. & Thomas, E. D. (1980). Viability of human marrow after long-term cryopreservation. *Cryobiology* **17**, 424–428.

Sandler, S. & Andersson, A. (1984). The significance of culture for successful cryopreservation of isolated pancreatic islets of Langerhans. *Cryobiology* **21**, 503–510.

Santarius, K. A. & Giersch, C. (1983). Cryopreservation of spinach chloroplast membranes by low-molecular-weight carbohydrates. II. Discrimination between colligative and non-colligative protection. *Cryobiology* **20**, 90–99.

Schaefer, U. W. (1980). Bone marrow stem cells. In *Low Temperature Preservation in Medicine and Biology* (ed. M. T. Ashwood-Smith & J. Farrant), pp. 139–154. Bath: Pitman Medical.

Schlafer, M. (1981). Pharmacological considerations in cryopreservation. In *Organ Preservation for Transplantation* (ed. A. K. Karow, Jr & D. E. Pegg), pp. 177–212. New York: Marcel Dekker Inc.

Schlafer, M. & Karow, A. M., Jr (1972). Ultrastructure-function correlative studies for cardiac cryopreservation. III. Hearts frozen to −10°C and −17°C with and without dimethyl sulfoxide (DMSO). *Cryobiology* **9**, 38–50.

Schlunk, T., Rüber, E. & Schleyer, M. (1981). Survival of human bone marrow progenitor cells after freezing: Improved detection in the colony-formation assay. *Cryobiology* **18**, 111–118.

Schwarz, A., Lindenmayer, G. E. & Allen, J. C. (1972). The Na^+, K^+-ATPase membrane transport system: Importance in cellular function. In *Current Topics in Membranes and Transport* (ed. F. Bronner & A. Meinzeller), vol. 3, pp. 1–82. New York and London: Academic Press.

Shikama, K. (1965). Effect of freezing and thawing on the stability of double helix of DNA. *Nature, Lond.* **207**, 529–530.

Southard, J. H., Rice, M. J. & Belzer, F. O. (1985). Preservation of renal function by adenosine-stimulated ATP synthesis in hypothermically perfused dog kidneys. *Cryobiology* **22**, 237–242.

STEPONKUS, P. L. & WIEST, S. C. (1979). Freeze-thaw induced lesions in the plasma membrane. In *Low Temperature Stress in Crop Plants* (ed. J. M. Lyons, D. G. Graham & J. K. Raisen), pp. 231–254. New York: Academic Press.

STRAUSS, G. & INGENITO, E. P. (1980). Stabilization of liposome bilayers to freezing and thawing: Effects of cryoprotective agents and membrane proteins. *Cryobiology* 17, 508–515.

SUDA, I., KITO, K. & ADACHI, C. (1974). Bioelectric discharges of isolated cat brain after revival from years of frozen storage. *Brain Res.* 70, 527–531.

TAKAHASHI, T., BROSS, J. B., SHABER, R. E. & WILLIAMS, R. J. (1985). Effect of cryoprotectants on the viability and function of unfrozen human polymorphonuclear cells. *Cryobiology* 22, 336–350.

TAKEHARA, I. & ROWE, A. W. (1971). Increase in ATPase activity in red cells membranes as a function of freezing regimen. *Cryobiology* 8, 559–565.

TAMIYA, T., OKAHASHI, N., SAKUMA, R., AOYAMA, T., AKAHANE, T. & MATSUMOTO, J. J. (1985). Freeze denaturation of enzymes and its prevention with additives. *Cryobiology* 22, 446–456.

TAYLOR, M. J. & HUNT, C. J. (1981). Dual staining of corneal endothelium with trypan blue and alizarin red S: Importance of pH for the dye-lake reaction. *Br. J. Ophthalmol.* 65, 815–819.

TAYLOR, M. J. & PEGG, D. E. (1983). The effect of ice formation on the function of smooth muscle tissue stored at −21 or −60°C. *Cryobiology* 20, 36–40.

TAYLOR, M. J. & PIGNAT, Y. (1982). Practical acid dissociation constants, temperature coefficients, and buffer capacities for some biological buffers in solutions containing dimethylsulfoxide between 25 and −12°C. *Cryobiology* 19, 99–109.

TSVETKOV, T., TSONEV, L., MERANZOV, N. & MINKOV, I. (1985). Functional changes in mitochondrial properties as a result of their membrane cryodestruction. I. Influence of freezing and thawing on succinate-ferricyanide reductase of intact liver mitochondria. *Cryobiology* 22, 47–54.

VAN DER MEULEN, F. W., REISS, M., STRICKER, E. A. M., VAN ELVEN, E. & VON DEM BORNE, A. E. G. (1981). Cryopreservation of human monocytes. *Cryobiology* 18, 337–343.

WEBER, T. R., DENT, T. L., SALLES, C. A., RAMSBURGH, S. R., FONSECA, F. P. & LINDENAUER, S. M. (1975). Cryopreservation of venous homografts. *Surg. Forum* 26, 291–293.

WEBER, T. R., LINDENAUER, S. M., DENT, T. L., ALLEN, E., SALLES, C. A. & WEATHERBEE, L. (1976). Long-term patency of vein grafts preserved in liquid nitrogen in dimethyl sulfoxide. *Ann. Surg.* 184, 709–712.

WEISENTHAL, L. M., DILL, P. L., KUZNICK, N. B. & LIPPMAN, M. E. (1983). Comparison of dye exclusion assays with a clonogenic assay in the determination of drug-induced cytotoxicity. *Cancer Res.* 43, 258–264.

WELLS, S. A. & CHRISTIANSEN, C. (1974). The transplanted parathyroid gland: Evaluation of cryopreservation and other environmental factors which affect its function. *Surgery* 75, 49–55.

WEZEMAN, F. H. & GUZZINO, K. M. (1986). Retention of bone cell viability in mouse calvarial explants after cryopreservation. *Cryobiology* 23, 81–87.

WHITTINGHAM, D. G. (1980). Principles of embryo preservation. In *Low Temperature Preservation in Medicine and Biology* (ed. M. J. Ashwood-Smith & J. Farrant), pp. 66–83. Bath: Pitman Medical.

WILKINSON, J. H. & ROBINSON, J. M. (1974). Effect of ATP on release of intracellular enzymes from damaged cells. *Nature, Lond.* 249, 662–663.

WILSON, A. P. (1986). Cytotoxicity and viability assays. In *Animal Cell Culture, A Practical Approach* (ed. R. I. Freshney), pp. 183–216. Oxford, Washington D.C.: IRL Press.

Printed in Great Britain © *Society for Experimental Biology 1987*   429

TEMPERATURE EFFECTS ON DIFFERENT ORGANIZATION LEVELS IN ANIMALS

KARI Y. H. LAGERSPETZ

Laboratory of Animal Physiology, Department of Biology, University of Turku, Finland

Summary

One of the central concepts in present biology is the recognition of different organization levels and their hierarchical array. Complex multicellular animals are constituted of organ systems, the organs of cells, the cells of organelles, membranes, and molecules. The primary effects of many environmental factors (e.g. light, concentrations of ions and molecules) can be delimited mainly to one level. Temperature, being the macroscopic physical measure of the random motion of smallest material particles, affects directly the animal life at all organization levels.

The special physical nature of temperature means also, that during the history of life, organisms have always been subjected to temperature variations. Many different ways to evade the pervasive effects of temperature have been evolved during the course of evolution. The study of the temperature relations of organisms can therefore give models for other branches of environmental biology.

The temperature limits and relations of an animal cannot be explained by the temperature relations and limits of its cells without taking into account such interactions between different types of cells, which are found only through the study of the organs. Also, the temperature limits and relations of animal cells cannot be explained just through the study of the constituent molecules. The possible interactions of the molecules (e.g. lipids and proteins in a cell membrane) are so manifold and complex that in order to ascertain the relative importance of them in the temperature relations of the cells we must rely in part on studies done on organelles (e.g. on the plasma membrane).

The study of the thermal biology of animal cells thus exemplifies the situation often found in biology: the attainment of a reductive explanation is not always a one-way deduction, but it may involve modifications of the lower level concepts according to the knowledge derivable only from studies of the higher level systems.

Temperature limits of animal life

The temperature range of active (self-reproducing) animal life extends from about 55°C which is the living temperature of *Thermobathynella adami*, a crustacean found in hot springs in central Africa (Capart, 1951), to −1·86°C, the living temperature of Antarctic fishes. Among plants, some blue-green algae live and reproduce in hot springs at 89°C.

Many multicellular animals have resting stages, which resist much higher and lower temperatures than the active animals. The cysts of the salt shrimp *Artemia salina* are early embryos covered with a chitinous shell. Air-dried cysts tolerate well −190°C, and also a vacuum of about 1 pm Hg pressure (Whitaker, 1940). About 25 % of them are viable after a stay of 90 min at 104°C, and some survive even a treatment at 110°C (Hinton, 1954, 1968). The larvae of the chironomid midge *Polypedilum vanderplancki*, studied also by Hinton (1951, 1960) live in shallow rock pools in central Africa. In dry condition, with a body water content of 3 to 8 %, these larvae tolerate temperatures as high as *Artemia* cysts, and also the immersion in liquid helium (−270°C). The dormant anabiotic or cryptobiotic forms of some nematodes, rotifers, and of the arthropod group of tardigrades resist temperatures as low as 0·008°K (Becquerel, 1951).

These records are useful because they show the importance of liquid water for active life, and also the importance of the dehydration for the survival of passive life at its extreme thermal limits. The water content of the cysts of *Artemia* and of the anabiotic stages of nematodes, rotifers and tardigrades is very low. It is obviously the profound dehydration which enables them to survive both in heat and in cold.

The recent reports of viable bacteria from the deep-sea waters of hydrothermal vents at 306°C (Baross, Lilley & Gordon, 1982; Baross & Deming, 1983) are interesting, since water at the pressure of 265 atm, from which these bacteria were collected, stays liquid up to 460°C, and 306°C is hence well below the boiling point.

The life in any form we know, seems to have its upper temperature limit close to the boiling point of water and its lower limit close to 0°K. At the natural conditions on earth, cold is the worst enemy of life (Franks, 1985) in its active forms, but when combined with dehydration, cold is the best preserver of passive life.

Why study animal cells?

Animal cells are either (1) free-living self-reproducing single-cell organisms (protozoans), (2) cultured self-reproducing cells of multicellular animals, (3) cells of multicellular animals living in temporary isolation, or (4) cells of multicellular animals living *in situ*. The gametes and zygotes of many animals, and cells isolated for study or for subsequent transfer to other animals are cells of the type 3.

Much of the studies on animal cells are conducted on the two types of self-reproducing cells or on cells in temporary isolation. One great advantage in such studies is the possibility of accurate monitoring and control of the origin and the life conditions of the objects.

The study of the effects of temperature on animal cells has three important objectives. First, the results of such studies show us how the various molecular structures within the cells and the manifold functions of them respond to changes in temperature. In this way these results add to our knowledge about the general biology, biochemistry, and biophysics of cellular temperature responses.

Secondly, the work done on animal cells has practical aspects, important for the enhancement of life of highly developed animals, even humans. Studies concerning the heat tolerance of tumour cells compared with normal ones, and those on the cold tolerance of cells, on their storage in cold and on their recovery for insemination, transfusion and transplantation, have as their ultimate aim to further the life of animal organisms.

Thirdly, much of the work done with animal cells has as its aim to explain the temperature relations of life functions of whole animals. I will start with this aspect.

Temperature limits on different organization levels

One of the central concepts in present biology is the recognition of different organization levels. Ecologists work on the ecosystem, population, and organism levels. To physiologists the organism, organ, cell, and molecule levels are the most familiar. The relations between the levels are not clear, as witnessed by the differing popular and even expert opinions as to what extent cell and molecule research can replace whole organism and organ system studies.

Temperature limits of survival and life functions can be studied on different organization levels. These limits are referred to as tolerance or resistance; these terms are usually considered as synonyms, e.g. by Prosser (1986). Franks (1985) makes a distinction between tolerance and resistance referring to the use of these terms in mechanics. In tolerance the system would be changed by the stress, but remain partly functional, while in resistance there would be within the system forces counteracting the stress. As combinations of so defined tolerance and resistance are common as conditions of survival, the distinction between them is not always possible (Franks, 1985), and the two terms are considered in the following as interchangeable. However, no cell does actually tolerate freezing in the sense of crystallization of ice inside it, although cells may avoid or resist freezing (Franks, 1985).

All animals and animal cells are always acclimated to a certain temperature or a range of temperatures, or they are in transition between two states of acclimation. There are no such 'controls', which have not been subjected to acclimation. This means that it is not possible to distinguish between the hereditary ('genotypic') and acquired ('phenotypic') components of temperature resistance or tolerance. However, those differences which are found in comparisons of two similar organisms, or of two similar non-dividing populations of cells in which no selection occurs, and which are subjected to different temperature treatments, must depend on acquired properties, i.e. on differences in acclimation. Such differences may be found out also if the same individual animal or cell can be studied repeatedly after different temperature treatments.

As all 'genotypic' effects on temperature relations of life functions must be expressed in the phenotype in order to be found out, and all 'phenotypic' effects occur within the limits set by the genotype, it is preferable to speak about heritable or genetic or evolutionary adaptation, and acclimation (acquired, non-heritable adaptation).

Brief cold and heat shocks (from a few minutes to a few hours) often shift the resistance limits of animals and animal cells to lower and higher temperatures, respectively. These effects are transient and non-heritable. They are referred to as cold hardening, and heat hardening or thermotolerance (for references, see Laudien, 1986; Burdon, this volume).

The resistance to cold or heat is not the same on different organization levels. The heart of an adult rat stops beating when its body temperature has decreased to about 15 °C. The isolated heart of a rat, when maintained in perfusion, stops at about 10 °C. But isolated myocardial cells continue their spontaneous phasic and electrically triggered fast contractions still at about 0 °C (Tirri, Lehto & Karttunen, 1983). The failure of the heart at low body temperature does not depend on the cold tolerance of the myocytes, but on the cold tolerance of the higher level control systems.

The late Professor Boris Ushakov and his group in Leningrad have for a number of years studied the heat resistance of various functions and proteins in a number of animals from different climatic conditions. For example, the reflex activity in the frog *Rana temporaria* stops after 30 min at 30–32 °C, while the muscle contraction tolerates 36·3 °C for the same time, the motility of spermatozoa 39·2 °C and the activity of cilia 41·7 °C. Succinate dehydrogenase from the muscle is 50 % inactivated after 30 min at 42 °C, and muscle actomyosin ATPase after 15 min at 40·5 °C (Ushakov, 1964).

Similar results have been obtained in a number of other animals. In all these, the proteins selected for study seem to be more thermostable *in vitro* than the cellular functions in which they participate, and cells studied more resistant than multicellular control functions (e.g. Prosser, 1986). Of course, it could hardly be the other way round, the proteins being less thermostable than the cells, and the cells less thermostable than the functional system in its entirety, and it would be quite a coincidence and probably a not well adaptable system, if all the components would break down at the same temperature. On the other hand, the correlations between the thermostabilities of different proteins from a number of animal species are remarkably high, i.e. the protein thermostabilities in each species are 'coupled' (Ushakov, 1967).

Animals and their cells: temperature limits

Sometimes it has been said, that since unicellular organisms are also injured by heat and show heat resistance acclimation, the function of control systems of multicellular animals cannot be decisive for their heat resistance. However, all cells of multicellular animals need not to be equal in their importance for the resistance of the whole animal. The control systems are themselves composed of cells of various types. The search for the mechanisms of cold and heat death and of cold and heat tolerance of the whole organism is therefore a search for the 'weakest link', which may well be a cellular structure and/or function. Knowledge about the temperature responses of various types of animal cells is therefore indispensable for the explanation of the thermal tolerances of whole animals.

The more complex the multicellular, systemic functions are, the less resistant to heat they seem to be. Kivivuori (1980) found in her studies on the crayfish *Astacus astacus* in our laboratory, that its isolated nerve cord shows neuronal activity still at about 36°C, while simple reflexes like those maintaining the respiratory movements of scaphognathites stop at 32°C, walking at 30°C, and the still more complex righting reflex already at about 26°C. Several other studies show that simple reflexes are generally more heat resistant than the more complex (e.g. conditioned) reflexes. This seems to indicate, that the more synapses there are in the control pathway, the less resistant is the function.

The work of Precht (1960), Jensen (1972) and Grainger (1973) is relevant in this respect. They studied the heat and cold tolerance of the neuromuscular function in frogs, and found that the transmission between the motor nerve cells and the muscle cells, the endplate or synaptic function, was more sensitive to heat and cold than the functions of either nerve or muscle cells. The information transfer between the two types of cells was the most sensitive part of this system.

We may ask further what is the thermally most sensitive part of the neuromuscular transmission process? In the case of the neuromuscular junctions of the crayfish and the frog, the release of the neurotransmitter from the presynaptic cell is probably the weakest link (White, 1976, 1983). In other cases of intercellular communication the receptor mechanisms may be the most sensitive (e.g. the insulin receptor, see Calderwood, this volume).

In the study by White (1983) on the crayfish *Procambarus clarkii*, the temperatures for the neuromuscular block were within the range of the heat death temperature, when the duration of the experiments was taken into account. The resting membrane potential of muscle fibres increased in normal crayfish physiological solution at least up to 38°C, at which temperature the average times to heat death of whole animals were 15 to 40 min, depending on the previous acclimation temperature.

A different picture of the mechanisms of heat death in crayfish emerges from the studies of Bowler and his collaborators (Bowler, 1963*a*, 1963*b*, 1981; Gladwell, Bowler & Duncan, 1976; Gladwell, 1976) on *Austropotamobius pallipes*. The heat tolerance in this species agrees well in its temperature and time relations with the increase of potassium concentration in the haemolymph. This occurs when the permeability of membranes of muscle cells to potassium increases and intracellular potassium ions leak out from them. Accordingly, the muscle fibres are depolarized at 32°C after an initial small hyperpolarization. The increased concentration of potassium in the haemolymph upsets the normal function of the nervous system and thus causes heat death. The responsiveness of nerve-muscle preparations in a crayfish saline solution, adjusted to resemble the haemolymph of heat injured animals, ceases more rapidly at sublethal temperature than in the normal crayfish saline.

These analyses of the mechanisms of the heat death in crayfish are among the most thorough-going which have been done on any multicellular animal. The 'weakest link' in both of them is the plasma membrane of a cell, the synaptic membrane of a neurone or the membrane of a muscle fibre. There is an accumulating body of evidence which supports the conclusion that certain cell membranes are the primary targets of heat and cold injury. Plenty of such evidence is presented in this volume by Bowler, Chapman, Cossins, Ellory, Watson & Morris, Willis, and Yatvin.

Returning to the issue about the relations of different organization levels of living nature, an often discussed problem is: can the phenomena encountered on one level (e.g. the whole organism level) be explained as certain properties found on another level (e.g. the cell level)? Such cross-level explanations are called reductive explanations, and there is no doubt that many advances in biology depend on them. This view comes well out, for instance, in the recent book on adaptational biology by a pioneer of thermal physiology, C. Ladd Prosser (1986).

This does not mean, that the whole-animal or organ system studies should be abandoned in favour of studies on animal cells or be replaced by them. Rather, many 'cell level' studies on the mechanisms of heat death use not cells but multicellular function systems like neuromuscular preparations, and derive the idea of which cells should be studied from the knowledge about whole organisms and their organ systems. The interactions of different types of cells are often more important than their properties in isolation. Attempts to the reductive explanation of whole animal properties on cellular level are valuable but they have to be complemented by background data from studies concerning the higher organization levels.

Cells in isolation and culture: temperature limits

Certain cells or their interactions are the 'weakest links' of vitally important functions. The mechanisms of the heat and cold death of the other cells, which form the bulk of the body of a multicellular animal, may be different. Those cells are more heat and cold resistant than the 'weakest link' cells. In poikilotherms, they often tolerate temperatures 5 to 10°C above the lethal temperature of the animal. In homeotherms, the limit of heat tolerance of cultured cells is slightly above the lethal limit of hyperthermia in the animal, and the limit of cold tolerance is close to freezing (isolated rat cardiac myocytes tolerate about 0°C; Tirri et al. 1983).

The question, whether the resistance limits of isolated or cultured cells can be shifted by a change of temperature, has not received much attention. However, apart from the possibility of finding out the mechanisms of heat and cold death of cells, such studies could perhaps also answer the question, to what extent the temperature acclimation phenomena observed at the cell and the molecule levels are caused indirectly by changes in the superposed regulatory mechanisms (e.g.

neural, hormonal, and circulatory), and to what extent by the direct effects of temperature on all cells.

Apparently this problem was first approached by studies concerning the thermal adaptability of cell functions of isolated organs. The ciliary activity of isolated surviving gills of the fresh water mussel *Anodonta cygnea* was shown to gain a better heat tolerance, if the gills were kept for 1–3 days at 21–24 °C than if kept at 4–5 °C (Lagerspetz & Dubitscher, 1966). The shift in the heat resistance was similar as that found in the gills taken from whole animals acclimated to these temperatures (Senius & Lagerspetz, 1974). In this case, the heat resistance acclimation of a cellular function occurred also in an isolated organ, and not only in the whole animal.

This 'direct' acclimation effect has since been studied further. No such effect was found in a related species, *Anodonta anatina*, or in the marine mussel *Mytilus edulis* (Senius, 1975a, 1975b, 1977). Instead, in these two species, the isolation (including, of course, the denervation) of gills in itself caused an increase of the heat resistance of ciliary activity, which did not depend on the subsequent storage temperature of the surviving gills. On the other hand, the cholinergic antagonists atropine and eserine abolished the acclimation effect, when present in the acclimation medium of isolated gills of *Anodonta cygnea* (Senius & Lagerspetz, 1974). This and other evidence suggested the involvement of an activation of cholinergic receptors as a possible mechanism of the organ level acclimation in this case (Senius, 1978).

Bowler, Laudien & Laudien (1983) have shown that cultured cells of the fish *Pimephales promelas* (fat head minnow) tolerate 5 min treatments at temperatures of 41 to 44 °C better if previously cultured at 34 °C than if cultured at 18 °C. The poster presented by Merz and Laudien at this Symposium gives further evidence on this point.

Chinese hamster ovary (CHO) cells are more heat resistant if grown at higher temperatures (39 and 41 °C) than if cultured at 37 or 32 °C (Anderson, Minton, Li & Hahn, 1981). The membranes of the cells cultured at higher temperatures are also less fluid, as shown by the higher steady state fluorescence polarization values of the probe molecule of membrane core, 1,6-diphenyl-1,3,5-hexatriene (DPH). This indicates homeoviscous adaptation, which will be discussed later.

These results show that isolated and cultured cells of multicellular animals may gain heat resistance acclimation. There is also evidence about the occurrence of cold resistance acclimation of the ciliary activity in isolated surviving tentacles of the snail *Planorbis corneus* (Precht & Christophersen, 1965). Recently, it has been shown that the stability of microtubules in cold is increased by the culture of rainbow trout gonad cells at a low temperature (Tsugawa & Takahashi, in press).

Tsugawa (1976) has earlier observed a relative decrease of anodic isozymes of lactate dehydrogenase by cold acclimation in cultured kidney and liver cell lines of the South African clawed toad, *Xenopus laevis*, and found a similar phenomenon in the liver of toads acclimated *in toto* to cold. The anodic isozymes are also more cold-sensitive than the more cathodic isozymes (Tsugawa, 1980). These results point out, that the acclimation of cultured animal cells may have functional importance also at physiological temperatures.

Cells and cell membranes: temperature limits

Many functions essential for life occur in the watery solution inside the cell. The environment of the cells is also mainly water. A non-aqueous phase constituted mainly of phospholipids, the plasma membrane, delimits the cells. All substances affecting the organism must either pass through this barrier or act upon it. Changes in temperature affect directly the molecular functions inside the cell, but by changing the state of the membrane they cause changes also in the relations between the cell and its environment.

A bilayer constituted only of lipids would form a virtually impermeable barrier to electrolytes and hydrophilic non-electrolytes. Cell membranes also contain proteins, which shunt the aquatic phases outside and inside the cell. It is obvious that the proteins in the membrane are affected by the dynamic state of membrane lipids, by their packing and movements, which depend, among other things, on their thermal disorder. The membrane lipids may thus restrict more or less the changes in the conformations of membrane proteins.

Some average parameters of thermal disorder or 'fluidity' of the lipid matrix of cell membrane preparations can be measured e.g. with spectroscopic techniques using different probe molecules. 'The term "membrane fluidity" refers to the physical state of the fatty acyl chains comprising the membrane bilayer structure' (Stubbs & Smith, 1984) is one definition of it.

The membrane fluidity parameters are directly proportional to the temperature at which they are measured. However, a longer stay of the organisms or cells at a changed environmental temperature often produces a change in membrane fluidity towards the direction opposite to the immediate temperature effect. This has been called homeoviscous adaptation (Sinensky, 1974). The acute and the (sub)chronic effects of a temperature change are thus opposite.

Differences in membrane fluidity have been found in cell membranes of various animals, and they correlate well with their thermal environment and tolerance (Cossins & Prosser, 1978; Cossins, 1983). Thus, the fluidity of synaptic membranes of vertebrates, from a fish living at about $-1 \cdot 9\,°C$ to mammals and birds (having a body core temperature of 37 and 42 °C, respectively) shows evolutionary homeoviscous adaptation.

The fluidity of cell membranes is also affected by the temperature to which the organisms have been acclimated, heat acclimation usually making the membranes less fluid and the cells more resistant to high temperatures, and cold acclimation changing them in the opposite way. Most of this work has been done on prokaryotes and protozoans (Thompson, 1980), and on fish (Cossins, 1983, and this volume).

In the previous section, the heat death of crayfish was shown to be correlated with the breakdown of membrane functions of certain cells, and with the death of these cells. Those membrane functions, the release of neurotransmitter and the permeability to potassium, do not in the first place depend on the state of lipid matrix of membrane but rather on the proteins embedded in it: receptors, enzymes and passive transporter proteins. The inactivation of Mg^{2+}-ATPase of plasma membranes of muscle fibres of crayfish by heat correlates well with the leak of potassium to the haemolymph (Gladwell, 1976), and the inactivation of two other membrane-bound enzymes, Na^+-K^+-ATPase of plasma membrane and Ca^{2+}-ATPase of the sarcoplasmic reticulum occur at the temperatures of the heat death of the crayfish (Bowler, Gladwell & Duncan, 1973; Cossins & Bowler, 1976). This similarity between the inactivation temperatures of membrane-bound enzymes of crayfish muscles suggests the importance of the lipid-protein interactions in cell membranes for cellular heat resistance.

Synaptic membranes of goldfish brain are more fluid in fish acclimated to 7°C than in fish acclimated to 28°C. The activity of Na^+-K^+-ATPase is also less thermostable in synaptic membrane preparations from cold acclimated goldfish. A decrease in the thermostability of this enzyme is also observed when the synaptic membranes are under the influence of a membrane fluidizing substance, *n*-hexanol (Cossins, Bowler & Prosser, 1981). The termination of the activity of this membrane-bound enzyme by heat seems to depend on the fluidity of lipids in the membrane.

In the fresh water mussel *Anodonta cygnea*, when whole animals are subjected to different temperatures for a week or longer, the heat resistance of ciliary function in the epithelial cells of gills is higher in animals acclimated to 24°C than in those kept at 4°C (Lagerspetz & Dubitscher, 1966; Senius & Lagerspetz, 1974; Senius, 1978). The activities, K_m values and thermal stabilities of ATPases in membrane preparations from gills are not changed by temperature acclimation, but the fluidity of the membranes is lower in warm acclimated animals (Lagerspetz, 1985). *n*-Hexanol, which increases membrane fluidity also in these preparations, shortens the resistance time of ciliary activity in a concentration dependent manner (Fig. 1), and calcium and magnesium ions have the opposite effect on both variables at high temperatures (Lagerspetz, 1985). In this case, there is a correlation between the membrane order and the heat resistance of a directly measurable cellular function, ciliary activity.

There is apparently often a close relation between the fluidity of membranes of cells and their thermal tolerance. However, a parameter as important as membrane fluidity probably affects cell functions also within the physiological temperature range and not only at such extreme temperatures, which are never encountered by these cells in their natural environment. Homeoviscous adaptation of cell membranes is probably also physiologically important, although this aspect of it has not been much studied (Lagerspetz & Laine, 1986).

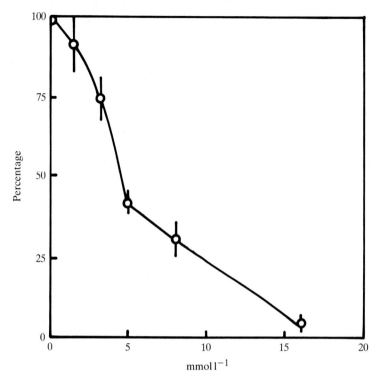

Fig. 1. The effect of a membrane fluidizing substance, *n*-hexanol on the heat resistance of the ciliary activity of frontal cilia in the gills of *Anodonta cygnea*. Heat resistance time is measured in minutes at 39·0°C, and expressed as percentage of the resistance time of cilia of gills in the medium without *n*-hexanol. Test animals have been acclimated to 24°C. The resistance time in animals acclimated to 4°C is equal to those subjected to $5 \, \text{mol} \, \text{l}^{-1}$ *n*-hexanol. Both treatments cause also a similar depolarization of DPH fluorescence. According to Lagerspetz, 1985.

Cells and cell membranes: functions at physiological temperatures

On the other hand, the interactions of membrane lipids and membrane proteins have been studied extensively during the last ten years. The techniques of preparing well characterized lipid bilayers and of incorporating purified membrane proteins with well known functions into them have much increased our knowledge of these interactions and of their functional significance.

The activity of Na^+-K^+-ATPase incorporated in phospholipid bilayers is positively correlated with the fluidity of them (Kimelberg & Papahadjopoulos, 1974; Sinensky, Pinkerton, Sutherland & Simon, 1979; Abeywardena, Allen & Charnock, 1983). The fluidity of bilayers does not affect the activity of Mg^{2+}-ATPase (Abeywardena, Allen & Charnock, 1983), although negative correlations between the fluidity and the activity of Mg^{2+}-ATPase have also been observed (Riordan, 1980). The turnover number of the passive sugar transporter glycoprotein of human erythrocytes is mainly affected by the headgroups and the hydrocarbon chain length of phospholipids in bilayers and not by the bulk fluidity (Carruthers & Melchior, 1986). In these studies, the bilayer fluidity was varied either by using different types of lipids or by adding cholesterol.

The acclimation to different temperatures can often be used to change the fluidity of cell membranes *in vivo*. The activity of Mg^{2+}-ATPase is higher in microsomes from the brain and from the epidermal cells of warm acclimated frogs than in similar preparations from the cold acclimated (Lagerspetz, 1977; Lagerspetz & Skyttä, 1979; Lagerspetz & Ellmén, 1984). If the enzyme is solubilized from the microsomal membranes of epidermal cells, the differences between the acclimation groups in the activity and in its temperature dependence disappear (Fig. 2). In this case, the differences in enzyme activity caused by temperature acclimation seem to depend on the lipid environment of the enzyme (Lagerspetz & Ellmén, 1984), the fluidity of which is affected by temperature acclimation (Lagerspetz & Laine, 1984).

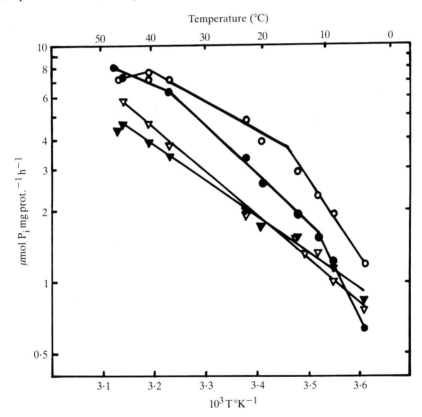

Fig. 2. The solubilization of Mg^{2+}-ATPase from microsomes of epidermal cells of the frog *Rana temporaria* abolishes the differences in enzyme activity and in its temperature dependence caused by the acclimation of animals to 6 °C (filled symbols) and to 23 °C (open symbols). Circles: enzyme activity of untreated microsomal preparations; triangles: enzyme activity in the supernatant after solubilization with Triton X-100 and washing. No activity was found in the sediment. SEM in 3 cases exceeds the double diameter of the symbol. According to Lagerspetz & Ellmén, 1984.

One of the important functions of cell membrane is to maintain the sodium and potassium ion gradients between the outside and the inside of the cell. The sodium pump system, identified with the activity of membrane-bound Na^+-K^+-ATPase, is implicated in this, as are the passive membrane permeabilities of these ions. The isolated frog skin is the classical model for the study of these ion movements.

According to the generally accepted theory, the whole transport process occurs in the epidermal cells. The entry of sodium ions to epidermal cells from the external medium across the outward directed (apical) cell membrane occurs by passive diffusion through sodium channels of the membrane. This is the step usually limiting the overall rate of sodium transport across the skin. From the epidermal cells, sodium is pumped out to the tissue fluid through the basal and basolateral cell membranes by the activity of Na^+-K^+-ATPase (Ussing, Erlij & Lassen, 1974; Rick, Doloff, Dorge, Beck & Thurau, 1984).

The net sodium transport rate across isolated frog skin can be measured as the short circuit current (SCC). SCC is generally directly proportional to the net rate of sodium ion transport, also at different temperatures. In isolated frog skins, SCC is temperature dependent. However, if a group of frogs (*Rana temporaria*) has been kept for 2–4 weeks at 4°C and another at 24°C, the average level of the SCC will be higher in those acclimated to the lower temperature. This can be observed over a wide range of measurement temperatures (Lagerspetz & Skyttä, 1979). Another observation made on frogs acclimated to 4°C and to 24°C is that membrane preparations made from the isolated epidermal cells are more fluid in the cold acclimated than in the warm acclimated group over a range of measurement temperatures from 5 to 45°C (Lagerspetz & Laine, 1984). There is thus a correlation between the SCC across the frog skin and the membrane fluidity of epidermal cells.

Additional correlations between these two variables have been sought for. First, *n*-hexanol increases the fluidity of microsomal preparations made from epidermal cells and also the SCC across the skin. Secondly, calcium decreases both these variables (Lagerspetz & Laine, 1984; Lagerspetz & Ellmén, 1984). In these experiments, the fluidity of cell membrane is in positive correlation with the sodium permeability of membrane. The homeoviscous acclimation of frog epidermal cell membrane seems then to be adaptive for the sodium transport function of the cell also at normal living temperatures, not only at its thermal limits.

More direct evidence on the relation of membrane fluidity and ion transport is given by Dr Raynard in a poster presented at this Symposium. His work shows that the sodium pump turnover rate is increased by cold acclimation in the red blood cells of rainbow trout, as is the fluidity of membranes of these cells.

If the sodium and potassium transport across cell membranes is generally affected by their fluidity, this phenomenon would probably be important in the function and temperature acclimation of nerve cells. In the crayfish *Astacus astacus* there occurs during temperature acclimation to 5°C and 20°C a compensatory shift in the temperature dependence of the conduction velocity and of the duration of the declining phase of the intracellularly recorded action potential of the medial giant axon in the nerve cord (Kivivuori & Lagerspetz, 1982). These effects are found at temperatures below about 20 °C, which is close to the ecological upper temperature limit of this species.

The clearest acclimation effect is seen between 5 and 20 °C in the maximal velocity of decline of action potential, which represents the maximum flow of potassium ions into the axon. The fluidity of cell membranes prepared from the nerve cord of *Astacus* is higher between 5 and 30 °C in crayfish acclimated to 5 °C than in those acclimated to 20 °C. Thus, the rate of change in potassium conductance across the axonal membrane and its fluidity seem to show positive correlation (Kivivuori, Laine & Lagerspetz, 1984, and unpublished observations).

These examples show that the thermal order of cell membranes is important for the temperature relations of functions of epidermal and nerve cells at physiological temperatures, and not only for the resistance of cells to high and low temperatures.

Cell membranes and the concept of membrane fluidity

The relations between membrane fluidity and ion permeability are not as simple as presented above. In artificial bilayers, the permeability is usually highest at the phase transition temperature (see e.g. Thompson, 1980). In addition, between molecules as complex as phospholipids and proteins, there are interactions which cannot be explained as depending only on the effects on the membrane lipid fluidity. Membrane proteins are in contact with the water phase, the headgroups and the backbones of membrane phospholipids, as well as with the non-polar hydrocarbon chains. The lipid environment of proteins embedded in it affects them, and the effects are complex and therefore variable. The effects often seem to depend on the immediate lipid environment of the proteins rather than on the bulk fluidity. The integral proteins may also order the membrane locally (e.g. Lee, this volume).

Membrane fluidity is decreased by alcohols and other 'membrane perturbants' (Harris & Hitzemann, 1981). The effects of these may differ depending on the biological system studied. Thus, for instance, in the synaptosomal membranes prepared from mouse brain, membrane fluidizing substances, like *n*-hexanol, close the voltage-activated sodium channels (Harris, 1984). In frog epidermal cells, *n*-hexanol seems to make the sodium channels of apical cell membrane more passable for sodium ions (Lagerspetz & Laine, 1984).

The acute effects of membrane perturbing substances are not always similar to those caused by the more chronic effects of the diet, temperature acclimation, and thermal shocks (Yatvin, Cree, Elson, Gipp, Tegmo & Vorpahl, 1982). There may be differences between an acute and a chronic fluidization of the target cell membrane.

The concentration of amiloride, which inhibits 50 % of the net sodium transport across the frog skin (measured as the SCC) does not differ significantly in cold and warm acclimated animals, but is $24 \pm 4\,\mu\text{mol}\,l^{-1}$ and $37 \pm 4\,\mu\text{mol}\,l^{-1}$, respectively. The mean values in frog skins under $1\,\text{mmol}\,l^{-1}$ n-hexanol and their controls are similar, 25 and $36\,\mu\text{mol}\,l^{-1}$, respectively, and do not differ significantly. But $5\,\text{mmol}\,l^{-1}$ n-hexanol, which decreases membrane fluidity only as much as cold acclimation, decreases the average 50 % inhibitory concentration of amiloride to $7\cdot4\,\mu\text{mol}\,l^{-1}$ (Lagerspetz & Laine, 1987). The acute fluidizing effect caused by n-hexanol and the chronic effect of cold acclimation are therefore not always similar in their actions on sodium channels.

Adrenalin is the major hormone of the adrenal medulla also in frogs, and its concentration in the adrenals is increased during cold acclimation (Harri, 1972), and several effects of cold acclimation can be mimicked by repeated injections of adrenalin to frogs (Lagerspetz, Harri & Okslahti, 1974). The sensitivity of sodium transport across frog skin to adrenalin is not different in frogs acclimated to 6°C and 23°C. However, the maximum increase of SCC caused by adrenalin is significantly higher in skins from cold-acclimated animals, and also in skins under the fluidizing influence of $2\,\text{mmol}\,l^{-1}$ n-hexanol (Lagerspetz, unpublished).

Amiloride acts directly on the sodium channels of the apical membrane of epithelial cells. The natriferic effect of low concentrations of adrenalin is mediated by beta-adrenergic receptors (e.g. Lindemann & Voûte, 1976), with a subsequent activation of adenylate cyclase residing in the membrane. Cyclic AMP, either itself or through further mediators then increases the passive sodium permeability of the apical cell membrane by acting on the sodium channels. Although the final target of these actions of amiloride and adrenalin is the same, the direct effect of the latter is not on the channel protein, but on a receptor protein of the same cell membrane.

As cell membranes are composed of lipids with different phase transition temperatures, all lipids of a cell membrane may not be in the same physical state, liquid-crystalline or gel. The lipid molecules situated at the boundaries between two physically distinct microdomains may be especially mobile due to packing faults, which may increase the transport capability or enhance the enzyme activity (Thompson, 1980, p. 10; Watson & Morris, this volume).

The cooling of a membrane usually leads to lateral migration of the lipids readily entering the gel state to form more rigid assemblages in the liquid-crystalline environment. This phenomenon is called phase separation (e.g. Thompson, 1980). The different temperature-time relations of the phase separation and of its reversal may cause hysteresis phenomena in membrane functions, such as observed in the time constants of sodium currents of frog muscle membranes at cooling and rewarming (Schwarz, 1979), and in the membrane potential of stretch receptor neurone of crayfish (Moser, Ottoson & Rydqvist, 1979; see Fig. 3). There is some evidence about the modifiability of the form of hysteresis loop (the reversal of it) in the stretch receptor neurones by temperature acclimation (Lehti & Kivivuori, 1986, and unpublished results).

Apparently, the workings of cells cannot be explained only by what we know about isolated molecules, lipids or proteins, but knowledge about their behaviour

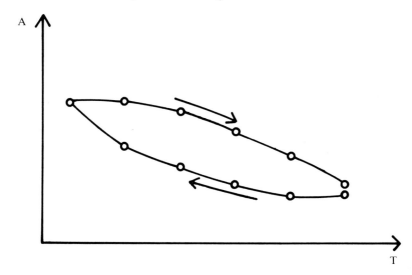

Fig. 3. A scheme exemplifying a thermal hysteresis loop. At cooling, the parameter studied (A) changes with temperature (T) along the lower curve, at rewarming along the upper curve.

in the combinations occurring in cell membranes are indispensable. Intracellular membranes delimit organelles, which are in their relative importance and physical dimensions well comparable to most organs of a multicellular animal. The plasma membrane of the cell can also be regarded as an organelle with several important functions.

Thermal disorder is a physical concept on the molecular level. It can be more easily applied to artificial lipid bilayers than to more complex fluid-mosaic membranes or to the even more complex and dynamic native cell membranes. The concept of membrane fluidity has been criticized as not taking into account the all possible changes in molecules caused by thermal disorder. However, membrane fluidity is a concept which can be defined only operationally, by referring to the methods of its measurement, as ESR, NMR, and fluorescence polarization, and the different probe molecules used.

Membrane fluidity is perhaps not as useful concept in bilayer studies as, e.g. the lipid packing density (Carruthers & Melchior, 1986). It is not related in a simple manner to the unsaturation degree of the fatty acids of lipids of a bilayer or of many membrane preparations (see Stubbs & Smith, 1984), and can therefore not be estimated on the basis of chemical analyses of the concentrations of different molecules in the membrane. However, membrane fluidity has proved to be a useful physical measure on the organelle (cell membrane) level.

Attempts to explain the temperature relationships of animals by references to their cells, and of animal cells by the properties of certain molecules, must mostly depend on the correlation approach. This may be problematic, since we do not usually know which parameters are the biologically most important and of what type the correlations would be. A 50 % inhibition of the activity of an enzyme may not be biologically more important than a 10 % or a 90 % inhibition of it. The relationships between membrane fluidity and enzyme activities or ion permeabilities are not necessarily linear. A change of the relative fluidity of a bilayer often corresponds to a 10–20-fold change in the functions (Hanski, Rimon & Levitzki, 1979; Pang, Chang & Miller, 1979; Harris & Schroeder, 1981).

Cell proteins and temperature

The pervasiveness of temperature as a factor influencing cells means that it has also effects not mediated by the cell membranes. As the activation energies of different enzyme catalysed reactions vary, a change in cell temperature may change not only the reaction rates, but affect also the balances between different reactions and reaction products. These may, in turn, affect reactions through such well-known biochemical mechanisms as end-product inhibition and enzyme induction and repression. As pointed out by Calderwood and Willis in this volume, the balances between reaction rates and between concentrations of ions and molecules are often more important than their absolute values.

This comes out also from the work on the effects of temperature acclimation on activities of enzymes in ectothermic animals. Hazel & Prosser (1974) enumerated about 60 enzymes, 3/4 of which show partial temperature compensation in their maximum activity in some preparations from poikilothermic animals after acclimation, while most of the remaining quarter of them show only the inverse effect, the activity being lower in cold-acclimated than in warm-acclimated animals. Many of the enzymes in this latter group are associated with the breakdown of metabolic products, and since the temperature compensation of metabolism is seldom perfect, more of such products would accumulate in animals living in a warm than in a cold environment. The increase of the activity of such enzymes in warm-acclimated animals would thus be adaptive (Hazel & Prosser, 1974).

This means, that changes in enzyme activities during acclimation are probably not independent from each other, but rather represent the attainment of a new steady state by the cell metabolism. Experiments, in which a temperature change has produced a series of oscillations of enzyme activities (e.g. Künnemann, Laudien & Precht, 1970) perhaps also speak for this.

In many types of multicellular animals and their cells in culture, brief heat shocks induce the production of certain types of proteins, and often suppress the synthesis of some others. Although the favoured proteins are called heat shock proteins (HSP), an increased synthesis of them is also induced by other agents causing structural damage to protein molecules. Alcohols, heavy metal ions, amino acid analogues, and the starvation, anoxia, and acidosis of the animal in question are some of these (Laudien, 1986; Burdon, this volume).

There is evidence about the enhancing effects of HSP on subsequent heat tolerance of cells, and they have therefore been considered as possible mediators of heat hardening or thermotolerance (Laudien, 1986), but the effects of heat shocks differ from those of warm acclimation in their time relations, and as changes in the cell membrane fluidity are not involved (Konings & Ruifrok, 1985).

After a heat shock, the pattern of protein synthesis in cells is changed, in some cases in a way resembling the patterns found during embryonic development (Burdon, this volume). HSP may have a role in the repair of injured organelles and in the degradation of damaged proteins.

Although the motor functions of cells are easily observable, very little work has been done on the effects of temperature on them. The motor functions of cells depend on the activity of contractile proteins. Some attention has been paid on the activity of cell cilia (see references above, and Lloyd & Kippert, in this volume), but such topics as, e.g. flagellar motion (important in spermatozoans), movements inside the muscle fibres, amoeboid or pseudopodial motion (e.g. in white blood cells and in cells of early embryos), phago- and pinocytosis, and cytoplasmic flow (e.g. in axons of nerve cells) have generally been neglected in the study of thermal biology of animal cells.

In spite of the importance of cytoskeleton as an organelle, the studies on the effects of temperature changes on it are also few (Coakley, this volume; Tsugawa & Takahashi, in press).

Cell survival in cold

Several contributions to this volume deal with the survival of animal cells and organs at low temperatures. The subject is of great practical importance, but it will be considered only briefly in this overview.

The separation of organs, tissues or cells from their natural supply of oxygen and nutrients and from the transport of metabolic end products makes them liable to ischemic damage. This can be reduced to some extent by lowering the temperature, which decreases the metabolism. During a prolonged storage at subnormal temperatures, phase separation may occur in cell membranes, they may lose their selective permeability, and the membrane ion pump mechanisms will be slowed down. This causes a change in the ion balances of cells. Such effects can be diminished by appropriate changes in the ion concentrations of the medium.

As stated at the beginning of this article, the problem of the survival of animal life at the lowest temperatures found in the universe has been solved several times during the course of the evolution, and in each case through the development of tolerance to extreme dehydration. No animal cell can withstand the formation of ice crystals within it (Franks, 1985). Intracellular ice crystals break the membrane structures of the cell. This has deleterious consequences on the molecular level.

The freezing point of water solutions is lowered by the increase of their osmotic concentration. The increase in the concentrations of, e.g. amino acids and glycerol is a common phenomenon in animals and plants at the onset of wintering. However, only relatively moderate decreases in the freezing point are attained at the concentrations tolerable to the cells.

One possibility to avoid intracellular freezing is to dehydrate the cells by an increase of the osmotic concentration of the extracellular medium. Some dehydration of cells occurs when extracellular ice is formed, as the osmotic concentration of the medium increases. This can be enhanced by the addition of relatively inert solute molecules (e.g. glycerol) to the medium of the cells, or by the natural occurrence of such molecules in it. Another type of substances may set on the ice crystal lattice and prevent it from growing. In addition to the cryoprotectant substances, also the rate of temperature change is important for the survival of cells in cold.

If the cooling occurs very rapidly, ice crystals are not formed, but the cell water is vitrified to an amorphous solid. The growth of ice crystals is much retarded at temperatures below $-40\,^{\circ}$C. The rewarming of the vitrified, brittle material must also be as fast as not to allow the crystal formation. Such rapid and uniform changes of temperature can be produced only in small volumes of cells or tissue.

Although the formation of ice crystals, whether inside or outside the cell membrane is a phenomenon on the molecule level, its deleterious effects on cells are not caused on effects on any particular molecules, but on the rupture of membranes limiting the cell and its organelles, with manifold secondary consequences on the molecular level. Crystallization of water sounds as a straightly reductive, molecular level explanation of the death of an animal cell in cold. But such explanation could never have been given without preceding extensive morphological and physiological cell level studies.

We may, as a mental experiment, think about the situation which prevails in such conditions of the universe, in which there is hydrogen and oxygen but no water. Probably the important properties of the combination of these elements could not be deduced from such properties of hydrogen and oxygen which would be known without making water. There is as little mystery in this notion as there is, e.g. in that of $0\,^{\circ}$K – it shows something about the structure of the universe.

References

ABEYWARDENA, M. Y., ALLEN, T. M. & CHARNOCK, J. S. (1983). Lipid-protein interactions of reconstituted membrane-associated adenosinetriphosphatases. *Biochim. biophys. Acta* **729**, 62–74.

ANDERSON, R. L., MINTON, K. W., LI, G. C. & HAHN, G. M. (1981). Temperature-induced homeoviscous adaptation of Chinese hamster ovary cell. *Biochim. biophys. Acta* **641**, 334–348.

BAROSS, J. A. & DEMING, J. W. (1983). Growth of 'black smoker' bacteria at temperatures of at least 250°C. *Nature, Lond.* **303**, 423–426.

BAROSS, J. A., LILLEY, M. D. & GORDON, L. I. (1982). Is the CH_4, H_2 and CO venting from submarine hydrothermal systems produced by thermophilic bacteria? *Nature, Lond.* **298**, 366–368.

BECQUEREL, P. (1951). La suspension de la vie au confins du zero absolu entre 0·0075°K et 0·047°K. Rôle de la synérèse reversible cytonucléoplasmique. *Proc. 8th int. Congr. Refrig.*, p. 326.

BOWLER, K. (1963*a*). A study of the factors involved in acclimatization to temperature and death at high temperatures in *Astacus pallipes*. I. Experiments on intact animals. *J. cell. comp. Physiol.* **62**, 119–132.

BOWLER, K. (1963*b*). A study of the factors involved in acclimatization to temperature and death at high temperatures in *Astacus pallipes*. II. Experiments at the tissue level. *J. cell. comp. Physiol.* **62**, 133–146.

BOWLER, K. (1981). Heat death and cellular heat injury. *J. therm. Biol.* **6**, 171–178.

BOWLER, K., GLADWELL, R. T. & DUNCAN, C. J. (1973). Acclimatization to temperature and death at high temperatures in the crayfish *Austropotamobius pallipes*. In *Freshwater Crayfish* (ed. S. Abrahamsson), pp. 121–131. Lund: Studentlitteratur.

BOWLER, K., LAUDIEN, H. & LAUDIEN, I. (1983). Cellular heat injury. *J. therm. Biol.* **8**, 426–430.

CAPART, A. (1951). *Thermobathynella adami* gen. et sp. nov., Anaspidacé du Congo Belge. *Bull. Inst. r. Sci. nat. Belg.* **27**:10, 1–4.

CARRUTHERS, A. & MELCHIOR, D. L. (1986). How bilayer lipids affect membrane protein activity. *Trends in biochem. Sci.* **11**, 331–335.

COSSINS, A. R. (1983). The adaptation of membrane structure and function to changes in temperature. In *Cellular Acclimatisation to Environmental Change*. Soc. Exp. Biol. Seminar Series 17 (ed. A. R. Cossins & P. Sheterline), pp. 3–32. Cambridge: Cambridge University Press.

COSSINS, A. R. & BOWLER, K. (1976). Resistance adaptation of the freshwater crayfish and thermal inactivation of membrane-bound enzymes. *J. comp. Physiol.* **111**, 15–24.

COSSINS, A. R., BOWLER, K. & PROSSER, C. L. (1981). Homeoviscous adaptation and its effect upon membrane-bound proteins. *J. therm. Biol.* **6**, 183–187.

COSSINS, A. R. & PROSSER, C. L. (1978). Evolutionary adaptation of membranes to temperature. *Proc. natn. Acad. Sci. U.S.A.* **75**, 2040–2043.

FRANKS, F. (1985). *Biophysics and Biochemistry at Low Temperatures.* x+210pp. Cambridge: Cambridge University Press.

GLADWELL, R. T. (1976). Heat death in the crayfish *Austropotamobius pallipes*: thermal inactivation of muscle membrane-bound ATPases in warm and cold adapted animals. *J. therm. Biol.* **1**, 95–100.

GLADWELL, R. T., BOWLER, K. & DUNCAN, C. J. (1976). Heat death in the crayfish *Austropotamobius pallipes* – ion movements and their effects on excitable tissues during heat death. *J. therm. Biol.* **1**, 79–94.

GRAINGER, J. N. R. (1973). A study of heat death in nerve-muscle preparations of *Rana temporaria* L. *Proc. R. Ir. Acad.* **B 73**, 283–290.

HANSKI, E., RIMON, G. & LEVITZKI, A. (1979). Adenylate cyclase activation by the β-adrenergic receptors as a diffusion-controlled process. *Biochemistry* **18**, 846–853.

HARRI, M. N. E. (1972). Effect of season and temperature acclimation on the tissue catecholamine level and utilization in the frog, *Rana temporaria*. *Comp. gen. Pharmac.* **3**, 101–112.

HARRIS, R. A. (1984). Differential effects of membrane perturbants on voltage-activated sodium and calcium channels and calcium-dependent potassium channels. *Biophys. J.* **45**, 132–134.

HARRIS, R. A. & HITZEMANN, R. J. (1981). Membrane fluidity and alcohol actions. *Currents in Alcoholism* **8**, 379–404.

HARRIS, R. A. & SCHROEDER, F. (1981). Ethanol and the physical properties of brain membranes. Fluorescence studies. *Molec. Pharmac.* **20**, 128–137.

HAZEL, J. R. & PROSSER, C. L. (1974). Molecular mechanisms of temperature compensation in poikilotherms. *Physiol. Rev.* **54**, 620–676.

HINTON, H. E. (1951). A new chironomid from Africa, the larva of which can be dehydrated without injury. *Proc. zool. Soc. Lond.* **121**, 371–380.

HINTON, H. E. (1954). Resistance of the dry eggs of *Artemia salina* L. to high temperatures. *Ann. Mag. nat. Hist.* **7**, 158–160.

HINTON, H. E. (1960). A fly larva that tolerates dehydration and temperatures from $-270\,°C$ to $+102\,°C$. *Nature, Lond.* **188**, 336–337.

HINTON, H. E. (1968). Reversible suspension of metabolism and the origin of life. *Proc. R. Soc Lond.* **B 171**, 43–57.

JENSEN, D. W. (1972). The effect of temperature on transmission at the neuromuscular junction of the sartorius muscle of *Rana pipiens*. *Comp. Biochem. Physiol.* **41**A, 685–695.

KIMELBERG, H. K. & PAPAHADJOPOULOS, D. (1974). Effects of phospholipid acyl chain fluidity, phase transitions, and cholesterol on (Na^++K^+)-stimulated adenosine triphosphatase. *J. biol. Chem.* **249**, 1071–1080.

KIVIVUORI, L. (1980). Effects of temperature and temperature acclimation on the motor and neural functions in the crayfish *Astacus astacus* L. *Comp. Biochem. Physiol.* **65A**, 297–304.

KIVIVUORI, L. & LAGERSPETZ, K. Y. H. (1982). Temperature acclimation of axonal functions in the crayfish *Astacus astacus* L. *J. therm. Biol.* **7**, 221–225.

KIVIVUORI, L., LAINE, A. & LAGERSPETZ, K. Y. H. (1984). Effect of acclimation temperature on the axonal functions and on the fluidity of neuronal membranes in the crayfish (*Astacus astacus* L.). *First Congr. Comp. Physiol. Biochem., Liege*, p. C 35.

KONINGS, A. W. T. & RUIFROK, A. C. C. (1985). Role of membrane lipids and membrane fluidity in thermosensitivity and thermotolerance of mammalian cells. *Radiat. Res.* **102**, 86–98.

KÜNNEMANN, H., LAUDIEN, H. & PRECHT, H. (1970). Der Einfluss von Temperaturanderungen auf Enzyme der Fischmuskulatur. Versuche mit Goldorfen *Idus idus*. *Mar. Biol.* **7**, 71–81.

LAGERSPETZ, K. Y. H. (1977). Effect of temperature acclimation on the microsomal ATPases of the frog brain. *J. therm. Biol.* **2**, 27–30.

LAGERSPETZ, K. Y. H. (1985). Membrane order and ATPase activity as correlates of thermal resistance acclimation of ciliary activity in the gills of *Anodonta*. *J. therm. Biol.* **10**, 21–28.

LAGERSPETZ, K. Y. H. & DUBITSCHER, I. (1966). Temperature acclimation of the ciliary activity in the gills of *Anodonta*. *Comp. Biochem. Physiol.* **17**, 665–671.

LAGERSPETZ, K. Y. H. & ELLMÉN, J. (1984). Epidermal ATPases and thermal acclimation of Na$^+$ transport in frog skin. *Molec. Physiol.* **6**, 201–209.

LAGERSPETZ, K. Y. H., HARRI, M. N. E. & OKSLAHTI, R. (1974). The role of the thyroid in the temperature acclimation of the oxidative metabolism in the frog *Rana temporaria*. *Gen. comp. Endocrin.* **22**, 169–176.

LAGERSPETZ, K. Y. H. & LAINE, A. (1984). Fluidity of epidermal cell membranes and thermal acclimation of Na$^+$ transport in frog skin. *Molec. Physiol.* **6**, 211–219.

LAGERSPETZ, K. Y. H. & LAINE, A. (1986). The functional significance of homeoviscous adaptation of cell membranes in multicellular animals. In *Temperature Relations in Animals and Man*. Biona-Report 4 (ed. H. Laudien), pp. 101–108. Stuttgart: Gustav Fischer.

LAGERSPETZ, K. Y. H. & LAINE, A. M. (1987). Changes in cell membrane fluidity affect the sodium transport across frog skin and its sensitivity to amiloride. *Comp. Biochem. Physiol.* (in press).

LAGERSPETZ, K. Y. H. & SKYTTÄ, M. (1979). Temperature compensation of sodium transport and ATPase activity in frog skin. *Acta physiol. Scand.* **106**, 151–158.

LAUDIEN, H. (1986). Heat-hardening in animals. In *Temperature Relations in Animals and Man*. Biona-Report 4 (ed. H. Laudien), pp. 147–153. Stuttgart: Gustav Fischer.

LEHTI, S. & KIVIVUORI, L. (1986). Temperature acclimation affects the thermal hysteresis of resting potential of stretch receptor neuron in the crayfish (*Astacus astacus* L.). *8th Conf. Eur. Soc. Comp. Physiol. Biochem., Strasbourg*, p. 113.

LINDEMANN, B. & VOÛTE, C. (1976). Structure and function of epidermis. In *Frog Neurobiology* (ed. R. Llinas & W. Precht), pp. 169–210. Berlin: Springer.

MOSER, H., OTTOSON, D. & RYDQVIST, B. (1979). Step-like shifts in membrane potential in the stretch receptor neuron of the crayfish (*Astacus fluviatilis*) at high temperatures. *J. comp. Physiol.* **133**, 257–265.

PANG, K.-Y., CHANG, T.-L, & MILLER, K. W. (1979). On the coupling between anesthetic induced membrane fluidization and cation permeability in lipid vesicles. *Molec. Pharmac.* **15**, 729–738.

PRECHT, H. (1960). Uber die Resistenzadaptation gegenuber extremen Temperaturen bei einigen Organfunktionen des Grasfrosches (*Rana temporaria* L.). *Z. wiss. Zool.* **164**, 336–353.

PRECHT, H. & CHRISTOPHERSEN, J. (1965). Temperaturadaptation des Cilienepithels isolierter Kiemen und Fühlerspitzen von Mollusken. *Z. wiss. Zool.* **171**, 197–209.

PROSSER, C. L. (1986). *Adaptational Biology. Molecules to Organisms*. xii + 784 pp. New York: John Wiley & Sons.

RICK, R., DOLOFF, C., DÖRGE, A., BECK, F. X. & THURAU, K. (1984). Intracellular electrolyte concentrations in the frog skin epithelium: effect of vasopressin and dependence on the Na concentration in the bathing media. *J. Membrane Biol.* **78**, 129–145.

RIORDAN, J. R. (1980). Plasma membrane Mg^{2+} ATPase activity is inversely related to lipid fluidity. In *Membrane Fluidity* (ed. M. Kates & A. Kuksis), pp. 119–129. Clifton, New Jersey: The Humana Press.

SCHMIDT, J., LAUDIEN, H. & BOWLER, K. (1984). Acute adjustments to high temperature in FHM-cells from *Pimephales promelas* (Pisces, Cyprinidae). *Comp. Biochem. Physiol.* **78A**, 823–828.

SCHWARZ, W. (1979). Temperature experiments on nerve and muscle membranes of frogs. Indications for a phase transition. *Pflügers Arch. ges. Physiol.* **382**, 27–34.

SENIUS, K. E. O. (1975a). The thermal resistance and thermal resistance acclimation of ciliary activity in the *Mytilus* gills. *Comp. Biochem. Physiol.* **51A**, 957–961.

SENIUS, K. E. O. (1975b). Thermal resistance acclimation of ciliary activity inn the gills of *Anodonta anatina* and *Anodonta cygnea*. *Comp. Biochem. Physiol.* **51C**, 157–160.

SENIUS, K. E. O. (1977). Thermal resistance of the ciliary activity in the gills of the fresh water mussel *Anodonta anatina*. *J. therm. Biol.* **2**, 233–238.

SENIUS, K. E. O. (1978). Control of the thermal resistance of ciliary activity in bivalve gills. *Rep. Dept Zool. Univ. Turku* **7**, 1–35.

SENIUS, K. E. O. & LAGERSPETZ, K. Y. H. (1974). The role of cholinergic receptors in the thermal resistance and thermal resistance acclimation of the ciliary activity in the *Anodonta* gills. *Comp. gen. Pharmac.* **5**, 169–179.

SINENSKY, M. (1974). Homeoviscous adaptation – a homeostatic process that regulates the viscosity of membrane lipids in *Escherichia coli*. *Proc. natn. Acad. Sci. U.S.A.* **71**, 522–525.

SINENSKY, M., PINKERTON, F., SUTHERLAND, E. & SIMON, F. (1979). Rate limitation of $Na^{+}+K^{+}$-stimulated adenosinetriphosphatase by membrane acyl chain ordering. *Proc. natn. Acad. Sci. U.S.A.* **76**, 4893–4897.

STUBBS, C. D. & SMITH, A. D. (1984). The modification of mammalian membrane polyunsaturated fatty acid composition in relation to membrane fluidity and function. *Biochim. biophys. Acta* **779**, 89–137.

THOMPSON, G. A., JR (1980). *The Regulation of Membrane Lipid Metabolism*. 218pp. Boca Raton, Florida: CRC Press.

TIRRI, R., LEHTO, H. & KARTTUNEN, P. (1983). The ability of isolated myocardial cells of the rat to contract at temperatures near freezing point. *J. therm. Biol.* **8**, 337–342.

TSUGAWA, K. (1976). Direct adaptation of cells to temperature: similar changes of LDH isozyme patterns by *in vitro* and *in situ* adaptations in *Xenopus laevis*. *Comp. Biochem. Physiol.* **55B**, 259–263.

TSUGAWA, K. (1980). Thermal dependence in kinetic properties of lactate dehydrogenase from the African clawed toad, *Xenopus laevis*. *Comp. Biochem. Physiol.* **66B**, 459–466.

TSUGAWA, K. & TAKAHASHI, K. P. (1987). Direct adaptation to temperature: cold-stable microtubule in rainbow trout cells cultured *in vitro* at low temperature. *Comp. Biochem. Physiol.* (in press).

USHAKOV, B. (1964). Thermostability of cells and proteins of poikilotherms and its significance in speciation. *Physiol. Rev.* **44**, 518–560.

USHAKOV, B. P. (1967). Coupled evolutionary changes in protein thermostability. In *Molecular Mechanisms of Temperature Adaptation* (ed. C. L. Prosser), pp. 107–129. Am. Ass. Adv. Sci., Washington, D.C.

USSING, H. H., ERLIJ, D. & LASSEN, U. (1974). Transport pathways in biological membranes. *A. Rev. Physiol.* **36**, 17–49.

WHITAKER, D. M. (1940). The tolerance of *Artemia* cysts for cold and high vacuum. *J. exp. Zool.* **83**, 391–399.

WHITE, R. (1976). Effects of high temperature and low calcium on neuromuscular transmission in frog. *J. therm. Biol.* **1**, 227–232.

WHITE, R. L. (1983). Effects of acute temperature change and acclimation temperature on neuromuscular function and lethality in crayfish. *Physiol. Zool.* **56**, 174–194.

YATVIN, M. B. CREE, T. C., ELSON, C. E., GIPP, J. J., TEGMO, I.-M. & VORPAHL, J. W. (1982). Probing the relationship of membrane "fluidity" to heat killing of cells. *Radiat. Res.* **89**, 644–646.

Printed in Great Britain © *Society for Experimental Biology 1987*   451

POSTER COMMUNICATION

INCREASED HEAT TOLERANCE IN FHM-CELLS: HEAT-SHOCK *VERSUS* ELEVATED CULTURING TEMPERATURE

REINHARD MERZ and HELMUT LAUDIEN

Department of Zoology, CAU, Olshausenstr. 40, 2300 Kiel, Federal Republic of Germany

Heat-hardening, caused by a short exposure to sub-lethal temperatures, results in a transient increase in heat resistance and is closely associated with the synthesis of heat-shock proteins (hsp). Heat resistance acclimation develops following a rise in environmental temperature, within viable limits, over a relatively long period of time and endures as long as the elevated temperature persists.

In FHM-cells, a cell line derived from the fat-head minnow *Pimephales promelas*, both phenomena have been described. The purpose of this work was to determine whether heat-hardening and resistance acclimation exhibit similar characteristics or can be distinguished as different phenomena.

We used standardized resistance assays to compare the occurrence of heat tolerance after hardening shock (35·5°C, 10 min) and a shift of culturing temperature (CT) from 16°C to 32°C. Therefore batches of cells with different thermal histories were divided after resistance treatment with one half plate out at 16°C, the other one at 32°C.

A hardening shock applied 60 min before the resistance treatment increases the rate of attachment (experimental temperature (ET) 16°C) from 23% to 53% (hardening effect). Heat-resistance acclimation becomes obvious, if 32°C is used as ET. (CT = ET: 16°C: 23% attachment; 32°C: 88% attachment after resistance treatment.) After shifting the culturing temperature from 16°C to 32°C both effects are detectable. The hardening effect occurs 1 h after shift, reaches its maximum after 6 h and decreases afterwards, although the inducing temperature persists. So the hardening effect is specific to a temperature shift, not to the elevated temperature *per se*. Opposed to this, heat resistance acclimation is not detectable earlier than 6 h after temperature shift and is fully developed after 7 days.

We investigated protein patterns of FHM-cells made thermotolerant by short- or longterm exposure to elevated temperatures using 2D-electrophoresis. Immediately after hardening shock several hsp were detectable: hsp84; hsp70; hsp39; hsp36 and hsp22.

One hour after shifting the culturing temperature, the same heat-shock proteins were synthesized as after heat shock. Six hours after CT-shift hsp-synthesis returned to the level of control cells, only hsp39 remained at an elevated level.

These data exhibit good correlation between the cellular concentration of hsp and the hardening effect, because heat-shock proteins in FHM-cells have an average half-time of 8h. There is no doubt that hsps play no role in heat resistance acclimation.

We also compared protein patterns of cells from different CT. Two proteins (P.I. 6·2, 50 kD and 52 kD) show greatly increased concentrations in cells grown at 16°C, 3 proteins (P.I. 5·8, 14 kD; P.I. 7·2 and 7·5, 50 kD) in cells grown at 32°C. At the moment it is not possible to draw any conclusion about whether one or some of these proteins play a role in heat-resistance acclimation or not.

Printed in Great Britain © *Society for Experimental Biology 1987*

POSTER COMMUNICATION

TEMPERATURE ACCLIMATION OF CARP INTESTINAL MORPHOLOGY

J. A. C. LEE

Department of Zoology, University of Liverpool, Liverpool L69 3BX, UK

Changes in cellular performance during acclimation to temperature are generally directed to offset or compensate for the direct effects of the temperature change. Most emphasis has been placed on studies at the cellular level of organization, such as modifications in enzyme performance (Hochachka & Somero, 1984) and 'homeoviscous adaptation' of membranes (Cossins & Sinensky, 1984). However, it has recently become clear that the adjustment of cellular and subcellular morphology may be a particularly important part of the overall adaptive response.

The effects of temperature acclimation upon the morphology of carp intestinal mucosa has been studied using morphometric techniques. Carp intestine showed an absence of anatomical regionalization with a gradient of villus dimensions along the tract which was more developed proximally. The decrease in villi dimensions was greatest in the anterior half. Temperature acclimation had no effect on intestinal-somatic indices. Total villus numbers remain unchanged by thermal acclimation though there were large changes in villi dimensions. Thus cold acclimation produced significant increases in mean villus height and breadth along the entire intestine. These villus shape changes resulted in a 58 % increase in total mucosal surface area and a 102 % increase in total villi volume in cold-acclimated fish relative to warm-acclimated fish. Surface area of the unmodified intestinal tube increased with cold acclimation by 28 %.

Studies of transmucosal transport rates in thermally-acclimated fish requires the measurement of a nominal surface area for the mucosa. These measurements usually do not take account of the possibility of differentially developed mucosal surfaces in the differently acclimated animals. Thus experiments, using unstripped preparations will overestimate the differences in area-specific transport capacity by up to 50 %. Estimates of surface area using stripped preparations probably do not suffer from this problem (Gibson *et al.* 1985), but even then it is necessary to take acount of the sampling position of tissue in the anterior–posterior gradient.

References

Hochachka, P. W. & Somero, G. N. (1984). *Biochemical Adaptations*. Princeton, New Jersey: Princeton University Press. Pp. 355–449.
Cossins, A. R. & Sinensky, M. (1984). In *Physiology of Membrane Fluidity*, vol. II (ed. M. Shinitzky). Boca Raton, Florida: CRC Press Inc.
Gibson, J. S., Ellory, J. C. & Cossins, A. R. (1985). *J. exp. Biol.* **114**, 355–364.

POSTER COMMUNICATION

THERMAL COMPENSATION OF THE Na$^+$ PUMP OF RAINBOW TROUT (*SALMO GAIRDNERI*) ERYTHROCYTES

ROBERT S. RAYNARD

Department of Zoology, University of Liverpool, Liverpool, L69 3BX, UK

Many poikilotherms show adaptive responses at the cellular level which compensate for the disturbance of membrane fluidity caused by a change in temperature. This cellular response is termed 'homeoviscous adaptation' and plays a crucial role in temperature acclimation. There is good evidence which indicates an important effect of membrane fluidity upon a variety of membrane functions such as permeability, active transport and the activity of membrane-bound enzymes. Thus it seems reasonable to expect homeoviscous adaptation would lead to compensatory responses in these various functional processes.

This study examines the effects of homeoviscous adaptation upon a cellular transport function, namely the Na$^+$ pump of the trout erythrocyte cell membrane. The approach consisted of measuring the maximal transport capacity and number of ouabain-binding sites in order to compare the absolute pump turnover rates of erythrocytes from cold (6°C) and warm (21°C) acclimated fish. The transport capacity of the Na$^+$ pump was determined by measuring the ouabain-sensitive influx of ^{42}K in erythrocytes using the method of Bourne & Cossins (1984). In order to measure maximal K$^+$ influx both extracellular K$^+$ concentration ($[K]_0$) and intracellular Na$^+$ concentration ($[Na]_i$) had to be at saturating concentrations. The Kms were determined as 1·5 mM for extracellular K$^+$ and 15 mM for intracellular Na$^+$. The ionophoric antibiotic, nystatin, was used to elevate $[Na]_i$ to approximately 80 mmol l^{-1} and $[K]_0$ was set at 6 mM, thus saturating concentrations for both were provided.

Measured at 10°C, the ouabain-sensitive influx of K$^+$ was 1·81 ± 0·08 (mean ± S.E.) mmol h^{-1}.10^{-6} cells in cells from warm acclimated fish and 2·24 ± 0·11 mmol h^{-1}.10^{-6} in cells from cold acclimated fish, an increase of 24 % ($P < 0.005$; $n = 10$). Intracellular ATP concentrations ($[ATP]_i$) were 370 ± 30 μmol lpcv^{-1} (μmol per litre packed cell volume) and 437 ± 93 μmol lpcv^{-1} for warm and cold acclimated fish respectively, these values were not significantly different ($P = 0.477$; $n = 5$) so that different Na$^+$ pump activities were not due to different $[ATP]_i$. Using radiolabelled ^{3}[H] ouabain the number of Na$^+$ pump sites in the erythrocyte membranes was determined. Two methods were used, equilibrium binding and correlation of pump inhibition with ouabain binding (Erdmann, 1982). Both methods resulted in similar values for the number of ouabain binding sites per cell, these being 22·9×10^3 ± 0·6×10^3 and 23·5×10^3 ± 0·5×10^3 ouabain binding sites per cell for warm and cold fish respectively. These values were not significantly different ($P = 0.479$; $n = 10$). The turnover number of the Na^+ pump was calculated from the ouabain sensitive K^+ influx and the number of ouabain binding sites per cell. Cells from warm acclimated fish had turnover numbers of $6.48 \pm 0.31 \, s^{-1}$ ($n = 10$) and the corresponding value of cold acclimated fish was 8.31 ± 0.59. It seems, therefore, that the increased total pumping capacity of cold acclimated fish erythrocytes is indeed due to an increase in the rate at which individual pumps operate. This implies a difference in conformational freedom of constituent polypeptide chains of the molecules which comprise the Na^+ pump perhaps as a result of homeoviscous adaptation.

To examine this possibility the membrane fluidity of trout erythrocytes was measured by steady state fluorescence polarization, using 1,6,diphenyl-1,3,5 hexatriene (DPH) as membrane probe, in vesicles prepared by cell lysis and sonication. Polarization values obtained at 13°C were 0.290 ± 0.003 ($n = 4$) for membranes from 21°C acclimated fish and 0.264 ± 0.003 for membranes from 6°C acclimated fish. The homeoviscous adaptation was found to be translatory with a homeoviscous efficacy of 44 % (Cossins, 1983). Thus by inference, cells from cold acclimated fish have reduced membrane order compared to warm acclimated fish. This reduction of order in the bilayer lipids may promote the conformational flexibility of membrane-bound proteins and enhance the catalytic and transport rates because the bilayer provides an effective solvent environment for membrane-bound enzymes.

The corresponding changes in membrane order and number of the Na^+ pump turnovers as a result of thermal acclimation are consistent with the idea that bilayer lipids play an important regulatory role in cellular transport.

References

BOURNE, P. K. & COSSINS, A. R. (1984). Sodium and potassium transport in trout (*Salmo gairdneri*) erythrocytes. *J. Physiol., Lond.* **347**, 361–375.

COSSINS, A. R. (1983). The adaptations of membrane structure and function to changes in temperature. In *Cellular Acclimatization to Environmental Change* (ed. A. R. Cossins & P. Sheterline), pp. 3–32. Cambridge: Cambridge University Press.

ERDMANN, E. (1982). Binding studies with ³H-ouabain to red cell membranes. In *Red Cell Membranes – a Methodological Approach* (ed. J. C. Ellory & J. D. Young), pp. 251–262. London: Academic Press.

Printed in Great Britain © *Society for Experimental Biology 1987*   457

POSTER COMMUNICATION

HEAT SENSITIVITY OF TUMOUR TISSUE

K. BOWLER[1], *A. M. S. KASHMEERY*[2] *and C. J. BARKER*[3]

[1] Department of Zoology, University of Durham, Durham DH1 3LE, UK
[2] Faculty of Applied Science and Engineering, Umm-al-Qura University, PO Box 3711, Mecca, Saudi Arabia
[3] Department of Biochemistry, University of Birmingham, PO Box 363, Birmingham, UK

Two transplantable tumours D23 hepatoma and MC7 sarcoma have been studied. Heating for 1 h at 44°C *in vivo* delayed tumour growth in all cases and in 25 % of D23 and 15 % of MC7 tumours no regrowth had occurred after 200 days. The administration of 1 mM tetracaine during heating potentiated the curative effect in D23 with about 75 % showing no regrowth after 200 days.

The sensitivity of the tumour tissue and liver slices to *in vitro* heating has also been studied. D23 and MC7 tissue respiration is sensitive to a 1 h preincubation at temperatures above 43°C when measured subsequently at 37°C. Liver slice respiration was less sensitive to heat. The inclusion of 1 mM tetracaine during heating sensitized the tissue to the hyperthermic treatment. However *in vivo* heating at 44°C for 1 h did not cause inhibition of D23 tissue respiration when subsequently measured at 37°C.

Mitochondria isolated from the D23 hepatoma tissue showed coupled respiration, however mitochondria isolated from *in vivo* heated tumour did not show coupled respiration. In contrast to mitochondria isolated from unheated tissue, these mitochondria lacked cristae and contained electron-dense granules, indicators of damage. The lack of effect of an *in vivo* heat-dose, known to cause tumour regression, on respiration and the reports that ATP levels are unaffected by such heating, suggests that cell respiration is not a primary lesion in cellular heat injury. This implies that the observed impairment of mitochondrial function following *in vivo* heating is best explained if the heating sensitized the mitochondria to subsequent damage during isolation.

Incubation of tumour tissue at 44°C caused a marked rise in K^+ leakage from the cells compared with incubation at 37°C. The inclusion of 1 mM tetracaine caused a 3 to 4-fold increase in the rate of K^+ leakage from the tumours and a 1·3-fold increase from liver.

The 'fluidity' of plasma membranes derived from the tumours and from normal liver was measured by fluorescence polarization using 1,6-diphenyl 1,3,5-hexatriene. Tumour plasma membranes were more 'fluid' (decreased order) than liver membranes at all temperatures tested. Tetracaine decreased membrane order in both tumours and in liver. Collectively these data support the hypothesis that the plasma membrane is a primary target in heat injury and that the tumour cell membrane may be particularly susceptible to hyperthermia as a result of the lower order of its lipid moiety.

POSTER COMMUNICATION

TEMPERATURE DEPENDENT MEMBRANE CHANGES IN HUMAN AND EEL RED BLOOD CELLS

D. J. HORNSEY and M. A. EL-MISSIRY

School of Biological Sciences, University of Bath, Bath BA2 7AY, UK

Human red blood cells, as typical examples of endothermic membrane systems, show an apparent anomalous temperature effect for the extent of haemolysis after 5 min exposure to hypo-osmotic salines. Greater lysis (90 %) occurs at lower temperatures (15°C) than at physiological ones (30 %) when subjected to a NaCl solution of 150 mosm Kg^{-1}. The final percentage haemolysis achieved after 5 min appears to be dependent only on the initial temperature of the haemolysing solution, for lowering to 15°C the 30 % haemolysed cells maintained at 37°C does not result in any increase in haemolysis.

Washing the cells free from plasma reduces this temperature effect significantly. At 37°C washed cells under osmotic stress are approximately 60 % haemolysed.

In contrast, ectothermic cells (eel red blood cells) show a temperature independent profile, but are still influenced in the same way by washing. Plasma factors obviously influence the osmotic integrity of the membrane but their identity is unknown. Certainly, albumin or divalent cations do not appear to be involved.

Endothermic cells with relatively high levels of saturated fatty acids in their membranes would be expected to show temperature-induced phase transitions (mobile $\rightleftharpoons$ crystalline states) at 20–30°C, whereas ectothermic cells, attempting to maintain their membranes mobile over a wider environmental temperature range and with higher level of unsaturated fatty acids to accomplish this may be expected to show little or no such transitions. It was therefore considered that such differences, if they exist, may account for variations observed in the temperature response to membrane fragility.

Using the haemolysis temperature data, Arrhenius plots were constructed relating haemolysis rates (derived from 1st or 2nd order kinetics) against 1/T with the following deductions:

(1) Phase transitions at 25°C were found in both washed and unwashed human cells with activation energies of 11–15 kJ mol^{-1} above the break (at high temperatures) and 4–8 kJ mol^{-1} below the break. All values were significantly below the water diffusion activation energy of 20 kJ mol^{-1} suggesting an assisted passage of water, particularly at low temperatures, through a channel lined with molecules of lower frictional coefficients than water. The blockage of such water channels with *p*-chloromercuribenzoate (*p*CMB) resulted not only in the expected increase in activation energy to the diffusional level (20 kJ mol^{-1}) as water

resorted to passage through the lipid phase but also a loss of the break point. This latter observation suggests that the observed phase change may involve the protein rather than the lipid moiety of the membrane.

(2) No phase transition was seen for eel cells and activation energies were $25\,kJ\,mol^{-1}$ for unwashed and $15{\cdot}6\,kJ\,mol^{-1}$ for washed cells. These values, close to the diffusional value, suggest that water movement is more energy demanding in these cells than in the human. Of note is the fact that pCMB, a known channel blocker for human cells, fails to increase the activation energy and in fact significantly reduces it implying a channel opening phenomenon.

Conclusion

Human cells undergo a phase transition possibly in their protein moiety at $25\,°C$ resulting in a more water impermeable state of the membranes at physiological temperatures as measured by the activation energy of water transport. Eel cells show no such phase change but significantly higher activation energies than those for human. Such phase changes would support the observation that endothermic cells have lower fragility at higher temperatures whereas ectothermic cells are temperature independent.

INDEX OF SUBJECTS